T0313758

Tropospheric and Ionospheric Effects on Global Navigation Satellite Systems

Tropospheric and Ionospheric Effects on Global Navigation Satellite Systems

Timothy H. Kindervatter
SciTec Inc.
Princeton, NJ, US

Fernando L. Teixeira
The Ohio State University
Columbus, OH, US

IEEE PRESS

WILEY

Registered Office
John Wiley & Sons, Inc., 111 River Street, Hoboken, NJ 07030, USA

Editorial Office
111 River Street, Hoboken, NJ 07030, USA

For details of our global editorial offices, customer services, and more information about Wiley products visit us at www.wiley.com.

Wiley also publishes its books in a variety of electronic formats and by print-on-demand. Some content that appears in standard print versions of this book may not be available in other formats.

Library of Congress Cataloging-in-Publication Data

Names: Kindervatter, Timothy H., author. | Teixeira, Fernando L. (Fernando Lisboa), 1969- author.
Title: Tropospheric and ionospheric effects on global navigation satellite systems / Timothy H. Kindervatter, Fernando L. Teixeira.
Description: Hoboken, NJ : Wiley-IEEE Press, 2022. | Includes bibliographical references and index.
Identifiers: LCCN 2022017545 (print) | LCCN 2022017546 (ebook) | ISBN 9781119863038 (cloth) | ISBN 9781119863045 (adobe pdf) | ISBN 9781119863052 (epub)
Subjects: LCSH: Global Positioning System. | Troposphere. | Ionosphere.
Classification: LCC G109.5 .K56 2022 (print) | LCC G109.5 (ebook) | DDC 910.285–dc23/eng20220720
LC record available at https://lccn.loc.gov/2022017545
LC ebook record available at https://lccn.loc.gov/2022017546

Cover Design: Wiley
Cover Image: © Andrey Armyagov/Shutterstock

Set in 9.5/12.5pt STIXTwoText by Straive, Chennai, India

Contents

Preface

Global Navigation Satellite Systems (GNSS) have a range of applications that include not only navigation, but also mapping and surveying, satellite orbit determination, and a precisely synchronized global time reference. This book is focused on the atmospheric effects which impact the propagation of the signals transmitted from GNSS satellites in orbit to receivers on the ground. The book also covers various predictive models which seek to quantify such effects.

In order to fully appreciate the limitations imposed by atmospheric propagation effects on GNSS performance and the design considerations that aim to address these effects, the first two chapters of this book provide an overview of GNSS. Chapter 1 introduces the Global Positioning System (GPS) in the role of an exemplary GNSS. System-level design and logistics of GPS are presented as a concrete example of how a typical GNSS operates. In particular, this chapter discusses the roles of the three major segments of the GPS system and a discussion of the GPS communication system, including the carrier frequencies and modulation scheme, ranging codes, and the so-called navigation message (i.e. the data modulated onto the GPS signal).

Chapter 2 broadens the focus to include all modern GNSS signals, and discusses the fundamentals of GNSS positioning. There are several ways to determine the position and velocity of a GNSS receiver based on the signals it receives from a GNSS constellation. Three main quantities, known as GNSS observables – pseudorange, carrier phase, and Doppler shift – are used in positioning and this chapter touches upon each of them. Each GNSS observable is modeled in a straightforward fashion, which makes mathematical analysis simple, but inaccurate – at least without error correction. Various sources of error, from atmospheric to relativistic effects, contribute some error to GNSS positioning. The most important of these effects, as well as common mitigation techniques, are discussed in this chapter. Finally, a derivation of GNSS positioning using pseudorange is presented, building upon the earlier concepts in the chapter.

The remainder of the book is dedicated to an in-depth discussion of atmospheric effects on the propagation of electromagnetic waves transmitted by GNSS satellites. In order to provide a strong conceptual foundation for the propagation phenomena being discussed, the authors have made an effort to provide detailed derivations whenever possible. From this more general standpoint, the relative importance of each particular propagation phenomenon to GNSS positioning becomes clear.

The first of these chapters is Chapter 3, which discusses propagation effects in the troposphere, the lowest region of the atmosphere that is comprised of non-ionized particles. In this region, the refractivity is the main parameter of interest since it determines the magnitude of a signal's group delay. Group delay is the largest contributor to tropospheric error in GNSS positioning, and several predictive models have been developed to offset this error. Two of these models – the Saastamoinen model and the Hopfield model – are covered in Chapter 4. The remaining sections of Chapter 3 discuss other tropospheric propagation phenomena such as scattering, absorption, and scintillations, while the final section of Chapter 4 discusses the US Standard Atmosphere, a general reference model for characteristics of the non-ionized atmosphere, such as temperature and densities of various molecular species.

Chapters 5–8 discuss the ionosphere: upper layers of the atmosphere where molecules have been ionized, creating a plasma that induces a variety of propagation effects on electromagnetic signals. The structure and dynamics of the ionosphere itself are quite complex, and these complexities are often necessary to understand in order to discuss ionospheric propagation. Therefore Chapters 5 and 6 are dedicated to a detailed discussion of the ionosphere's structural properties, including the mechanisms which drive ionization and plasma transport, its morphology, and a number of anomalies that represent deviations from the simplest mathematical models used to understand the ionosphere.

Chapter 7 builds upon these topics to explore ionospheric propagation. The Earth's magnetic field causes the ionosphere to behave like an anisotropic medium, giving rise to a number of propagation effects such as group delay, dispersion, and Faraday rotation. These effects are all dependent on the total electron content (TEC) along a signal's propagation path, and since the Sun is the primary driver of ionization in the atmosphere, these effects exhibit strong geographical variations and diurnal, seasonal, and solar-cycle periodicities. The diurnal changes are of particular importance, since the recombination of ions and free electrons shortly after sundown gives rise to strong fluctuations in TEC, which cause ionospheric scintillations that can lead to GNSS signal outages.

Finally, Chapter 8 presents a few of the major models used to quantify and forecast the ionospheric propagation effects presented in Chapter 7. Just as in the troposphere, the group delay introduced into the signal by the ionosphere is a major source of error in GNSS positioning. Chapter 8 discusses two prominent predictive models, the Klobuchar model and NeQuick, which can be used to mitigate the effect of ionospheric delay. Later in the chapter, the Global Ionospheric Scintillation Model (GISM) is briefly covered, which aims to predict the severity of ionospheric scintillations. In addition, this chapter provides an in-depth overview of the International Reference Ionosphere, a general reference model for characteristics of the ionosphere such as electron density, electron and ion temperatures, and ion composition.

This book is aimed at the level of a graduate student or an advanced undergraduate student. As prerequisites, the reader is expected to have a mathematical background typical of an undergraduate engineering or physics program. In addition, some exposure to electrodynamics at the level of Griffiths [1] is expected. Appendix A contains a review of relevant topics in electrodynamics, and Appendix B provides further exploration of concepts that will be referenced in this book, but are not typically part of a first course in electrodynamics.

31 March 2022
Timothy H. Kindervatter
SciTec, Inc., Princeton, NJ, USA
Fernando L. Teixeira
The Ohio State University, Columbus, OH, USA

1

Overview of the Global Positioning System

1.1 Introduction

A global navigation satellite system (GNSS) is a complex global network consisting of three segments: the space segment (satellite constellation), the control segment (ground-based tracking stations), and the user segment (receiver equipment). The purpose of such a system is to obtain highly accurate positioning information of a receiver on the surface of the Earth.

The system operates via the use of a constellation of satellites. These satellites orbit in a configuration such that several of them are visible from any point on Earth at any given time. Each satellite continuously transmits a coded signal, which includes information that uniquely identifies the transmitting satellite as well as its location in space. A receiver on the surface of the Earth may pick up one of these signals and use it to determine the distance between its corresponding satellite and the receiver. Since an electromagnetic signal travels almost exactly at the speed of light, c, through Earth's atmosphere, it is possible to calculate this distance by simply multiplying the time between transmission and reception by c.

Using multiple signals simultaneously, the receiver may trilaterate its position by determining the x, y, and z coordinates of its location on the surface of the Earth. In theory, only three satellites are needed to determine a receiver's position, since we appear to have only three unknowns: x, y, and z. However, the clock in the receiver that is used to determine the time of reception is not synchronized to the satellite clocks. Since the distance calculation is dependent on the signal's travel time, any error in the timestamp at the moment of reception will translate to an error in the calculated distance as well. This receiver clock error is a fourth unknown in our system of equations, so we actually need a fourth satellite to determine our solution. A more rigorous mathematical analysis of this process will be explored in Section 2.4.

Tropospheric and Ionospheric Effects on Global Navigation Satellite Systems, First Edition.
Timothy H. Kindervatter and Fernando L. Teixeira.
© 2022 The Institute of Electrical and Electronics Engineers, Inc. Published 2022 by John Wiley & Sons, Inc.

Several independent GNSS networks are currently in operation, including the United States' Global Positioning System (GPS), Russia's Globalnaya Navigazion-naya Sputnikovaya Sistema (GLONASS), and the European Union's Galileo. Several regional navigation satellite systems are also currently operational, such as China's BeiDou (planned to be expanded to a global network), India's Navigation with Indian Constellation (NAVIC), and Japan's Quasi-Zenith Satellite System (QZSS).

Since its inception, GNSS has become a ubiquitous technology with far-reaching applications. Among other things, GNSS has been applied to navigation, surveying and mapping, precision agriculture, logistical supply chain management, weather prediction, emergency and relief services, seismology, and even facilitation of stock trades. A brief discussion of some applications is presented in Section 1.2. It is clear that GNSS has played a significant role in shaping the modern world, and will likely continue to expand in utility as further improvements are made and new applications are devised.

Maintenance and improvement of GNSS is a tremendous global endeavor, resulting from the collaboration of numerous scientists and engineers in disparate fields, including aerospace engineering, electrical engineering, astronomy, electromagnetics, and geophysics. As a highly multidisciplinary field of study, the literature on GNSS is correspondingly broad. It is infeasible for an individual to be an expert in every facet of GNSS, and one may find the literature of a neighboring field impenetrable without at least a cursory understanding of its fundamental concepts.

The primary focus of this book is the atmospheric effects on the radiowave signals used by GNSS. In order to appreciate the challenges posed to GNSS positioning by atmospheric propagation effects, a basic understanding of GNSS fundamentals is necessary. To that end, this chapter and Chapter 2 aim to provide an overview of important concepts, many of which will be referenced throughout the remainder of this book.

This chapter's focus is a concrete example of an existing GNSS system: the United States' GPS. In particular, the system-level design and logistics of operating the GPS service will be covered. Although the implementation details of each GNSS differ, the broad design considerations relevant to GPS are typical and so serve as a suitable example for GNSS design in general. Section 1.3 discusses the three major subsystems of GPS, also known as GPS segments. They are the space segment (the constellation of satellite vehicles), the control segment (a network of ground stations for monitoring and control), and the user segment (receivers which decode GPS signal data and provide services such as positioning and timing).

Section 1.4 briefly introduces the terminology related to the Keplerian orbits (a type of trajectory that includes GPS satellite orbits). This is relevant when discussing satellite ephemeris data, which is information encoded into a GPS signal regarding current and predicted trajectories of satellites in the constellation. Ephemeris data is part of the so-called Navigation Message, which is discussed in Section 1.5 alongside other aspects of the GPS communication subsystem such as broadcast frequencies, modulation scheme, and ranging codes.

1.2 Applications of GNSS

The applications of GNSS are multitudinous and varied, and over the last few decades numerous industries have seen tremendous innovation by leveraging the positioning and timing information GNSS provides [2].

1.2.1 Applications of Standard GNSS Positioning

One of the most common uses of GNSS is for navigation. The average person is no doubt familiar with the utilization of GNSS for driving directions, but there are many other navigation applications as well. For example, aviation has been dramatically improved by GNSS. Accurate positioning information allows for safer takeoffs and landings by providing air traffic controllers with real-time information on the locations of multiple aircraft. Pilots may also use positioning information to land in remote locations with limited ground-based services, such as a medical evacuation helicopter attempting to rescue people from a mountain. GNSS has also been used to optimize flight routes, reducing travel time, and costs. Analogous marine applications have also been sought, providing efficient routes for ships and ensuring safe arrival and departure at ports. GNSS may even be used to aid spacecraft launch, and is helpful for monitoring the orbits of other space-based vehicles such as telecommunication satellites.

Another major application of GNSS is surveying and mapping. Quickly and readily available positioning information has allowed surveyors to model the Earth to an unprecedented degree of accuracy and completeness. Prior to the advent of GNSS, surveying techniques required a line of sight from a ground station to the receiver, limiting their functionality to areas with good vantage points (such as hilltops or mountains), or wide open areas free of obstructions (such as plains). Conversely, as long as a GNSS receiver has a good view of the sky, it is able to communicate with satellites to determine its position. This expanded functionality allows surveyors to map many more areas, such as valleys, cities with tall buildings, and even coastal areas with limited access to land-based reference stations.

GNSS has even aided in mapping the ocean floor, though indirectly. Ships utilize GNSS information to determine their location on the surface of the ocean, and sonar is used to map the local topography of the ocean floor and any submersed structures. This information is tremendously important for mariners, as it can alert them to the location of shoals or hazardous obstacles that must be avoided. Additionally, underwater construction projects such as bridge piers or oil rigs rely heavily on GNSS.

1.2.2 Applications of Centimeter and Millimeter-Level Positioning Accuracy

A single GNSS receiver is able to determine its location on the surface of the Earth to an accuracy of less than 10 m. Through the use of differential positioning or real-time kinematic (RTK) positioning, in conjunction with error correction techniques, it is possible to achieve centimeter or even millimeter-level accuracy. Sources of error as well as mitigation techniques will be discussed in Chapter 2.

Centimeter-level accuracy has allowed farmers to employ precision agriculture, in which entire fields of crops may be planted, watered, and harvested by autonomous vehicles. Additionally, this level of accuracy allows for better control of pesticide, herbicide, and fertilizer dissemination, producing a higher crop yield for the farmer as well as mitigating damage to the environment.

Construction has also seen an increase in automation from centimeter-level GNSS positioning. Autonomous vehicles are able to bulldoze or grade an area to specified dimensions, or lay down concrete such as for curbs or sidewalks. This level of automation reduces time, cost, and the number of workers required to complete a construction project.

Self-driving cars would also be infeasible without centimeter-accurate GNSS. These positioning measurements, in conjunction with light detection and ranging (LIDAR) and various sensors deployed throughout the vehicle, provide the vehicle with a detailed picture of its location and surroundings, allowing it to navigate efficiently and safely on the road.

For some applications, even centimeter-level accuracy may not be sufficient. For example, measurement stations can be used to determine tectonic drift. However, tectonic plates drift very slowly – on the order of millimeters per year [3]. In order to accurately determine this drift, GNSS-based measurement stations utilize numerous correction techniques to remove as many sources of error as possible and measure their location to within a millimeter.

1.2.3 Applications of GNSS Timing Information

As we will see later in this chapter, the operation of GNSS is dependent on extremely precise timing information. To achieve this, each satellite has an atomic clock onboard which timestamps every signal upon transmission. In addition to its intended use for positioning, a side benefit of GNSS is that these atomic clocks provide a valuable reference time which may be used for timing and synchronization in a number of applications.

National laboratories across the world use GNSS timing information as a reference in order to establish a globally agreed-upon standard time called the Coordinated Universal Time (abbreviated UTC). This standard facilitates travel and commerce on a global scale, ensuring that flight schedules are not confused by conflicting timing information, and allowing banks and corporations to keep consistent records of transactions. The stock market has become increasingly automated, and computers are able to make trades on time scales of microseconds or even nanoseconds. GNSS timing is standard for all stock trading now, to ensure that these deals go through properly.

Telecommunication networks utilize GNSS timing to synchronize base stations, efficiently allocate bandwidth, and allow users to tune radio stations with minimal delay. Power grids require precise synchronization to ensure that geographically separated generators all run in phase. Distributed sensor networks may also be synchronized; for example, seismic monitoring networks can be used to quickly determine the epicenter of an earthquake and provide forewarning for natural disasters like tsunamis. Similarly, a network of doppler radars across the United States are synchronized to allow prediction and reporting of hazardous weather such as tornadoes and hurricanes.

The broad and diverse applications of GNSS have cemented the technology as a ubiquitous and fundamental aspect of modern society, and the motivation to understand and improve this technology is clear.

1.3 GPS Segments

Much of the information in this section is adapted from Chapter 3 of [4] and the descriptions of the GPS segments available from [2].

1.3.1 Space Segment

The space segment of the GPS consists of a constellation of satellites orbiting the Earth in a specific pattern. This constellation is continuously maintained and frequently upgraded by the United States Air Force (USAF). The constellation is comprised of six orbital planes, each with an inclination angle of 55° with respect to the equator. Each of these orbital planes will contain a minimum of four operational satellites at any given time. These satellites are positioned such that there are between 4 and 10 satellites visible from any point on Earth at any given time. Since four satellites are required to trilaterate a position on Earth, this ensures that a receiver anywhere on Earth will be able to determine its position. If more than four satellites are available, the receiver may choose the best four signals to improve the accuracy of the position calculation.

The USAF works to ensure that this baseline constellation has at least 24 satellites operating 95% of the time. The six orbital planes also have additional slots for extra satellites which are not necessary for the system's operation but may increase performance or act as backups in the case of a core satellite's failure. In June 2011, the Air Force expanded the baseline constellation from 24 to 27 satellite vehicles, increasing coverage. At the time of this writing, there are a total of 31 satellites considered part of the GPS constellation: 27 which comprise the baseline constellation and 4 additional vehicles.

The entire constellation orbits in a medium Earth orbit (MEO) at an altitude of approximately 20,200 km. These orbits are very nearly circular, having an eccentricity of 0.02. The orbital period is half of one sidereal day (approximately 11 hours and 58 minutes). The vehicles that comprise the constellation are continually phased out and replaced by improved designs with additional capabilities, increased lifespans, and improved performance. A summary of the different generations of satellite hardware is shown in Table 1.1, which has been adapted from information provided in [2].

Navigation Payload

The navigation payload is a collection of subsystems contained in every GPS satellite. These subsystems are responsible for the generation and transmission of the positioning signals utilized by the user segment. Figure 1.1 shows a block diagram of the navigation payload.

Every GPS satellite transmits an L-band (see Table 1.2) downlink that is utilized for trilateration by the user segment. The control segment also uses these signals to monitor the satellite constellation and ensure that it is transmitting properly. All generations of satellite hardware transmit signals on the L1 and L2 carrier frequencies, and more recent hardware revisions (Block IIF and newer) also transmit on the L5 carrier frequency. Block III satellites are designed with the ability to transmit on the L1C carrier frequency, but have not yet begun broadcasting this signal. The exact structure and content of these signals is essential to GPS operation, so Section 1.5 covers this information in greater detail.

Tracking, telemetry, and control (TT&C) is accomplished by an S-band uplink from the control segment. A dedicated subsystem is responsible for receiving TT&C data and managing functions such as orbit correction. A separate ultra-high frequency (UHF)-band (see Table 1.3) crosslink subsystem allows satellites to communicate with others in the constellation. Therefore, if a satellite is unable to communicate with the control segment, it may still receive important data updates from neighboring satellites. All current GPS satellites also have the ability to perform autonomous navigation via crosslink ranging, a function known as

Table 1.1 Overview of current and planned GPS satellite blocks.

Block IIR satellites
8 operational
C/A code on L1
P(Y) code on L1 and L2
On-board clock monitoring
AutoNav allows autonomous operation for 180 days
7.5 year design lifespan
First launched in 1997
Block IIR-M satellites
7 operational
All Block IIR signals and features
Second civil signal on L2 (L2C)
New military code (M-code) on L1 and L2; provides enhanced jam resistance
7.5 year design lifespan
First launched in 2005
Block IIF satellites
12 operational
All Block IIR-M signals and features
Third civil signal on L5 (L5)
Advanced atomic clocks
Improved accuracy, signal strength, and quality
12 year design lifespan
First launched in 2010
Block III satellites
4 operational, more in production[a]
All Block IIF signals and features
Fourth civil signal on L1 (L1C)
Enhanced signal reliability, integrity, and accuracy
No selective availability
IIIF satellites equipped with laser retroreflectors
15 year design lifespan

a) Up to 32 satellites are planned, split among two series: 10 Block IIIA vehicles and up to 22 Block IIIF vehicles. As of this writing, 4 of the 10 Block IIIA vehicles have been launched, and the remaining 6 are expected to launch in the next few years, with the tenth and final launch projected for 2023. The Block IIIF vehicles are expected to launch beginning in 2026 and continue through 2034.

Source: Based on GPS.gov [2].

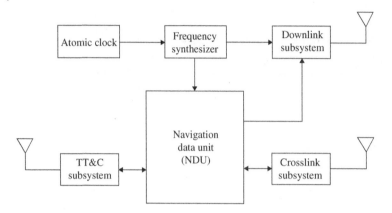

Figure 1.1 Block diagram of navigation payload.

Table 1.2 IEEE frequency band designations.

Band name	Frequency range (GHz)	Wavelength range (cm)
L	1.0–2.0	15–30
S	2.0–.0	7.5–15
C	4.0–8.0	3.75–7.5
X	8.0–12.0	2.5–3.75
K_u	12.0–18.0	1.67–2.5
K	18.0–27.0	1.11–1.67
K_a	27.0–40.0	0.75–1.11
V	40.0–75.0	0.40–0.75
W	75.0–110	0.27–0.40

Source: Based on IEEE Standard Letter Designations for Radar-Frequency Bands [6].

AutoNav. Uplink and crosslink data is handled by the Navigation Data Unit's (NDU) processor.

Since timing is an essential factor in proper GPS operation, each satellite is also equipped with multiple onboard atomic clocks. Only one of these clocks operates at a given time; the remaining clocks are redundant systems included to ensure consistent system operation in the case of the main clock's failure. These clocks use either Rubidium (Rb) or Cesium (Cs) and are very stable, making them very reliable for long-term, precise timing. Rb clocks typically drift no more than 10 ns/day, and Cs clocks drift even less: no more than 1 ns/day.

Atomic clocks oscillate at the natural resonant frequency of the atom used to build the clock (Rb or Cs). Both of these frequencies are on the order of GHz,

Table 1.3 ITU frequency band designations.

Band name	Abbreviation	Frequency range	Wavelength range
Very low frequency	VLF	3–30 kHz	10–100 km
Low frequency	LF	30–300 kHz	1–10 km
Medium frequency	MF	300 kHz–3 MHz	100 m–1 km
High frequency	HF	3–30 MHz	10–100 m
Very high frequency	VHF	30–300 MHz	1–10 m
Ultra-high frequency	UHF	300 MHz–3 GHz	0.1–1 m
Super high frequency	SHF	3–30 GHz	1–10 cm
Extremely high frequency	EHF	30–300 GHz	1–10 mm

Source: Based on ITU-R V.431-8 [5].

making them impractical for modulation. A phase-locked frequency synthesizer is used to generate a 10.23 MHz frequency, which is then passed to the NDU. In the NDU, the C/A and P(Y) ranging codes (further discussed in Section 1.5) are generated and passed to the L-band subsystem. There, they are modulated onto the L-band carrier frequencies for transmission.

1.3.2 Control Segment

The GPS control segment is a globally distributed network of ground stations that monitors and controls the satellite constellation. The control segment is comprised of three elements: monitoring stations, ground antenna stations, and the master control station (MCS).

Monitoring Stations

At the time of this writing, the control segment includes 16 monitoring stations [2] located in a number of locations worldwide. Six are operated by the Air Force, and the other 10 are under the purview of the National Geospatial-Intelligence Agency (NGA). These stations are equipped with a dual-frequency receiver, a choke-ring antenna that provides multipath signal rejection, two on-site cesium clocks, and meteorological sensing equipment.

Each station collects pseudorange and carrier phase measurements from satellites as they pass overhead. The receiver then decodes the signal to obtain the navigation message, and forwards all of this information to the MCS for processing. Additionally, each station utilizes sensors to obtain surface pressure, temperature, and humidity measurements that are used to develop models of the troposphere. We will later see in Chapter 4 that such models may be used to estimate errors introduced by propagation phenomena in the troposphere. The meteorological data is also sent to the MCS for processing.

While user segment receivers are entirely passive and designed only to pick up signals in compliance with the intended specifications, monitoring stations must be able to detect non-compliant signals so that they can report these abnormalities to the MCS. To facilitate this, every monitoring station may be controlled remotely by the MCS to acquire and track signals from unhealthy satellites.

Ground Antenna Stations

As of this writing, there are currently 11 ground antenna stations [2] included in the control segment network. These S-band antennas provide TT&C for the satellites based on commands from the MCS. The ground antennas establish control sessions with a single satellite at a time, but the MCS is capable of coordinating control sessions for multiple satellites simultaneously using the various ground antennas.

In addition to TT&C, ground antenna stations are responsible for transmitting the navigation message generated by the MCS. The stations also support S-band ranging, which allows the MCS to quickly detect and resolve orbital anomalies.

Master Control Station

The MCS is located at Schriever Air Force Base in Colorado Springs, Colorado and is operated by the USAF Space Command, Second Space Operation Squadron (2SOPS). A backup MCS in Gaithersburg, Maryland provides redundancy in the system. The MCS is responsible for a number of tasks, including:

- Monitoring the location and health status of satellites.
- Computing estimated satellite clock and ephemeris parameters.
- Synchronizing GPS time to UTC.
- Generating and uploading the navigation message.
- Monitoring integrity of navigation message.
- Verifying and logging navigation data supplied to user segment.
- Maneuvering and repositioning satellites to correct orbital anomalies or restructure constellation in case of satellite failure.

The primary function of the MCS is to create the navigation message (discussed in detail in Section 1.5) and transmit it to the satellite constellation. There are several steps involved in this process.

The control segment's monitoring stations receive signals transmitted by GPS satellites and are able to calculate the basic GPS observables, pseudorange, and carrier phase, by demodulating the encoded information. This information includes the location of each satellite, the timing of their onboard clocks, and the health of each vehicle. All of this data is sent to the MCS, where it is organized and processed.

The MCS uses a Kalman filter [4, p. 89] to "smooth" the raw measured data and produce precise estimates of the satellite ephemeris. Each satellite's motion is modeled and, in conjunction with the smoothed ephemeris data, a reference trajectory is computed for each satellite. This reference trajectory forms the basis for the ephemeris parameters included in the navigation message.

In order to ensure synchronization among all the satellites in the constellation, the MCS must maintain a coherent time scale as a reference. Since a pseudorange measurement is derived from the travel time of a signal, monitoring stations are able to use the measured pseudorange of a satellite to provide the timing information of every clock in the satellite constellation. Using an ensemble of all the clocks on every satellite and every monitoring station, the MCS sets up a composite reference time, called GPS time. To retain consistency with other global time scales, GPS time is steered to align with UTC.

The raw timing data provided by the monitoring stations is smoothed by a Kalman filter and referenced to GPS time to obtain parameters representing the bias, drift, and drift rate of every clock in the satellite constellation. These parameters are included as clock correction data in the navigation message.

The processed ephemeris and clock correction data for each individual satellite, along with some other information, is compiled and encoded in the manner described in Section 1.5. Each satellite's complete navigation message is backed up locally, matched to the proper vehicle, and then transmitted. Every message is updated and re-sent several times per day to ensure that satellites have up-to-date information.

To ensure that the navigation message is delivered correctly, each satellite is designed to rebroadcast the message they receive. The control segment's monitoring stations receive these rebroadcasted messages and forward them to the MCS, where they are compared with the local copies. If there are significant differences in any of the received messages, an alert is sent to 2SOPS, which then intervenes to manually correct the problem.

Finally, the MCS is able to utilize the ground antenna stations to manually control GPS satellite vehicles. In the case of orbital anomalies that are unable to be corrected autonomously, 2SOPS may actively steer a vehicle back into its proper orbit. Additionally, in the case of a vehicle's failure, several satellites may need to be repositioned to ensure full coverage. Using multiple ground antennas, it is possible to steer several satellites simultaneously.

1.3.3 User Segment

The GPS user segment refers to all receivers capable of receiving and decoding the L-band signals transmitted by the GPS satellite constellation. Reception by these receivers is passive, allowing for an unlimited number of users. Many different types of receivers exist on the market today, boasting different features intended for various applications. Improved functionality and accuracy comes at a greater price, so the purchaser of a GPS receiver must consider the context in which it will be used.

At the time of writing, receivers with the most rudimentary functionality are available for less than $50 US. On the other hand, high-end receivers boasting centimeter-level accuracy will cost thousands of US dollars. In some applications, less accurate positioning may be acceptable, so an inexpensive receiver is desirable, while applications with more strict positioning requirements require more sophisticated and expensive hardware.

Receiver Design Considerations

A number of considerations must be taken into account when designing a GPS receiver. For example, some receivers are designed to be portable, so they must be small, lightweight, and potentially sturdy enough to withstand drops, shocks, vibrations, or other physical impact. Some receivers are intended for outdoor applications, so they must be able to withstand inclement weather, as well as the temperature and humidity extremes of its environment. Receivers for maritime applications must also take atmospheric salt content and corrosive effects into account. Receivers mounted on vehicles may also need to take into account velocity and acceleration. For example, a receiver mounted on a fighter jet will need to withstand multiple Gs of acceleration, and will need to account for the doppler shift induced by the jet's high velocity in order to acquire signals properly.

The geometry of the receiver antenna is a major factor for a GPS receiver as well. Vehicles designed to be aerodynamic will require low-profile antennas, high gain antennas may be used to improve signal reception, phased arrays allow for digital beam steering and multipath or interference rejection, and choke ring antennas greatly mitigate multipath errors. The proper choice of antenna is paramount to the successful operation of a GPS receiver.

The internal software and front-end hardware of the receiver platform are also an important facet of receiver design. A user-friendly user interface is always desirable, but it is especially important in the case of receivers targeted at the general public, such as Google's Maps service, or Garmin's dashboard receiver modules. In some applications, GPS receivers will be used in conjunction with a variety of other sensors to accomplish a specific task. For example, a drone used for aerial imaging may utilize inertial measurement units (IMUs), computer vision, and GPS positioning simultaneously, so the GPS receiver must be able to coordinate with the software that orchestrates this complex process.

The design parameters outlined here are but a few of the myriad challenges facing engineers designing GPS receivers. As GPS technology continues to mature, the seemingly limitless applications will introduce still further problems to be solved.

Types of Receivers

There is significant variation in the feature sets of receivers on the market today. Feature-rich designs are generally accompanied by a higher price tag. A few of the most important feature distinctions are the carrier frequencies and codes supported, the satellite constellations accessible to the receiver, and whether the receiver can utilize augmentation services or reference networks to improve performance.

All current GPS satellites broadcast L1 and L2 carrier frequencies, while Block IIF satellites and future generations have been upgraded to transmit the L5 carrier frequency as well. However, not all receivers are designed to pick up all of these signals. In fact, most receivers are single-frequency, and are only tuned to pick up L1. Dual-frequency receivers are more expensive, but they allow the receiver to mitigate the ionospheric error term from GPS positioning measurements (these concepts will be explored later in this book). Triple-frequency receivers provide even further advantages over dual-frequency, including reduction of multipath errors and phase ambiguities.

Similarly, not all GPS receivers are designed to make use of all modulation codes. Some codes are authorized for civilian use, while others are designed specifically for military applications. Currently, there are three available civil signals: L1 C/A, L2C, L5. Capability for a fourth civil signal, L1C (not to be confused with L1 C/A) is included in the Block III generation of satellites, but at the time of this writing, the L1C signals are not yet being broadcast. All GPS receivers are capable of receiving L1 C/A, and multi-frequency receivers are designed to receive L2C and L5 as well.

The non-civilian signals currently in operation are the L1 P(Y) code and the L2 P(Y) code. These codes were intended solely for military applications, and as such are encrypted and not directly accessible for commercial use. However, many receiver manufacturers have developed proprietary solutions to indirectly measure pseudorange and carrier phase using these codes.

Many modern receivers are designed to receive signals not only from the GPS constellation, but also the other GNSS constellations (GLONASS, Galileo, and BeiDou). Utilization of satellites from multiple constellations affords benefits such as improved availability, increased accuracy and system integrity, better dilution of precision (DOP), and improved robustness against spoofing.

Finally, receivers may also be designed to take advantage of a number of augmentation systems and techniques to improve accuracy. The two major categories of augmentation systems are Satellite Based Augmentation Systems (SBAS) and Ground Based Augmentation Systems (GBAS). An example of an SBAS is Japan's QZSS, which is a regional constellation of satellites designed to improve system performance. A GBAS is a collection of ground stations that receive GNSS signals, perform corrections, and send updated measurements to other receivers near Earth's surface. This is very useful for performing real-time positioning improvements for receivers in range of the GBAS.

Some receivers are designed to make use of a network of reference stations to improve their positioning accuracy. Common techniques include differential global navigation satellite system (DGNSS) – in which measurements from two or more receivers are differenced to reduce error – and Precise Point Positioning (PPP) in which additional data from nearby reference stations is forwarded to the receiver in order to calculate and remove some sources of error.

Receiver Block Diagram

Figure 1.2 shows a simplified block diagram of a generic GPS receiver.

Upon reception, RF signals are sent through a bandpass filter to remove interference. After this, the signal is boosted by a low-noise preamplifier. At this point, the signal is downconverted to an intermediate frequency (IF) and subsequently sampled by an analog-to-digital converter (ADC). To satisfy the Nyquist criterion, the sampling frequency of the ADC must be at least twice the chipping rate of the encoded signal. However, oversampling reduces quantization noise and increases resolution, reducing the number of bits required in the ADC, so sampling frequencies up to 20 times the chipping rate are frequently employed. The chipping rate of L1 C/A code is 1.023 MHz, and the chipping rate of both L1 P(Y) and L2 P(Y) is 10.23 MHz.

The sampled data is fed into a digital signal processor (DSP) with N parallel channels. The processor isolates each signal received by a satellite and assigns it to a channel. Each channel continuously tracks the satellite assigned to it, meaning that up to N satellites may be tracked simultaneously (typically N ranges from 8 to 12). Each channel has loops to track and calculate pseudorange and carrier phase measurements. Additionally, each signal's encoded navigation message is demodulated at this stage. All of this information is then forwarded to the navigation processor.

The navigation processor uses the pseudorange and/or carrier phase measurements to determine positioning information, as shown in Section 2.4. Additional information included in the navigation message (and in the case of DGNSS/PPP,

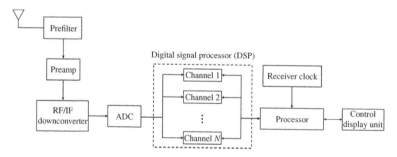

Figure 1.2 Simplified block diagram of a GPS receiver.

information received from nearby reference stations) may be used to further refine the positioning calculation.

The control display unit (CDU) is an I/O device allowing the user to interact with the receiver, issue commands, and receive feedback information. For example, a car dashboard navigation receiver is typically equipped with a touchscreen and a graphical user interface (GUI) allowing the user to input a desired destination. The screen will display roadmap information and real-time navigation information, and a built-in synthetic speech synthesizer will vocalize directions so a driver can keep his or her eyes on the road. In more complex systems, such as an aircraft, the CDU may be integrated directly into a network of instruments so that the receiver may exchange information with them directly.

1.4 Keplerian Orbits

GPS satellite motion is based on so-called "Keplerian orbits," which are briefly discussed in this section. Johannes Kepler, a German mathematician and astronomer, was a pioneer in the study of planetary motion. He developed three basic laws that accurately describe orbital motion, and this work would eventually lay the foundation for Newton's universal law of gravitation. We still use Kepler's laws as the basis for planning satellite trajectories today.

1.4.1 Shape of Orbit

Keplerian orbits are, in general, elliptical. When a small object elliptically orbits a large object, the large object will be located at one of the foci of the ellipse. The nearest point of an elliptical orbit is called the periapsis, and the farthest point is called the apoapsis. For an object orbiting the Earth, these points are given special names: the periapsis is called the perigee and the apoapsis is called the apogee.

One may draw a straight line through the two foci, connecting the periapsis and the apoapsis. This line is the largest distance between any two points on the ellipse, and is called the major axis. The line perpendicular to this is called the minor axis, and is the shortest distance between any two points on the ellipse. Dividing these two quantities by two gives the semi-major axis (a) and the semi-minor axis (b), respectively (see Figure 1.3).

1.4.2 Vernal Point

The Earth orbits around the Sun in a plane known as the ecliptic plane, but the Earth's axis is tilted at an angle of 23.5° with respect to this plane. Thus, the equatorial plane of the Earth is also tilted at this same angle. The intersection of these

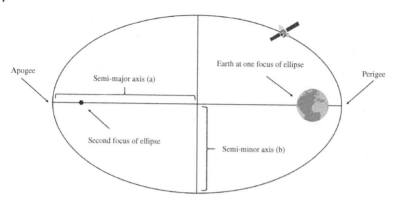

Figure 1.3 Elliptical orbit of satellite around Earth (not to scale).

two planes creates a stationary line. On the vernal equinox, the Sun lies directly along this line at a point termed the vernal point. Six months later on the autumnal equinox, the Sun intersects with this line once again at the autumnal point. The direction of the vernal point is arbitrarily chosen as the zero degree reference, so the autumnal point is at an angle of 180°.

Greek astronomer Hipparchus first discovered the vernal point in 130 BCE by observing the apparent location of the Sun on the vernal equinox. The Sun appeared to be in the constellation of Aries, so he termed the vernal point the First Point of Aries. This terminology is reflected in the notation for the vernal point, ♈, which is the astrological symbol for Aries. The Earth's axis precesses with a period of about 26,000 years, so the vernal point does not remain stationary over such timescales. In fact, the vernal point is currently within the constellation of Pisces, but the notation has been kept the same.

The position of an object in 3D space may be described using three parameters. For astronomical observations a spherical coordinate system – in which two angles and a radial distance are the three parameters – is most convenient. In astronomical coordinates, one of these angles is called the *right ascension* (α) of the object, and is measured eastward along the equator with respect to the vernal point. The second angle, called the *declination* (δ), is measured with respect to the equatorial plane, with 90° representing an object at the north pole and 270° representing one at the south pole. The radial distance is self-explanatory, and is given by the altitude of the object, typically measured in kilometers for satellite-Earth systems. This spherical coordinate system is referred to as the equatorial coordinate system, and is shown in Figure 1.4.

1.4.3 Kepler Elements

A Keplerian orbit has six degrees of freedom: the position and velocity components each in three spatial dimensions. Cartesian coordinates are cumbersome to work

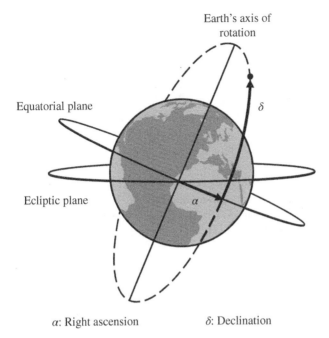

Equatorial plane

Ecliptic plane

Earth's axis of rotation

δ

α

α: Right ascension δ: Declination

Figure 1.4 Astronomical coordinates in the equatorial coordinate system.

with, so instead six parameters known as Kepler elements are used to describe an orbital trajectory. The Kepler elements are:

- *Semi-major axis (a)*: Half of the distance between the periapsis and apoapsis of an elliptical orbit.
- *Eccentricity (e)*: A measure of the shape of an ellipse, given by $e = \sqrt{1 - \frac{b^2}{a^2}}$. For an ellipse, $0 < e < 1$, where $e = 0$ represents a circle and larger values indicate a more elongated ellipse.
- *Inclination (i)*: The angle between the orbital plane and the reference plane. The reference plane for a satellite-Earth system is the equator.
- *Longitude of the ascending node (Ω)*: The angle within the reference plane between the reference direction (the vernal point, ♈) and the ascending node. The ascending node is defined as the point where the orbital trajectory passes upward through the reference plane. For a satellite-Earth system, "up" refers to northward travel.
- *Argument of periapsis (ω)*: The angle within the orbital plane between the ascending node and the periapsis.
- *Periapsis passing time (T_0)*: The epoch at which the satellite passes the point of closest approach (periapsis). The position of the satellite within the orbit may be given by one of three quantities known as "anomalies."

- *True anomaly (v)*: The angle within the orbital plane between the periapsis and the current location of the orbiting body at a specific time. The time chosen to take a "snapshot" of the system is referred to as the epoch. The value of the true anomaly will vary depending on the choice of epoch.
- *Eccentric anomaly (E)*: To determine the eccentric anomaly, consider a circle of radius a, concentric with the elliptical orbit. A line perpendicular to the major axis is drawn through the satellite until it intersects the circle. A line is then drawn between this point and the center of the orbit. The angle between the major axis and this line is defined as the Eccentric anomaly. See Figure 1.5 [7].
- *Mean anomaly (M)*: The mean anomaly considers a fictitious satellite moving with the same orbital period as the real satellite, but instead in a circular orbit at constant speed. The mean anomaly is the angle between the semi-major axis of the real orbit and the location of the fictitious satellite along its circular orbit.

Each of the three anomalies may be represented mathematically as a function of time:

$$M(t) = \frac{2\pi}{T}(t - T_0) = \sqrt{\frac{\mu}{a^3}}(t - T_0),$$

$$E(t) = M(t) + e \sin E(t), \qquad \qquad (1.1)$$

$$v(t) = 2 \arctan \left[\sqrt{\frac{1+e}{1-e}} \right] \tan \frac{E(t)}{2},$$

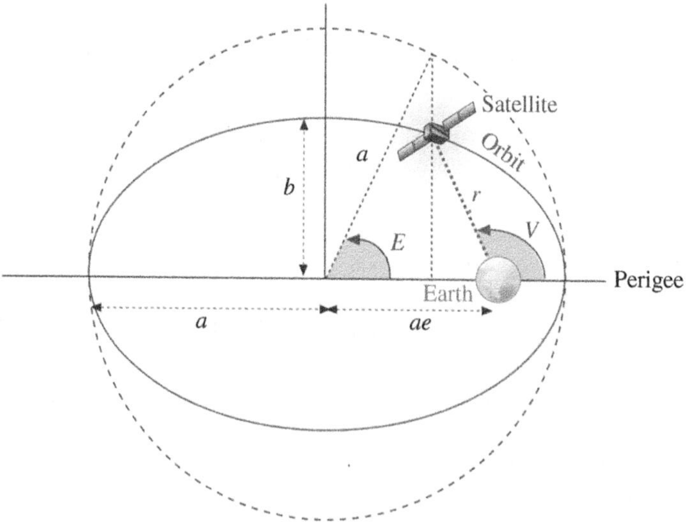

Figure 1.5 Eccentric and true anomalies of a satellite orbiting Earth. Source: Sanz Subirana et al. [7]. Reproduced with permission of the European Space Agency (ESA).

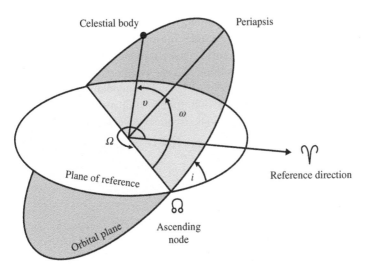

Figure 1.6 Ecliptic and equatorial planes of the Earth.

Ω: Longitude of ascending node υ: Argument of periapsis

ω: Argument of periapsis i: Inclination

where T is the period of the orbit and Earth's standard gravitational parameter $\mu = 3{,}986{,}004.418 \times 10^{14}$ m^3/s^2.

The semi-major axis and eccentricity of an orbit may be observed in Figure 1.3. The remaining Kepler elements are shown in Figure 1.6.

1.5 Satellite Broadcast

The L-band signals broadcast by the GPS satellite constellation are the cornerstone of the entire system's operation, so it is worthwhile to explore them in some depth. A GPS signal is comprised of three major components:

Carrier signal: An electromagnetic wave oscillating at a single frequency known as the carrier frequency. Modulation techniques may be used to encode information onto this signal.

Ranging codes: Also known as pseudorandom noise (PRN) codes. A ranging/PRN code is a unique sequence of bits that may be encoded onto the carrier signal. Upon transmission, the code is timestamped, so a GPS receiver may use this information to determine the signal's time of travel and thus the distance between the satellite and receiver. Each satellite is assigned a unique PRN code, allowing receivers to distinguish signals received from different satellites.

Navigation message: A lengthy bit string modulated onto the carrier frequency of every L-band GPS signal. The string includes information on the orbital trajectory (ephemeris data), health status and clock parameters of the transmitting satellite, rough estimates of the positions of *every* satellite in the constellation (almanac data), and some other helpful data such as model parameters used to (partially) mitigate errors caused by ionospheric propagation effects on the positioning information provided by single-frequency receivers.

In the remainder of this section, each of these three components is analyzed in more detail.

1.5.1 Carrier Frequencies

In a wireless communication link, information is typically encoded on a sinusoidal electromagnetic wave. This wave is known as the carrier wave and its frequency of oscillation is termed the carrier frequency. Various propagation effects will alter the behavior of the carrier wave as it travels, and these effects are primarily determined by the carrier frequency. This means that the choice of carrier frequency is of utmost importance when designing a wireless communication system.

The medium of propagation for an Earth-space link such as GPS is the atmosphere. We will see in later chapters that the atmosphere may be split into two distinct regions for the purposes of electromagnetic wave propagation. These two regions are called the troposphere and the ionosphere, and the effects they have on propagation are drastically different. When the carrier frequencies for GPS signals were chosen, engineers needed to consider all of these effects in order to determine the range of frequencies that would minimize error and maximize signal reception.

Major ionospheric effects such as absorption, group delay, refraction, and others are proportional to $1/f^2$, where f denotes the carrier frequency, meaning they become more significant at lower frequencies. To avoid deleterious ionospheric effects we avoid lower frequencies; those above 0.1 GHz are typically sufficient. Unfortunately, tropospheric effects behave in an opposite way, becoming more pronounced at higher frequencies. In particular, gas attenuation becomes significant above 10 GHz, and heavy rain can cause meaningful attenuation for frequencies above 5 GHz. We are thus left with a practical frequency range of 0.1–5 GHz. More thorough analyses of tropospheric and ionospheric propagation effects are presented in Chapters 2 and 4, respectively.

Since ionospheric effects tend to induce larger positioning errors and can be more difficult to predict and mitigate, especially for single-frequency receivers, carrier frequencies in the upper end of the practical range are chosen. Importantly, higher frequencies also enable the use of smaller antennas at the receiver. The control segment transmits TT&C signals on the S-band (2–4 GHz), while the satellite constellation broadcasts positioning signals in the L-band (1–2 GHz). The L-band positioning signals will be the focus of the remainder of this section.

Currently, three distinct L-band carrier frequencies are broadcast by the GPS constellation. Each frequency is an integer multiple of a fundamental frequency $f_0 = 10.23$ MHz:

$$L1 = 154 \times f_0 = 1575.42\,\text{MHz} \quad \Rightarrow \quad \lambda_{L1} = 19.0\,\text{cm},$$

$$L2 = 120 \times f_0 = 1227.60\,\text{MHz} \quad \Rightarrow \quad \lambda_{L2} = 24.4\,\text{cm},$$

$$L5 = 115 \times f_0 = 1176.45\,\text{MHz} \quad \Rightarrow \quad \lambda_{L5} = 25.5\,\text{cm}.$$

L1 and L2 are broadcast by all generations of GPS satellites, while L5 is only broadcast by Block IIF and newer satellites.

It is worth noting that each GNSS constellation has its own unique signal plan consisting of carrier frequencies, modulating codes, and navigation messages. Engineers designing multi-constellation receivers will need an understanding of the differences between the GPS signal plan and those of other constellations. The interested reader may consult [7, pp. 18, 19] for a detailed discussion of these topics.

1.5.2 Digital Modulation

Information is encoded onto a carrier signal by modulating it. In GNSS systems information is encoded digitally, meaning the information is represented by a bit string. Using various methods, parameters of the carrier signal may be *keyed*, or switched between discrete values.

Modulation Methods

The three parameters of a carrier wave that may be modulated to encode information are its amplitude, frequency, and phase. While there are numerous sophisticated modulation techniques, for our purposes we will only focus on the simplest of them: binary modulation. In binary modulation, the chosen parameter may only switch between one of two values.

For example, a binary amplitude shift keyed (BASK) signal will oscillate at the same amplitude as the carrier signal if the modulating bit is a 1, and will not oscillate at all if the modulating bit is a 0. This is illustrated in Figure 1.7b. Similarly, a binary frequency shift keyed (BFSK) signal will oscillate at one frequency for a "1" bit, and a different frequency for a "0" bit, as seen in Figure 1.7c.

Binary Phase Shift Keying (BPSK) is the chosen modulating technique for nearly every GPS ranging code. In this modulation scheme, whenever two consecutive bits have opposite values the carrier signal is phase-reversed (i.e. a phase shift of

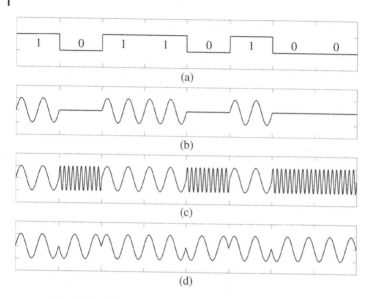

Figure 1.7 (a) Modulating bit string represented as a square wave. Carrier modulated using: (b) Amplitude Shift Keying (ASK); (c) Frequency Shift Keying (FSK); (d) Phase Shift Keying (PSK).

180° is introduced). Put another way, whenever there is an edge transition in the square wave representing the bit string (see Figure 1.7a) the phase of the carrier is shifted by 180°. An example of a BPSK signal is shown in Figure 1.7d.

Code Division Multiplexing

Via a process called multiplexing, it is possible to transmit multiple coded signals on the same carrier frequency without interference. There are multiple types of multiplexing, including time division, frequency division, and code division. GNSS signals employ the latter technique, code division multiplexing. This is especially desirable for GNSS signals because rather than needing a distinct carrier signal for each satellite in the constellation, all satellites may broadcast on the same frequency (L1 for example). Each satellite is assigned its own unique ranging code, which the receiver references in order to determine which satellite it is receiving from. Since a receiver is able to access multiple satellites using only one frequency, this technique is called code division multiple access (CDMA).

To understand how this works, let's consider a concrete example. Consider a satellite in the GPS constellation, say space vehicle number 1 (SV-1). SV-1 is assigned a unique bit string that it modulates onto the L1 carrier frequency. This bit string is SV-1's C/A code. Since each satellite is assigned its own C/A code, it is possible for a receiver to distinguish between signals received from multiple satellites by figuring out which C/A codes it has received.

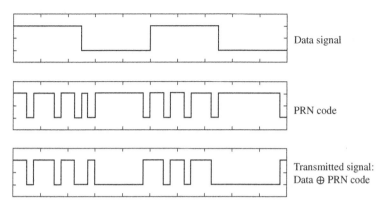

Figure 1.8 Data combined with PRN code to create a CDMA signal.

The set of all C/A ranging codes is stored locally in every GPS receiver. When an L1 signal is received, the encoded C/A code is retrieved. The receiver then compares this retrieved code with each stored code one by one until it achieves a match. The "identifier" code is called a PRN code, and does not carry any data itself. In fact some GPS signals, called pilot signals, are broadcast without any data at all (see Section 1.5.3 for more details). For signals that do transmit data, the PRN code must be combined with the data signal via an exclusive or (XOR) operation. Figure 1.8 demonstrates this.

The frequency of the PRN code is called the *chipping rate*. In general, this frequency is not the same as the data rate, so to distinguish the two, we typically measure the chipping rate in chips per second (cps) and the data rate in bits per second (bps).

The degree of similarity between two signals corresponds to the *correlation* of two signals. Correlation of two different signals is called cross-correlation, and correlating a signal with a time-displaced copy of itself is called autocorrelation. To ensure that a GPS receiver properly identifies which satellite it is receiving from, and does not mistake it for another one, each satellite must transmit a code that is as different as possible from each other satellite's. In other words, the cross-correlation between the two signals must be minimized.

Code Correlation

The algorithm that a GPS receiver uses to cross-correlate two digital signals is as follows:

1. Receiver creates replica code from local data.
2. Signals are multiplied bit-by-bit (analogous to an inner product).
3. Bits of resulting string are summed to obtain correlation.
4. Replica code is shifted right one bit and algorithm is repeated.

Table 1.4 Cross-correlation of two digital signals.

		1	1	0	1	0	
Step 1:	*	1	0	1	1	0	
	=	1	0	0	1	0	$\Sigma = 2$
		0	1	1	0	1	
Step 2:	*	1	0	1	1	0	
	=	0	0	1	0	0	$\Sigma = 1$
		1	0	1	1	0	
Step 3:	*	1	0	1	1	0	
	=	1	0	1	1	0	$\Sigma = 3$
		0	1	0	1	1	
Step 4:		1	0	1	1	0	
	=	0	0	0	1	0	$\Sigma = 1$
		1	0	1	0	1	
Step 5:	*	1	0	1	1	0	
	=	1	0	1	0	0	$\Sigma = 2$

The correlation is normalized – that is, the correlation value is divided by the sum of the received signal's bits. This results in a value between 0 and 1, where 0 means there is no similarity between the two signals at all, and 1 means they are a perfect match. If the receiver obtains a normalized correlation of 1, it knows that the replica code it has generated is identical to the code on the received signal.

An example is given in Table 1.4, in which the top string in every step represents the replica code. Notice that this replica code is shifted right one bit after every step. The second string, which represents the received code, remains unshifted, and the sum of its bits is 3.

The correlation value for each step is represented by Σ, and the normalized correlation value for this example would be $\Sigma/3$. It can be seen that in Step 3, when the two codes line up perfectly, the correlation value is 3. This means that the normalized correlation is 1, and the receiver concludes that the two codes match. The number of shifts required to achieve this match tells the receiver the time delay between transmission and reception of the signal. Since the signal travels at approximately the speed of light, the receiver may estimate its distance from the satellite by multiplying the time delay by c. If the receiver shifts all the way through a replica code without achieving a normalized correlation of 1, it concludes that the replica code does not match the received code. It then selects the next code in the memory bank and repeats the correlation process. Replica codes are considered one-by-one until a match is found.

Orthogonal Codes

To ensure that the receiver never mixes up two satellites, it is important that each satellite is "speaking a different language" from all the others. In other words, the correlation value between two different codes must be as small as possible no matter how they are aligned. Similarly, to ensure that the receiver determines the correct time delay, the autocorrelation must be very small unless it is aligned perfectly with itself.

A special class of signals called orthogonal signals is ideal for this purpose. Orthogonal signals have the following special properties:

1. The cross-correlation of two orthogonal signals is 0 regardless of their relative phase.
2. The autocorrelation of an orthogonal signal is 0 except when it is in-phase with itself, in which case it is 1.

By using orthogonal signals, GNSS systems ensure that positioning errors never arise due to false alarms or improperly identified satellites. An example of a set of orthogonal codes is the Gold codes, named after their discoverer Robert Gold [8]. There are 37 Gold codes that match the specifications for the L1 C/A code. Each of the 31 satellites in the GPS constellation is assigned one of these codes, which it utilizes for its entire lifespan. When a satellite ceases to function or is decommissioned, its code is recycled by the satellite replacing it.

Binary Offset Carrier (BOC) Modulation

In addition to the modulation techniques previously mentioned, GPS signals make use of a more sophisticated technique called binary offset carrier, or BOC. The details are beyond the scope of this text, but a curious reader may find a detailed discussion in [9, pp. 88–100, 148]. We will simply discuss a few key points to provide a general picture of its purpose.

Most GPS signals are modulated using BPSK, meaning they have a power spectrum shaped like a sinc function. Therefore, the power is mostly concentrated around the carrier frequency. In contrast, BOC-modulated signals have a relatively low power density near the carrier, and instead have two large sidelobes some distance away on either side of the carrier. This *split-spectrum* technique allows more modern GPS codes to be modulated onto the same carrier frequencies as older codes without interfering.

BOC is accomplished by modulating with an additional sub-carrier frequency, f_{sc}, which is equal or higher than the chip frequency f_c. A shorthand notation for BOC modulation is BOC(m, n), where $m = f_{sc}/f_{ref}$ and $n = f_c/f_{ref}$ are integers. For GPS, $f_{ref} = 1.023$ Mcps. The M-codes on L1 and L2 have $f_{sc} = 10f_{ref}$, and $f_c = 5f_{ref}$, so the notation for their modulation is BOC(10,5).

It should be noted that the same convention is used for BPSK; however, since BPSK does not employ a subcarrier frequency, there is only one ratio to consider: $n = f_c/f_{ref}$. So for example, consider C/A code, which is modulated at $f_c = 1.023$ Mcps. This modulation would be denoted BPSK(1). Similarly, for P(Y) code, where $f_c = 10.23$ Mcps, the modulation is denoted BPSK(10).

There are also several variants of BOC modulation, including sine BOC (BOC_{sin}), cosine BOC (BOC_{cos}), alternative BOC (BOC_{alt}), multiplexed BOC (MBOC), etc. When listed without a subscript, BOC(m, n) implicitly refers to BOC_{sin}. The two variants used in GPS are BOC_{sin} and MBOC.

As discussed earlier, multiplexing may be performed in multiple different ways, one of which is time-multiplexing. The L1C civil signal, launched with Block III satellites, is modulated using a time-multiplexed binary offset carrier (TMBOC) scheme. As discussed in Section 1.5.3, the L1C signal is comprised of two components, the pilot signal and the data signal. The pilot signal, $G_P(f)$ is time-multiplexed such that 29 of every 33 symbols are modulated using BOC(1,1) and the remaining four are modulated using BOC(6,1). Every symbol of the data signal, $G_D(f)$ is modulated with BOC(1,1):

$$G_P(f) = \frac{29}{33}G_{BOC(1,1)}(f) + \frac{4}{33}G_{BOC(6,1)}(f),$$

$$G_D(f) = G_{BOC(1,1)}(f).$$

75% of the signal's total power is allocated to the pilot signal, while 25% is reserved for the data signal:

$$G_{MBOC(6,1,1/11)}(f) = \frac{3}{4}G_P(f) + \frac{1}{4}G_D(f) = \frac{10}{11}G_{BOC(1,1)}(f) + \frac{1}{11}G_{BOC(6,1)}(f).$$

Therefore the notation MBOC(6, 1, 1/11) may be used to denote this TMBOC signal, where the 1/11 refers to the fraction of the power dedicated to the component with the higher frequency subcarrier, BOC(6,1). The remainder of the power is concentrated in the BOC(1,1) component.

Forward Error Correction

Forward error correction (FEC) is a technique commonly used to mitigate error in communication systems and is employed by all modern GPS signals. The primary benefit of FEC is that an error in a given transmission may be corrected using only the transmitted data. In other words, no reverse channel is required to request retransmission of the data. This is extremely useful for GPS since the satellite downlink is a passive, one-way communication channel – receivers have no way of contacting satellites.

FEC is accomplished by including redundancy into the transmitted data, which means that this technique comes at the cost of a higher forward channel bandwidth. A very simple, but illustrative example of FEC is presented next to aid the

Table 1.5 (3,1) Repetition code FEC.

Triplet received	Interpreted as
000	0
001	0
010	0
100	0
111	1
110	1
101	1
011	1

reader in understanding the concept. However, such rudimentary implementations are not used in practice.

On a noisy channel, bits may become corrupted and switch to the opposite value. Ordinarily, there would be no way for a receiver to know that this error had occurred, and the accuracy of the encoded information would be degraded. Now consider an FEC implementation using (3,1) repetition code – meaning every bit is transmitted three times in a row (0 becomes 000 and 1 becomes 111). Due to errors, a receiver may receive one of eight distinct triplets, as shown in Table 1.5.

If one bit in a triplet happens to be corrupted, there are still two bits carrying the correct value. Therefore, when the receiver decodes a triplet, it may "democratically" choose whether to interpret it a 0 or a 1 depending on which value is in the majority in the triplet. Of course, it is possible that two or even all three bits in a triplet may be simultaneously corrupted. However, this will happen far less frequently than only one bit being corrupted, so there will be a net decrease in the number of errors overall.

GPS signals use a more sophisticated FEC technique known as convolutional codes. This topic will not be discussed here, but a general discussion may be found in [10, pp. 617–637] or any comparable undergraduate communication systems text.

1.5.3 Ranging Codes

Every ranging code is what is known as a PRN code. A PRN code is a deterministic sequence of bits that appears random to an observer who does not know the sequence. GPS receivers have access to these codes, so they are able to recognize and decode them. Any other receiver would interpret the code as noise and ignore it.

There are a number of different ranging codes included in GPS signals. Three legacy codes – L1 C/A, L1 P(Y), and L2 P(Y) – have been broadcast by all satellites since the inception of GPS. An ongoing modernization effort has phased in additional codes over time, including L2C, L5C, and the L1 and L2 M-codes. The upcoming Block III satellites will be designed to broadcast another new code: L1C. These codes all have quite different structures and applications. This section will provide an overview of all of these codes and their differences.

Legacy Codes

Coarse/acquisition code or C/A code: This code is only broadcast on the L1 carrier frequency. In this text, we commonly refer to this code as L1 C/A, though in most literature it is simply denoted as C/A. This was the first and originally the only signal available for civil use.

The PRN codes used in this signal are orthogonal Gold codes, each containing 1023 bits repeated with a period of 1 ms. This implies a *chipping rate* of 1.023 Mcps. Thus, the duration of one code chip is about 1 μs, which can be multiplied by the speed of light c to receive the *chip width* (analogous to wavelength) of approximately 293.1 m. Modern receivers typically have a resolution of about 1% of chip width, or about 2.931 m using C/A code.

Precision code or P-code: This code is restricted and is only accessible to military users and select authorized civilian users. It is transmitted on both the L1 and L2 carrier frequencies.

Each satellite is assigned a PRN code that is 6.1871×10^{12} bits long (approximately 720.213 GB). The chipping rate is 10.23 Mcps, meaning the entire sequence repeats once every week. Each satellite's unique PRN code is actually just a segment of a master P-code 235,469,592,765,000 bits ($\sim 2.35 \times 10^{14}$ bits or 26.716 TB) in length. The master code is 38 weeks long, so 38 unique week-long segments are available to assign to satellites.

The chipping rate is 10 times faster than that of C/A code, so its chip duration is 10 times smaller, at 0.1 μs. This gives a chip width of 29.31 m and a positioning resolution of 0.2931 m. This is 10 times the positioning resolution of C/A code, hence the name precision code.

Synchronizing with P-code directly would be a very time-consuming process for a receiver. The expectation was that receivers would first lock onto the simpler C/A code, obtain the time and their approximate position, and then use that information to synchronize with P-code for more accurate positioning.

This master P-code is actually the product of two codes, called X1 and X2 codes. The X1 code is 15,345,000 bits long, while the X2 code is 37 bits longer, at 15,345,037 chips ($15,345,000 \times 15,345,037 = 235,469,592,765,000$). At a chipping rate of 10.23 Mcps, the X1 code is exactly 1.5 seconds long. We may define a 1.5 s interval as an X1 epoch, and as we will see in Section 1.5.4, this is essential in allowing a receiver to transition from C/A code to P-code.

Anti-spoofing or A/S: Spoofing is a type of attack on GPS whereby a malicious party attempts to deceive a GPS receiver by broadcasting signals that resemble those transmitted by a GPS satellite, or by rebroadcasting genuine signals captured at a different location or time. In this way, the GPS receiver is given misleading data, causing it to estimate that it is somewhere other than its true position, or that it is in the correct position but at a different time.

This type of attack can be problematic or even dangerous. For example, an aircraft GPS could be spoofed into navigating into enemy airspace, where it may be shot down. Alternatively, spoofing could be used to delay competitors' timing systems in automated stock trading, putting the attacker at an unfair advantage. To protect against such attacks, P-code is encrypted with a second, confidential code called W-code. When W-code is added to P-code, the result is called Y-code, hence the common abbreviation P(Y). The encryption key for decrypting W-code is available only to authorized users. Few details are openly available regarding the nature of this code, although it is known that it is applied at a frequency of about 500 kHz. Using this information, some manufacturers have built semi-codeless receivers that are capable of tracking P(Y) signals, even without detailed knowledge of W-code.

Selective availability or S/A: Selective availability refers to the intentional degradation of satellite clock and ephemeris data. This effect used to be implemented by all GPS satellites and would induce tens of meters of positioning error. S/A is a holdover from the earliest days of GPS, when it was used exclusively for military purposes. It was intended as a feature to confuse US adversaries attempting to make use of GPS positioning.

Selective availability quickly became a nuisance as civil applications of GPS began to grow rapidly. President Bill Clinton announced that at midnight on 1 May 2000 selective availability would be shut off, allowing more accurate GPS positioning by civil users. This made GPS an even more attractive technology, and spurred the boom of innovation in the field we see today.

The technology allowing selective availability is still included in satellites up to (but not including) Block IIF vehicles, and it is within the power of the US government to turn it back on. This uncertainty was a cause for concern for civil GPS users. However, in September 2007 the government reaffirmed the decision to permanently abandon S/A by announcing that Block III satellites would be designed without the feature.

On L1, the C/A code, $C(t)$, and the P(Y) code, $P(t)$, are modulated in quadrature (i.e. with a phase difference of 90°), with P(Y) on the in-phase channel and C/A on the quadrature channel. The navigation message $D(t)$ is modulated on top of each signal via modulo 2 addition. These two signals are summed to result in the L1 signal

$$s_{L1} = a_p P_i(t) D_i(t) \sin(\omega_1 t + \phi_{L1}) + a_c C_i(t) D_i(t) \cos(\omega_1 t + \phi_{L1}),$$

where i represents the ith satellite.

The L2 signal is simpler, since it only has P(Y) code modulated onto it. Once again, the navigation message is added via modulo 2 addition, resulting in the signal

$$s_{L2} = b_p P_i(t) D_i(t) \sin(\omega_2 t + \phi_{L2}).$$

Figure 1.9, which has been adapted from [11], shows a block diagram of how the legacy GPS signals are constructed. A local oscillator synthesizes a fundamental frequency $f_0 = 10.23$ MHz using the atomic clock aboard the satellite. The L1 and L2 frequencies, the C/A and P(Y) codes, and the navigation message all operate at frequencies derived from f_0.

For quick reference, Table 1.6 [11] summarizes the parameters of legacy GPS signals.

Modern Codes

The original GPS design specifications reached full operational capability on 17 July 1995 [12]. With this goal completed, the US government began looking toward

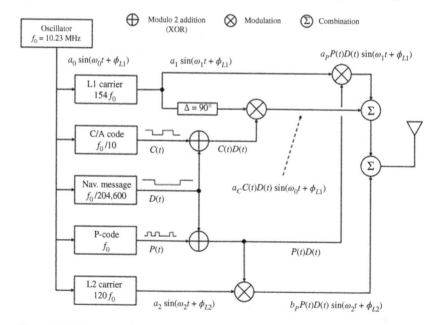

Figure 1.9 Block diagram of legacy signal construction. Source: Adapted from Figure 7.11 of [11]. Reproduced with permission of De Gruyter.

Table 1.6 Parameters of GPS legacy signals.

Atomic clock frequency	$f_0 = 10.23$ MHz
L1 frequency	$154 \times f_0 = 1575.42$ MHz
L1 wavelength	$\lambda_{L1} = 19.0$ cm
L2 frequency	$120 \times f_0 = 1227.60$ MHz
L2 wavelength	$\lambda_{L2} = 24.42$ cm
C/A code chipping rate	$f_0/10 = 1.023$ Mcps (MHz)
C/A code chip width	293.1 m
C/A code period	1 ms
P(Y) code chipping rate	$f_0 = 10.23$ Mcps (MHz)
P(Y) code chip width	29.31 m
P(Y) code period	266 days (38 weeks) total
	7 day block per satellite
Navigation message frequency	$f_0/204{,}600 = 50$ bps (Hz)
Frame length	30 s
Total message length	12.5 min

Source: Adapted from Seeber [11]. Reproduced with permission of De Gruyter.

the future. Advances in technology in conjunction with increasing civil demand for GPS prompted a modernization effort which simultaneously improved upon the existing system and introduced additional features.

One of the major features introduced by the modernization initiative was the introduction of various new GPS signals. These new signals have been phased in over the last few decades as part of each new generation of satellite hardware. The modern GPS signals include the second civil code L2C, two improved military codes L1 M and L2 M, and the third civil code L5C. Upcoming Block III satellites will also include a fourth civil code, L1C.

Modern signals feature several general improvements over legacy signals. Each modern signal now includes an additional, dataless signal called a pilot signal alongside the traditional data signal. The pilot signal is designed to be easier for the receiver to acquire. Upon acquisition, the receiver may use the pilot to lock onto the data signal. Additionally, all modern signals employ FEC when encoding the navigation message.

Second civil signal or L2C: Prior to 2005, the only signal authorized for civil use was the C/A code on L1. This made it difficult for civil users to take advantage of techniques afforded by dual-frequency receivers. In 2005, the first Block IIR-M satellite was launched, which included an additional civil signal on L2 for the express purpose of facilitating dual-frequency applications.

This signal, abbreviated as L2C, is comprised of two distinct PRN sequences: the civil-moderate code L2 CM and the civil-long code L2 CL. The CM code is

10,230 bits long, with a period of 20 ms while the CL code is 767,250 bits long, and has a period of 1500 ms. Both codes are modulated using BPSK, and each has a chipping rate of 511.5 kcps. The final L2C signal multiplexes both of these codes together for a total chipping rate of 1.023 Mcps.

The CM code is modulated with the navigation message, while the CL code remains dataless and acts as the pilot signal. The CL code has strong cross-correlation properties, providing an additional 24 dB of effective power (i.e. ~250 times greater) when compared with C/A code. This improves reception, making it easier for receivers to maintain a lock on a satellite while tracking. In addition, this makes it easier to receive L2C in heavily forested areas, indoors, or in other areas with typically poor reception.

New military code or M-code: Alongside the L2C signal, two new restricted codes were made available starting with Block IIR-M generation satellites. These new codes are intended for military use only and are termed M-codes. There is one M-code broadcast on L1, and another broadcast on L2. These two codes are identical, which both modulated using the BOC modulation technique and transmitting at a chipping rate of 5.115 Mcps.

Since the code is designed to be highly confidential, not much else is known about the structure of these signals. It is, however, known that M-codes are designed to allow for autonomous synchronization, in contrast to P(Y) codes, which require synchronization to a C/A code first. M-codes are also designed for improved jamming resistance, making them less susceptible to enemy disruption.

Third civil code or L5: Beginning with the Block IIF generation of satellites in 2010, a third civil signal became available. This signal was the first and is currently the only signal broadcast on the L5 carrier frequency. This signal is sometimes abbreviated to L5C, denoting its availability to civil users. However, since no military signals are broadcast on L5, the C is commonly dropped, and the signal is simply referred to as "L5."

The L5 frequency lies in the Aeronautical Radio Navigation Service (ARNS) band. The ARNS band is an internationally protected band of frequencies allocated specifically for life-critical applications called *Safety-of-Life* (SoL) applications, such as precision aircraft approach guidance. Since this band is protected, users can be assured that there will be no interference from other signals, ensuring the most secure and robust navigation possible.

Like all other modern signals, L5 broadcasts both a data code and a dataless pilot code. On L5, these two codes are broadcast in quadrature. The in-phase code, L5-I or I5, is the data channel which contains the navigation message, while the quadrature code, L5-Q or Q5, is the dataless pilot channel. Both codes are modulated using BPSK, are 10,230 bits long, have a chipping rate of 10.23 Mcps, and a period of 1 ms.

Fourth civil signal or L1C: This signal is not yet available, but is planned to be included with all Block III satellites once they launch. The first block III satellite launched in December 2018, and at the time of this writing at least 24 block III satellites are planned for launch by the late 2020s [2].

L1C will be another civil signal, which will make it the fourth civil signal available in total, and the second civil signal available on L1. Just like all other modern signals, L1C is comprised of two codes. These codes are broadcast in quadrature, similar to the L5 signal: the in-phase $L1C_D$ data channel upon which the navigation message is encoded, and the $L1C_P$ dataless pilot channel. Both codes are 10,230 bits long, and have a chipping rate of 1.023 Mcps, resulting in a period of 10 ms. The modulation scheme used is called multiplexed binary offset carrier (MBOC), which is designed to improve mobile GPS reception in environments such as cities. Some details of this modulation scheme are discussed in Section 1.5.2.

The L1C signal was designed in a collaborative effort between the United States and Europe, and the signal will be included in both the GPS and Galileo constellations, allowing direct interoperability. In addition, Japan's QZSS and China's BeiDou constellations are adopting L1C-like signals, further facilitating multi-constellation applications.

Finally, Table 1.7 [7] summarizes a few important parameters of every GPS code.

Table 1.7 Parameters of current and planned GPS signals.

Link	Carrier frequency (MHz)	PRN code	Modulation type	Code rate (Mcps)	Data rate (bps)	Service
L1	1575.420	C/A	BPSK(1)	1.023	50	Civil
		P	BPSK(10)	10.23	50	Military
		M	$BOC_{sin}(10,5)$	5.115	N/A	Military
		L1C-I data	MBOC(6,1,1/11)	1.023	50	Civil
		L1C-Q pilot			—	
L2	1227.600	P	BPSK(10)	10.23	50	Military
		L2C M	BPSK(1)	1.023	25	Civil
		L			—	
		M	$BOC_{sin}(10,5)$		N/A	Military
L5	1176.450	L5-I data	BPSK(10)	10.23	50	Civil
		L5-Q pilot			—	

Source: Sanz Subirana [7]. Reproduced with permission of the European Space Agency (ESA).

1.5.4 Navigation Message

Each GPS satellite has a unique set of data called the navigation message modulated onto the signals it broadcasts. The navigation message contains key information that receivers require in order to perform positioning, including:

- *Ephemeris data*: Orbital data (Kepler elements and perturbation parameters) used to predict the trajectory of a satellite.
- *Satellite clock information*: Coefficients of a second degree polynomial indicating the bias, drift, and drift rate of the atomic clock on board a satellite.
- *Satellite health information*: Service parameters indicating whether a satellite's signal is to be trusted or not.
- *Almanac data*: Ephemeris, clock, and health data for every satellite in the constellation, but with reduced accuracy.
- *Additional parameters*: Ionospheric correction parameters for single frequency receivers, data for conversion between GPS time and UTC

The control segment continuously monitors the GPS constellation and updates the parameters of the navigation message. Processing techniques such as Kalman filtering are implemented to provide the most accurate parameters possible. Ephemeris and clock parameters are assigned individually to each satellite, and must be updated frequently to maximize their precision. The control segment typically sends these updates every two hours. The almanac data does not need to be as precise, so the control segment only updates this data about once every six days. If a satellite is unable to establish contact with the control segment for some reason, satellites may use their crosslink channel to share updates among themselves.

It is worth exploring the types of data included in the navigation message a bit further to understand the role they play in GPS operation. Ionospheric correction parameters will be a subject of separate consideration in this work, so their discussion is delayed until Chapters 7 and 8.

Ephemeris Data

In Section 1.4, it was shown that a satellite's orbit may be modeled by an elliptical trajectory using six parameters called Kepler elements. However this is a highly idealized model, and in reality orbits are far more complex. A satellite is subject to a number of perturbations that may cause it to deviate from the Keplerian ellipse. A list of perturbing forces [11, p. 83] is as follows:

1. Non-uniform gravitational forces arising from non-sphericity and inhomogeneity of the mass distribution of Earth.
2. Gravitational forces from the moon, the Sun, planets, and other celestial bodies in our solar system.

3. Tidal forces which deform the shape of the Earth, altering its gravitational potential.
4. Radiation pressure from photons, both directly from the Sun and reflected from Earth.

The six Kepler elements defining an ideal trajectory, as well as a number of correction terms accounting for perturbing forces, are collectively referred to as ephemeris data. Ephemeris (plural: ephemerides) comes from the Latin word for diary, and refers to the apparent position of a celestial object in the sky at a given time. A GPS receiver uses ephemeris data to estimate the position of a satellite, which is essential to its ability to calculate its own position.

Every satellite in the GPS constellation broadcasts its own ephemeris data as part of its navigation message. The receiver extracts this data and stores it locally as a binary string. The ephemeris information is forwarded to the control segment for monitoring and processing, and updated data is sent back to the satellites via the S-band uplink every two hours. The binary data can also be downloaded and converted into an ASCII format called a Receiver Independent Exchange (RINEX) file.

Additionally, every satellite's navigation message includes almanac data, which includes ephemerides for all satellites in the constellation. The ephemerides included in the almanac use fewer parameters, so they are a coarser estimate of position than a satellite's individualized broadcast ephemeris. The almanac is updated by the control segment at least every six days.

Every GPS receiver has access to the ephemeris data encoded in the navigation message, but the parameters included in the message are coarsely calculated, and only provide positioning accuracy to within a few meters. Applications that require centimeter or millimeter level accuracy may benefit from even more accurate ephemeris data. A global federation of universities, research institutions, and other scientific agencies called the International GNSS Service (IGS) collects and shares GPS data. Data tracked by the IGS includes GPS satellite ephemerides, tracking station positions and velocities, Earth rotation parameters, clock information for satellites and tracking stations, zenith tropospheric delay estimates, and global ionosphere maps [13].

Post-processing techniques are used to further refine the accuracy of GPS ephemerides at various IGS analysis centers. The data is archived in regional IGS data centers, and the analysis center coordinator compiles the analysis centers' data into a concise, deliverable product for distribution.

There are various tiers of ephemeris data available, and increased accuracy comes at the cost of longer processing time. For example: "Ultra-Rapid" GPS ephemeris data is available in real time, but the most accurate "Final" ephemeris takes 12–18 days of post-processing. It is easy to see how the former data set is useful – applications such as navigation benefit from improved positioning

Table 1.8 Sets of ephemeris data available from the IGS.

Type	Accuracy		Latency	Update interval	Sample interval
Broadcast	Orbits	~ 100 cm	Real time	—	Daily
	Sat. clocks	~ 5 ns RMS ~2.5 ns SDev			
Ultra-rapid (predicted half)	Orbits	~ 5 cm	Real time	Every six hours (03, 09, 15, 21 UTC)	15 min
	Satellite clocks	~ 3 ns RMS ~1.5 ns SDev			
Ultra-rapid (observed half)	Orbits	~ 3 cm	3–9 h	Every six hours (03, 09, 15, 21 UTC)	15 min
	Satellite clocks	~ 150 ps RMS ~50 ps SDev			
Rapid	Orbits	~ 2.5 cm	17–41 h	Daily (At 17 UTC)	15 min
	Sat. and station clocks	~ 75 ps RMS ~25 ps SDev			5 min
Final	Orbits	~ 2.5 cm	12–18 d	Every Thursday	15 min
	Sat. and station clocks	~ 75 ps RMS ~20 ps SDev			Sat.: 30 s Stn.: 5 min

Source: International GNSS Service [13].

accuracy in real time. The usefulness of data that takes hours or even days to produce is less obvious. An example is a station designed to monitor tectonic plate drift. Since tectonic plates move extremely slowly (millimeters per year), the receiver must be able to achieve millimeter level accuracy to discern minute differences in position. A latency of days to weeks in processing data is not significant compared with the months or years over which meaningful movement of tectonic plates occurs. The need for the greatest accuracy possible, regardless of latency makes Final ephemeris data perfectly suited to such an application.

Table 1.8 compares the various ephemeris data sets provided by the IGS [13].

GPS Time System

The modern SI unit definition of a second is derived from the hyperfine transitions of electrons in certain atoms. Two such atomic frequency standards (AFS) are currently used: the cesium standard, which uses Cs-133, and rubidium standard, which uses Rb-87. The cesium standard is more accurate and is the primary AFS, but rubidium atomic clocks are less expensive and more compact that cesium clocks, so the rubidium standard has been adopted as a secondary AFS.

The chosen hyperfine transition of Cs-133 oscillates at a frequency of 9,192,631,770 Hz, meaning one second in the cesium standard is defined by 9,192,631,770 oscillations of Cs-133. Similarly, the chosen hyperfine transition of Rb-87 oscillates at 6,834,682,610.904 Hz, so one second in the rubidium standard occurs every 6,834,682,610.904 oscillations of Rb-87. The frequency of the hyperfine oscillations in the atoms of a given clock are known as the *nominal frequency* of that clock.

Atomic clocks are chosen as the basis for time systems because they are incredibly stable. Stability of a clock is defined as the drift in the nominal frequency over time. Rb clocks have a nominal stability of about 10 ns/day ($\sim 10^{-13}$ s/s), while Cs clocks are even more stable, exhibiting a drift of only about 1 ns/day ($\sim 10^{-14}$ s/s). Mathematically, a drift in time is equivalent to a drift in frequency, and can be expressed by the expression: $\delta t / T = \delta f / f_0$.

International Atomic Time (TAI) is a time system that uses AFS to provide a universal definition of the second that may be referenced by scientists worldwide. However, this definition differs very slightly from the prior definition of the second, which was based on the rotation of the Earth. In this time system, called *sidereal time*, one full period of Earth's rotation takes one day, and one second is defined as $1/86,400$ days ($24 \times 60 \times 60 = 86,400$).

A second modern time system called UTC was introduced alongside TAI to define a global time that is approximately in sync with the rotation of the Earth. This is done by periodically adding a leap second to TAI, meaning UTC always differs from TAI by an integer number of seconds. TAI and UTC currently differ by 37 seconds (TAI − UTC = 37).

GPS utilizes a time system based on AFS, very similar to TAI. This time system, called GPS time, is defined using an ensemble of all atomic clocks in GPS satellites and monitoring stations. Similar to TAI, GPS time does not utilize leap seconds, but GPS time was originally synchronized to UTC at its inception on 5 January 1980. Therefore, GPS time currently differs from UTC by only 18 seconds.

Since GPS time is an accurate, reliable, and easily accessible time system, many users simply utilize GPS for timing applications, disregarding the positioning capability entirely. Most of these applications function using UTC however, so the navigation message of all GPS satellites includes a conversion between UTC and GPS time.

Each satellite also broadcasts a set of parameters indicating the error in its onboard atomic clock. This information is represented as the coefficients of a second degree polynomial whose terms represent the bias, drift, and drift rate of the clock at the epoch of transmission (discussed further in Section 2.3.1). Each satellite broadcasts this information about itself every 30 seconds. Additionally, the almanac data included in every satellite's navigation message includes clock correction data with reduced accuracy, similar to ephemeris data.

GPS time is tracked using a number called the Z-count. The Z-count measures the amount of time elapsed since the zero time point at the midnight between 5 January and 6 January 1980. This number is represented by a 29-bit string: the 10 most significant bits representing GPS week and the 19 least significant bits representing the number of X1 epochs (1.5 seconds intervals) elapsed since the beginning of the week. The reason X1 epochs are used instead of seconds is that it helps receivers switch over from C/A code to P-code. This is explained further later in this section as part of a discussion on the handover word (HOW).

GPS week represents how many weeks have passed since GPS time began on 6 January 1980, starting with week zero. Only 10 bits are allocated for GPS week, meaning the largest number of weeks that can be represented in this format is 1023; week 1024 rolls back over to week 0. Note that 1024 weeks is approximately 19.6 years, so one rollover occurred on 22 August 1999 and the latest rollover occurred on 7 April 2019.

GPS week transitions occur at the transition between Saturday and Sunday. The time within a given week is given by the number of X1 epochs elapsed since the week transition, and is referred to as the time of week (TOW). The number of seconds in a week is 604,800, so the number of X1 epochs is 604,800/1.5 = 403,200. This may be represented by a 19-bit binary number, ranging from 0 to 403,199.

Almanac Data

The GPS almanac is a collection of data that includes information on every satellite in the GPS constellation. The information includes rough estimates of the satellites' orbital parameters and clock information. The almanac is encoded onto the navigation message of every satellite, is updated by the control segment at least every six days, and is considered accurate for 180 days after being updated.

GPS almanac data is presented in a text file called a YUMA file, available from the US Coast Guard website [14]. An example YUMA file for SV-01 at the epoch GPS week 859 (actually week 1883 due to rollover) is shown in Figure 1.10.

One of the primary functions of the almanac is to reduce the time to first fix (TTFF) of a receiver. A receiver may store almanac data in its non-volatile memory in order to reference it upon start-up. Almanac data allows the receiver to approximate the current time to within ~20 seconds and the approximate positions of every satellite in the constellation to within ~100 km. This information allows the receiver to reduce the TTFF. For example, a given receiver has a good idea of which satellites will be overhead and may begin searching for those right away, rather than wasting its time searching for satellites below the horizon. Utilizing almanac data in this fashion is known as a warm start, and may reduce the TTFF to as little as 30 seconds.

Conversely, a cold start is one in which almanac data is either unavailable or out of date. For example, upon factory start the receiver will have no data stored locally and will need to find a satellite with no prior knowledge of its position, velocity or the time. Once locked onto a satellite, the receiver may wait to receive the entire navigation message, which includes almanac data, and use it to acquire the remaining satellites. Downloading the entire navigation message takes 12.5 minutes. Another case of a cold start is when the receiver has been switched off for

```
******** Week 859 almanac for PRN-01 ********
ID:                          01
Health:                      000
Eccentricity:                0.5026340485E-002
Time of Applicability (s):   233472.0000
Orbital Inclination (rad):   0.9641153141
Rate of Right Ascen (r/s):   -0.7908900866E-008
SQRT (A) (m 1/2):            5153.609375
Right Ascen at Week (rad):   -0.2792631499E+000
Argument of Perigee (rad):   0.469108253
Mean Anom(rad):              0.6852946818E+000
Af0(s):                      0.1144409180E-004
Af1(s/s):                    0.0000000000E+000
week:                        859
```

Figure 1.10   Example YUMA file. Source: Based on United States Coast Guard [14].

a longer than 180 days or the receiver has been moved hundreds of miles without updating the almanac. In these cases, the satellites that the receiver thinks should be overhead are not there, so the receiver will have to search for satellites blindly, as if it had no almanac data at all.

Finally, a receiver may also perform a hot start if it has current ephemeris, almanac, time, and position data. Clock and ephemeris data are updated every 30 seconds, so a receiver may only perform a hot start if it is restarted within seconds of switching off. A hot start takes anywhere from 0.5 to 20 seconds.

Navigation Message Structure

The navigation message is modulated onto a carrier frequency at 50 bps. It contains 25 "pages" of data called frames, each subdivided into five subframes. Each subframe is 300 bits long so at 50 bps, each subframe is six seconds long and a whole frame lasts 30 seconds. The entire navigation message is 25 frames long, resulting in a total message duration of 12.5 minutes.

Each subframe is numbered 1–5 and each is designated a specific set of information. Subframe 1 contains clock correction parameters and health information for the broadcasting satellite as well as the 10-bit number representing GPS week. Subframes 2 and 3 contain the satellite's ephemeris data. These three subframes contain fairly accurate parameters, as represented by the "Broadcast" row of Table 1.8. To minimize the TTFF of a receiver, these three subframes are reproduced in every frame of the navigation message. That way, a receiver never has to wait more than 30 seconds to receive full ephemeris and clock data from a satellite.

Subframes 4 and 5 are different from the first three. The data transmitted in these subframes are common to all satellites in the constellation, as opposed to subframes 1–3, which transmit satellite-specific data. Secondly, rather than repeating the same information every frame, these subframes each contain a small piece of a much larger set of data. Every frame transmits 1/25 of the total information, and the full data set is completely transmitted over the course of 12.5 minutes. Subframe 4 contains ionospheric correction parameters, information for conversion between GPS time and UTC, and part of the almanac data. Subframe 5 is dedicated entirely to almanac data. Figure 1.11 (adapted from [15]) provides a visualization of the navigation message's structure.

Each subframe is divided into 10 words, each 30 bits in length. The first two words in every subframe are the telemetry word (TLM) and the HOW.

The TLM begins with an 8-bit preamble (10001011) that never changes, regardless of the contents of the subframe. This preamble is used to allow the receiver to distinguish the beginning of a new frame. Following the preamble are 16 bits of reserved data and 6 parity bits. The parity bits are a checksum allowing the receiver to confirm that it is receiving the beginning of a new subframe and not a very similar bit string instead. To do this, the receiver decodes the 16 reserved data bits, creates a parity, and checks to ensure it corresponds to the last 6 bits of the

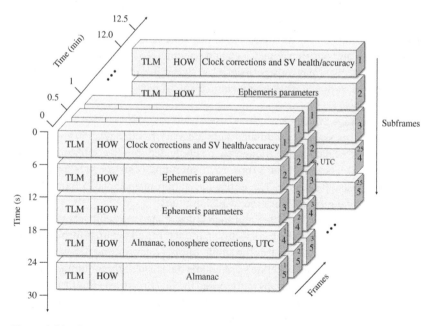

Figure 1.11 Structure of the navigation message.

TLM. If it does, the receiver moves on to the HOW. If not, the receiver restarts its search for the beginning of a new subframe.

The HOW begins with a 17-bit number. This number is the 17 most significant bits of the 19-bit TOW. Recall that TOW represents how many X1 epochs (1.5 seconds intervals) have elapsed since the beginning of the GPS week. The truncated 17-bit number ranges from 0 to 100,799 and represents the number of 6 second subframes elapsed since the beginning of the week. The number contained in the HOW corresponds to the arrival time of the leading edge of the next subframe.

Since the receiver knows the number of subframes elapsed since the beginning of the week, it may simply multiply by 4 to receive the number of elapsed X1 epochs. Each X1 epoch corresponds to 15,345,000 chips of P-code, so it is possible to determine the exact point of the P-code that is currently being broadcast. This allows a receiver to acquire C/A code, use the HOW to determine the status of the P-code and then switch to it, hence the name HOW.

The HOW also contains two flag bits: an alert flag which indicates that the satellite may be giving an inaccurate measurement, and an anti-spoof flag indicating whether anti-spoofing is currently enabled or not. Following these are three subframe identifier bits indicating the number of the subframe, 1–5. The final six bits are parity bits, and function exactly the same as in the TLM. If this second parity check is passed, the receiver will go on to decode and use the data encoded in the remainder of the frame. If the parity check fails, the receiver will start all over, searching once again for the TLM preamble indicating the beginning of a frame.

Figure 1.12 (adapted from [15]) shows the basic structure of the telemetry and HOWs. After the TLM and the HOW, the remaining eight words in a subframe are

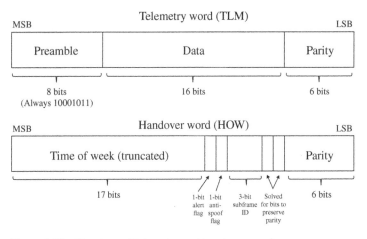

Figure 1.12 Structure of telemetry and handover words.

dedicated to the data specified by the subframe (e.g. ephemeris data for frames 2 and 3).

Modern GPS signals broadcast different versions of the navigation message. All of these messages broadcast the same data (ephemeris, almanac, clock correction, etc.), but in different formats. Each modern signal has its own unique navigation message with different names: L2C uses CNAV, both M-codes use MNAV, L5C uses L5-CNAV, and L1C uses CNAV-2. To distinguish it from modern messages, the original navigation message is called the "legacy navigation message" or LNAV. Some of the features included in the modern navigation messages include:

- FEC and cyclic redundancy checks (CRC) to detect and correct erroneous transmissions.
- Rather than using frames, which must always be transmitted in the same pattern, modern navigation messages (excluding CNAV-2) are "packetized," enabling specific types of information (e.g. ephemeris, almanac, etc.) to be broadcast in a flexible order, allowing greater versatility.
- The time offset between GPS and other GNSS systems is included, allowing for greater interoperability between GNSS constellations.
- GPS week is broadcast using 13 bits instead of just 10. This increases the maximum week number to 8192, meaning that GPS week rollover will only occur once every 157.1 years rather than every 19.6 years.

A detailed discussion of these modern messages is beyond the scope of this text, but further information may be found in the GPS Interface Specification documentation. Documentation for L1 C/A, L2C and P(Y) codes may be found in [16]; for L5C in [17]; and for L1C in [18].

2

Principles of GNSS Positioning

2.1 Introduction

In this chapter, we broaden our focus as compared with Chapter 1; rather than concerning only on GPS, the topics in this chapter will apply to global navigation satellite systems (GNSS) more broadly. In particular, this chapter covers the process by which GNSS receivers perform positioning based on the signals they receive from a satellite constellation. There are several physical observables that can be used to perform positioning, each with their own advantages and disadvantages. Section 2.2 covers the three major observables of interest: pseudorange, carrier phase, and Doppler shift.

Although GNSS observables can be modeled simply, these models do not account for the numerous practical complexities introduced by the real world. For example, a GNSS receiver calculates pseudorange by comparing the transmission and reception times of a received GNSS signal, and then multiplying this travel time by the speed of light. However, this simplistic model fails to account for the slower signal speed and longer propagation path induced by refraction in the atmosphere. In fact, many of the atmospheric phenomena discussed in later chapters contribute significantly to GNSS positioning error.

Since atmospheric sources of error are covered throughout the rest of this book, Section 2.3 is dedicated to discussion of other prominent sources of error. For example, relativistic effects and clock synchronization issues would degrade the accuracy of GNSS timestamps if left unaccounted for. Additionally, effects such as multipath interference and diffraction can cause loss of phase lock (an undesirable effect that negatively affects positioning when using the carrier phase observable, as discussed in Section 2.2.2). All sources of GNSS positioning error can be collected into a single error term that is added to the observable of interest. Precise positioning then becomes a matter of minimizing this error term.

Section 2.4 presents a derivation of the equations used for GNSS positioning based on the pseudorange observable. A system of four equations and four

Tropospheric and Ionospheric Effects on Global Navigation Satellite Systems, First Edition.
Timothy H. Kindervatter and Fernando L. Teixeira.
© 2022 The Institute of Electrical and Electronics Engineers, Inc. Published 2022 by John Wiley & Sons, Inc.

unknowns is required to ascertain a receiver's position, which requires signals from four distinct satellites. One might assume that only three equations would be needed, since the receiver's coordinates in 3D space appear to be all that's needed. However, as noted in the Chapter 1, a fourth unknown, namely, the bias of the receiver's clock with respect to the GNSS time reference (e.g. GPS time), must also be obtained. This can be done simply by including the signal from a fourth satellite in the calculation.

This section goes on to discuss further topics related to GNSS positioning. These include an iterative algorithm that rapidly converges to the solution of the system of equations, use of the method of least squares to account for random error, and a phenomenon known as dilution of precision (DOP) wherein uncertainties in individual GNSS observables are magnified by the geometric configuration of satellites.

Finally, this chapter closes with a discussion on data combinations in Section 2.5.1. By combining GNSS observables, it is often possible to completely remove the contribution of a certain source of error. For example, since the ionosphere is dispersive, waves of different frequencies travel at different speeds. Thus the pseudorange measurement obtained on, say, the L1 GPS frequency would be different than the pseudorange measurement obtained on the L2 GPS frequency. By solving the two resulting equations simultaneously, it is possible to mostly remove the ionosphere's contribution to positioning error. This technique is called the ionosphere-free combination. A range of similar techniques is discussed in the rest of this section.

2.2 Basic GNSS Observables

There are three basic quantities available to a GNSS receiver for ranging, also known as *GNSS observables*. The three quantities are the code pseudorange (commonly just "pseudorange"), the carrier phase measurement, and the Doppler frequency shift of the received signal. All three of these measurements may be used to deduce the distance between a satellite and a receiver, and the Doppler shift may also be used to determine the receiver's velocity relative to a satellite.

Throughout this section, a quantity related to a satellite will be denoted with a superscript s, as in t^s. Similarly, a quantity related to a receiver will be given the subscript r, as in t_r. A quantity relating satellite and receiver terms may be represented using both a subscript r and a superscript s, as in R_r^s.

2.2.1 Pseudorange

In Section 1.5.2, it was shown that the time delay between transmission and reception of a signal could be determined via correlation of the modulated code at the

GNSS receiver. Since electromagnetic signals travel at the speed of light c, it is possible to determine the distance between a satellite and receiver by multiplying a signal's time delay by c.

Consider a satellite that transmits a signal when its internal clock reads T^s, which then travels toward Earth where a receiver picks it up. The reception time according to the reading on receiver's internal clock is T_r. If the clocks on the satellite and the receiver were perfectly synchronized, this would be the only information required to perform ranging. In practice, however, it is impossible to perfectly synchronize the clocks. To solve the synchronization problem, the clocks in both the satellite and the receiver are referenced to a common time system – in this case, GNSS time (see Section 1.5.4). Each clock will exhibit an error, or *bias*, with respect to GNSS time. The satellite clock bias is denoted δ^s and the receiver clock bias is denoted δ_r. Therefore in GNSS time, $T^s = t^s + \delta^s$ and $T_r = t_r + \delta_r$ where t^s and t_r are the absolute transmission and reception times. Subtracting the transmission time from the reception time gives the total travel time of the signal. Multiplying this quantity by c results in:

$$R_r^s = c(T_r - T^s) = c(t_r + \delta_r) - c(t^s + \delta^s) = c\tau + c(\delta_r - \delta^s), \qquad (2.1)$$

where $\tau = t_r - t^s$ and R_r^s is the pseudorange measurement. This is not the true range between the satellite and the receiver, because the clock biases introduce error into the measurement! The true range is given by $\rho = c(t_r - t^s)$. In addition to clock biases, numerous other sources of error degrade the accuracy of the pseudorange measurement. Many of these sources of error will be discussed in Section 2.3. For now, these additional sources of error may be combined into one term, e_r^s. Combining all of this information gives us the most basic form of the pseudorange equation:

$$R_r^s = \rho + c(\delta_r - \delta^s) + e_r^s. \qquad (2.2)$$

By utilizing the clock parameters broadcast by the satellite's navigation message, the satellite clock bias can be accounted for, so $\delta^s \to 0$. Similarly, the ephemeris data from the navigation message gives the approximate location of the satellite. This location can be represented in Cartesian coordinates, where x^s, y^s, and z^s designate the satellite's position in three dimensions. In addition, the receiver is situated at Cartesian coordinates x_r, y_r, and z_r. The displacements in each direction are given by $\Delta x = x^s - x_r$, $\Delta y = y^s - y_r$ and $\Delta z = z^s - z_r$. Using the Pythagorean theorem, the distance (i.e. the true range) between the satellite and the receiver is $\rho = \sqrt{(x^s - x_r)^2 + (y^s - y_r)^2 + (z^s - z_r)^2}$. Therefore, Equation 2.2 can be rewritten as

$$R_r^s = \sqrt{(x^s - x_r)^2 + (y^s - y_r)^2 + (z^s - z_r)^2} + c\delta_r + e_r^s. \qquad (2.3)$$

Recall that x^s, y^s, and z^s are known a priori by the receiver from the ephemeris data in the navigation message. Neglecting the error term, this leaves four unknowns in Equation 2.3: x_r, y_r, z_r, and δ_r. To solve for these four unknowns, we need four linearly independent equations. These can be obtained by observing the pseudorange of four different satellites simultaneously. The details of this positioning calculation are provided in Section 2.4.

Typical GPS receivers are capable of measuring pseudorange to within 1% of a code's chip width. For C/A code the chip width is roughly 300 m, resulting in a resolution of about 3 m. P(Y) code is 10 times more precise, having a chip width of about 30 m and a resolution of about 0.3 m.

Basic Configurations

For either a stationary or a moving receiver, the number of satellites required to perform positioning at a single epoch is four. Additionally, for a stationary receiver, fewer satellites may be used if measurements from several epochs are considered. The following derivations (as well as a similar examination using the carrier phase measurement in Section 2.2.2) are adapted from [19].

At a given epoch, n_s satellites are available for observation. At subsequent epochs, the same satellites may be observed again to provide additional n_s measurements. Thus over n_t epochs, the total number of observations available to a receiver from n_s satellites is $n_s n_t$.

Consider a stationary receiver. For a single epoch, the number of unknowns is four: the three components of the receiver's position and the receiver clock bias. Over several epochs, the coordinates of the receiver remain constant, since the receiver is stationary. However, due to clock drift, the receiver clock bias will not remain constant with time so a new receiver clock bias must be calculated at every epoch. Therefore, over n_t epochs, the number of unknowns to be calculated is $3 + n_t$. In order for the receiver to calculate its position, the number of observations must be equal to or greater than the number of unknowns. Thus we are given the expression $n_s n_t \geq 3 + n_t$. We may solve for n_t, which gives us:

$$n_t \geq \frac{3}{n_s - 1}. \tag{2.4}$$

We see that for a single epoch, i.e. $n_t = 1$, we must have at least $n_s = 4$ satellites to obtain a solution. The minimum number of satellites which provides a positioning solution is $n_s = 2$, in which case n_t must equal 3. This means that only two satellites may be used for positioning as long as they are observed over at least three epochs. In theory, using fewer satellites over several epochs seems like an attractive solution; however, in practice this solution requires the epochs to be spaced by several hours to ensure that the satellites had moved sufficiently far from their previous position. Otherwise, the successive measurements of a given satellite would be too similar and the positioning measurement would provide unacceptable results or even fail entirely. Due to this limitation, observations from four satellites are used simultaneously, which always ensures an accurate positioning measurement.

Now consider the case of a moving receiver. The number of observations available from n_s satellites over n_t epochs is still $n_s n_t$. However, the number of unknowns has now changed. Since the receiver is no longer stationary, the components of its position now change with time. These components must be recomputed at every epoch, providing $3n_t$ unknowns over n_t epochs. The receiver clock bias must also be computed at every epoch, as before, so the total number of unknowns over n_t epochs is now $4n_t$. Relating the number of observations and the number of unknowns now gives us $n_s n_t \geq 4n_t$, which simply reduces to $n_s \geq 4$. Note that the n_t term cancels out, meaning we cannot solve for it in this case. This makes sense because none of the unknowns remain constant, so a previous epoch's observations are useless for calculating the receiver's current position. However, at any given epoch the position may always be computed if there are at least four satellites available.

2.2.2 Carrier Phase

The other main GNSS observable is the carrier phase. Every GNSS signal oscillates with a certain carrier frequency. The period, or the time in seconds that it takes for the signal to complete one full cycle, is given by $T = 1/f$, where f is the carrier frequency. Phase is commonly measured in radians, taking values between 0 and 2π; however, in GNSS applications, phase is typically measured in cycles instead.

To convert phase to cycles, the value in radians is divided by 2π, resulting in values that range from 0 to 1. Therefore, the phase of the wave denotes what fraction of a full cycle has passed. Since phase is a unitless quantity, it may be multiplied by, say, the wavelength of the carrier signal to obtain a measurement in meters. This can be used to determine the distance between a satellite and receiver.

Phase must be measured with respect to some reference value. At some arbitrary initial time t_0, the phase of the wave will be ϕ_0. At some time later, t, the phase value will be ϕ. The phase difference between these two points is $\Delta\phi = \phi - \phi_0 = (t - t_0)f$, and its value is modulo 2π since 2π represents one full cycle. This is depicted in Figure 2.1. When a satellite transmits a signal, it will have some phase $\phi^s(t) = \phi_0^s + \Delta\phi = \phi_0^s + (t - t_0)f^s$. Upon reception, the receiver will generate a replica signal at the same frequency, but which in general has a different initial phase. The phase of this wave will be $\phi_r(t) = \phi_{0,r} + \Delta\phi = \phi_{0,r} + (t - t_0)f_r$. The satellite and receiver clocks will each be referenced to GNSS time, as discussed previously. Therefore, there will be an additional phase offset of $f^s \delta^s$ in the satellite-broadcasted signal, and $f_r \delta_r$ in the replica signal. When the satellite-broadcasted signal arrives at the receiver, it will have undergone a time

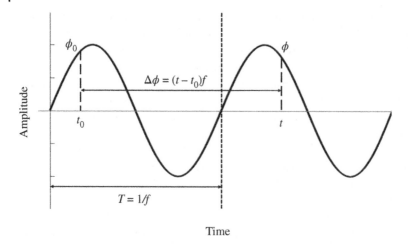

Figure 2.1 Phase relationships for a sinusoidal signal.

delay τ, equal to the travel time between the satellite and receiver. Therefore, the phase difference between the received signal and the replica signal will be

$$\Delta\phi_r^s = \phi_r(t) - \phi^s(t - \tau)$$
$$= [(t - t_0)f_r + \phi_{0,r} + f_r\delta_r] - [(t - \tau - t_0)f^s + \phi_0^s + f^s\delta_s]$$
$$= (f_r - f^s)(t - t_0) + f^s\tau + f_r\delta_r - f^s\delta^s + \Delta\phi_0, \tag{2.5}$$

where $\Delta\phi_0 = \phi_{0,r} - \phi_0^s$. For simplicity, we may choose ϕ_0^s and $\phi_{0,r}$ as reference points for the phase broadcast and replica signals, respectively. In other words, this allows us to set $\phi_0^s = \phi_{0,r} = 0$, effectively removing the term $\Delta\phi_0$ from the expression.

The frequencies f^s and f_r are referenced to a nominal frequency f_0 which is defined in terms of the ensemble atomic frequency standard used by GNSS time. Typically, f^s and f_r only deviate from the nominal frequency by some tiny fraction of 1 Hz. Therefore, the phase contribution of the term $(f_r - f_s)(t - t_0)$ is $\mathcal{O}\left(10^{-4}\right)$ radians and may be neglected [19, p. 107]. In the remaining terms, f^s and f_r may be approximated as f_0, giving us

$$\Delta\phi_r^s = f_0\tau + f_0(\delta_r - \delta^s). \tag{2.6}$$

Integer Ambiguity

When a GNSS receiver measures carrier phase, it is only capable of reading the fraction of the current full cycle. It is impossible for the receiver to know how many full cycles were traversed prior to reception. As a simple analogy, consider a solid disc with a single notch in the top (refer to Figure 2.2). If the disc is rotating, a full cycle occurs when the notch returns to its original position. If one

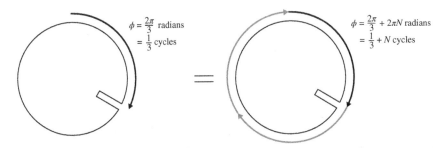

Figure 2.2 Integer ambiguity represented by a spinning disc.

wishes to measure the present fraction of a cycle that has been traversed by the notch, they may take a snapshot of the disc and observe its location. The angle between the top of the disc and the current location of the notch is analogous to the phase of the carrier signal upon reception. However, by only looking at the snapshot of the disc, it is impossible to determine how many full revolutions the disc has undergone since it started spinning. Thus, when a GNSS receiver observes the phase of the carrier signal, the measurement is ambiguous. Adding any integer number of cycles N to this equation does not change the phase at all. This phenomenon is appropriately termed *integer ambiguity*. Therefore, the *total* carrier phase, Φ_r^s, equals the observed fractional carrier phase given by Equation 2.6 plus an unknown integer N.

$$\Phi_r^s = \Delta\phi_r^s + N = f_0\tau + f_0(\delta_r - \delta^s) + N. \tag{2.7}$$

Multiplying Φ_r^s by the wavelength of the carrier signal results in the total number of wavelengths traversed between the satellite and the receiver. This is equivalent to the distance between the two, and can thus be used for ranging. Recalling that $c = f_0\lambda$ for an electromagnetic signal with frequency f_0, we get

$$\lambda\Phi_r^s = c\tau + c(\delta_r - \delta^s) + \lambda N.$$

Recall that τ is simply the travel time of the signal, so $c\tau$ is the true range ρ. Using Equation 2.1, we see that

$$\lambda\Phi_r^s = R_r^s + \lambda N. \tag{2.8}$$

In other words, the carrier phase measurement only differs from the pseudorange measurement by an integer number of wavelengths.

As for pseudorange, carrier phase measurements are degraded by various error sources. These error sources will be elaborated upon in Section 2.3, but for now we will denote their overall net effect as e_r^s. Thus the carrier phase equation finally becomes

$$\boxed{\lambda\Phi_r^s = R_r^s + \lambda N + e_r^s.} \tag{2.9}$$

The carrier phase is able to be measured far more precisely than the pseudo-range, offering a resolution of up to 1% of the carrier wavelength. For L1 this corresponds to a resolution of 2.0 mm, and for L2 a resolution of 2.4 mm. Though these measurements are very precise, they are ambiguous due to the unknown integer N.

Positioning with Carrier Phase

We see from Equation 2.9 that the carrier phase measurement only differs from the pseudorange measurement by an integer number of wavelengths. Thus, once the receiver has calculated the pseudorange (as per the methods outlined in Section 2.4), the only remaining task is to solve for the integer ambiguity N. Each satellite has its own individual carrier phase measurement and thus its own integer ambiguity N^j, where the superscript j refers to the jth satellite. Because of this, each different carrier phase measurement introduces a new unknown to the system of equations, so simply observing more satellites does not help solve the ambiguity problem. Additionally, N is different for each carrier frequency, so for example N_{L1} and N_{L2} are not equal.

Fortunately, N is constant for a given satellite and a given frequency over time. Therefore, if the receiver establishes a phase lock with a satellite, the carrier phase on that frequency may be measured over successive epochs to solve for N. Consider a receiver switched on at epoch t_0; at this epoch a measurement of the carrier phase is taken (denoted by ϕ_r^s). If the carrier phase is measured again some time later, it will equal the carrier phase at t_0 plus some additional elapsed phase $\Delta\phi_r^s$. Therefore, the carrier phase measurement as a function of time is given by:

$$\phi_r^s(t) = \Delta\phi_r^s(t) + N, \tag{2.10}$$

where $\Delta\phi_r^s = \phi_r^s(t) - \phi_r^s(t_0)$. Figure 2.3 [19] illustrates this geometrically, where $\Delta\phi_i$ is shorthand for $\phi_r^s(t_i) - \phi_r^s(t_0)$. For simplicity, we may assume that $\Delta\phi_0 = 0$. The receiver may use the measurements from several epochs to calculate the ambiguity term.

When considering measurements from multiple satellites, as we must to perform positioning, there are a number of unknowns to be considered. Let's take stock of them. Consider the carrier phase measurement for a stationary receiver on a single frequency, say L1. For a single epoch, the number of unknowns is $4 + n_s$, where n_s is the number of satellites under observation. These unknowns are:

1. Receiver coordinates: x_r, y_r, z_r.
2. Receiver clock bias: δ_r.
3. Ambiguities on L1: $N_1^j; j = 1, 2, \ldots, n_s$.

Each satellite provides one observation, so the number of observations available is n_s. Every measurement made at successive epochs provides an additional n_s

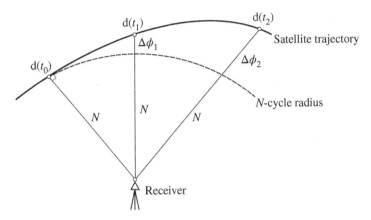

Figure 2.3 Geometric representation of carrier phase measurement over several epochs. The N-cycle radius is the set of points which are N full wavelengths away from the receiver, so a signal that traverses that distance will accrue a phase of $N\lambda$. At time t_i, the signal will accrue an additional phase of $\delta\phi_i$, which depends on the distance between the satellite's position at t_i and the N-cycle radius.

observations, so for n_t epochs, the number of observations is $n_s n_t$. The ambiguities always remain constant, and since the receiver is stationary, the receiver coordinates remain constant as well. Conversely, due to clock drift, the receiver clock bias is not constant with time. Therefore a new clock bias needs to be calculated at every epoch. Therefore, for n_t epochs, the number of unknowns is now $3 + n_s + n_t$. These unknowns are:

1. Receiver coordinates: x_r, y_r, z_r.
2. Receiver clock biases: $\delta_{r,i}$; $i = 1, 2, \ldots, n_t$.
3. Ambiguities on L1: N_1^j; $j = 1, 2, \ldots, n_s$.

The design matrix for this system has a rank deficiency of 1 (see [19, p. 164] for details), making our model unsolvable in its current form. To solve this, we must arbitrarily choose a value for one of our unknowns. If we fix a value for the receiver clock bias at the first epoch, we now have $3 + n_s + (n_t - 1)$ unknowns. To solve for all unknowns, the number of observables must be greater than or equal to the number of unknowns. For n_s satellites observed over n_t epochs, we have $n_s n_t \geq 3 + n_s + (n_t - 1)$. Solving for n_t gives us:

$$n_t \geq \frac{n_s + 2}{n_s - 1}. \tag{2.11}$$

We see that if measurements from four satellites are considered, $n_t \geq 2$ satisfies Equation 2.11. In other words, it takes at least two epochs for a receiver to calculate its position using the carrier phase measurements from four different satellites. In general, any combination of n_s and n_t which satisfies Equation 2.11 will provide a positioning solution. The minimum number of satellites which provides a solution is $n_s = 2$, in which case $n_t \geq 4$.

Now consider the case of a moving receiver. In this case, the receiver coordinates change with time, so over n_t epochs and fixing the first receiver clock bias, we have: $n_s n_t \geq 3n_t + n_s + n_t - 1$. Solving for n_t gives:

$$n_t \geq \frac{n_s - 1}{n_s - 4}. \tag{2.12}$$

In this case, the minimum number of satellites to obtain a solution is $n_s = 5$, and they must be tracked for $n_t \geq 4$ epochs. Note that a solution for only one epoch is available in neither the stationary nor the kinematic case.

Cycle Slips

The phase difference $\Delta\phi_i$ is known because the receiver remains phase locked with the satellite. If the receiver ever loses lock with the satellite, it will need to recalculate N upon re-establishing lock because the integer ambiguity will not necessarily remain the same, even if phase lock is only lost very briefly. This phenomenon is known as a *cycle slip*. Cycle slip detection and correction is a large topic by itself and is beyond the scope of this text. Discussion of these topics may be found in [7, 11, 19, 20].

Loss of lock occurs when the receiver drops the GNSS signal, so it is important for satellites to broadcast their signals at a sufficiently high power to minimize the number of dropped signals. Physical phenomena that may cause significant fading at GNSS frequencies include attenuation from atmospheric gases and water vapor, scintillations in the troposphere and ionosphere, and multipath interference. As a result, these phenomena must be modeled and accounted for in order to ensure consistent GNSS operation. A further exploration of propagation phenomena and modeling techniques is provided later in this text.

2.2.3 Doppler Shift

Until now, we have only discussed how GNSS is able to determine position and time, but it is also possible for a GNSS receiver to determine its velocity. This is accomplished through the use of the Doppler effect, so it is worth a brief discussion of this phenomenon.

Doppler Effect

In general, the phase of a signal is a function of time, $\phi(t)$. The instantaneous frequency of a signal is given by $1/2\pi$ times the derivative of the phase with respect to time. This can be easily observed via a few basic expressions for phase. For example, a signal whose phase varies linearly has

$$\phi(t) = 2\pi(f_0 t + \phi_0),$$
$$f(t) = \frac{1}{2\pi} \frac{d\phi(t)}{dt} = f_0.$$

This signal is a pure sinusoid, oscillating at only a single frequency f_0.

Consider a transmitter broadcasting a signal of wavelength λ. A receiver at a distance R from the transmitter will observe a phase reading of $\phi = 2\pi R/\lambda$. Now imagine the transmitter begins to move toward the receiver with a constant velocity along the line of sight. As the transmitter moves toward the receiver, it continues to transmit the signal. However, the transmitter is now moving in the same direction as the wavefronts it sends in front of it, so each successive wavefront is closer together than if the transmitter had remained stationary. The wavelength of a signal is defined by the distance between wavefronts, so the motion of the transmitter has actually decreased the wavelength of the signal that the receiver will observe. Recalling that $c = f\lambda$ and that c is constant in all frames of reference, this means that as λ decreases, f must increase. Conversely, if the transmitter were moving away from the receiver, the distance between wavefronts would be increased as compared with a stationary receiver. Thus λ increases and f decreases. Figure 2.4 illustrates this visually.

One of the fundamental postulates of the theory of relativity is that physical phenomena behave the same regardless of the frame of reference. Therefore, although our previous example was from the frame of reference of a stationary receiver and a moving transmitter, the result is identical if, for example, we change our frame of reference to that of a stationary transmitter, in which the receiver is now the one that is moving. Indeed, regardless of the chosen frame of reference, the same result will always be obtained. Thus, what matters is the *relative* motion of the transmitter and receiver.

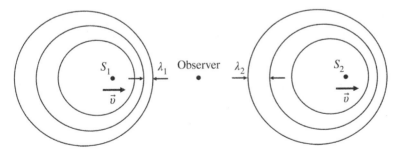

Figure 2.4 Doppler effect.

Due to the relative motion of the transmitter and receiver, the distance between them is now a function of time, $R(t)$. Consequently, the phase is also a function of time, so that

$$\frac{d\phi(t)}{dt} = \frac{d}{dt}\left(2\pi\frac{R(t)}{\lambda}\right) = \frac{2\pi}{\lambda}\frac{dR(t)}{dt} = 2\pi\frac{v(t)}{\lambda},$$

where $v(t)$ is the relative velocity of the transmitter and receiver. Since $d\phi(t)/dt = 2\pi f(t)$, we see that

$$f_D(t) = \frac{1}{2\pi}\frac{d\phi(t)}{dt} = \frac{v(t)}{\lambda}, \tag{2.13}$$

where $f_D(t)$ is called the Doppler frequency. Note that Equation 2.13 is measured in radians. If $\phi(t)$ were instead measured in cycles, the $1/2\pi$ term would not be needed. For objects moving at a constant relative velocity, f_D will be constant. We will use the convention that a positive Doppler frequency represents objects moving toward each other. The frequency seen at the receiver, f_r, will be the transmitted frequency f^s plus the Doppler frequency. Therefore,

$$f_r = f^s + f_D \quad \Rightarrow \quad f_D = f_r - f^s.$$

This phenomenon is referred to as the Doppler effect, and f_r is said to be *Doppler shifted* due to the relative motion of the transmitter and receiver. The Doppler effect is commonly observed in sound waves when vehicles pass by at high speeds. As the vehicle approaches the sound of the engine will be high pitched, but the pitch will quickly decrease as it speeds past the listener.

Previously, we assumed that all relative motion occurred only along the line of sight between the transmitter and the receiver. In general this will not be the case, and only the component of the velocity that lies along the line of sight contributes to the Doppler effect. Thus, if we say that $\hat{r}$ is the unit vector pointing along the line of sight, and $\vec{v}$ is the relative velocity vector of the transmitter and receiver, then

$$v_\perp = \vec{v} \cdot \hat{r}$$

is the projection of $\vec{v}$ onto $\hat{r}$. This quantity is sometimes called the *range rate* or the *radial velocity*, and is the velocity used to determine the Doppler shift.

Determining Velocity

Using all of the information in this section, we may return to Equation 2.8 to see how GNSS receivers determine velocity. By taking the time derivative of both sides of Equation 2.7, we see that

$$\boxed{\lambda D_r^s = \lambda\dot{\Phi}_r^s = \dot{R}_r^s + \dot{e}_r^s = \dot{\rho} + c(\dot{\delta}_r - \dot{\delta}^s) + \dot{e}_r^s,} \tag{2.14}$$

where time derivatives are denoted by a dot, D_r^s is the Doppler observable, $\dot{\rho}$ is called the pseudorange rate, and $\dot{\delta}_r$ and $\dot{\delta}^s$ represent the drifts of the receiver clock and satellite clock, respectively. Recall that $\lambda\phi_r^s$ contains an integer ambiguity term λN. This term is constant with respect to time, so it vanishes when the time derivative of ϕ_r^s is taken.

The derivative of the carrier phase (measured in cycles) with respect to time is the raw Doppler shift f_D. Using Equation 2.13, we see that multiplying $\dot{\Phi}_r^s$ (measured in cycles so that $1/2\pi$ is removed) by λ gives us the relative velocity of the transmitter and receiver. This is simply the range rate $v_\perp$. As such, λD_r^s is sometimes referred to as the "Doppler shift scaled to range rate." Thus, by measuring the Doppler frequency, the receiver is able to develop an expression for the velocity of a receiver relative to a satellite. Much like the pseudorange, this equation has four unknowns: the x, y and z components of the receiver's velocity, and the receiver clock drift. The x, y, and z components of the satellite's velocity and the satellite clock drift are provided in the navigation message. Using the Doppler frequency measurements from four satellites simultaneously, the receiver may solve the system of equations for the four unknowns and determine the receiver's velocity. This velocity calculation is nearly identical to the pseudorange calculation presented in Section 2.4; however, every parameter is now differentiated with respect to time (including the error parameters).

The Doppler shift is measurable to a precision of about $0.001\,\text{Hz}$ [19]. Using Equation 2.13, this corresponds to a velocity resolution of $2 \times 10^{-4}\,\text{m/s}$, based on a carrier frequency of $1.5\,\text{GHz}$. However, as with positioning measurements, the velocity determination is subject to errors which reduce its accuracy. Since Doppler frequency is the time derivative of carrier phase, the corresponding error terms are the time derivatives of the carrier phase error terms.

Differential and Integrated Doppler

Consider two carrier phase measurements at different epochs t_j and t_k. Differencing these two measurements and dividing by $\Delta t = t_k - t_j$ gives:

$$\frac{\lambda \Delta \Phi_r^s}{\Delta t} = \frac{\Delta \rho}{\Delta t} + c\frac{\left(\Delta \delta_r - \Delta \delta^s\right)}{\Delta t} + \lambda\frac{\Delta N}{\Delta t} + \frac{\Delta e_r^s}{\Delta t}. \tag{2.15}$$

Note that $\Delta N = 0$ because the integer ambiguity remains constant with time. Therefore, the term containing ΔN vanishes, and we see that the remaining equation is nearly identical to Equation 2.14. The difference is that Equation 2.14 represents the observed instantaneous Doppler shift, while Equation 2.15 represents a numerically differenced Doppler shift formed from carrier phase measurements. The latter quantity is referred to as the *differential Doppler shift*, or simply differential Doppler. Whereas the instantaneous Doppler measurement may be used for instantaneous velocity determination, the differential Doppler may be used to determine the average velocity between two epochs.

Integrating the instantaneous Doppler over the interval Δt gives:

$$\lambda \int_{t_j}^{t_k} D_r^s dt = \Delta \rho + c \left(\Delta \delta_r - \Delta \delta^s \right) + \Delta e_r^s. \tag{2.16}$$

This is known as the integrated Doppler shift, or just integrated Doppler. Multiplying Equation 2.15 by Δt results in:

$$\lambda \Delta \Phi_r^s = \Delta \rho + c \left(\Delta \delta_r - \Delta \delta^s \right) + \lambda \Delta N + \Delta e_r^s, \tag{2.17}$$

which is simply the difference of two carrier phase measurements at different epochs. Comparing Equations 2.16 and 2.17, we see that they only differ by the ΔN term. We may write:

$$\Delta N = \Delta \Phi_r^s - \int_{t_j}^{t_k} D_r^s dt, \tag{2.18}$$

where λ has been canceled from every term. This expression can be used in cycle slip detection. As long as there are no cycle slips, N will remain constant and ΔN will be zero. Therefore, if Equation 2.18 provides a nonzero value, a cycle slip must have occurred.

Positioning with Integrated Doppler

In addition to its role in velocity determination, the Doppler shift may be used for positioning. To accomplish this, a quantity known as the integrated Doppler count is introduced. When a receiver acquires a Doppler shifted GNSS signal, its frequency, f_r will differ slightly from a stable reference frequency f_g generated by the receiver. These two frequencies may be mixed, causing a beating effect with a beat frequency of $f_g - f_r$. The zero crossings of the beat frequency may be tracked. The number of zero crossings in a certain time interval is referred to as the integrated Doppler count, and may be expressed as follows:

$$N_{jk} = \int_{T_j}^{T_k} D_r^s \, dt = \int_{T_j}^{T_k} (f_g - f_r) dt, \tag{2.19}$$

where T_j and T_k represent the beginning and end of the counting interval. A satellite transmitting a signal at epoch t^j will arrive at the receiver at epoch T_j, meaning that the two epochs may be related by the expression:

$$T_j = t^j + \frac{\rho_i^j}{c}, \tag{2.20}$$

where $\rho_i^j = \sqrt{(x^j - x_i)^2 + (y^j - y_i)^2 + (z^j - z_i)^2}$ represents the true range between a receiver (denoted by index i) and the satellite at epoch t^j. An analogous expression may be written relating the transmission epoch t^k to the reception epoch T_k:

$$T_k = t^k + \frac{\rho_i^k}{c}, \tag{2.21}$$

where $\rho_i^k = \sqrt{(x^k - x_i)^2 + (y^k - y_i)^2 + (z^k - z_i)^2}$ is the true range between the receiver and the satellite at epoch t^k. Substituting Equations 2.20 and 2.21 into the bounds of the integral in Equation 2.19 gives:

$$N_{jk} = \int_{t^j + \frac{\rho_i^j}{c}}^{t^k + \frac{\rho_i^k}{c}} (f_g - f_r) dt. \tag{2.22}$$

Note that the integral of a frequency over some time interval is simply equal to the number of cycles traversed during that time interval. The number of cycles transmitted between epochs t^j and t^k must equal the number of cycles received between epochs T_j and T_k, so:

$$\int_{t^j}^{t^k} f^s dt = \int_{T_j}^{T_k} f_r dt = \int_{t^j + \frac{\rho_i^j}{c}}^{t^k + \frac{\rho_i^k}{c}} f_r dt, \tag{2.23}$$

where f^s is the frequency transmitted by the satellite (prior to any Doppler shift). Substituting Equation 2.23 into Equation 2.22 yields the observation equation:

$$N_{jk} = (f_g - f_s)(t^k - t^j) + \frac{f_g}{c}(\rho_i^k - \rho_i^j). \tag{2.24}$$

In this way, the integrated Doppler count may be used to determine the change in a satellite's position over the counting interval, namely $(\rho_i^k - \rho_i^j)$. The integrated Doppler count is used less commonly than pseudorange or carrier phase measurements because it requires very stable receiver oscillators and long observations times (several hours) for the satellite to move a sufficient distance. This method of positioning is, however, useful for resolving ambiguities. This technique will not be covered in this text, but a discussion of ambiguity resolution is presented in [11, pp. 269–276].

2.3 GNSS Error Sources

The expressions developed in Section 2.2 for calculating GNSS observables are merely simplified models. These models fail to account for several significant physical effects, causing the predicted values of the observables to differ from measured values. The discrepancy between the predicted and measured values is known as GNSS error.

Numerous physical effects contribute to the error in the measurement of GNSS observables. In Section 2.2, the total contribution of all error sources was condensed into a single term, e_r^s. However, by studying each source of error individually, each source's contribution to the total error may be isolated and given its own term in the observable equation. Sources of error in GNSS may be

Table 2.1 Categorization of GNSS error sources.

Category	Error source
Satellite	Clock error
	Ephemeris errors
Signal propagation	Ionospheric effects
	Tropospheric effects
Receiver	Clock error
	Multipath
	Antenna phase center variation
	Receiver noise

Source: Based on Hofmann-Wellenhof et al. [19].

grouped into three categories: satellite-related errors, propagation-related errors, and receiver-related errors. Table 2.1 [19] categorizes a few important sources of GNSS error.

Commonly, satellite clock and ephemeris errors are referred to as signal-in-space range error (SISRE). The remaining error sources (i.e. ionospheric and tropospheric effects, multipath, receiver-related errors, and noise) are collectively termed user equipment error (UEE). Together, all sources of error are referred to as the user equivalent range error (UERE).

Another common distinction is between systematic and random errors. Systematic errors, also called biases, are those which have consistent and predictable values (i.e. small statistical variance). Most sources of GNSS error are systematic, including clock and ephemeris errors, atmospheric effects, multipath and antenna phase center (APC) variation. Random errors have large statistical variance and are difficult to predict and model. Receiver noise is the most prominent source of random error in GNSS positioning.

Error terms may be either approximated and accounted for by using correction terms, or may be nearly eliminated using various positioning and design techniques. Parameters broadcast in the navigation message, provided by the international GNSS service (IGS) or calculated using atmospheric models, are examples of correction terms. They may be used to remove some, but not all, of the error contribution from a given source. For example the Klobuchar model, a model of ionospheric effects, provides parameters that only remove about half of the error contribution from the ionosphere. Clock and ephemeris parameters are available to various degrees of precision, as discussed in Section 1.5.4. Additionally, through the use of relative positioning and dual-frequency receivers, it is possible to remove certain error terms almost entirely. Differencing techniques (discussed in detail in

Section 2.5.2) may be used to solve for certain systematic errors. For example, differencing measurements of two receivers to the same satellite removes error terms related to that satellite, and differencing measurements from two satellites at the same receiver removes that receiver's error terms. Double differencing (involving two receivers and two satellites) can therefore effectively be used to remove hardware-related error terms.

In the remainder of this section, each major source of error affecting GNSS observables is briefly discussed. Ionospheric and tropospheric propagation errors will be treated in more detail in the following chapters of this book.

2.3.1 Clock and Ephemeris Errors

GNSS positioning depends strongly on accurate determination of the travel time of coded GNSS signals. The clocks in both satellites and receivers are referenced to a common time system; for example, GPS clocks are referenced to so-called GPS time, as discussed in Section 1.5.4. Consider a receiver receiving a signal from a single satellite. If either or both of the clocks deviate from GPS time, the receiver will produce an incorrect estimate of the range to the satellite. Ranging is determined by multiplying a signal's travel time by the speed of light, so even very small deviations, or *biases*, in a clock's reading can cause large positioning errors. For example, a clock bias of 10 ns corresponds to a positioning error of 3 m. A clock's bias will generally change with time. The rate of change of the clock bias is known as the clock drift, and is usually measured as the number seconds added to the initial clock bias per second (s/s). Similarly, the clock drift may itself change with time. The rate of change of clock drift is called the drift rate and is measured in seconds per second squared (s/s^2).

Satellite Clock Error

The bias, drift, and drift rate of every GNSS satellite's onboard atomic clock are included in its own individualized navigation message. This information is presented as a second-order polynomial:

$$\delta^s = a_0 + a_1(t - t_0) + a_2(t - t_0)^2,$$
(2.25)

where δ^s is the satellite clock error (s), a_0 is the clock bias (s), a_1 is the clock drift (s/s), a_2 is the clock drift rate (s/s^2), and t_0 is the reference epoch (s).

In the case of GPS, the control segment continuously monitors these parameters and sends updated values to the satellite constellation every two hours. The reference epoch t_0 is set to the time of the last update. Atomic clocks are chosen for satellite timing due to their high stability (i.e. small clock drift). Small drift means slow change in the clock bias. This ensures that once a satellite's bias is determined, it can be guaranteed to remain close to that value for a long time. Thus, a correction term to offset the satellite clock's bias need not be constantly recalculated. Typical values for a satellite clock are $a_0 < 1$ μs, $a_1 \approx 10^{-11}$ s/s, $a_2 \approx 0$ s/s^2 [11].

When a receiver calculates position using, say, pseudorange, it extracts the value of δ^s from the navigation message. A correction term equal to $-\delta^s$ is added to the pseudorange calculation, mostly removing the effect of satellite error on positioning. Clock parameters broadcast in the navigation message typically remove satellite clock error to within ~ 5 ns. As shown in Table 1.8, more accurate data provided by the IGS may improve this precision up to within approximately 75 ps.

Receiver Clock Error

Atomic clocks are expensive, so many GNSS receivers instead use inexpensive alternatives such as quartz clocks. Quartz clocks are less stable than atomic clocks, exhibiting a clock drift on the order of 10^{-8}–10^{-10} s/s. Despite the larger drift value, a GNSS receiver is able to estimate its own clock bias and drift. Using these estimates, the receiver may employ certain techniques to mitigate the effect of the receiver clock error [21]. One technique involves "steering" the clock's oscillator, continuously driving the clock drift toward zero. This keeps the clock bias constant to within the noise level, allowing for simple correction. A second, more common technique allows the receiver to introduce discrete jumps in the clock's time reading to offset accumulated bias. Typically, a jump of 1 ms is introduced whenever the clock's bias exceeds 1 ms in magnitude.

The receiver clock bias can be estimated and compensated to within 1 μs, which corresponds to a ranging error of about 300 m. This error is obviously unacceptable for the vast majority of GNSS applications, so methods must be employed to remove the remaining bias. Generally, receiver clock error. δ_r is included as an unknown in the observable equations. Alternatively, differencing techniques may be employed to eliminate the receiver clock error term entirely.

Ephemeris Errors

The ephemeris data broadcast by the navigation message may differ slightly from the true orbit of the satellite. More accurate ephemeris data is available from IGS, as seen in Table 1.8. In the past, inaccuracies in the broadcast ephemeris amounted to several meters of positioning error. Numerous efforts have been made to reduce this error, which have been largely successful. A 2012–2013 report found that the GPS constellation had an average SISRE (satellite clock plus ephemeris error) of 0.7 m [22].

For short baselines, differential positioning may also be used to remove orbital errors. An approximate relationship between baseline length and orbital (ephemeris) error is given by:

$$\frac{db}{b} = \frac{dr}{\rho}, \tag{2.26}$$

where b is the baseline, ρ is the true range between the satellite and receiver, dr is the orbital error, and db is the baseline error.

2.3.2 Relativistic Effects

Einstein's theories of special and general relativity dramatically reframed the way we view the universe. In the early twentieth century, the work of Einstein, Lorentz, Minkowski, and others overturned the prevailing understanding of space and time, demonstrating that the two quantities are neither distinct nor absolute, as previously thought. Instead, space and time are merely aspects of the same physical entity – a malleable four-dimensional continuum termed *spacetime*.

The consequences of special relativity (SR) and general relativity (GR) are surprising and unintuitive. Physical phenomena related to relativity include:

- *Length contraction*: The length of an object moving with some velocity relative to an observer is decreased along the velocity direction.
- *Motion time dilation*: An object moving relative to an observer will experience the passage of time more slowly than the observer will.
- *Gravitational time dilation*: An object in the presence of a stronger gravitational field will also experience the passage of time more slowly.
- *Relativity of simultaneity*: Events that appear to occur simultaneously to a stationary observer will appear to occur at different times to an observer with a relative velocity.

These phenomena occur for *any* object moving relative to an observer, but at slow speeds the effects are too small for a human to notice. When the relative velocity between the object and the observer is a significant fraction of the speed of light (i.e. at a *relativistic speed*), the effects become much more pronounced.

Relativistic effects may not be ignored when designing a GNSS. Time dilation effects on GPS, for example, would cause positioning error at a rate of approximately 10 km/day. This accumulated error would make the system completely useless if not accounted for. What follows is a discussion of relativistic effects that must be considered to ensure proper GNSS operation.

Time Dilation of the Satellite Clock

GNSS receivers typically use an Earth-centered, Earth-fixed (ECEF) reference system which accounts for the rotation of the Earth. In other words, in such a reference frame, receivers are at rest. In this frame of reference, GNSS satellites are moving at a high velocity relative to receivers, so the atomic clocks on board the satellites are subject to time dilation as per special relativity.

Consider a rest frame $S(x, y, z, t)$ and another frame $S'(x', y', z', t')$ moving relative to S with constant velocity v. S represents the reference frame of a GNSS receiver, while S' represents the reference frame of a satellite. Assume that the two systems coincide at the epoch $t = t' = 0$. According to special relativity, time will move more slowly for the moving reference frame, S'. Thus at some later epoch, the reading of a clock in S' will differ from the reading of a clock in S by some factor.

The time in the moving frame S' is called proper time and is denoted t'. The time in the rest frame S is called coordinate time and is denoted t. We may write the proportionality factor between proper time and coordinate time as γ, thus giving us the relationship $t = \gamma t'$. The derivation of γ is not presented here, but may be found in [23]. It may be shown that

$$\gamma = \frac{1}{\sqrt{1 - \frac{v^2}{c^2}}}, \tag{2.27}$$

where v is the velocity of the moving reference frame S', and c is the speed of light. It follows that the time interval $\Delta t' = t_2' - t_1'$ will be changed with respect to the time interval $\Delta t = t_2 - t_1$. This may be expressed using the formula

$$\Delta t = \gamma \Delta t' = \frac{\Delta t'}{\sqrt{1 - \frac{v^2}{c^2}}}. \tag{2.28}$$

It can be seen that as v increases, the denominator of Equation 2.27 decreases, so γ increases. Therefore Δt is larger than $\Delta t'$, telling us that time has elapsed more quickly in S than S', as expected. We know that frequency is inversely proportional to time, so we can immediately deduce that

$$f = \frac{f'}{\gamma} = f'\sqrt{1 - \frac{v^2}{c^2}}. \tag{2.29}$$

We may use a Taylor series expansion to rewrite the expressions for γ and $1/\gamma$:

$$\frac{1}{\sqrt{1 - \frac{v^2}{c^2}}} = 1 + \frac{1}{2}\left(\frac{v}{c}\right)^2 + \dots,$$
$$\sqrt{1 - \frac{v^2}{c^2}} = 1 - \frac{1}{2}\left(\frac{v}{c}\right)^2 + \dots. \tag{2.30}$$

For small values of v^2/c^2 (i.e. small values of v), higher order terms of the Taylor series expansion are negligible and may be ignored. GNSS satellites move at a very small fraction of c, so this assumption is valid for our purposes. Using the reduced Taylor series expansions in Equations 2.28 and 2.29, we may write the relations:

$$\frac{\Delta t' - \Delta t}{\Delta t} = \frac{f' - f}{f} = -\frac{1}{2}\left(\frac{v}{c}\right)^2. \tag{2.31}$$

Equation 2.31 summarizes the effect of time dilation due to special relativity. Note that the term on the right hand side is the kinetic energy of a unit mass ($\frac{1}{2}v^2$) scaled by c^2. Thus the motion time dilation effect predicted by special relativity may be thought of in terms of the kinetic energy of an object.

Analogously in general relativity, the potential energy due to a gravitational field gives rise to the gravitational time dilation effect. The expression describing gravitational time dilation is:

$$\frac{\Delta t' - \Delta t}{\Delta t} = -\frac{f' - f}{f} = -\frac{\Delta U}{c^2}. \tag{2.32}$$

In this equation, ΔU represents the potential energy difference between two reference frames (such as that of a GNSS receiver on Earth's surface and that of an orbiting satellite). Equations 2.31 and 2.32 may be linearly superimposed to give a single equation that accounts for the time dilation effects of both special and general relativity:

$$\frac{\Delta t' - \Delta t}{\Delta t} = -\frac{f' - f}{f} = -\frac{1}{2}\left(\frac{v}{c}\right)^2 - \frac{\Delta U}{c^2}. \tag{2.33}$$

As discussed in Section 1.5.4, GNSS satellites' atomic clocks are based on certain atomic frequency standards. It is easily seen in Equation 2.33 that time dilation will cause the frequency of the clock to deviate from its nominal value. This deviation must be accounted for when designing satellite hardware, or else major positioning errors would be induced. An error term describing the deviation from the nominal frequency is given by:

$$\boxed{\frac{f_0' - f_0}{f_0} = \frac{1}{2}\left(\frac{v}{c}\right)^2 + \frac{GM}{c^2}\left[\frac{1}{R_e + h} - \frac{1}{R_e}\right],} \tag{2.34}$$

where ΔU has been expanded, R_e is the radius of the Earth, and h is the height of an orbiting satellite.

As an example, GPS is designed for a nominal frequency of $f_0 = 10.23\,\text{MHz}$. A satellite's atomic clock is obviously designed at rest on the surface of Earth, so once launched time dilation effects will change this nominal frequency to some value f_0'. In order to calculate f_0', we must substitute in values for every other term of Equation 2.34. Though v is generally not constant for an orbiting satellite, it is sufficient to use the mean orbital velocity for this calculation. This quantity is given by

$$v = \sqrt{\frac{GM}{R_e + h}}. \tag{2.35}$$

The gravitational constant G has a value of $6.673 \times 10^{-11}\,\text{N} \cdot \text{m}^2/\text{kg}^2$. The mass of the Earth, M is approximately $5.98 \times 10^{24}\,\text{kg}$, and the radius of the Earth, R_e, is approximately 6.38×10^6 m. GPS satellites orbit at a height h of approximately $20,200$ km. Substituting all of these values into Equation 2.35 results in a mean orbital velocity of approximately 3.874 km/s. Recalling that $c = 3 \times 10^8$ m/s, we may now substitute all of the aforementioned parameters into Equation 2.34 to see that

$$\frac{f_0' - f_0}{f_0} = -4.464 \times 10^{-10}. \tag{2.36}$$

This quantity is the fraction by which a satellite's clock runs more slowly than the nominal frequency. Using $f_0 = 10.23$ MHz, we may solve for f_0'. We find that $f_0' - f_0 = -4.57 \times 10^{-3}$ Hz, which means that $f_0' = 10.22999999543$ MHz. This frequency difference seems inconsequential, but the effect of time dilation is cumulative, so timing errors quickly become very large. We may determine how much a satellite's clock will drift with respect to a clock on Earth in the span of one day. The clock drift is given by $-4.464 \times 10^{-10} \times 60 \times 60 \times 24 = -0.00003856$ s/day, or approximately -38.6 µs/day. When calculating position, the receiver multiplies the time delay by c, so a timing error of this magnitude results in a positioning error of about 11.6 km every day. This error would make GPS completely useless, which is why it is so important that relativistic effects are considered when designing satellite hardware.

Relativistic Effects Due to Orbit Eccentricity

A relativistic effect caused by the eccentricity of a satellite's orbit introduces a periodic bias in the satellite clock. This bias is in addition to the time dilation effects previously discussed, and may be calculated using the formula:

$$\delta^{rel} = \delta_0^{rel} + \frac{2e}{c^2} \sqrt{\mu a} \sin E, \tag{2.37}$$

where $\mu = GM$ is the gravitational constant of the Earth, a and e are respectively the semi-major axis and the eccentricity of the orbital ellipse, E is the eccentric anomaly of the orbit, and δ_0^{rel} is a constant factor. The constant μ is known and equal to 3.986×10^{14} N · m^2/kg, whereas a, e, and E are ephemeris parameters broadcast in the navigation message. An equivalent expression for the relativistic clock bias is given by:

$$\delta^{rel} = -2\frac{\vec{r} \cdot \vec{v}}{c^2}. \tag{2.38}$$

In this expression, $\vec{r}$ is the position vector of a given satellite, and $\vec{v}$ is that same satellite's velocity vector.

The bias δ^{rel} is included in the navigation message by adding it to the clock polynomial shown in Equation 2.25. The receiver may then subtract this value from the timing estimate to remove this source of error. The value of δ^{rel} is less than 25 m, so this term is not included in the almanac data (which has an accuracy of less than 135 m) [24].

Relativistic Acceleration of Satellites
Relativistic effects also induce perturbations in the acceleration of a satellite. The details of these effects are beyond the scope of this text, but details may be found in [24, 25]. The International Earth Rotation and Reference Systems Service (IERS) is an organization responsible for maintaining global time and reference frame standards. The IERS standard correction for relativistic acceleration effects [20, p. 66] is given by:

$$\Delta \vec{a} = \frac{\mu}{c^2 r^3} \left[\left(4\frac{\mu}{r} - v^2 \right) \vec{r} + 4 \left(\vec{r} \cdot \vec{v} \right) \vec{v} \right],$$

where $\vec{a}$, $\vec{v}$, and $\vec{r}$ are the acceleration, velocity, and position vectors of the satellite, respectively, with $v = |\vec{v}|$ and $r = |\vec{r}|$.

Path Range Error
From general relativity, the spacetime curves in the presence of matter, and more massive objects cause more curvature. Due to this curvature, an object will travel along a slightly different path than if spacetime were flat. Electromagnetic waves are subject to this effect, meaning it must be accounted for in order to properly calculate the travel time of a GNSS signal. For a satellite to ground link on Earth, the difference between the length of a path in curved spacetime and the length of a path in flat spacetime is given by:

$$\Delta r^{rel} = \frac{2\mu}{c^2} \ln \left(\frac{r^s + r_r + r_r^s}{r^s + r_r - r_r^s} \right), \tag{2.39}$$

where r^s and r_r are the geocentric distances of the satellite and receiver, respectively. Similarly, r_r^s is the distance between the satellite and receiver. r_r is simply the radius of the Earth, R_e, and $r^s = R_e + h$, for a satellite orbiting at height h. r_r^s is, in general, given by the Pythagorean theorem: $r_r^s = \sqrt{(R_e + h)^2 + R_e^2}$. R_e is approximately 6.38×10^6 m, and h is about 20,200 km for GPS satellites, giving r_r^s a maximum value of about 25,600 km. Combining all of this information, a maximum range error $\Delta r^{rel} = 18.6$ mm results from Equation 2.39. This positioning error is important to account for in applications which require millimeter-level positioning. Fortunately, relative positioning may be used to reduce this effect to as little as 0.001 ppm [25].

Earth Rotation Correction
The rotation of the Earth induces an error that must be accounted for. This error may be interpreted in two different ways, depending on the frame of reference of the observer. An Earth-centered inertial (ECI) frame is one which does not

rotate along with the Earth. The frame remains stationary while the Earth rotates within it. In this frame, rotation is accounted for through a "transmission path correction," which determines for the movement of a receiver in the time between transmission and reception of a GNSS signal. An ECEF frame *does* rotate along with the Earth. In this frame, the Earth appears to be stationary while the rest of the universe spins around it. Since an ECEF frame is a rotating frame, it is non-inertial, and therefore experiences a relativistic effect known as the Sagnac effect. The Sagnac effect is a manifestation of the relativity of simultaneity, and so influences the time at which a receiver picks up a signal. The receiver must apply a timing correction to account for this effect. Though the interpretation of the correction differs between the two frames, the numerical value remains the same in both [26]. The two interpretations will be briefly discussed next.

ECI frame: Consider a GNSS receiver on the surface of the Earth receiving from a single satellite. At the instant the satellite transmits a signal, the receiver is at some location. As the signal propagates, the receiver will rotate along with the Earth, and will thus be in a slightly different location by the time the signal arrives at the receiver. The distance the signal travels is given by:

$$c\Delta t = \left| \vec{r}_r + \vec{v}_r \Delta t - \vec{r}^s \right|, \tag{2.40}$$

where $\vec{r}^s$ is the satellite's geocentric distance vector, $\vec{r}_r$ is the receiver's geocentric distance vector, and $\vec{v}_r$ is the velocity vector of the receiver. $\vec{v}_r$ accounts for both the velocity of the receiver due to the rotation of the Earth as well as $\vec{v}_k$, the kinematic velocity of the receiver with respect to the Earth's surface. Therefore, $\vec{v}_r$ is given by the expression:

$$\vec{v}_r = \vec{\omega}_e \times \vec{r}_r + \vec{v}_k, \tag{2.41}$$

where $\vec{\omega}_e$ is the angular velocity vector of the Earth. To obtain the transmission path correction, $\Delta\rho$, the initial separation between the satellite and the receiver, $\left| \vec{r}_r - \vec{r}^s \right|$ must be subtracted.

$$\Delta\rho = \left| \vec{r}_r + \vec{v}_r \Delta t - \vec{r}^s \right| - \left| \vec{r}_r - \vec{r}^s \right|. \tag{2.42}$$

Since $|\vec{v}_r \Delta t| \ll \left| \vec{r}_r - \vec{r}^s \right|$, this expression may be simplified using a Taylor series expansion [27] to:

$$\Delta\rho = \frac{(\vec{r}_r - \vec{r}^s) \cdot \vec{v}_r}{c}. \tag{2.43}$$

The magnitude of the path correction can be up to 30 m [24].

ECEF frame: The Sagnac effect affects the timing of the receiver clock. The motion of the receiver from point a to point b induces a time difference equal to [26]:

$$\Delta t_{Sagnac} = \frac{1}{c^2} \int_a^b (\vec{\omega} \times \vec{r}) \cdot d\vec{r} = \frac{2\omega A}{c^2} = \frac{\omega}{c^2} \left(x_a y_b - x_b y_a \right), \tag{2.44}$$

where $\vec{\omega}$ is the angular velocity of the Earth, A is the projection onto the equatorial plane of the area swept out by the center of rotation and the endpoints of the light path, and (x_a, y_a) and (x_b, y_b) are the coordinates of the endpoints of the light path projected onto the equatorial plane. Sagnac corrections are typically on the order of hundreds of nanometers. A study performed in 1984 measured the time delays induced by the Sagnac effect in a global experiment using timing centers in Boulder, Colorado; Braunschweig, West Germany; and Tokyo, Japan. The experiment reported time delays of 240–350 ns [28]. A more detailed discussion of the Sagnac effect may be found in [26, 28].

2.3.3 Carrier Phase Wind-Up

Wind-up is a phenomenon which only affects carrier phase measurements, and arises in circularly polarized waves. Since all GNSS signals are right-hand circularly polarized (RHCP), wind-up must be taken into account. The effect is small, so it may be ignored for applications which do not require high accuracy.

Wind up is dependent on the relative orientation of the satellite and receiver antennas as well as the line-of-sight vector between the two. As a satellite orbits the Earth, it must continually reorient itself so that its solar panels are receiving maximum energy from the Sun. This requires the satellite to rotate, changing the orientation of its antenna with respect to the (typically stationary) antenna of the receiver. This rotation creates a phase variation which is erroneously interpreted by a receiver as a range variation.

The carrier phase wind up correction w, measured in radians, is equivalent to the change in carrier phase induced, $\Delta\Phi$. $\Delta\Phi$ is given by the expression:

$$\Delta\Phi = \delta\phi + 2\pi n, \tag{2.45}$$

where $\delta\phi$ is the fractional part of the phase measurement. This quantity may be further expressed as:

$$\delta\phi = \text{sgn}(\zeta) \arccos\left(\frac{\vec{d}^s \cdot \vec{d}_r}{|\vec{d}^s||\vec{d}_r|}\right), \tag{2.46}$$

where $\text{sgn}(\cdot)$ represents the sign function, $\zeta = \hat{\rho} \cdot (\vec{d}^s \times \vec{d}_r)$, $\hat{\rho}$ is the unit vector along the line-of-sight from the satellite to the receiver, and $\vec{d}^s$ and $\vec{d}_r$ are the dipole vectors of the satellite and receiver, respectively, which are given by the expressions:

$$\begin{aligned} \vec{d}_r &= \hat{a}_r - \hat{\rho}\left(\hat{\rho} \cdot \hat{a}_r\right) + \hat{\rho} \times \hat{b}_r, \\ \vec{d}^s &= \hat{a}^s - \hat{\rho}\left(\hat{\rho} \cdot \hat{a}^s\right) - \hat{\rho} \times \hat{b}^s. \end{aligned} \tag{2.47}$$

Earlier, $\hat{a}_r$ and $\hat{b}_r$ are orthogonal unit vectors defining the local coordinate system of the receiver. In east, north, up (ENU) coordinates, $\hat{a}_r$ and $\hat{b}_r$ represent the unit vectors pointing east and north, respectively. In addition, $\hat{a}^s$ and $\hat{b}^s$ are orthogonal unit vectors defining a satellite-fixed coordinate frame. These vectors are defined in terms of $\hat{r}^s_{Earth}$ and $\hat{r}^s_{Sun}$, the unit vectors pointing from the satellite's center of mass to the center of the Earth and to the Sun, respectively, so that

$$\hat{b}^s = \hat{r}^s_{Earth} \times \hat{r}^s_{Sun},$$
$$\hat{a}^s = \hat{b}^s \times \hat{r}^s_{Earth}. \tag{2.48}$$

Note that both $\hat{r}^s_{Earth}$ and $\hat{r}^s_{Sun}$ can be determined from ephemeris data.

Returning to Equation 2.45, the term $2\pi n$ refers to the integer number of cycles elapsed in addition to the fractional phase. The integer number n is given by:

$$n = \text{nint}\left[\frac{\Delta\Phi_{prev} - \delta\phi}{2\pi}\right], \tag{2.49}$$

where $\text{nint}[\cdot]$ is the nearest integer function, and $\Delta\Phi_{prev}$ is the previous value of the wind up correction. The value of n may be initialized to zero, meaning it is simply incorporated into the unknown carrier phase ambiguity. In other words, when the ambiguity term is solved for, it will include the value of n.

To scale the wind-up correction w to a range correction, it must be multiplied by the wavelength of the carrier signal, λ. Further details on carrier phase wind-up can be found in [7].

2.3.4 Atmospheric Effects

The troposphere and ionosphere are each responsible for a host of propagation effects that influence GNSS operation. The remainder of this book discusses these effects and methods of mitigation. Chapter 3 discusses propagation effects in the troposphere, and Chapter 4 covers various models that may be used to account for the errors these effects induce in GNSS positioning. Chapters 5–8 explore analogous topics, this time for ionospheric effects.

2.3.5 Multipath, Diffraction, and Interference Effects

Multipath

Multipath propagation, or simply multipath, is a phenomenon which can occur if there is more than one possible path that a signal can take from the satellite to the receiver. For example, multipath occurs when a surface near the receiver, such as the ground or a wall, reflects an incident signal. In general, the direct and any number of reflected signals will arrive at the receiver at slightly differing times due to different path lengths as depicted in Figure 2.5a. In some cases, the direct signal

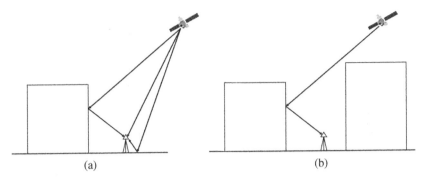

Figure 2.5 Multipath propagation. Source: Adapted from Figure 7.53 of [11]. Reproduced with permission of De Gruyter.

may be obstructed entirely, in which case only reflected signals will arrive at the receiver, as illustrated in Figure 2.5b.

The path of a reflected signal will be longer than that of a direct signal, which introduces a positioning error at the receiver. Multipath error is larger when using pseudoranges, typically amounting to about 10–20 m [29], though under extreme circumstances the error may exceed 100 m [30]. Multipath error in carrier phase measurements is much smaller, typically no larger than 1 cm [19, p. 155], though this error may still need to be considered for certain applications.

There is no generalized analytical model of multipath propagation, since it varies dramatically depending on the local environment as well as the incident signal's angle of arrival (which varies with time as the satellite passes overhead). In the presence of multiple reflectors, such as in a dense urban environment, multipath propagation quickly becomes too complex to predict in a deterministic sense and therefore are typically modeled statistically. Discussion of such models is beyond the scope of this text, but the interested reader may consult [31] for details.

Another method for estimating multipath error is via differential positioning. For short baselines, a double-differenced code range or carrier phase measurement will be free of all sources of error other than those dependent on frequency (i.e. ionospheric effects and multipath). Dual-frequency receivers can be used to remove the ionospheric error. Apart from the noise level, the remaining error is approximately the contribution of multipath. This technique is highly effective for stationary receivers, since their environment is unlikely to change. As such, the station may be calibrated to offset multipath error. Similar calibration for moving receivers is much more difficult because the multipath error is constantly changing along with the surrounding environment.

Additional methods for multipath mitigation include: (i) choice of environment, (ii) antenna design, and (iii) receiver and software design.

Choice of environment: When possible, an environment with few or no nearby reflectors should be chosen. An absorbing material may be deployed on the ground near the receiver to reduce reflections. Multiple antenna arrays, controlled antenna motion, and long observation times may all be utilized to average out multipath variation near an antenna.

Antenna design: Proper antenna design can significantly reduce reception from undesired directions and hence mitigate multipath effects. For example, a ground plane will reject any signals arriving from below the antenna and a "choke ring" design will reject signals arriving at shallow angles (which often arise due to multipath).

Receiver and software design: All GNSS signals are RHCP. Upon reflection, RHCP waves change polarization, instead becoming left-hand circularly polarized (LHCP). A receiver may be designed to reject all LHCP signals, thus removing their contribution to multipath error. Reflected signals also arrive at a receiver later than direct signals. A strobe correlator may be used to discriminate between the direct and reflected signals.

Diffraction

In some cases, the receiver may be completely obstructed from all direct and reflected signals. This effect, known as shadowing, is demonstrated in Figure 2.6. Due to diffraction, a signal will still arrive at a shadowed receiver. The diffracted signal will travel a slightly longer path than the straight-line distance between the receiver and the satellite. Thus, a positioning error of several centimeters to decimeters will be introduced [11, p. 319]. Models of diffraction effects on GNSS may be used to remove this source of error, though details of these models are not discussed here. One such model, the stochastic SIGMA-Δ model, is discussed in detail in [32].

Interference

When two wave signals superimpose in space, their amplitudes will sum at every point. This phenomenon is called interference. When the peak of one wave lines

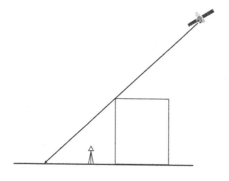

Figure 2.6 Receiver in the shadow of an obstacle.

up with the peak of the other, the amplitude of the resultant wave increases; this is called *constructive interference*. Conversely, when the peak of one wave aligns with the trough of the other, they produce a diminished amplitude. This is known as *destructive interference*.

Destructive interference is of particular importance in wireless communications. A signal with high power relative to the noise at the receiver, i.e. a signal with a high signal-to-noise ratio (SNR), is able to be received. Signals with a low SNR cannot be adequately distinguished from noise. Since the power of a signal is dependent upon its amplitude, destructive interference reduces the power of the incident signal, making it susceptible for being dropped. Dropped signals caused by interference are especially problematic for GNSS, specifically for carrier phase measurements. A carrier phase measurement from a given satellite is initially ambiguous, but the receiver may establish a lock with the satellite. Over time, the receiver may use successive measurements from the same satellite to calculate and remove the ambiguity from the carrier phase measurement (this technique is discussed further in Section 2.2.2). If the receiver loses lock with the satellite signal at any point, this process will need to be repeated. For this reason, it is especially important that GNSS receivers be designed to minimize signal drops. Interference may be reduced by filtering out all frequencies except those very near the carrier. In principle, GNSS could function by broadcasting only at the carrier frequency. However, if all of a GNSS signal's power were concentrated at the carrier, any source of interference at that frequency would cause major reception problems. For example, a malicious party may intentionally broadcast at GPS frequencies to jam a receiver, which is potentially dangerous in military applications. To avoid this problem, the power of a GPS signal is spread over a specified bandwidth (2.046 MHz for C/A code and 20.46 MHz for P-code). In this *spread-spectrum* technique, even if one of the constituent frequencies experiences interference, it will not dramatically reduce the power of the overall signal. This increases the system's robustness against jamming as well as other sources of interference.

2.3.6 Hardware-Related Errors

A number of error sources arise within a GNSS receiver due to the receiver's hardware. Receiver clock error was already covered in Section 2.3.1, but there are a number of other error sources that must be considered as well. These include APC variation, biases in receiver instruments, and other sources of noise.

Antenna Phase Center Variation

Discussion of signal transmission and reception has thus far assumed that the transmitter and receiver antennas are both point sources. In reality, any antenna has volume and geometry, so it is not immediately obvious which point of the antenna should be considered when measuring the distance between the transmitter and receiver. The point used for such measurements is called the APC, and is the apparent source of radiation of the antenna. The distance between a transmitter and receiver is defined as the distance between their antennas' respective phase centers. For the remainder of this section, only the APC of the receiver will be discussed, though examination of the satellite's APC uses the same concepts and terminology.

No two antennas have the same APC, and the APC of a given antenna is not constant, since it is a function of the elevation angle, azimuth angle, intensity, and frequency of the signal. In other words, every signal has its own phase center. The mean position of the APC is used as a constant reference point, and the difference between this point and the APC for a given measurement is called the phase center variation (PCV). Typically, the PCV is given for each carrier frequency separately, each represented as an azimuth- and elevation-dependent surface in 3D space. PCVs are typically on the order of centimeters to millimeters [11, p. 321]. The APC usually resides inside the geometry of an antenna. This makes it unsuitable for defining the height of an antenna, so instead a geometrical point called the antenna reference point (ARP) is used to define the height. The ARP is defined as the intersection of the antenna's vertical mechanical axis of symmetry with the lowest part of the antenna housing. The distance between the mean APC and the ARP is called the phase center offset (PCO). The PCO for each carrier frequency is typically provided by the antenna manufacturer. The total APC correction is dependent upon both the azimuth- and elevation-dependent PCV and the PCO. Let $\vec{a}$ be a vector representing the PCO, and $\hat{r}_0$ be the unit vector between the satellite and receiver (see Figure 2.7 [19]). The contribution of the PCO to positioning error is given by the projection of $\vec{a}$ onto $\vec{r}_0$:

$$\delta_{PCO} = \vec{a} \cdot \hat{r}_0. \qquad (2.50)$$

The contribution of the PCV to the positioning error, denoted δ_{PCV} is simply given by the distance between the PCV contour and the mean APC. The total phase center correction is the sum of these two quantities, i.e.

$$\delta_{APC} = \delta_{PCO} + \delta_{PCV}. \qquad (2.51)$$

When this correction is applied to the positioning measurement by the receiver, the range measurement refers to the ARP of the receiving antenna rather than its APC. The height of the antenna may then be subtracted from this measurement to obtain the receiver's position on the surface of the Earth.

Instrumental Biases

As a signal passes through the hardware of either a satellite or a receiver, delays known as instrumental biases are introduced. These biases differ for each GNSS

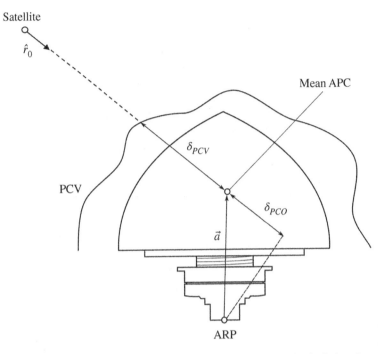

Satellite

$\hat{r}_0$

Mean APC

δ_{PCV}

PCV

δ_{PCO}

$\vec{a}$

ARP

Figure 2.7 Diagram depicting the quantities of interest when calculating the antenna phase center correction.

observable, and are dependent on frequency. For a given observable at a given frequency, each instrumental bias is constant. Instrumental biases may be generically modeled as follows [20]:

$$d_c = d_{c,i}(f) + d_c^j(f), \tag{2.52}$$

$$d_p = d_{p,i}(f) + d_p^j(f), \tag{2.53}$$

$$d_d = d_{d,i}(f) + d_d^j(f), \tag{2.54}$$

where the subscripts c, p, and d refer to the code, phase, and Doppler observables, respectively. In addition, the subscript i refers to the ith receiver, and the superscript j refers to the jth satellite. Each instrument bias is a function of frequency f.

Instrument biases can be removed via relative positioning. Double differencing removes all instrumental biases from each receiver and satellite used to perform the difference.

Receiver Noise
The positioning measurement is subject to a number of influences from random sources of error. In contrast to the *systematic errors* that have been the focus of this

section thus far, random errors (denoted ϵ_r^s) are difficult to model, and therefore harder to remove from positioning measurements. Random error sources include disturbances in the receiver's antenna, cables, amplifiers, thermal noise, dynamic stress, etc. Noise is impossible to avoid completely, but methods such as low-pass filtering can be used to reduce the noise level. Modern receivers are capable of reducing noise positioning error down to submillimeter levels [11, p. 323].

In addition to the receiver hardware noise, a concept known as *process noise* must be considered. Process noise arises from the fact that there is some freedom in the approach to data analysis. Various software packages exist for performing positioning calculations, each with their own particular options. Additionally, a different set of reference stations may be used to define the reference frame. Because of these freedoms, it is possible for the same observation to be calculated slightly differently by different software packages. Alternatively, it is possible that the same software package will give slightly different results in different laboratories. These variations may amount to positioning errors of approximately 1 cm horizontally and 2 cm vertically [33].

2.3.7 Dilution of Precision

The term UERE is oftentimes used to denote the total range error from the mean caused by the various factors discussed earlier. The geometry of the satellites used for positioning has an effect on the UERE. If the satellites are grouped too closely together, the UERE will be multiplied by some factor. This factor is called the *dilution of precision* or DOP. Theoretically, an ideal DOP is equal to 1 since it would not increase the UERE at all. However in practice DOP is rarely below 2 [34], meaning the UERE is multiplied by a factor of 2. DOP can assume values greater than 20, which would amplify the UERE to such an extent that the positioning measurement may become impractical. It is possible to mathematically model DOP and design receivers to choose satellites such that DOP is minimized. This concept is elaborated upon further in Section 2.4.3.

2.3.8 Additional Error Sources

The most important error sources have been covered in this section, but this list is not fully comprehensive. There are many other physical effects that contribute a small amount of error to positioning measurements. One example is deformation of the Earth's crust due to tidal effects such as solid tides and oceanic loading, which can change both the horizontal and vertical position of a receiver by up to a few centimeters. A detailed discussion of these phenomena may be found in [7, 20]. In addition to all known error sources, it is possible that there may be sources of error yet to be identified. The contribution of these hypothetical sources would be very small, and ignoring their effect is acceptable for present applications. However, as GNSS technology progresses, applications requiring even more precise positioning may become feasible. In principle, there could be sources contributing as little as tens of micrometers of error to positioning measurements. These sources would need to be considered to make such precise positioning possible.

2.4 Point Positioning

GNSS positioning is very simple conceptually. Consider a receiver that computes the range R_1 to a single satellite, whose location is known from the ephemeris data it broadcasts. The receiver knows that it must lie somewhere on the surface of a sphere of radius R_1 around the satellite, and cannot narrow down its location without additional information. If the range to a second satellite, R_2, is introduced, then the receiver knows it must be on the surface of not only the sphere of radius R_1, but also a sphere of radius R_2 as well. The intersection of these two spheres forms a circle, so at this point the receiver can narrow down its location to somewhere on that circle. Finally, if a third satellite at a distance R_3 is introduced, a third sphere will intersect the first two. There will only be two points on the surface of all three spheres, as shown in Figure 2.8. One of these two points is the location of the receiver. One of these two points gives a nonsensical answer, for example the point may be too far from the surface of the Earth (or moving at an impossible velocity in the case of velocity determination). Therefore, this point may be discarded and the receiver determines that the remaining point is the location of the receiver.

In theory, only three satellites are needed to perform positioning, as in the previous example. However this only applies to errorless measurements. Most sources of error may be modeled and accounted for, but the receiver clock bias must instead be solved as a fourth unknown variable. Thus, in practice, an additional fourth satellite is required to perform positioning as noted before.

GNSS receivers calculate their range to a given satellite using one of several GNSS observables. In Section 2.2, three main GNSS observables were identified: pseudorange, carrier phase, and Doppler shift. This section will explore the mathematics behind the pseudorange calculation. The carrier phase and Doppler shift equations are dependent upon the pseudorange, so calculation of these quantities is very similar and will not be shown explicitly here. At the end of this section, a brief discussion of DOP is provided.

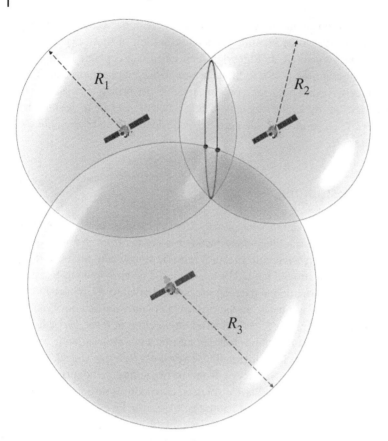

Figure 2.8 Visualization of GNSS positioning.

2.4.1 Positioning Using Pseudorange

GNSS satellites continuously transmit positioning signals, and receivers on Earth are designed to pick up these signals, decode the information encoded on them, and use that information to determine its position. To summarize, a basic outline of this process is as follows:

1. Satellite generates a carrier frequency and modulates the ranging code and navigation message onto it. The ranging code includes a timestamp of the moment of transmission.
2. The transmitted signal travels at approximately the speed of light toward the surface of the Earth.
3. A GNSS receiver picks up the signal and measures the time of reception.

4. The receiver may decode the ranging code to receive the transmission time, and may decode the navigation message to obtain the ephemeris data (and thus the approximate position of the satellite).
5. Utilizing all of this information, the receiver is able to calculate the pseudorange of the satellite.
6. This measurement is performed for at least four satellites simultaneously, providing enough information for the receiver to determine its position.

Recall that the equation for pseudorange is:

$$R_r^s = \rho_r^s + c(\delta_r - \delta^s) + e_r^s, \tag{2.55}$$

where ρ_r^s is the true geometric range between the satellite and receiver, c is the speed of light, δ_r is the receiver clock error, δ^s is the satellite clock error, and e_r^s is a term accounting for all other sources of systematic error. Recall also that ρ_r^s may be expanded as:

$$\rho_r^s = \sqrt{(x^s - x_r)^2 + (y^s - y_r)^2 + (z^s - z_r)^2}, \tag{2.56}$$

where x^s, y^s, and z^s are the components of the geocentric position of a satellite, and x_r, y_r, and z_r are the ECEF coordinates of a (stationary) receiver. The satellite coordinates may be calculated from ephemeris data, and a satellite clock correction is broadcast in the navigation message, allowing the receiver to remove δ^s from the pseudorange calculation. The resulting pseudorange equation is:

$$R_r^s = \sqrt{(x^s - x_r)^2 + (y^s - y_r)^2 + (z^s - z_r)^2} + c\delta_r + e_r^s. \tag{2.57}$$

By considering important sources of error as discussed in Section 2.3, it is possible to subtract the error term from both sides of this equation. The resulting quantity on the left hand side is known as the "observed minus computed" or "O minus C" observation. Thus we may write:

$$^*R_r^s = \sqrt{(x^s - x_r)^2 + (y^s - y_r)^2 + (z^s - z_r)^2} + c\delta_r, \tag{2.58}$$

where $^*R_r^s$ represents the O minus C pseudorange observation. Henceforth, all references to pseudorange will assume the O minus C observation, but the asterisk will be omitted for notational simplicity. There are four unknowns in this equation: the three components of the receiver position – x_r, y_r, and z_r – and the receiver clock error δ_r. Since there are four unknowns, the receiver needs pseudorange observations from four different satellites to solve the system of equations. The system of equations is as follows:

$$
\begin{aligned}
R_r^{(1)} &= \sqrt{(x^{(1)} - x_r)^2 + (y^{(1)} - y_r)^2 + (z^{(1)} - z_r)^2} + c\delta_r, \\
R_r^{(2)} &= \sqrt{(x^{(2)} - x_r)^2 + (y^{(2)} - y_r)^2 + (z^{(2)} - z_r)^2} + c\delta_r, \\
R_r^{(3)} &= \sqrt{(x^{(3)} - x_r)^2 + (y^{(3)} - y_r)^2 + (z^{(3)} - z_r)^2} + c\delta_r, \\
R_r^{(4)} &= \sqrt{(x^{(4)} - x_r)^2 + (y^{(4)} - y_r)^2 + (z^{(4)} - z_r)^2} + c\delta_r,
\end{aligned}
\tag{2.59}
$$

where a superscript (i) refers to the ith satellite being observed. This system of equations is nonlinear and difficult to solve in its current form. Fortunately, we may use a Taylor series expansion to linearize the system, which will allow us to use matrix methods to quickly find a solution. Recall that the general form of a Taylor series expansion of a function $f(x)$ about a point x_0 is:

$$f(x) = f(x_0) + \frac{f'(x_0)}{1!}(x - x_0) + \frac{f''(x_0)}{2!}(x - x_0)^2 + \frac{f'''(x_0)}{3!}(x - x_0)^3 + \dots .$$

(2.60)

If $|x - x_0|$ is small, we can drop the quadratic and higher order terms to obtain a linear approximation of $f(x)$ in terms of x.

The pseudorange equation is a function of four variables, so a Taylor series expansion must be performed for each variable separately, using the partial derivative with respect to that variable. We must also make an initial prediction of the position and clock error of the receiver. The elements of this prediction vector $\vec{x}_0 = [x_{r,0}\ y_{r,0}\ z_{r,0}\ \delta_{r,0}]^T$ will serve as the center points about which each Taylor series expansion will be evaluated. The Taylor series expansion of the pseudorange equation is given by:

$$R_r^s = R_{r,0}^s + \frac{\partial R_{r,0}^s}{\partial x_r}(x_r - x_{r,0}) + \frac{\partial R_{r,0}^s}{\partial y_r}(y_r - y_{r,0}) + \frac{\partial R_{r,0}^s}{\partial z_r}(z_r - z_{r,0})$$
$$+ \frac{\partial R_{r,0}^s}{\partial \delta_r}(\delta_r - \delta_{r,0}),$$

(2.61)

where $R_r^s = \sqrt{(x^s - x_r)^2 + (y^s - y_r)^2 + (z^s - z_r)^2} + c\delta_r$ is the pseudorange measurement taken by the receiver, and $R_{r,0}^s = \sqrt{(x^s - x_{r,0})^2 + (y^s - y_{r,0})^2 + (z^s - z_{r,0})^2} + c\delta_{r,0}$ is a calculated parameter based on our prediction vector. If we now define $\Delta R_r^s = R_r^s - R_{r,0}^s$, $\Delta x = (x_r - x_{r,0})$, $\Delta y = (y_r - y_{r,0})$, $\Delta z = (z_r - z_{r,0})$ and $\Delta\delta_r = (\delta_r - \delta_{r,0})$, we may now write the system of equations 2.59 in matrix form:

$$
\begin{bmatrix} \Delta R_r^{(1)} \\ \Delta R_r^{(2)} \\ \Delta R_r^{(3)} \\ \Delta R_r^{(4)} \end{bmatrix}
=
\begin{bmatrix}
\dfrac{\partial R_{r,0}^{(1)}}{\partial x_r} & \dfrac{\partial R_{r,0}^{(1)}}{\partial y_r} & \dfrac{\partial R_{r,0}^{(1)}}{\partial z_r} & \dfrac{\partial R_{r,0}^{(1)}}{\partial \delta_r} \\[2mm]
\dfrac{\partial R_{r,0}^{(2)}}{\partial x_r} & \dfrac{\partial R_{r,0}^{(2)}}{\partial y_r} & \dfrac{\partial R_{r,0}^{(2)}}{\partial z_r} & \dfrac{\partial R_{r,0}^{(2)}}{\partial \delta_r} \\[2mm]
\dfrac{\partial R_{r,0}^{(3)}}{\partial x_r} & \dfrac{\partial R_{r,0}^{(3)}}{\partial y_r} & \dfrac{\partial R_{r,0}^{(3)}}{\partial z_r} & \dfrac{\partial R_{r,0}^{(3)}}{\partial \delta_r} \\[2mm]
\dfrac{\partial R_{r,0}^{(4)}}{\partial x_r} & \dfrac{\partial R_{r,0}^{(4)}}{\partial y_r} & \dfrac{\partial R_{r,0}^{(4)}}{\partial z_r} & \dfrac{\partial R_{r,0}^{(4)}}{\partial \delta_r}
\end{bmatrix}
\begin{bmatrix} \Delta x \\ \Delta y \\ \Delta z \\ \Delta\delta_r \end{bmatrix} .
$$

Note that:

$$\frac{\partial R_{r,0}^s}{\partial x_r} = \frac{-(x^s - x_{r,0})}{\rho_{r,0}^s}, \quad \frac{\partial R_{r,0}^s}{\partial y_r} = \frac{-(y^s - y_{r,0})}{\rho_{r,0}^s}, \quad \frac{\partial R_{r,0}^s}{\partial z_r} = \frac{-(z^s - z_{r,0})}{\rho_{r,0}^s},$$

$$\frac{\partial R_{r,0}^s}{\partial \delta_r} = c, \tag{2.62}$$

where $\rho_{r,0}^{(i)} = \sqrt{(x^s - x_{r,0})^2 + (y^s - y_{r,0})^2 + (z^s - z_{r,0})^2}$. In other words:

$$\begin{bmatrix} \Delta R_r^{(1)} \\ \Delta R_r^{(2)} \\ \Delta R_r^{(3)} \\ \Delta R_r^{(4)} \end{bmatrix} = \begin{bmatrix} \dfrac{-(x^{(1)} - x_{r,0})}{\rho_{r,0}^{(1)}} & \dfrac{-(y^{(1)} - y_{r,0})}{\rho_{r,0}^{(1)}} & \dfrac{-(z^{(1)} - z_{r,0})}{\rho_{r,0}^{(1)}} & c \\ \dfrac{-(x^{(2)} - x_{r,0})}{\rho_{r,0}^{(2)}} & \dfrac{-(y^{(2)} - y_{r,0})}{\rho_{r,0}^{(2)}} & \dfrac{-(z^{(2)} - z_{r,0})}{\rho_{r,0}^{(2)}} & c \\ \dfrac{-(x^{(3)} - x_{r,0})}{\rho_{r,0}^{(3)}} & \dfrac{-(y^{(3)} - y_{r,0})}{\rho_{r,0}^{(3)}} & \dfrac{-(z^{(3)} - z_{r,0})}{\rho_{r,0}^{(3)}} & c \\ \dfrac{-(x^{(4)} - x_{r,0})}{\rho_{r,0}^{(4)}} & \dfrac{-(y^{(4)} - y_{r,0})}{\rho_{r,0}^{(4)}} & \dfrac{-(z^{(4)} - z_{r,0})}{\rho_{r,0}^{(4)}} & c \end{bmatrix} \begin{bmatrix} \Delta x \\ \Delta y \\ \Delta z \\ \Delta \delta_r \end{bmatrix}.$$

For the sake of conciseness, let's write this equation as:

$$\vec{y}_0 = A\vec{x}_{\Delta,0}, \tag{2.63}$$

where $\vec{y}_0$ is called the observation vector, A is the design matrix, and $\vec{x}_{\Delta,0}$ is the unknown vector. We may simply left-multiply both sides by A^{-1} to solve for the unknown vector $\vec{x}_{\Delta,0}$. Note that A is always invertible because the pseudorange equations corresponding to different satellites are linearly independent.

Each element of $\vec{x}_{\Delta,0}$ is a differential quantity, e.g. $\Delta x = x_r - x_{0,r}$. These differentials represent the difference between the true position (or clock error) of the receiver and the predicted value. We may now add this vector to our initial prediction vector, $\vec{x}_1 = \vec{x}_0 + \vec{x}_\Delta$, that is

$$\begin{bmatrix} x_{1,r} \\ y_{1,r} \\ z_{1,r} \\ \delta_{1,r} \end{bmatrix} = \begin{bmatrix} x_{0,r} \\ y_{0,r} \\ z_{0,r} \\ \delta_{0,r} \end{bmatrix} + \begin{bmatrix} \Delta x \\ \Delta y \\ \Delta z \\ \Delta \delta_r \end{bmatrix}. \tag{2.64}$$

The new vector on the left hand side, which we may call $\vec{x}_1$ can be used as a new prediction of the receiver's position (and clock error). We may plug the elements of this vector into Equation 2.61 as the center point of a new Taylor series expansion. This gives us a new observation vector $\vec{y}_1$ and thus a new unknown vector $\vec{x}_{\Delta,1}$. The elements of $\vec{x}_{\Delta,1}$ will be smaller than those of $\vec{x}_{\Delta,0}$, meaning we are closer to finding the true position of the receiver. We can generalize the relationship shown in Equation 2.64 as follows:

$$\vec{x}_n = \vec{x}_{n-1} + \vec{x}_{\Delta,n}. \tag{2.65}$$

We may continue to iterate in this fashion until the elements of $\vec{x}_{\Delta,n}$ are zero, at which point $\vec{y}_n = \vec{x}_n$ simply represents the true location of the receiver.

Practically speaking, the elements of $\vec{x}_{\Delta,n}$ will never be exactly zero, but the receiver algorithm may be set to terminate when $|\vec{x}_{\Delta,n}|$ (i.e. the sum of the squares of the elements of $\vec{x}_{\Delta,n}$) is less than some predetermined threshold, say, 1 cm. A MATLAB script that implements this algorithm is shown in the following text.

```
%Pseudorange measurements in meters, provided in RINEX file
P14 = 21389121.8317;
P20 = 22977368.2926;
P25 = 21967290.2846;
P31 = 20067290.1517;

%Satellites' x, y, and z coordinates in meters, provided in SP3 file
X14 = 16392906.651;
X20 = -15160016.715;
X25 = 15808190.995;
X31 = -1928273.889;

Y14 = -14085726.897;
Y20 = -4829877.001;
Y25 = -7296305.740;
Y31 = -19479815.577;

Z14 = 15552716.208;
Z20 = 21080170.652;
Z25 = 19955096.500;
Z31 = 17956947.627;

%Satellite clock biases in seconds, provided in SP3 file
ds14 = 188.819990e-6;
ds20 = 171.855965e-6;
ds25 = 19.210329e-6;
ds31 = 333.907169e-6;

%Initial Taylor series expansion point
Xr = 0;
Yr = 0;
Zr = 0;
dr = 0;

%Speed of light in meters/second
c = 299792458;

%Removing satellite clock biases from pseudorange measurements
P14 = P14 - c*ds14;
P20 = P20 - c*ds20;
```

```
P25 = P25 - c*ds25;
P31 = P31 - c*ds31;

E = 1; %Initializing error variable
i = -1; %Initializing iteration variable

while E >.01 % Positioning convergence threshold of 1 centimeter

%Predicted range from receiver to satellite
ri014 = sqrt((X14-Xr)^2 + (Y14 - Yr)^2 + (Z14 - Zr)^2) + c*dr;
ri020 = sqrt((X20-Xr)^2 + (Y20 - Yr)^2 + (Z20 - Zr)^2) + c*dr;
ri025 = sqrt((X25-Xr)^2 + (Y25 - Yr)^2 + (Z25 - Zr)^2) + c*dr;
ri031 = sqrt((X31-Xr)^2 + (Y31 - Yr)^2 + (Z31 - Zr)^2) + c*dr;

%Design matrix
A = [(Xr - X14)/ri014, (Yr - Y14)/ri014, (Zr - Z14)/ri014, c;...
(Xr - X20)/ri020, (Yr - Y20)/ri020, (Zr - Z20)/ri020, c; ...
(Xr - X25)/ri025, (Yr - Y25)/ri025, (Zr - Z25)/ri025, c;...
(Xr - X31)/ri031, (Yr - Y31)/ri031, (Zr - Z31)/ri031, c];

%Observation vector
y = [P14 - ri014; P20 - ri020; P25 - ri025; P31 - ri031];

xdelta = A^-1*y; %Unknown vector

E = xdelta'*xdelta; %Positioning discrepancy

%Equation 1.55
Xr = Xr + xdelta(1);
Yr = Yr + xdelta(2);
Zr = Zr + xdelta(3);
dr = dr + xdelta(4);

i = i+1; %Increment counter by 1
end
```

In this code, data for satellite vehicles numbered 14, 20, 25, and 31 were used. The epoch of measurement was 20 March 2014 at 19:00:00, converted from GPS time. Parameters were obtained from the Receiver Independent Exchange (RINEX) file and the Standard Product # 3 (SP3) file for the aforementioned epoch. Both of these file formats are available from the IGS. The parameters in the SP3 file are computed in an ECEF frame [35]. The user is free to change the elements of the initial Taylor series expansion point. The closer these values are to the true position and clock bias values, the fewer iterations are needed. In this case, the vector representing the expansion point was simply initialized to zero, and the algorithm required six iterations to converge within the specified threshold of one centimeter. An alternative approach to the one presented in this section is Bancroft's algorithm, a direct positioning solution which does not require a priori knowledge of the receiver's position. The details of this algorithm are beyond the scope of this text, but the interested reader may find a derivation in [36].

2.4.2 Accounting for Random Error

Previously we removed the error term from the pseudorange equation to form the O minus C observation. This was done under the assumption that we could compute the contribution of error sources and account for them. This approach works for systematic errors, but random error cannot be modeled as easily, so we cannot simply remove it. When random error is considered, Equation 2.63 becomes:

$$\vec{y}_0 = \mathbf{A}\vec{x}_{\Delta,0} + \vec{e}_r^s, \tag{2.66}$$

where $\vec{e}_r^s$ represents the random error vector. With this term added in, solving for $\vec{x}_{\Delta,0}$ is no longer trivial. The original system is now inconsistent, i.e. $\vec{y} \neq \mathbf{A}\vec{x}_{\Delta,0}$. The solution to this problem is to utilize the measurements of more than four satellites at a time. In general, more than four satellites will be visible at any given time, and it is possible for a receiver to choose the four satellites that will give the best positioning measurement. Assume a signal is received from m different satellites at once. As before, we may linearize the pseudorange measurement from each satellite to obtain a matrix equation. However, now the observation vector $\vec{y}$ and the design matrix $\mathbf{A}$ have dimensions $m \times 1$ and $m \times 4$, respectively, as shown in the following equation

$$
\begin{bmatrix} \Delta R_r^{(1)} \\ \Delta R_r^{(2)} \\ \vdots \\ \Delta R_r^{(m)} \end{bmatrix} =
\begin{bmatrix}
\frac{-(x^{(1)} - x_r)}{\rho_r^{(1)}} & \frac{-(y^{(1)} - y_r)}{\rho_r^{(1)}} & \frac{-(z^{(1)} - z_r)}{\rho_r^{(1)}} & c \\
\frac{-(x^{(2)} - x_r)}{\rho_r^{(2)}} & \frac{-(y^{(2)} - y_r)}{\rho_r^{(2)}} & \frac{-(z^{(2)} - z_r)}{\rho_r^{(2)}} & c \\
\vdots & \vdots & \vdots & \vdots \\
\frac{-(x^{(m)} - x_r)}{\rho_r^{(m)}} & \frac{-(y^{(m)} - y_r)}{\rho_r^{(m)}} & \frac{-(z^{(m)} - z_r)}{\rho_r^{(m)}} & c
\end{bmatrix}
\begin{bmatrix} \Delta x \\ \Delta y \\ \Delta z \\ \Delta \delta_r \end{bmatrix} +
\begin{bmatrix} \epsilon_r^{(1)} \\ \epsilon_r^{(2)} \\ \vdots \\ \epsilon_r^{(m)} \end{bmatrix}.
$$

This system of equations is overdetermined, i.e. it has more equations than unknowns. Overdetermined systems do not have an exact solution, but one can seek a solution that minimizes the effect of the remaining error. It is possible to determine this solution through a standard technique known as the method of least squares. The remainder of this section explores this method.

Method of Least Squares

We may simplify the notation a bit and drop the subscripts and superscript to rewrite Equation 2.66 as:

$$\vec{y} = \mathbf{A}\vec{x} + \vec{\epsilon}. \tag{2.67}$$

We may rearrange Equation 2.67 to solve for $\vec{\epsilon}$

$$\vec{\epsilon} = \vec{y} - \mathbf{A}\vec{x}. \tag{2.68}$$

The squared magnitude of $\vec{e}$ is equal to $\vec{e}^{\mathrm{T}}\vec{e}$ and can be expressed as

$$J(\vec{x}) = \vec{e}^{\mathrm{T}}\vec{e} = (\vec{y} - \mathbf{A}\vec{x})^{\mathrm{T}}(\vec{y} - \mathbf{A}\vec{x}). \tag{2.69}$$

Expanding $J(\vec{x})$ gives:

$$\begin{aligned} J(\vec{x}) &= (\vec{y}^{\mathrm{T}} - \vec{x}^{\mathrm{T}}\mathbf{A}^{\mathrm{T}})(\vec{y} - \mathbf{A}\vec{x}) \\ &= \vec{y}^{\mathrm{T}}\vec{y} - \vec{y}^{\mathrm{T}}\mathbf{A}\vec{x} - \vec{x}^{\mathrm{T}}\mathbf{A}^{\mathrm{T}}\vec{y} + \vec{x}^{\mathrm{T}}\mathbf{A}^{\mathrm{T}}\mathbf{A}\vec{x}. \end{aligned} \tag{2.70}$$

The squared magnitude of $\vec{e}$ is minimized when $J(\vec{x})$ is also at a minimum. $J(\vec{x})$ has a unique minimum when its derivative equals zero. This gives

$$J'(\vec{x}) = -2\vec{y}^{\mathrm{T}}\mathbf{A} + 2\vec{x}^{\mathrm{T}}\mathbf{A}^{\mathrm{T}}\mathbf{A} = 0 . \tag{2.71}$$

We may then transpose both sides and rearrange to obtain:

$$\mathbf{A}\vec{y}^{\mathrm{T}} = \mathbf{A}^{\mathrm{T}}\mathbf{A}\vec{x}. \tag{2.72}$$

Equation 2.72 represents a system of equations termed the normal equations. Even if $\mathbf{A}$ is a rectangular matrix, $\mathbf{A}^{\mathrm{T}}\mathbf{A}$ will always be a square matrix. If $\mathbf{A}^{\mathrm{T}}\mathbf{A}$ is invertible, the normal equations may be solved for a vector $\hat{x}$, which is called the least squares solution:

$$\hat{x} = \left(\mathbf{A}^{\mathrm{T}}\mathbf{A}\right)^{-1}\mathbf{A}^{\mathrm{T}}\vec{y}, \tag{2.73}$$

i.e. $\hat{x}$ is the solution for which the square magnitude of $\vec{e}$ is minimized. The least squares solution is merely an approximation – it is not the exact solution that we would have gotten if the system were consistent (i.e. not overdetermined). This solution may be thought of as the slope and vertical axis intercept of a trendline that best fits a scatter plot of data. In fact, the method of least squares is commonly used in regression analysis for this very purpose.

In the aforementioned derivation, we have neglected the effect of observation uncertainties. We must account for these uncertainties by weighting our least squares solution. To understand how this is done, we must first understand the concept of a covariance matrix.

Covariance Matrices

Consider two random vectors $\vec{x}$ and $\vec{y}$. The uncertainties, or standard deviations, in $\vec{x}$ and $\vec{y}$ are given by $\vec{\sigma}_x$ and $\vec{\sigma}_y$ respectively. A matrix known as the *covariance matrix* gives the variances and covariances for these two vectors. The covariance matrix of $\vec{x}$ and $\vec{y}$ is given by:

$$\mathrm{cov}(\vec{x}, \vec{y}) = \left\langle \left(\vec{x} - \langle\vec{x}\rangle\right)\left(\vec{y} - \langle\vec{y}\rangle\right)^{\mathrm{T}} \right\rangle = \left\langle \vec{x}\vec{y}^{\mathrm{T}} \right\rangle - \langle\vec{x}\rangle\langle\vec{y}\rangle^{\mathrm{T}}, \tag{2.74}$$

where $\langle\cdot\rangle$ represents the expectation value (i.e. the mean) of a random variable. We also introduce the notation $\mathrm{cov}(\vec{x}, \vec{y}) = \mathbf{C}_{\vec{x}\vec{y}}$ for the covariance matrix. Let us examine the case where both $\vec{\mathbf{x}}$ and $\vec{\mathbf{y}}$ vectors consist of a single element, i.e. $\vec{\mathbf{x}} = x$ and $\vec{\mathbf{y}} = y$. In this case, the explicit form of the covariance matrix is

$$\mathbf{C}_{\vec{x}\vec{y}} = \begin{bmatrix} \sigma_x^2 & \sigma_{xy}^2 \\ \sigma_{xy}^2 & \sigma_y^2 \end{bmatrix}. \tag{2.75}$$

The diagonal elements are the variances of each parameter. The off-diagonal elements represent the covariance of the two parameters. Note that $\mathbf{C}_{\vec{x}\vec{y}}$ is always a symmetric matrix.

The *correlation* between two parameters is given by the ratio of their covariance and the product of their standard deviations. For our example, the correlation between $\vec{\mathbf{x}}$ and $\vec{\mathbf{y}}$ is:

$$\rho_{xy} = \frac{\sigma_{xy}^2}{\sigma_x \sigma_y}. \tag{2.76}$$

The value of the correlation ranges from -1 to 1, i.e. complete negative correlation to complete positive correlation. If the two parameters are completely independent of each other, their correlation, and thus their covariance, will be 0 and we say that the variables are decorrelated.[1] In that case, all off-diagonal entries of the covariance matrix are 0, making it a diagonal matrix.

It should be noted that the covariance matrix of a random vector with itself, e.g. $\text{cov}(\vec{\mathbf{x}}, \vec{\mathbf{x}})$, is called the variance–covariance matrix of $\vec{\mathbf{x}}$. We may denote this matrix more concisely as $\mathbf{C}_{\vec{x}}$. The variance–covariance matrix of a random vector generalizes the concept of variance to multiple dimensions. A useful quantity which encapsulates the information provided by the variance–covariance matrix is the *total variation*. The total variation of a random vector is the combined effect of all of its elements' variances. Thus, the total variation is given by the trace of the variance–covariance matrix.

For the GNSS pseudorange observations, the covariance matrix is given by:

$$\mathbf{C}_{\vec{y}} = \begin{bmatrix} \sigma_x^2 & \sigma_{xy} & \sigma_{xz} & \sigma_{x\delta} \\ \sigma_{yx} & \sigma_y^2 & \sigma_{yz} & \sigma_{y\delta} \\ \sigma_{zx} & \sigma_{zy} & \sigma_z^2 & \sigma_{z\delta} \\ \sigma_{\delta x} & \sigma_{\delta y} & \sigma_{\delta z} & \sigma_\delta^2 \end{bmatrix}. \tag{2.77}$$

The covariance matrix $\mathbf{C}_{\vec{x}_\Delta}$ is of the same form. Recall that we wish to properly weight the positioning solution. The weights for each observation are proportional to their variance. We may define a *cofactor* as:

$$q_{ij} = \frac{\sigma_{ij}}{\sigma_0^2}, \tag{2.78}$$

---

1 Note that the concept of statistical independence is stronger than decorrelation. Independent variables are decorrelated but the reverse is not always true.

where σ_0^2 is an arbitrary reference variance chosen a priori. We may define the *cofactor matrix* as follows:

$$\mathbf{Q} = \frac{1}{\sigma_0^2} \mathbf{C}_{\bar{y}}. \tag{2.79}$$

The weight matrix $\mathbf{W}$ is the inverse of $\mathbf{Q}$:

$$\mathbf{W} = \mathbf{Q}^{-1} = \sigma_0^2 \mathbf{C}_{\bar{y}}^{-1}. \tag{2.80}$$

The elements w_i of $\mathbf{W}$ are the weights. The weight for the ith observation is given by:

$$w_i = \frac{\sigma_0^2}{\sigma_i^2}, \tag{2.81}$$

where σ_0^2 is used to define the unit weight, i.e. an observation i will have a weight of one when its variance equals the reference variance (i.e. $\sigma_i^2 = \sigma_0^2$). It is most convenient to choose $\sigma_0^2 = 1$, in which case the weight of an observation is simply the reciprocal of its variance, $1/\sigma_i^2$.

While the general form of $\mathbf{W}$ is given by Equation 2.80, there are several conditions which simplify it considerably. Pseudorange observations are made independent of one another, so we may make the assumption that they are uncorrelated. In this case, all the off-diagonal elements of the covariance matrix vanish:

$$\mathbf{C}_{\bar{y}} = \begin{bmatrix} \sigma_x^2 & 0 & 0 & 0 \\ 0 & \sigma_y^2 & 0 & 0 \\ 0 & 0 & \sigma_z^2 & 0 \\ 0 & 0 & 0 & \sigma_\delta^2 \end{bmatrix}. \tag{2.82}$$

Choosing $\sigma_0^2 = 1$ gives the following weight matrix:

$$\mathbf{W} = \sigma_0^2 \begin{bmatrix} \frac{1}{\sigma_x^2} & 0 & 0 & 0 \\ 0 & \frac{1}{\sigma_y^2} & 0 & 0 \\ 0 & 0 & \frac{1}{\sigma_z^2} & 0 \\ 0 & 0 & 0 & \frac{1}{\sigma_\delta^2} \end{bmatrix}. \tag{2.83}$$

Note that in the case where every measurement has the same variance, say, σ^2, then the weight matrix is simply given by:

$$\mathbf{W} = \frac{1}{\sigma^2} \mathbf{I}, \tag{2.84}$$

where $\mathbf{I}$ is the 4×4 identity matrix. Regardless of the form of $\mathbf{W}$, the weighted least squares solution is given by:

$$\hat{x} = \left(\mathbf{A}^T\mathbf{W}\mathbf{A}\right)^{-1}\mathbf{A}^T\mathbf{W}\vec{y}. \tag{2.85}$$

GNSS Positioning and Least Squares

We will now apply the method of least squares to the GNSS pseudorange observation. Instead of the simple model presented in Equation 2.66, we will use the linearized pseudorange equation:

$$R_r^s = R_{r,0}^s - \frac{(x^s - x_r)}{\rho_r^s}\Delta x - \frac{(y^s - y_r)}{\rho_r^s}\Delta y - \frac{(z^s - z_r)}{\rho_r^s}\Delta z + c\Delta\delta_r + e_r^s + \epsilon_r^s. \tag{2.86}$$

As before, we may remove the systematic error term e_r^s to receive the O minus C pseudorange (henceforth R_r^s will refer to O minus C). The random error term ϵ_r^s may not be removed. We may rewrite this as a matrix equation:

$$\mathbf{\Delta R}_r^s = \mathbf{A}\vec{\mathbf{x}}_\Delta + \vec{\epsilon}_r^s = \begin{bmatrix} -\frac{(x^s-x_r)}{\rho_r^s} & -\frac{(y^s-y_r)}{\rho_r^s} & -\frac{(z^s-z_r)}{\rho_r^s} & c \end{bmatrix}\begin{bmatrix} \Delta x \\ \Delta y \\ \Delta z \\ \Delta\delta_r \end{bmatrix} + \epsilon_r^s. \tag{2.87}$$

This represents the pseudorange observation from a single satellite. By considering additional satellites, the number of rows in these matrices increases. Considering observations from m satellites gives:

$$\begin{bmatrix} \Delta R_r^{(1)} \\ \Delta R_r^{(2)} \\ \vdots \\ \Delta R_r^{(m)} \end{bmatrix} = \begin{bmatrix} \frac{-(x^{(1)} - x_r)}{\rho_r^{(1)}} & \frac{-(y^{(1)} - y_r)}{\rho_r^{(1)}} & \frac{-(z^{(1)} - z_r)}{\rho_r^{(1)}} & c \\ \frac{-(x^{(2)} - x_r)}{\rho_r^{(2)}} & \frac{-(y^{(2)} - y_r)}{\rho_r^{(2)}} & \frac{-(z^{(2)} - z_r)}{\rho_r^{(2)}} & c \\ \vdots & \vdots & \vdots & \vdots \\ \frac{-(x^{(m)} - x_r)}{\rho_r^{(m)}} & \frac{-(y^{(m)} - y_r)}{\rho_r^{(m)}} & \frac{-(z^{(m)} - z_r)}{\rho_r^{(m)}} & c \end{bmatrix}\begin{bmatrix} \Delta x \\ \Delta y \\ \Delta z \\ \Delta\delta_r \end{bmatrix} + \begin{bmatrix} \epsilon_r^{(1)} \\ \epsilon_r^{(2)} \\ \vdots \\ \epsilon_r^{(m)} \end{bmatrix}.$$

We may now solve this for the least squares solution:

$$\hat{x}_\Delta = \left(\mathbf{A}^T\mathbf{W}\mathbf{A}\right)^{-1}\mathbf{A}^T\mathbf{W}\vec{y}, \tag{2.88}$$

where $\vec{y}$ is the observation vector on the left hand side, and $\mathbf{W} = \mathbf{Q}_{obs}^{-1}$ is the weighting matrix. In this case, $\mathbf{Q}_{obs}$ is a diagonal 4×4 matrix whose entries are the variances of the x, y, z, and δ_r observations. As before, the solution obtained may be iterated until it converges to the true location of the receiver.

2.4.3 Further Considerations on Dilution of Precision

The geometric configuration of a set of satellites has an effect on positioning error through a phenomenon called DOP, as noted before. It is important to note that positioning error differs from observation error. Observation error is the uncertainty in the measurement of a GNSS observable, caused by error sources such as those covered in Section 2.3. These observation errors may be magnified to a varying degree by the geometric configuration of the observed satellites. This phenomenon is illustrated for the two-dimensional case in Figure 2.9. Dots represent satellites, black semicircles represent the true range to the receiver, and the lighter semicircles represent observation error.

In Figure 2.9a, there is no uncertainty in the observations, so the location of the receiver is at the intersection of the two semicircles. In Figure 2.9b,c, each observation is subject to some uncertainty. This translates into a positioning uncertainty, represented by the shaded region. The observation uncertainties are unchanged between cases (b) and (c), but it is easy to see that the positioning uncertainty has changed considerably, simply due to the arrangement of the satellites. Extending this understanding to GNSS positioning, we may model our position error vector $\vec{x}_\Delta$ as a random vector to account for the uncertainty. We may assume that $\vec{x}_\Delta$ is normally distributed, and thus $\langle \vec{x}_\Delta \rangle = 0$. If $\vec{x}_\Delta$ is normally distributed, then the observation vector $\vec{y}$ is also normally distributed, so it is also true that

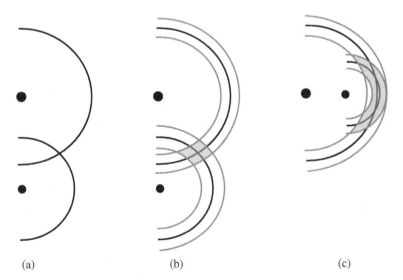

(a) (b) (c)

Figure 2.9 (a) Observations with no uncertainty, and therefore no positioning error. (b) and (c) Observations with some uncertainty; respective positioning errors are shaded.

$\langle \vec{y} \rangle = 0$. The following derivation has been adapted from [37, 38]. Let us consider the case where $\vec{x}_\Delta = \left(\mathbf{A}^T \mathbf{A} \right)^{-1} \mathbf{A}^T \vec{y}$, as per our discussion on least squares solutions in Section 2.4.1. The variance–covariance matrix of $\vec{x}_\Delta$ is given by:

$$\mathbf{C}_{\vec{x}_\Delta} = \left\langle \left(\vec{x}_\Delta - \langle \vec{x}_\Delta \rangle \right) \left(\vec{x}_\Delta - \langle \vec{x}_\Delta \rangle \right)^T \right\rangle, \tag{2.89}$$

but as we said, $\langle \vec{x}_\Delta \rangle = 0$, so Equation 2.89 reduces to

$$\mathbf{C}_{\vec{x}_\Delta} = \left\langle \vec{x}_\Delta \vec{x}_\Delta^T \right\rangle. \tag{2.90}$$

Now we may substitute Equation 2.88 into Equation 2.90 to obtain:

$$\mathbf{C}_{\vec{x}_\Delta} = \left\langle \left(\left(\mathbf{A}^T \mathbf{W} \mathbf{A} \right)^{-1} \mathbf{A}^T \mathbf{W} \vec{y} \right) \left(\left(\mathbf{A}^T \mathbf{W} \mathbf{A} \right)^{-1} \mathbf{A}^T \mathbf{W} \vec{y} \right)^T \right\rangle. \tag{2.91}$$

$\mathbf{A}$ is not a random matrix, so it may be factored out to result in:

$$\mathbf{C}_{\vec{x}_\Delta} = \left(\mathbf{A}^T \mathbf{W} \mathbf{A} \right)^{-1} \mathbf{A}^T \mathbf{W} \left\langle \vec{y} \vec{y}^T \right\rangle \mathbf{W}^T \mathbf{A} \left(\left(\mathbf{A}^T \mathbf{W} \mathbf{A} \right)^{-1} \right)^T. \tag{2.92}$$

Expanding all inverted and transposed quantities and recalling that $(\mathbf{A}^T)^{-1} = (\mathbf{A}^{-1})^T$ allows us to cancel nearly every term, leaving only:

$$\mathbf{C}_{\vec{x}_\Delta} = \mathbf{A}^{-1} \left\langle \vec{y} \vec{y}^T \right\rangle \left(\mathbf{A}^T \right)^{-1}. \tag{2.93}$$

Since $\langle \vec{y} \rangle = 0$, we see via Equation 2.74 that $\mathbf{C}_{\vec{y}} = \left\langle \vec{y} \vec{y}^T \right\rangle$. This allows us to write Equation 2.93 as:

$$\mathbf{C}_{\vec{x}_\Delta} = \mathbf{A}^{-1} \mathbf{C}_{\vec{y}} \left(\mathbf{A}^T \right)^{-1}. \tag{2.94}$$

Equation 2.94 is a specific case of a general law known as the *covariance law*. The general form of the covariance law for random matrices $\mathbf{X}$ and $\mathbf{Y}$, where $\mathbf{X} = \mathbf{BY}$, is:

$$\mathbf{C}_\mathbf{X} = \mathbf{B} \mathbf{C}_\mathbf{Y} \mathbf{B}^T. \tag{2.95}$$

We may assume that pseudorange observations are uncorrelated, so $\mathbf{C}_{\vec{y}}$ is diagonal. If each pseudorange observation has the same variance σ_{obs}^2, then $\mathbf{C}_{\vec{y}} = \sigma_{obs}^2 \mathbf{I}$, where $\mathbf{I}$ is the 4×4 identity matrix. Consequently, Equation 2.94 becomes:

$$\mathbf{C}_{\vec{x}_\Delta} = \sigma^2 \left(\mathbf{A}^T \mathbf{A} \right)^{-1} = \sigma_{obs}^2 \mathbf{D}. \tag{2.96}$$

$\mathbf{D}$ is a matrix whose elements d_{ij} are purely geometric in nature. We may write both $\mathbf{C}_{\vec{x}_\Delta}$ and $\mathbf{D}$ explicitly to obtain:

$$\begin{bmatrix} \sigma_x^2 & \sigma_{xy} & \sigma_{xz} & \sigma_{x\delta} \\ \sigma_{yx} & \sigma_y^2 & \sigma_{yz} & \sigma_{y\delta} \\ \sigma_{zx} & \sigma_{zy} & \sigma_z^2 & \sigma_{z\delta} \\ \sigma_{\delta x} & \sigma_{\delta y} & \sigma_{\delta z} & \sigma_\delta^2 \end{bmatrix} = \sigma^2 \begin{bmatrix} d_x^2 & d_{xy} & d_{xz} & d_{x\delta} \\ d_{yx} & d_y^2 & d_{yz} & d_{y\delta} \\ d_{zx} & d_{zy} & d_z^2 & d_{z\delta} \\ d_{\delta x} & d_{\delta y} & d_{\delta z} & d_\delta^2 \end{bmatrix}. \tag{2.97}$$

The standard deviation σ_{obs} is simply the UERE (introduced before in Section 2.3), so σ^2 is also known. The design matrix $\mathbf{A}$ is also known, so we may solve for $\mathbf{C}_{\bar{x}_\Delta}$. The total variation of $\vec{x}_\Delta$ is given by the trace of $\mathbf{C}_{\bar{x}_\Delta}$. This quantity is related to the positioning error, so we will denote it σ_{pos}^2.

$$\sigma_{pos}^2 = \text{tr}\left(\mathbf{C}_{\bar{x}_\Delta}\right) = \sigma_x^2 + \sigma_y^2 + \sigma_z^2 + \sigma_\delta^2. \tag{2.98}$$

Notice that $\mathbf{D}$ is very similar to $\mathbf{C}_{\bar{x}_\Delta}$. However, whereas $\mathbf{C}_{\bar{x}_\Delta}$ represents statistical information, $\mathbf{D}$ is geometric in nature. A parameter analogous to total variation may be defined for $\mathbf{D}$, and its square root represents the geometric dilution of precision (GDOP).

$$(\text{GDOP})^2 = (d_x^2 + d_y^2 + d_z^2 + d_\delta^2). \tag{2.99}$$

Thus we see that the positioning error is related to the observation factor by the GDOP.

$$\sigma_{pos}^2 = \sigma_{obs}^2 (\text{GDOP})^2 \quad \Rightarrow \quad \sigma_{pos} = \sigma_{obs} \text{GDOP}. \tag{2.100}$$

Thus, depending on the geometric configuration of the satellites, the UERE (σ_{obs}) will be multiplied by the GDOP to result in a larger positioning error. Typical values for the UERE are 0.2–1 m for P-code, and several meters for C/A code [15]. Smaller values of GDOP are desirable – typically a GDOP of less than 7 is considered good. GNSS receivers can observe more than four satellites simultaneously and choose the configuration which produces the lowest GDOP.

The position vectors between a GNSS receiver and the four satellites used for positioning form a tetrahedron. The volume of this tetrahedron is inversely related to the value of the GDOP. Therefore, the satellite configuration that provides the best positioning measurement is the one with the largest tetrahedron volume.

In theory, the largest possible tetrahedron has one satellite at zenith and three satellites equally spaced below the horizon at an elevation angle of $-19.47°$. This configuration produces a GDOP of 1.581. However, this configuration is impossible, because signals can not be received from satellites below the horizon. Therefore, the best realizable configuration is one with a satellite at zenith and three satellites equally spaced on the horizon. This configuration produces a GDOP of 1.732 [38].

In addition to GDOP, several other quantities are often referred to in the literature, so they bear mentioning. The GDOP is often separated into position dilution of precision (PDOP) and time dilution of precision (TDOP). PDOP can also be further separated into horizontal dilution of precision (HDOP) and vertical dilution of precision (VDOP). The expressions for all five DOP quantities are shown in Equation 2.101

$$
\begin{aligned}
\text{GDOP} &= \frac{\sqrt{\sigma_x^2 + \sigma_y^2 + \sigma_z^2 + \sigma_\delta^2}}{\sigma_{obs}} = \sqrt{d_x^2 + d_y^2 + d_z^2 + d_\delta^2}, \\[2mm]
\text{PDOP} &= \frac{\sqrt{\sigma_x^2 + \sigma_y^2 + \sigma_z^2}}{\sigma_{obs}} = \sqrt{d_x^2 + d_y^2 + d_z^2}, \\[2mm]
\text{TDOP} &= \frac{\sqrt{\sigma_\delta^2}}{\sigma_{obs}} = \sqrt{d_\delta^2}, \\[2mm]
\text{HDOP} &= \frac{\sqrt{\sigma_x^2 + \sigma_y^2}}{\sigma_{obs}} = \sqrt{d_x^2 + d_y^2}, \\[2mm]
\text{VDOP} &= \frac{\sqrt{\sigma_z^2}}{\sigma_{obs}} = \sqrt{d_z^2}.
\end{aligned}
\tag{2.101}
$$

Note that $\text{GDOP}^2 = \text{PDOP}^2 + \text{TDOP}^2$ and $\text{PDOP}^2 = \text{HDOP}^2 + \text{VDOP}^2$.

2.5 Data Combinations and Relative Positioning

Some of the most powerful techniques available for removing GNSS positioning errors involve taking linear combinations of GNSS observables. By combining observables in various ways, it is possible to suppress various sources of error, resulting in more accurate positioning calculations. This section will cover two main classes of data combinations. The first are multi-frequency combinations, which require the use of multi-frequency receivers. The second are differences of observables measured at multiple receivers, from multiple satellites, or at different epochs. This second class of combinations is used in so-called relative positioning. The two classes of combinations may also be used in tandem.

2.5.1 Multi-Frequency Combinations

Multi-frequency combinations are techniques involving multiple GNSS observables measured on different carrier frequencies at the same receiver. Up to nine GNSS observables are available to a single receiver, depending on how many frequencies it is designed to receive. In this section, we will adopt a notation in which the subscript of an observable refers to the frequency it is measured on.

A single frequency receiver has access to the GNSS observables on that frequency. For example, a single frequency GPS receiver could use the L1 pseudorange R_1, the L1 carrier phase Φ_1 and the L1 Doppler shift D_1. A dual-frequency GPS receiver could access the observables on two frequencies. So a receiver capable of receiving both L1 and L2 would not only receive the L1 observables previously mentioned, but also the L2 observables R_2, Φ_2 and D_2. Finally, a triple-frequency GPS receiver could receive signals from all three current GPS carrier frequencies – L1, L2, and L5. Therefore, in addition to the six observables already mentioned, a triple-frequency receiver can also measure the L5 observables R_5, Φ_5 and D_5. Multi-frequency combinations are linear combinations of two or more observables which provide certain benefits when performing GNSS positioning, such as canceling the effects of certain error terms or aiding in cycle slip detection. Typically only observables of the same type are combined – i.e. code–code, carrier–carrier, and Doppler–Doppler combinations – though code-carrier combinations are sometimes used as well.

The remainder of this subsection, which has been adapted from [7], discusses a few common combinations and the benefits they provide. Only combinations of pseudorange and carrier phase observables will be considered, though in principle Doppler shift measurements could be combined in a similar manner. Special attention is given to the ionosphere-free combination, since it is an essential tool for dealing with ionospheric error.

Ionosphere-Free Combination
Due to various time-dependent effects, it is difficult to accurately model and therefore mitigate, ionospheric error when performing GNSS positioning. The ionospheric modeling parameters included in the navigation message are only capable of removing about half of the error contribution of the ionosphere. This degree of mitigation typically suffices for standard point positioning; however, the remaining ionospheric error is unacceptably large for precise point positioning. To remove the remaining contribution, a data combination known as the ionosphere-free combination is employed.

The ionosphere is dispersive, meaning that signals of different frequencies travel at different velocities. This implies that the amount of time delay, and therefore the induced ionospheric error, differs between GNSS observables on different frequencies. Consider the pseudorange observables on L1 and L2:

$$R_1 = \rho + c\delta_r + I_1,$$
$$R_2 = \rho + c\delta_r + I_2. \tag{2.102}$$

ρ is the distance between the satellite and receiver, δ_r is the receiver clock bias, and I_i is the ionospheric error term for frequency i. All other sources of error may be neglected for now, as they are not relevant for the present discussion.

As mentioned previously, ionospheric error is frequency dependent, and may be approximated as:

$$I_i = \frac{40.3}{f_i^2}\text{TEC},\tag{2.103}$$

where f_i is frequency i and TEC is the (slant) total electron content, a spatial- and time-dependent parameter that depends on the conditions of the ionosphere along the path between the satellite and the receiver. Details on the TEC and a derivation of Equation 2.103 are presented in Sections 7.3.1 and 7.3.3, respectively. We may now rewrite the expressions in Equation 2.102 as:

$$R_1 = \rho + c\delta_r + \frac{40.3}{f_1^2}\text{TEC},$$
$$R_2 = \rho + c\delta_r + \frac{40.3}{f_2^2}\text{TEC}.\tag{2.104}$$

Multiplying both sides of these expressions by the squares of their respective frequencies gives:

$$f_1^2 R_1 = f_1^2\left(\rho + c\delta_r\right) + 40.3\text{TEC},$$
$$f_2^2 R_2 = f_2^2\left(\rho + c\delta_r\right) + 40.3\text{TEC}.\tag{2.105}$$

If we now difference these two expressions, we obtain:

$$f_1^2 R_1 - f_2^2 R_2 = \left(f_1^2 - f_2^2\right)\left(\rho + c\delta_r\right).\tag{2.106}$$

We see that the ionospheric error term cancels completely. Thus, this linear combination of R_1 and R_2 is called the ionosphere-free combination. Rearranging slightly and letting $R_{IF} = \left(\rho + c\delta_r\right)$, we get:

$$\boxed{R_{IF} = \frac{f_1^2 R_1 - f_2^2 R_2}{f_1^2 - f_2^2}.}\tag{2.107}$$

This is the standard form of the ionosphere-free pseudorange combination.

An ionosphere-free combination may also be obtained for carrier phase measurements. Consider the carrier phase observables on L1 and L2:

$$\lambda_1 \Phi_1 = \rho + c\delta_r + \lambda_1 N_1 - I_1,$$
$$\lambda_2 \Phi_2 = \rho + c\delta_r + \lambda_2 N_2 - I_2.\tag{2.108}$$

λ_i is the wavelength of a signal with frequency i and N_i is the integer ambiguity on frequency i. Once again, all other error terms have been omitted, since they are not relevant here. Note that the sign in front of the ionospheric error terms has changed. This is due to a phenomenon known as carrier-code divergence, and is caused by dispersion. Details of this effect are discussed in Section B.3.2.

Dividing both sides of the expressions in Equation 2.108 by λ_i and using the relation $c = f_i \lambda_i, i = 1, 2$ lets us rewrite this as:

$$\Phi_1 = \frac{f_1}{c}\rho + f_1\delta_r + N_1 - \frac{f_1}{c}I_1,$$
$$\Phi_2 = \frac{f_2}{c}\rho + f_2\delta_r + N_2 - \frac{f_2}{c}I_2. \tag{2.109}$$

Substituting Equation 2.103 gives:

$$\Phi_1 = \frac{f_1}{c}\rho + f_1\delta_r + N_1 - \frac{40.3}{cf_1}\text{TEC},$$
$$\Phi_2 = \frac{f_2}{c}\rho + f_2\delta_r + N_2 - \frac{40.3}{cf_2}\text{TEC}. \tag{2.110}$$

Multiplying both sides by their respective frequencies removes the frequency dependence of the ionospheric error term:

$$f_1\Phi_1 = \frac{f_1^2}{c}\rho + f_1^2\delta_r + f_1 N_1 - \frac{40.3}{c}\text{TEC},$$
$$f_2\Phi_2 = \frac{f_2^2}{c}\rho + f_2^2\delta_r + f_2 N_2 - \frac{40.3}{c}\text{TEC}. \tag{2.111}$$

Differencing these two expressions gives:

$$f_1\Phi_1 - f_2\Phi_2 = \left(f_1^2 - f_2^2\right)\left(\frac{\rho}{c} + \delta_r\right) + f_1 N_1 - f_2 N_2. \tag{2.112}$$

Note that $c = f_i \lambda_i$, so multiplying through by c results in:

$$f_1^2\lambda_1\Phi_1 - f_2^2\lambda_2\Phi_2 = \left(f_1^2 - f_2^2\right)\left(\rho + c\delta_r\right) + f_1^2\lambda_1 N_1 - f_2^2\lambda_2 N_2. \tag{2.113}$$

Dividing through by $f_1^2 - f_2^2$ gives:

$$\frac{f_1^2\lambda_1\Phi_1 - f_2^2\lambda_2\Phi_2}{f_1^2 - f_2^2} = \rho + c\delta_r + \frac{f_1^2\lambda_1 N_1 - f_2^2\lambda_2 N_2}{f_1^2 - f_2^2}. \tag{2.114}$$

Defining Φ_{IC} as the right hand side of this expression, we obtain the final form of our solution:

$$\boxed{\Phi_{IC} = \frac{f_1^2\lambda_1\Phi_1 - f_2^2\lambda_2\Phi_2}{f_1^2 - f_2^2}.} \tag{2.115}$$

Due to several approximations made in the formulation of the ionospheric error term I, the ionosphere-free combination does not actually completely eliminate ionospheric error. An improved model of I which accounts for higher order terms of the series expansion of the refractive index, the geomagnetic field effect, and bending of the ray path is discussed in [32].

Geometry-Free Combination

The geometry-free combination is the simplest data combination, merely requiring the difference of two observations. This eliminates all geometric terms, leaving only those that are dependent on frequency. The most notable terms remaining in a geometry-free combination are ionospheric delay and multipath errors, though less significant frequency-dependent terms such as instrumental biases and carrier phase wind-up remain as well. Additionally, the random noise term cannot be canceled.

Consider two pseudorange measurements on L1 and L2:

$$R_1 = \rho + c\left(\delta_r - \delta^s\right) + T_r^s + I_1 + K_1 + M_1 + \epsilon_1,$$
$$R_2 = \rho + c\left(\delta_r - \delta^s\right) + T_r^s + I_2 + K_2 + M_1 + \epsilon_2, \tag{2.116}$$

where ρ is the distance between the satellite and receiver, δ_r and δ^s are respectively the receiver and satellite clock biases, and T_r^s is the tropospheric delay. In addition, I_i is the ionospheric delay, $K_i = d_{c,i} - d_c^i$ where $d_{c,i}$ and d_c^i are respectively the code receiver and satellite instrumental biases, M_i is the code multipath error, and ϵ_i is the code random error. In all cases, the index i refers to a specific frequency. The geometry-free pseudorange combination is simply the difference of these two measurements.

$$R_{GF} = R_1 - R_2 = \left(I_1 - I_2\right) + \left(K_1 - K_2\right) + \left(M_1 - M_2\right) + \left(\epsilon_1 - \epsilon_2\right). \tag{2.117}$$

Similarly, consider two carrier phase measurements on L1 and L2:

$$\lambda_1 \Phi_1 = \rho + c\left(\delta_r - \delta^s\right) + \lambda_1 N_1 + T_r^s - I_1 + k_1 + \lambda_1 w + m_1 + \varepsilon_1,$$
$$\lambda_2 \Phi_2 = \rho + c\left(\delta_r - \delta^s\right) + \lambda_2 N_2 + T_r^s - I_2 + k_2 + \lambda_2 w + m_1 + \varepsilon_2, \tag{2.118}$$

where N_i is the integer ambiguity, $k_i = d_{p,i} - d_p^i$ where $d_{p,i}$ and d_p^i are respectively the phase receiver and satellite instrumental biases, w is the wind-up correction, m_i is the phase multipath error, and ε_i is the phase random error. Once again, note the sign change on the ionospheric error terms. The geometry-free carrier phase combination can be obtained from:

$$\lambda_1 \Phi_1 - \lambda_2 \Phi_2 = \left(\lambda_1 N_1 - \lambda_2 N_2\right) - \left(I_1 - I_2\right) + \left(k_1 - k_2\right)$$
$$+ \left(\lambda_1 - \lambda_2\right) w + \left(m_1 - m_2\right) + \left(\varepsilon_1 - \varepsilon_2\right). \tag{2.119}$$

In addition to the code–code and phase–phase combinations just shown, it is possible to create a code-phase geometry-free combination using the observables on the same frequency. This combination is given by:

$$\lambda_i \Phi_i - R_i = \lambda_i N_i - 2I_i + \left(k_i - K_i\right) + \lambda_i w + \left(m_i - M_i\right) + \left(\varepsilon_i - \epsilon_i\right), \tag{2.120}$$

where $i = 1$, 2, or 5, corresponding to L1, L2, or L5, respectively. Geometry-free combinations may be used to estimate TEC or detect cycle slips, among other applications [7].

Wide-Lane and Narrow-Lane Combinations

Wide-lane combinations are a type of combination designed to produce a signal with a larger wavelength than that of either constituent observable. These combinations are useful for cycle slip detection and ambiguity resolution [7].

Consider pseudorange measurements on L1 and L2, each multiplied by their respective frequencies. All sources of error except the ionosphere are neglected.

$$f_1 R_1 = f_1 \left(\rho + \delta_r \right) + f_1 I_1,$$
$$f_2 R_2 = f_2 \left(\rho + \delta_r \right) + f_2 I_2. \tag{2.121}$$

Differencing these two expressions gives:

$$f_1 R_1 - f_2 R_2 = \left(f_1 - f_2 \right) \left(\rho + \delta_r \right) + f_1 I_1 - f_2 I_2. \tag{2.122}$$

Dividing both sides by $f_1 - f_2$ results in:

$$\frac{f_1 R_1 - f_2 R_2}{f_1 - f_2} = \rho + \delta_r + \frac{f_1 I_1 - f_2 I_2}{f_1 - f_2}. \tag{2.123}$$

We may denote the right hand side of this expression as R_W, letting us write the wide-lane pseudorange combination as:

$$R_W = \frac{f_1 R_1 - f_2 R_2}{f_1 - f_2}. \tag{2.124}$$

A similar derivation may be performed for carrier phase measurements. Consider the L1 and L2 carrier phase observables multiplied by their respective frequencies. Again only the ionospheric error term is considered, and as before its sign is opposite of the pseudorange case.

$$f_1 \hat{\Phi}_1 := f_1 \lambda_1 \Phi_1 = f_1 \left(\rho + \delta_r \right) + \lambda_1 N_1 - f_1 I_1,$$
$$f_2 \hat{\Phi}_2 := f_2 \lambda_2 \Phi_2 = f_2 \left(\rho + \delta_r \right) + \lambda_2 N_2 - f_2 I_2. \tag{2.125}$$

Again, differencing these measurements and dividing by $f_1 - f_2$ give:

$$\frac{f_1 \hat{\Phi}_1 - f_2 \hat{\Phi}_2}{f_1 - f_2} = \rho + \delta_r + \frac{f_1 \lambda_1 N_1 - f_2 \lambda_2 N_2}{f_1 - f_2} - \frac{\left(f_1 I_1 - f_2 I_2 \right)}{f_1 - f_2}. \tag{2.126}$$

Denoting the right hand side as $\hat{\Phi}_W$ lets us finally write the wide-lane carrier phase combination:

$$\hat{\Phi}_W = \frac{f_1 \hat{\Phi}_1 - f_2 \hat{\Phi}_2}{f_1 - f_2}. \tag{2.127}$$

Narrow-lane combinations are combinations designed to produce a signal with a smaller wavelength than either constituent signal. Narrow-lane combinations are useful because the noise of the combination is lower than either of its constituent signals [7]. The derivation for narrow-lane combinations is nearly identical to that of wide-lane combinations, but rather than taking the difference of the

Table 2.2 Wide-lane and narrow-lane combinations of carrier frequencies.

Signals combined	Wide-lane wavelength (m)	Narrow-lane wavelength (m)
L1, L2	$\lambda_W = 0.862$	$\lambda_N = 0.107$
L1, L5	$\lambda_W = 0.751$	$\lambda_N = 0.109$
L2, L5	$\lambda_W = 5.861$	$\lambda_N = 0.125$

observables, their sum is taken instead. Therefore, the narrow-lane pseudorange combination may be written as:

$$R_N = \frac{f_1 R_1 + f_2 R_2}{f_1 + f_2} \tag{2.128}$$

and the narrow-lane carrier phase combination is:

$$\hat{\Phi}_N = \frac{f_1 \hat{\Phi}_1 + f_2 \hat{\Phi}_2}{f_1 + f_2}. \tag{2.129}$$

The wavelength of a wide-lane signal is given by $\lambda_W = c / \left(f_i - f_j \right)$, while a narrow-lane signal is given by $\lambda_W = c / \left(f_i + f_j \right)$. In both expressions, i and j can take values of 1, 2, or 5, corresponding to L1, L2 and L5, respectively. Recalling that L1 = 1575.420 MHz, L2 = 1227.600 MHz, and L5 = 1176.45 MHz we may determine the wide- or narrow-lane combination for any pair of carrier frequencies. Table 2.2 lists the possible combinations.

2.5.2 Relative Positioning

Relative positioning, also known as differential positioning, is a collection of techniques in which measurements from two or more receivers and/or satellites are differenced. Much like multi-frequency combinations, these differencing techniques are useful for eliminating certain sources of error. There are three different types of differential positioning: single differences, double differences, and triple differences. Single differences are performed at a single epoch, and they either compare the measurements from two satellites at one receiver or the measurement from the same satellite at two separate receivers. In the case where two receivers are used, the distance between them is known as the *baseline*. For short baselines (a few kilometers), ionospheric, tropospheric and ephemeris errors are all negligible and may be ignored [11]. These errors will grow along with an increasing baseline, so relative positioning is most useful for stations in

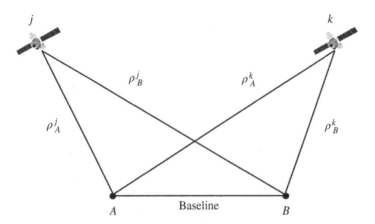

Figure 2.10 Relative positioning using two satellites and two receivers.

close proximity. A double difference can be formed by differencing two single differences. In other words, a double difference requires two satellites and two receivers. Furthermore, a triple difference is one in a double difference at one epoch is subtracted from the same double difference at a different epoch. Thus, a triple difference still only requires two satellites and two receivers, but measurements must be taken over successive epochs. The notation used in this section will denote satellites using superscripts, e.g. δ^s, $s = j, k, \ldots$ where j represents the jth satellite, k represents the kth satellite, and so on. Similarly, receivers will be denoted with subscripts, e.g. δ_r, $r = A, B, \ldots$ where A represents receiver A and so on. Figure 2.10 illustrates this.

Single Differences

It is possible to perform relative positioning using either pseudorange or carrier phase measurements. Only the carrier phase expressions will be considered going forward, but pseudorange expressions are affected in an analogous way. In the following expressions, the clock errors (δ_r, δ^s) and carrier phase instrumental biases ($d_{p,r}$, d_p^s) for receiver r and satellite s have been singled out in order to demonstrate how they cancel when differencing techniques are applied. The remaining error sources are grouped into a single term, e_A^j. There are two different types of single difference to consider: across-receiver and across-satellite.

Across-receiver single differences: First we will consider a single difference using one satellite and two receivers. This type of single difference is called an *across-receiver* single difference. The carrier phase measurements from a given satellite j are measured at a single epoch on a given frequency, say, L1. The expressions for the carrier phase measurement at receivers A and B are, respectively:

$$\Phi_A^j = \frac{1}{\lambda} \left[\rho_A^j + c \left(\delta_A - \delta^j \right) + \lambda N_A^j + d_{p,A} - d_p^j + e_A^j \right],$$

$$\Phi_B^j = \frac{1}{\lambda} \left[\rho_B^j + c \left(\delta_B - \delta^j \right) + \lambda N_B^j + d_{p,B} - d_p^j + e_B^j \right].$$

(2.130)

We will use the notation $\Phi_{AB}^j = \Phi_B^j - \Phi_A^j$ to represent the difference of these two quantities. The same notation applies to all differenced quantities. Therefore, by differencing the two expressions in Equation 2.130 we get:

$$\boxed{\Phi_{AB}^j = \frac{1}{\lambda} \left[\rho_{AB}^j + c\delta_{AB} + \lambda N_{AB}^j + d_{p,AB} + e_{AB}^j \right].}$$

(2.131)

Note that the terms related to the satellite hardware, δ^j and d_p^j have canceled out. If we had used satellite k instead of satellite j throughout this derivation, we would obtain the analogous equation:

$$\Phi_{AB}^k = \frac{1}{\lambda} \left[\rho_{AB}^k + c\delta_{AB} + \lambda N_{AB}^k + d_{p,AB} + e_{AB}^k \right].$$

(2.132)

Note that the only thing that has changed is the satellite index. The satellite hardware errors have still been removed, and the receiver hardware errors are identical to those in Equation 2.131.

Across-satellite single differences: Now consider a single difference using two satellites but only a single receiver, also known as a *across-satellite* single difference. Once again, only a single epoch and frequency are considered. The carrier phase measurements from satellites j and k are received by receiver A, giving the expressions:

$$\Phi_A^j = \frac{1}{\lambda} \left[\rho_A^j + c \left(\delta_A - \delta^j \right) + \lambda N_A^j + d_{p,A} - d_p^j + e_A^j \right],$$

$$\Phi_A^k = \frac{1}{\lambda} \left[\rho_A^k + c \left(\delta_A - \delta^k \right) + \lambda N_A^k + d_{p,A} - d_p^k + e_A^k \right].$$

(2.133)

The notation convention $\Phi_A^{jk} = \Phi_A^k - \Phi_A^j$ will be used for all quantities. Thus the across-satellite single difference is given by the expression:

$$\boxed{\Phi_A^{jk} = \frac{1}{\lambda} \left[\rho_A^{jk} - c\delta^{jk} + \lambda N_A^{jk} - d_p^{jk} + e_A^{jk} \right].}$$

(2.134)

In this case, the errors related to the receiver hardware, δ_r and $d_{p,A}$ have canceled out. If we had used receiver B instead of receiver A, we would have obtained the expression:

$$\Phi_B^{jk} = \frac{1}{\lambda} \left[\rho_B^{jk} - c\delta^{jk} + \lambda N_B^{jk} - d_p^{jk} + e_B^{jk} \right]. \tag{2.135}$$

The receiver hardware errors are also removed in this case, and the satellite hardware errors are identical to those in Equation 2.134.

Differenced measurements are not necessarily uncorrelated. We can use the covariance law given in Equation 2.95 to determine whether a set of differenced observables is correlated or not. For receivers A and B and satellites j and k, the four possible single differences can be expressed by the matrix equation:

$$\mathbf{S} = \mathbf{M}\mathbf{\Phi} \quad \Rightarrow \quad \begin{bmatrix} \Phi_{AB}^{j} \\ \Phi_{AB}^{k} \\ \Phi_{A}^{jk} \\ \Phi_{B}^{jk} \end{bmatrix} = \begin{bmatrix} -1 & 1 & 0 & 0 \\ 0 & 0 & -1 & 1 \\ -1 & 0 & 1 & 0 \\ 0 & -1 & 0 & 1 \end{bmatrix} \begin{bmatrix} \Phi_{A}^{j} \\ \Phi_{B}^{j} \\ \Phi_{A}^{k} \\ \Phi_{B}^{k} \end{bmatrix}. \tag{2.136}$$

If we call $\mathbf{C_S}$ the covariance matrix for $\mathbf{S}$ and $\mathbf{C_\Phi}$ the covariance matrix for $\mathbf{\Phi}$, from the covariance law we see that $\mathbf{C_S} = \mathbf{M}\mathbf{C_\Phi}\mathbf{M}^T$. Recall that if the standard deviation σ of every element of $\mathbf{\Phi}$ is the same, we may write $\mathbf{C_\Phi} = \sigma^2 \mathbf{I}$ where $\mathbf{I}$ is the identity matrix. Thus, the covariance law becomes:

$$\mathbf{C_S} = \mathbf{M}\mathbf{C_\Phi}\mathbf{B}^T = \mathbf{M}\sigma^2 \mathbf{I}\mathbf{M}^T = \sigma^2 \mathbf{M}\mathbf{M}^T. \tag{2.137}$$

Evaluating $\mathbf{M}\mathbf{M}^T$ gives us:

$$\mathbf{C_S} = \sigma^2 \begin{bmatrix} 2 & 0 & 1 & -1 \\ 0 & 2 & -1 & 1 \\ 1 & -1 & 2 & 0 \\ -1 & 1 & 0 & 2 \end{bmatrix}. \tag{2.138}$$

Using Equation 2.76, we see that the first two elements of $\mathbf{S}$ are uncorrelated, as are the third and fourth elements. In other words, single differences of the same type (across-receiver or across-satellite) are uncorrelated. Single differences of opposite types are correlated or anti-correlated to some degree. As we will see in the section on double differences, single differences of the same type can be combined to further eliminate errors. On the other hand, single differences of opposite types are typically not combined. Therefore, their correlations are irrelevant.

If only single differences of the same type are considered, Equation 2.136 reduces to either:

$$\mathbf{S} = \mathbf{M}\Phi \quad \Rightarrow \quad \begin{bmatrix} \Phi^j_{AB} \\ \Phi^k_{AB} \end{bmatrix} = \begin{bmatrix} -1 & 1 & 0 & 0 \\ 0 & 0 & -1 & 1 \end{bmatrix} \begin{bmatrix} \Phi^j_A \\ \Phi^j_B \\ \Phi^k_A \\ \Phi^k_B \end{bmatrix} \tag{2.139}$$

or

$$\mathbf{S} = \mathbf{M}\Phi \quad \Rightarrow \quad \begin{bmatrix} \Phi^{jk}_A \\ \Phi^{jk}_B \end{bmatrix} = \begin{bmatrix} -1 & 0 & 1 & 0 \\ 0 & -1 & 0 & 1 \end{bmatrix} \begin{bmatrix} \Phi^j_A \\ \Phi^j_B \\ \Phi^k_A \\ \Phi^k_B \end{bmatrix}. \tag{2.140}$$

In both cases, $\mathbf{C_S}$ reduces to:

$$\mathbf{C_S} = \sigma^2 \begin{bmatrix} 2 & 0 \\ 0 & 2 \end{bmatrix} = 2\sigma^2 \mathbf{I}. \tag{2.141}$$

Double Differences

A double difference is formed by simply subtracting one single difference from another. Note that these two single differences must be of the same type! For example, Equations 2.131 and 2.132 may be used to create a double difference, as can Equations 2.134 and 2.135. These two pairs alone will eliminate the remaining hardware-related errors, so while other pairs are mathematically possible, they are not useful.

Consider Equations 2.131 and 2.132. Recall that we made the assumptions that these differences were performed at a single epoch and a single frequency. From these two equations, we can form a double difference as follows:

$$\Phi^{jk}_{AB} = \frac{1}{\lambda} \left[\rho^{jk}_{AB} + \lambda N^{jk}_{AB} + e^{jk}_{AB} \right]. \tag{2.142}$$

Here, the notation convention $\Phi^{jk}_{AB} = \Phi^k_{AB} - \Phi^j_{AB}$ is used for all parameters. Note that the remaining receiver hardware errors have vanished. Note that had we started with Equations 2.134 and 2.135 instead, the satellite hardware errors would have vanished, leaving the exact same expression. Double differencing is a powerful technique because of this attractive property – hardware-related errors of all satellites and receivers involved are completely removed. Removal of these errors aids in ambiguity resolution.

Similarly to single differences, the correlations between double differences must be considered. Consider the two double differences Φ_{AB}^{jk} and Φ_{AB}^{jl}. The matrix equation describing these quantities is given by:

$$\mathbf{D} = \mathbf{MS} \quad \Rightarrow \quad \begin{bmatrix} \Phi_{AB}^{jk} \\ \Phi_{AB}^{jl} \end{bmatrix} = \begin{bmatrix} -1 & 1 & 0 \\ -1 & 0 & 1 \end{bmatrix} \begin{bmatrix} \Phi_{AB}^{j} \\ \Phi_{AB}^{k} \\ \Phi_{AB}^{l} \end{bmatrix}. \tag{2.143}$$

Note that this equation only deals with single differences of the same type that, as we found previously, are uncorrelated. Therefore, $\mathbf{C_S}$ reduces to the diagonal matrix $2\sigma^2 \mathbf{I}$. With this in mind, we may apply the covariance law to Equation 2.143 to obtain:

$$\mathbf{C_D} = \mathbf{M}\mathbf{C_S}\mathbf{M}^T = \mathbf{M}2\sigma^2\mathbf{I}\mathbf{M}^T = 2\sigma^2\mathbf{M}\mathbf{M}^T. \tag{2.144}$$

Evaluating $\mathbf{M}\mathbf{M}^T$ results in:

$$\mathbf{C_D} = 2\sigma^2 \begin{bmatrix} 2 & 1 \\ 1 & 2 \end{bmatrix}. \tag{2.145}$$

We see that the double differences are not uncorrelated. The corresponding weight matrix is the inverse of the covariance matrix:

$$\mathbf{W} = \mathbf{C_D}^{-1} = \frac{1}{6\sigma^2} \begin{bmatrix} 2 & -1 \\ -1 & 2 \end{bmatrix}. \tag{2.146}$$

This only applies to the case where two double differences are used. In general, for n double differences, the weight matrix becomes:

$$\mathbf{W} = \frac{1}{2\sigma^2(n+1)} \begin{bmatrix} n & -1 & -1 & \dots & -1 \\ -1 & n & -1 & \dots & -1 \\ -1 & -1 & n & & \vdots \\ \vdots & \vdots & & \ddots & -1 \\ -1 & -1 & \dots & -1 & n \end{bmatrix}. \tag{2.147}$$

Triple Differences

A triple difference may be formed by differencing a double difference at two separate epochs. For example, consider a double difference formed on a single frequency at two epochs t_1 and t_2:

$$\Phi_{AB}^{jk}(t_1) = \frac{1}{\lambda}\left[\rho_{AB}^{jk}(t_1) + \lambda N_{AB}^{jk} + e_{AB}^{jk}(t_1)\right],$$

$$\Phi_{AB}^{jk}(t_2) = \frac{1}{\lambda}\left[\rho_{AB}^{jk}(t_2) + \lambda N_{AB}^{jk} + e_{AB}^{jk}(t_2)\right]. \tag{2.148}$$

Using the notation $\Phi_{AB}^{jk}(t_{12}) = \Phi_{AB}^{jk}(t_2) - \Phi_{AB}^{jk}(t_1)$ for all parameters, we may form the triple difference:

$$\Phi_{AB}^{jk}(t_{12}) = \frac{1}{\lambda}\left[\rho_{AB}^{jk}(t_{12}) + e_{AB}^{jk}(t_{12})\right]. \tag{2.149}$$

Triple differences eliminate carrier phase ambiguities entirely, removing the need to solve for them. Additionally, triple differences may be used in cycle slip detection and repair.

Once again, the correlations between these differences must be addressed. Consider two triple differences $\Phi_{AB}^{jk}(t_{12})$ and $\Phi_{AB}^{jl}(t_{12})$. The matrix equation for these quantities is:

$$\mathbf{T} = \mathbf{MS} \quad \Rightarrow \quad \begin{bmatrix} \Phi_{AB}^{jk}(t_{12}) \\ \Phi_{AB}^{jl}(t_{12}) \end{bmatrix} = \begin{bmatrix} 1 & -1 & 0 & -1 & 1 & 0 \\ 1 & 0 & -1 & -1 & 0 & 1 \end{bmatrix} \begin{bmatrix} \Phi_{AB}^{j}(t_1) \\ \Phi_{AB}^{k}(t_1) \\ \Phi_{AB}^{l}(t_1) \\ \Phi_{AB}^{j}(t_2) \\ \Phi_{AB}^{k}(t_2) \\ \Phi_{AB}^{l}(t_2) \end{bmatrix}. \tag{2.150}$$

Double differences are always formed from uncorrelated single differences, so $\mathbf{C_S}$ still equals $2\sigma^2\mathbf{I}$. The covariance law gives:

$$\mathbf{C_T} = \mathbf{MC_S M^T} = \mathbf{M}2\sigma^2\mathbf{IM^T} = 2\sigma^2 \cdot \mathbf{MM^T} \tag{2.151}$$

Evaluating $\mathbf{MM^T}$ results in:

$$\mathbf{C_T} = 2\sigma^2 \begin{bmatrix} 4 & 2 \\ 2 & 4 \end{bmatrix}. \tag{2.152}$$

A systematic method for determining $\mathbf{MM^T}$ given triple differences across any two satellites or any two epochs exists. The ijth element of $\mathbf{MM^T}$ is given by the inner product of the ith column of $\mathbf{M}$ with the jth column of $\mathbf{M}$. This can be verified for our prior example by comparing Equations 2.150 and 2.152. Additional examples are also given:

$$\Phi = \begin{bmatrix} \Phi_{AB}^{jk}(t_{12}) \\ \Phi_{AB}^{jk}(t_{23}) \end{bmatrix}, \quad \mathbf{M} = \begin{bmatrix} 1 & -1 & 0 & -1 & 1 & 0 & 0 & 0 & 0 \\ 0 & 0 & 0 & 1 & -1 & 0 & -1 & 1 & 0 \end{bmatrix},$$

$$\mathbf{MM^T} = \begin{bmatrix} 4 & -2 \\ -2 & 4 \end{bmatrix},$$

$$\Phi = \begin{bmatrix} \Phi_{AB}^{jk}(t_{12}) \\ \Phi_{AB}^{jl}(t_{23}) \end{bmatrix}, \quad \mathbf{M} = \begin{bmatrix} 1 & -1 & 0 & -1 & 1 & 0 & 0 & 0 & 0 \\ 0 & 0 & 0 & 1 & 0 & -1 & -1 & 0 & 1 \end{bmatrix},$$

$$\mathbf{MM^T} = \begin{bmatrix} 4 & -1 \\ -1 & 4 \end{bmatrix},$$

$$\Phi = \begin{bmatrix} \Phi_{AB}^{kj}(t_{12}) \\ \Phi_{AB}^{jl}(t_{23}) \end{bmatrix}, \quad \mathbf{M} = \begin{bmatrix} -1 & 1 & 0 & 1 & -1 & 0 & 0 & 0 & 0 \\ 0 & 0 & 0 & 1 & 0 & -1 & -1 & 0 & 1 \end{bmatrix},$$

$$\mathbf{MM^T} = \begin{bmatrix} 4 & 1 \\ 1 & 4 \end{bmatrix}.$$

It should be noted that throughout this section, the assumption was made that all measurements were performed on the same frequency. This is not strictly necessary – single, double, and triple differences may be formed using different frequencies. In these cases, the aforementioned derivations must be modified to account for this change. Derivations of multi-frequency differences are not provided in this text, but a brief discussion is presented in [19, pp. 176–177].

Real Time Kinematic Positioning

Real time kinematic (RTK) positioning is a widely used and powerful positioning technique that makes use of the concepts of differential positioning. In RTK positioning, two receivers are used: one static reference station whose coordinates are already known, and one mobile receiver known as a rover. Rover receivers are small, lightweight, and suited for on-site surveying, among other applications. In addition, rovers are capable of calculating position to a high degree of accuracy in real time (i.e. without the need for post-processing). This capability makes RTK useful for a wide variety of applications, including navigation, autonomous vehicle direction, precision farming, construction, and more.

For a rover to perform RTK positioning, it must receive a positioning measurement from the base station; typically carrier phase measurements are used due to their higher accuracy. Measurements are sent from the base station to the rover in real time via the ultra high frequency (UHF) band. The rover then differences the base station's carrier phase measurement with its own, allowing it to determine its position relative to the base station with up to millimeter-level accuracy. The absolute positioning accuracy is dependent upon the accuracy of the base station measurements as well as the baseline length. RTK positioning is best used for baselines between 10 and 20 km [34].

One of the downsides of RTK is its dependence on short baselines. Due to this limitation, it is useful to set up a network of reference stations in order to maximize coverage for a rover receiver. It is also possible to combine measurements from three or more reference stations to create a virtual reference station (VRS) with an even shorter baseline to the rover than any of the reference stations. These topics are beyond the scope of this book, but the interested reader may find further information on these subjects, as well as RTK in general, in references [11, 19, 34] for example.

3

Tropospheric Propagation

3.1 Introduction

In this chapter, we turn our attention to the first of two atmospheric regions of interest for global navigation satellite systems (GNSS) signal propagation: the troposphere. The troposphere is the lowest region of the atmosphere, which consists of non-ionized gases and extends from the Earth's surface up to about 10–15 km,[1] depending on latitude. This is the region in which most weather phenomena occur, and where a typical commercial airplane will fly. As an electromagnetic wave travels through the troposphere, it will interact with gas molecules, giving rise to several effects such as refraction, absorption, scattering, and interference. The sections of this chapter will each discuss the relative importance of these effects to waves at GNSS frequencies.

One of the most impactful propagation effects imposed by the troposphere is that of group delay. As an electromagnetic wave travels through the troposphere, it propagates at a slower velocity and is refracted so that it traverses a longer path to the surface than if it had traveled along a straight path. The combination of these two effects – slower propagation velocity and longer path – introduces error when performing GNSS positioning known as tropospheric group delay. Since the group delay arises from refraction, the introduced error depends on the refractivity of the troposphere along the wave's propagation path. Tropospheric refractivity can in turn be determined from three quantities: temperature, pressure of dry gases, and pressure of water vapor. In Section 3.2, a derivation of tropospheric group delay is

1 In the context of GNSS applications, the major sources of delay error are the ionosphere and the non-ionized region below the ionosphere. Above the troposphere and below the ionosphere there are another non-ionized regions such as the stratosphere and the mesosphere.
The contribution of all of these non-ionized regions together is considered when computing the total tropospheric delay. Thus "tropospheric delay" is commonly used to simply mean "non-ionospheric delay" in GNSS literature [39]. Indeed, about 80% of the delay caused by the non-ionized atmosphere can be attributed to the troposphere per se [40].

Tropospheric and Ionospheric Effects on Global Navigation Satellite Systems, First Edition.
Timothy H. Kindervatter and Fernando L. Teixeira.

presented. Section 3.3 expands on tropospheric refraction, presenting a derivation of the curvature of a wave's propagation path. This information can be useful for determining the angle of arrival of an incident signal (which is a relevant consideration, for example, in antenna design).

Waves propagating through the troposphere are also subject to both absorption and scattering effects. The attenuation of a signal due to these combined effects is termed extinction, and is the subject of Section 3.4. At GNSS frequencies, extinction effects are typically not very pronounced, so this section consists of brief discussions that aim to explain the physical reasons as to why each effect is unimportant in the context of GNSS positioning.

Finally, the chapter concludes with Section 3.5, which discusses tropospheric scintillations. Scintillations are small fluctuations in a signal's amplitude over time. In the visible light range, scintillations appear as a twinkling effect that can commonly be observed when viewing stars. These amplitude fluctuations often arise due to atmospheric turbulence, which causes incident waves to interfere with randomly varying phase differences. Strong scintillations can attenuate a GNSS signal enough to cause loss of lock, an undesirable effect related to the carrier phase GNSS observable, which was discussed in Section 2.2.

3.2 Tropospheric Group Delay

GNSS signals are refracted as they travel through the troposphere, which causes them to travel along a curved path whose length is greater than that of a straight path. In addition, because the index of refraction of the troposphere is greater than one, the wave's propagation velocity, $v = c/n$, is smaller than the speed of light. Since the pseudorange observable is formulated by assuming a straight path and propagation at c, these deviations introduce a discrepancy between the measured range to a satellite and the true range. In particular, the effective radio path length of the path taken by a signal in the troposphere is[2]:

$$S = \int_S n(s)\,ds, \tag{3.1}$$

where $n(s)$ is the index of refraction as a function of position s along the actual path which is, in general, curved. The path length "expected" by the formulation of pseudorange is that of a straight path in a vacuum, where $n = 1$:

$$S_0 = \int_{S_0} ds. \tag{3.2}$$

---

2 Note that this does not include the $\cos \alpha$ term seen in Equation 7.49. This is because the troposphere does not exhibit the anisotropy seen in the ionosphere, which arises due to the interaction of free electrons with Earth's magnetic field. The troposphere does not have the abundance of free electrons that the ionosphere does, so this effect is not present.

The deviation from this path caused by tropospheric refraction is simply the difference between S and S_0:

$$\Delta S = S - S_0 = \int_S n(s)ds - \int_{S_0} ds. \tag{3.3}$$

This can be rewritten as

$$\Delta S = S - S_0 = \int_S [n(s) - 1]ds + \left(\int_S ds - \int_{S_0} ds \right). \tag{3.4}$$

The refractive index of the troposphere is only slightly higher than 1; for example, a typical value is approximately $n = 1.000350$. Since this difference is so slight, it is more convenient to instead consider a related quantity called *refractivity*, denoted N, which is given by the equation:

$$N = (n - 1) \times 10^6. \tag{3.5}$$

Both the refractive index and refractivity are unitless parameters. However, for ease of use, refractivity is commonly quantified in "N-units." As an example, the refractivity corresponding to $n = 1.000350$ is 350 N-units.

In terms of refractivity, Equation 3.3 becomes:

$$\Delta S = 10^{-6} \int_S N(s)ds + \left(\int_S ds - \int_{S_0} ds \right). \tag{3.6}$$

In GNSS applications, it is typically sufficient to consider only the first term, which corresponds only to the retarding of the signal by the troposphere, as opposed to the second term which describes the increased path length that the signal traverses (due to refractive bending of the ray). The second term only needs to be considered if the curvature of the ray path contributes appreciably to the total path length. For GNSS signals, this is only the case near the horizon [39], so we are justified in ignoring its contribution in most cases. Henceforth, when we refer to "tropospheric delay," we refer only to the first term and will denote it as follows:

$$\Delta^T = 10^{-6} \int_S N(s)ds. \tag{3.7}$$

Note that Δ^T is sometimes denoted T in the literature (for *tropospheric* delay) to distinguish it from ionospheric delay, which can similarly be derived from path differences due to refraction (see Section 7.3.2). We will not use the notation T for tropospheric delay in this section, since it can easily be confused with temperature.

As is the case for any medium, the refractive index of the troposphere depends on its molecular composition. In particular, it depends on how these molecules react to the presence of an electromagnetic field. The troposphere is primarily composed of so-called "dry gases" such as N_2 and O_2. At a macroscopic scale, these gases can be assumed to obey the ideal gas law [41]. For reasons beyond the scope of this book, the sum of all of the dry gases' reactions to an incident electromagnetic field can be expressed solely in terms of macroscopic parameters [42], namely the absolute temperature of the troposphere T and partial pressure p_d:

$$N_{dry} = K_1 \frac{p_d}{T}. \tag{3.8}$$

The proportionality constant K_1 is determined experimentally. The other major component of the troposphere is water vapor, which can also be assumed to obey the ideal gas law [41]. Just like dry gases, the refractivity of water vapor can be expressed in terms of its macroscopic parameters. However, water vapor differs from dry gases in that it is a polar molecule. Due to its polarity, the electric dipole of an H_2O molecule reacts to an incident electromagnetic field, providing an additional term to refractivity [42]:

$$N_{wet} = K_2 \frac{e}{T} + K_3 \frac{e}{T^2}. \tag{3.9}$$

Here, e is the partial pressure of water vapor, and K_2 and K_3 are experimentally determined constants. The total refractivity of the troposphere is simply the sum of the dry and wet components:

$$N = N_{dry} + N_{wet} = K_1 \frac{p_d}{T} + K_2 \frac{e}{T} + K_3 \frac{e}{T^2}. \tag{3.10}$$

The refractive index of a sample of air can be tested by determining the bending angle of incident light. The temperature and vapor pressure can be measured using a thermometer and hygrometer, respectively. One experiment, performed by Smith and Weintraub [43], obtained the values: $K_1 = 77.60 \pm 0.013$ K/mb, $K_2 = 71.6 \pm 8.5$ K/mb, and $K_3 = (3.747 \pm 0.031) \times 10^5$ K^2/mb. With these values (truncated to three significant figures), Equation 3.10 becomes:

$$N = 77.6 \frac{p_d}{T} + 72 \frac{e}{T} + 3.75 \times 10^5 \frac{e}{T^2}, \tag{3.11}$$

where T has units of Kelvins and both p and e have units of millibars. We can simplify this equation further by using the fact that the total pressure p is the sum of the partial pressures of the dry and wet components:

$$p = p_d + e. \tag{3.12}$$

With this, we can write:

$$N = 77.6 \frac{p - e}{T} + 72 \frac{e}{T} + 3.75 \times 10^5 \frac{e}{T^2}, \tag{3.13}$$

which simplifies to:

$$N = 77.6 \frac{p}{T} - 5.6 \frac{e}{T} + 3.75 \times 10^5 \frac{e}{T^2}. \tag{3.14}$$

We can even further simplify this equation by assuming that the temperature will always fall within the range of -50 to $+40$ °C, which is a reasonable assumption for the troposphere. If this is done, then the second and third terms may be combined without incurring much error ($< 0.5\%$ for frequencies below $100\,\text{GHz}$ [[44], p. 100]). To do this, we can set these two terms equal to a single new term proportional to e/T^2:

$$K_4 \frac{e}{T^2} = -5.6\frac{e}{T} + 3.75 \times 10^5 \frac{e}{T^2}. \tag{3.15}$$

At $T = 273\,\text{K}$, this results in $K_4 = 3.73 \times 10^5\,\text{K}^2/\text{mb}$. If we insert this new term in place of the second and third terms of Equation 3.14, we obtain:

$$N = 77.6\frac{p}{T} + 3.73 \times 10^5 \frac{e}{T^2}. \tag{3.16}$$

Finally, by factoring out $77.6/T$ from each term, we arrive at the tropospheric refractivity relation commonly found in the literature:

$$N = \frac{77.6}{T}\left(p + 4810\frac{e}{T}\right). \tag{3.17}$$

With this expression in hand, we can rewrite Equation 3.6 as

$$\Delta S = 10^{-6} \int_S \frac{77.6}{T(s)}\left(p(s) + 4810\frac{e(s)}{T(s)}\right)\,ds + \left(\int_S ds - \int_{S_0} ds\right), \tag{3.18}$$

and consequently, we can also rewrite Equation 3.7 as:

$$\boxed{\Delta^T = 10^{-6} \int_S \frac{77.6}{T(s)}\left(p(s) + 4810\frac{e(s)}{T(s)}\right)\,ds,} \tag{3.19}$$

where the quantities $T(s)$, $p(s)$, and $e(s)$ are the temperature, total pressure, and water vapor partial pressure along the signal's path, respectively.

Unlike the ionosphere, the troposphere is non-dispersive up to about $15\,\text{GHz}$ [7]. This means that there is no technique analogous to the ionosphere-free combination (see Section 2.5.1) which can be used to entirely remove tropospheric delay from GNSS observables. Instead, various models of tropospheric delay have been developed in order to predict the magnitude of the delay and remove it. We will discuss one such model, the Saastamoinen model, in Section 4.2.

The carrier phase measurement is affected by a non-unity refractive index in a manner analogous to the pseudorange. In a dispersive medium like the ionosphere, the carrier phase will be *advanced* rather than delayed. This is because of the difference in the effect of dispersion on the group velocity and phase velocity of a signal (see Sections 7.3.3 and A.2), an effect termed carrier-code divergence. In contrast to the ionosphere, the troposphere is non-dispersive at GNSS frequencies, so dispersion does not occur. For this reason, the carrier phase measurement does not advance like in the ionosphere, but instead is delayed by the same amount as the pseudorange measurement. This means that *both* observables overestimate the length of the path through the troposphere. By contrast, carrier phase underestimates the path length through the ionosphere, while pseudorange overestimates it.

3.2.1 Mapping Functions

The tropospheric path delay is directly dependent on the length of the path a signal takes through the troposphere. As a result, signals received at low elevation angles will experience greater delay than those at high elevation angles, as illustrated in Figure 3.1.

In principle, Equation 3.19 could be used to determine the tropospheric delay along any path through the atmosphere. In order to accomplish this, one would need precise information about the atmospheric profile along the path of interest. This information could be provided by experimental data input into an atmospheric model. In practice, however, it is impractical to specify all possible paths of interest this way. For example, the atmospheric parameters for a given location may be nonexistent or at least inaccessible to the person performing the calculation. Even if seasonal averages of atmospheric parameters were used, accurate atmospheric models are computationally intensive, so it may be undesirable to run a simulation for every path of interest [45, §12.4]. Instead, we can use *mapping functions* to relate the zenith path delay to paths at other elevation angles.

A mapping function $m(E)$ is simply the ratio of the zenith path delay to the delay along a path at some elevation angle[3] E. Once a general relationship has been found, the path delay at any elevation angle can be determined by knowing only the zenith path delay. For example, consider a simplified model in which we ignore the curvature of the Earth and assume a homogeneous atmosphere. We can see in

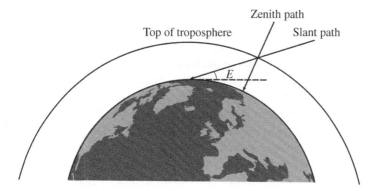

Figure 3.1 Signals traveling at low elevation angles traverse longer paths through the troposphere.

---

3 In reality, the mapping function also depends on the atmospheric profile of the refractive index as well as the elevation angle. However, for simplicity many mapping functions ignore the refractive index dependence and give an approximate relationship that depends only on elevation angle.

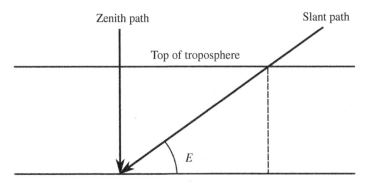

Figure 3.2 A simplified model of tropospheric refraction in which the zenith path and a slant path are related by a right triangle.

Figure 3.2 that the zenith path can be represented as a leg of a right triangle, and the slant path as the hypotenuse.

The relationship between the two is given by:

$$\Delta_s^T = \Delta_z^T \csc(E), \tag{3.20}$$

where Δ_s^T is the slant path delay, Δ_z^T is the zenith path delay, and E is the elevation angle of the slant path. Recall that the mapping function is simply the ratio of the slant path delay to the zenith path delay, so in this case, the mapping function is:

$$m(E) = \frac{\Delta_s^T}{\Delta_z^T} = \csc(E). \tag{3.21}$$

This model is merely meant to illustrate the concept of the mapping function. It is greatly simplified and does not produce very accurate results. For example, a typical value for the zenith path delay is 2.3 m, and a value at an elevation angle of 10° is about 20 m [11, p. 315]. Using Equation 3.20, we would get an estimate of 13.2 m for the slant path delay.

More accurate mapping functions are typically determined via fitting of experimental data. By examining measured path delays at a given location for various elevation angles, it is possible to observe the trend in the mapping function at that location with respect to E. For example, the Niell model [46] is a comprehensive model that uses empirical data to determine mapping functions globally with respect to latitude and altitude.

The first term on the right hand side of Equation 3.18 is also known as the dry delay. The dry delay is fairly stable with respect to time and accounts for about 90% of the total tropospheric delay. By contrast the wet delay, or the second term on the right hand side of Equation 3.18 involving $e(s)$, contributes only 10% of the total tropospheric delay, but is highly variable over time [11, pp. 314, 315]. Therefore, it is convenient to determine mapping functions for the dry and wet components of the tropospheric delay separately. The total slant path delay can therefore be given as:

$$\Delta_s^T = m_d(E)\Delta_{z,d}^T + m_w(E)\Delta_{z,w}^T, \tag{3.22}$$

where Δ_s^T is the total delay along a slant path at elevation angle E, $m_d(E)$ is the dry mapping function, $\Delta_{z,d}^T$ is the zenith dry delay, $m_w(E)$ is the wet mapping function, and $\Delta_{z,w}^T$ is the zenith wet delay.

3.3 Tropospheric Refraction

At low altitudes, temperature, pressure, and humidity all decrease with height. Pressure and humidity drop off more quickly than temperature [44], so there is an overall decrease in refractivity as altitude increases. This decrease is exponential, and is in general given by the equation:

$$N = N_s \exp\left(-\frac{h}{H}\right), \tag{3.23}$$

where N_s is the refractivity at Earth's surface, h is the altitude above the surface, and H is a quantity known as the scale height. A derivation of H is provided in Section 5.3.1. N_s and H vary by location, but measurements have shown [47] that the median change in refractivity over the first kilometer above the surface is $\Delta N = -40$ N-units/km. The International Radio Consultative Committee[4] defines an average exponential atmosphere with a refractivity profile:

$$N = 315 \exp\left(-\frac{h}{7.36}\right), \tag{3.24}$$

where $N_s = 315$ N-units and $H = 7.36$ km.

As discussed in Section 3.2, an electromagnetic wave traveling through the troposphere propagates along a path which is continuously bent due to the changing refractivity. Depending on the severity of the refractivity gradient ΔN, this bending may be severe enough to redirect the wave back toward the Earth. The following derivation will provide a general formula which will allow us to characterize the refraction of electromagnetic waves under differing refractivity conditions.

We will begin the derivation by considering two waves propagating at different altitudes. We will assume that the path they traverse is long enough that the curvature of the Earth becomes important. Therefore, we will express the altitude of the wave using a radius of curvature ρ, where ρ is greater than the radius of the Earth, r_e. We will assume that both waves traverse some Earth-centered angle $d\theta$

---

4 Commonly, the acronym CCIR is used to refer to this organization. This is based on its French name: Comité Consultatif International pour la Radio.

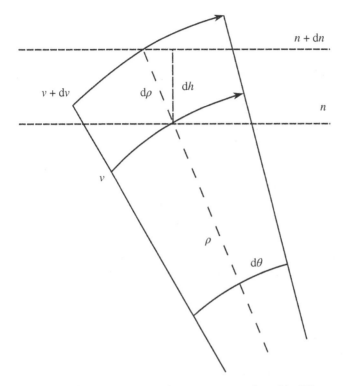

Figure 3.3 Two electromagnetic waves propagating with different radii of curvature through the troposphere.

in the same amount of time dt. One wave will propagate with a radius of curvature ρ and velocity v, while the other will propagate with some greater radius $\rho + d\rho$ at a larger velocity $v + dv$. This information is summarized in Figure 3.3.

First, consider the lower-altitude wave. This wave traverses an arc length $\rho \, d\theta$ in time dt, which is summarized by the equation:

$$\rho \, d\theta = v \, dt. \tag{3.25}$$

A similar observation for the higher-altitude wave results in the equation:

$$(\rho + d\rho)d\theta = (v + dv)dt. \tag{3.26}$$

If we divide Equation 3.26 by 3.25 and simplify slightly, we obtain:

$$\frac{d\rho}{\rho} = \frac{dv}{v}. \tag{3.27}$$

The index of refraction is defined as $n = c/v$, so it follows that:

$$\frac{dn}{dv} = \frac{-c}{v^2} = -\frac{n}{v}. \tag{3.28}$$

Furthermore, this tells us that:

$$\frac{dn}{n} = -\frac{dv}{v}. \tag{3.29}$$

Using Equation 3.27, we can replace the right hand side of Equation 3.29 and rearrange it so that:

$$\frac{1}{\rho} = -\frac{1}{n}\frac{dn}{d\rho}. \tag{3.30}$$

From Figure 3.3, we can see that:

$$d\rho = \frac{dh}{\cos\theta}, \tag{3.31}$$

so we may rewrite Equation 3.30 as:

$$\frac{1}{\rho} = -\frac{1}{n}\frac{dn}{dh}\cos\theta. \tag{3.32}$$

The quantity $1/\rho$ is termed curvature, and gives a measure of how much a wave's path is bent as it travels. Note that if refractivity were completely constant with height, i.e. if $dn/dh = 0$, then curvature would be zero. This indicates a straight-line propagation path, as one would expect if no refraction occurred.

The curvature of a wave's propagation path can be compared with the curvature of the Earth:

$$\frac{1}{r_e} - \frac{1}{\rho} = \frac{1}{r_e} + \frac{1}{n}\frac{dn}{dh}\cos\theta. \tag{3.33}$$

In terms of refractivity, this can be written as:

$$\boxed{\frac{1}{r_e} - \frac{1}{\rho} = \frac{1}{r_e} + 10^{-6}\frac{1}{n}\frac{dN}{dh}\cos\theta.} \tag{3.34}$$

In general, dN/dh can be either positive or negative, which means that a wave path's curvature can be larger or smaller than the curvature of the Earth, depending on the refractivity conditions.

Generally, refractivity conditions are classified as one of four categories. Conditions in which dN/dh is near the median value of -40 N-units/km are considered *normal refractivity*. A typical range for normal refractivity is $N = 0$ to -79 N-units/km. When $N > 0$, the atmosphere is said to be *subrefractive*, meaning waves are bent away from the surface of the Earth rather than back toward it. In the range $N = -79$ to -157 N-units/km, the atmosphere is said to be *superrefractive*. Under these conditions, waves are bent toward Earth's surface more strongly than usual, but not strongly enough to return them to Earth. At $N = -157$ N-units/km, the waves are refracted along a trajectory with curvature equal to the Earth's curvature, which causes them to remain at the same altitude as they travel. Finally, in cases where $N < -157$ N-units/km, waves are bent toward the Earth faster than the Earth "falls away" due to its curvature. In these cases, the wave cannot escape the atmosphere, so this category is referred to as a *trapping gradient*. The four categories of refractivity are summarized in Table 3.1.

Figure 3.4 depicts each of the refractivity categories for a signal arriving from a satellite.

When a trapping gradient is present, an electromagnetic wave incident at a low elevation angle will bounce repeatedly along the surface of the Earth. This behavior is known as *tropospheric ducting*. Ducts form whenever there is a sharp increase in refractivity with height (opposite the typical decrease). For example, consider

Table 3.1 Classification of refractivity conditions and their corresponding gradient values in N-units/km.

Classification	N
Subrefractive	$N > 0$
Normal refraction	$0 \geq N > -79$
Superrefractive	$-79 \geq N > -157$
Trapping gradient	$N < -157$

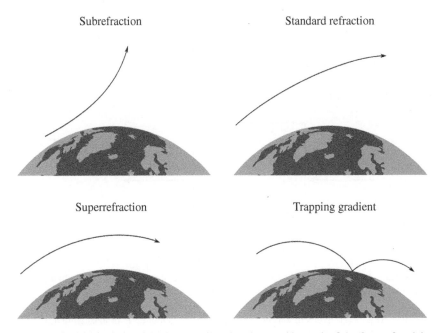

Figure 3.4 Bending of an incident satellite signal caused by each of the four refractivity gradient categories.

the case where a mass of warm air blows in above a region of colder air. Since temperature typically decreases with height, the region of warm air will create a refractivity gradient that is the opposite of the usual one. This is known as a temperature inversion, and is one of the common causes of tropospheric ducts. Ducting is uncommon for GNSS signals. One study found that among 5348 radio occultations,[5] only 536 showed ducting conditions [48].

For further information on tropospheric refraction, the reader may consult the ITU Recommendation ITU-R P.834 [49]. This document provides easy-to-use formulas and tables for the estimation of the ray bending and effective radio path length of the troposphere, among other effects.

3.4 Extinction

3.4.1 Beer–Lambert Law

As a plane electromagnetic wave travels through a medium such as the atmosphere, its amplitude will be lessened.[6] This effect is caused by two phenomena working in tandem: scattering and absorption. First, we will discuss these effects in terms of the bulk properties of a medium. Then, in the remainder of this section, we will re-examine each phenomenon in greater detail.

First, let's look at the bulk scattering properties of a medium. Consider a plane electromagnetic wave with intensity I traveling along a single axis, which we will label x. This wave travels through a uniform slab of material whose thickness is dx, and emerges on the other side with a lessened intensity $I - dI$. This situation is depicted in Figure 3.5.

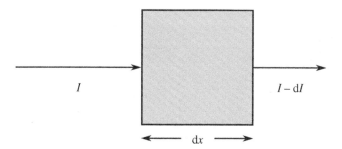

Figure 3.5 An electromagnetic wave travels through a slab of material and emerges with lessened intensity.

---

5 Radio occultation is a remote sensing technique wherein a low-Earth orbit satellite receives a transmission from a GNSS satellite crosses the horizon from its perspective. A rising occultation occurs just as the GNSS satellite appears above the horizon, and a setting occultation occurs just as it is obscured by the horizon.
6 Note that spherical waves will experience additional decrease caused by the $1/r^2$ geometric spreading factor for the power density (equivalent to a $1/r$ factor for the amplitude).

For now, we will assume that the energy absorbed by the medium is negligible, so that the drop in intensity is entirely due to scattering. Roughly speaking, scattering can be defined as the redirection of electromagnetic radiation from its initial trajectory. This means that scattering can be caused by reflection, refraction, diffraction, or even by absorption and subsequent spontaneous re-emission [50, p. 4]. Many of these processes will redirect the incident radiation away from the x-axis, its initial direction of travel. Therefore, the intensity along the x-axis will be diminished, resulting in the observed attenuation.

The proportion of light transmitted through the medium is the emergent intensity divided by the incident intensity:

$$\text{Proportion of light transmitted} = \frac{I + dI}{I}. \tag{3.35}$$

where, for mathematical convenience, the differential assumes a *negative* value when the wave intensity *decreases*. Therefore, it follows that the proportion of radiation that was scattered away from the x-axis is:

$$\text{Proportion of light scattered} = 1 - \frac{I + dI}{I} = -\frac{dI}{I}. \tag{3.36}$$

At a microscopic level, scattering occurs when an electromagnetic wave interacts with small particles. For example, in the atmosphere, a wave could scatter off of gas molecules, dust grains, or water droplets. These so-called *scatterers* can be modeled as spheres for simplicity. A spherical scatterer has some geometrical cross-section A. However, due to effects such as refraction and diffraction, the "shadow" cast by the scatterer, or its effective cross-section, may not be the same size as A. This phenomenon is illustrated in Figure 3.6.

The concept of an effective cross-section arises in situations other than scattering (it also applies to absorption, for example). But in the context of scattering, we refer to it as the *scattering cross-section*, denoted σ_s. The scattering cross-section has units of area, e.g. $\mu\,\text{m}^2$. The slab of material that we are considering also has a cross-sectional area s. The fraction of the area "shadowed" by a single scatterer is given by the ratio of its scattering cross section to the full area of the slab, i.e. σ_s/s.

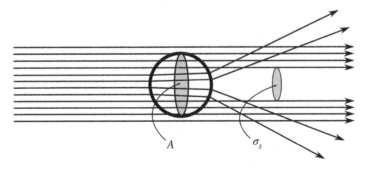

Figure 3.6 The scattering cross-section of a particle is not necessarily equal to its geometrical cross-section. Source: Adapted from Prahl and Jacques [51]. Reproduced with permission of Steven L. Jacques and Scott A. Prahl.

If we assume that the number density of scatterers in the slab is N, then the total number of scatterers is given by NV, where V is the volume of the slab. Further, we can rewrite the volume of the slab as:

$$V = s \, dx, \tag{3.37}$$

where s is the area of the slab's face and dx is the slab's thickness. Since each small scatterer within the volume has a cross-sectional ratio of σ_s/s, the total effect of all scatterers combined is:

$$\text{Fraction of light scattered} = \frac{\sigma_s}{s} Ns \, dx = \sigma_s N \, dx. \tag{3.38}$$

The quantity $\sigma_s N$ is known as the *scattering coefficient*, which we will denote μ_s. Note that since σ_s has units of area and N has units of inverse volume, μ_s has units of inverse length (e.g. μm^{-1}). We can rewrite Equation 3.38 as:

$$\text{Fraction of light scattered} = \mu_s \, dx. \tag{3.39}$$

Note that Equations 3.36 and 3.39 both refer to the fraction of light which was scattered off of the x-axis. Therefore, we can set them equal:

$$-\frac{dI}{I} = \mu_s \, dx. \tag{3.40}$$

Let's denote the incident intensity I_0. We may now integrate both sides of Equation 3.40 to determine the intensity I at any arbitrary point x along the x-axis:

$$\int_{I_0}^{I} -\frac{dI}{I} = \int_{0}^{x} \mu_s \, dx. \tag{3.41}$$

This evaluates to:

$$\ln(I) - \ln(I_0) = \ln\left(\frac{I}{I_0}\right) = -\mu_s x, \tag{3.42}$$

and therefore:

$$I = I_0 \, e^{-\mu_s x}. \tag{3.43}$$

This indicates that the loss of intensity due to scattering follows an exponential relationship. A nearly identical proof can be constructed to show that absorption follows the same relationship, except substituting the scattering coefficient μ_s for the absorption coefficient μ_a:

$$I = I_0 \, e^{-\mu_a x}. \tag{3.44}$$

So far we have only considered the effects of scattering and absorption in isolation. Their combined effect is given by the *extinction coefficient*:

$$\mu_e = \mu_s + \mu_a. \tag{3.45}$$

Therefore, the total extinction caused by the slab of material is given by:

$$I = I_0 \, e^{-\mu_e x}.$$

(3.46)

This relationship is known as the Beer–Lambert law.

In a complex medium like the atmosphere, which has many constituent molecular species, μ_e will be the sum of each species' extinction. For example, if the species of interest are N_2, O_2, NO_2, O_3, water vapor, and aerosols (i.e. particulate matter such as dust or volcanic ash), the extinction coefficient would be:

$$\mu_{e,tot} = \mu_{e,N_2} + \mu_{e,O_2} + \mu_{e,NO_2} + \mu_{e,O_3} + \mu_{e,H_2O} + \mu_{e,aer}.$$

(3.47)

Note that the extinction coefficient for each individual species is itself the sum of the scattering and absorption coefficients for that species.

The extinction coefficients are also functions of wavelength. A given species may scatter strongly at one wavelength and not at another, and the same is true of absorption. Scattering is also dependent on the size of the scatterer. A large sphere made of a given material may scatter light differently than a small sphere made of the exact same material. We will explore the reasons for this in greater detail in the upcoming sections.

3.4.2 Scattering

The character of light scattering by small particles is strongly dependent on the relationship between the particle's size and the wavelength of the incident light. For simplicity, we will assume that the scattering particle is spherical, so its size can be captured by its radius r. The wavelength of the incident light will be denoted λ. Roughly speaking, there are three regimes in which scattering occurs, as illustrated in Figure 3.7.

The first of these is the regime of *geometric optics* or *ray optics*, where $\lambda \ll r$. This is the regime that the reader is most familiar with from daily experience. The wavelength of visible light, which is on the order of hundreds of nanometers, is much

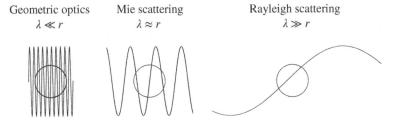

| Geometric optics | Mie scattering | Rayleigh scattering |
| $\lambda \ll r$ | $\lambda \approx r$ | $\lambda \gg r$ |

Figure 3.7 The three regimes in which spherical particles scatter light. Each regime is dictated by the relationship between the particle's radius and the incident light's wavelength.

smaller than the macroscopic objects humans typically interact with. For example, consider a glass marble with a radius of 0.5 cm. A green laser (with wavelength 500 nm) pointed at the marble will fall into the geometric optics regime, since r is 10,000 times larger than λ.

Under these conditions, incident light can be approximated as rays that follow the familiar laws of reflection and refraction. Chromatic dispersion also arises in this context, which explains the formation of rainbows when white light is incident on water droplets in the atmosphere. Light subject to reflection, refraction, or dispersion can be considered "scattered," though these phenomena are not typically discussed using that terminology. This regime is of little interest in the context of GNSS signal propagation, since the wavelength of these signals is typically on the order of tens of centimeters, and there are no particles in the atmosphere with radii several orders of magnitude larger than that.

The second regime occurs when $\lambda \approx r$. In the special case of spherical scatterers, this type of scattering is known as *Mie scattering*, named after the physicist Gustav Mie. This is the most complex of the three cases, and although Maxwell's equations can be solved exactly in the case of spherical scatterers, the derivation of the resulting scattered fields is beyond the scope of this book. The interested reader may refer to [50, 52, 53] for details. We are justified in ignoring this case for the purposes of GNSS signal propagation as well, because atmospheric particulate matter is measured on the order of tens of microns or smaller [54]. Even water droplets, which are typically in the range of 0.1 μm to 1 mm [55], are too small to affect GNSS signals in the Mie regime. The only case in which Mie scattering becomes relevant for GNSS applications is in the case of an ensemble of raindrops [44, p. 155].

The third and final case, known as Rayleigh scattering, occurs when $\lambda \gg d$, which is the regime of greatest interest for GNSS applications. For example, consider a water droplet in the atmosphere with a radius of 5 μm. An L1 GPS signal has a wavelength of about 19 cm, or 38,000 times larger than the radius of the water droplet. Since the wavelength is so much larger than the particle, the electric field which comprises the light wave changes very little over the width of the particle. Because of this, the electric field E inside the water droplet can be considered uniform. As the electric field varies in time, the constituent molecules of the water droplet will all oscillate in tandem (each as a small dipole), re-radiating an electric field. This re-radiated field is interpreted by a distant observer as scattered light.

The intensity of the re-radiated field can be obtained from its time-averaged Poynting vector. The equation for this quantity is given by Equation A.116. The most important thing to note from this equation is that the intensity of Rayleigh scattering is proportional to $1/\lambda^4$. This means that shorter wavelengths are scattered much more strongly than longer wavelengths.[7] For any wavelength greater than about 3.0 cm (i.e. frequency less than 10 GHz), very little of the incident wave's power will be scattered away from its propagation direction [44]. Therefore an L1 GPS signal, which has a wavelength of about 19 cm, will not suffer much attenuation from scattering as it traverses the troposphere.

3.4.3 Gaseous Absorption

In the presence of an electromagnetic wave, molecules will begin to oscillate. Each molecule has a set of frequencies that it "prefers" to oscillate at. Several terms are commonly used to refer to these frequencies, including *vibrational modes*, *resonance frequencies*, and *natural frequencies*. These terms are used interchangeably, but we will adopt "resonance frequencies" going forward. The nearer an electromagnetic wave's frequency is to one of the resonance frequencies of a molecule, the larger the fraction of the electromagnetic wave's energy that molecule will absorb. For further details on this phenomenon, see Sections B.2.2 and B.3.

The total absorption of an electromagnetic wave due to a given molecular species along an atmospheric path depends on:

- How close the electromagnetic wave's frequency is to a resonance frequency of the molecular species.
- The concentration of the molecular species along the path.
- The length of the path.

In the atmosphere, the major constituents to consider are Nitrogen, Oxygen, and water vapor. Trace gases such as Carbon Dioxide, Ozone, and Methane do not occur at concentrations sufficient to cause significant absorption, so their effect may be ignored [56]. At GNSS frequencies, even the major constituents cause minimal absorption, since none of their resonant frequencies lie near those of GNSS signals. For example, the nearest resonance frequencies to the GPS L1 signal (1.575 GHz) are: 22 GHz(H_2O), 60 GHz (O_2), 119 GHz (O_2), 180 GHz (H_2O), and 324 GHz (H_2O) [56–58]. The resonance lines of Nitrogen are even higher, in the 600 GHz to 1 THz range [59].

Total absorption needs to take into account absorption along a full path. Therefore, it is most useful to work with a quantity called the specific attenuation, denoted γ, which is typically measured in dB/km. The International Telecommunications Union (ITU) recommends a simple algorithm for calculating

---

7 This is the reason that sunsets appear red. Blue light has a shorter wavelength than red light, so it is scattered much more strongly as it propagates through the atmosphere. By the time the sunlight reaches the Earth's surface, its color is dominated by longer wavelength red light.

specific attenuation in the troposphere, which is summarized by the following equations:

$$\gamma = 0.1820f(N''_{O_2} + N''_{H_2O}),\tag{3.48}$$

where f is frequency in GHz, and N''_{O_2} and N''_{H_2O} are the imaginary parts of the complex refractivities of oxygen and water vapor, respectively. These quantities are themselves frequency dependent, and are given by the equations:

$$N''_{O_2}(f) = N''_D(f) + \sum_{i(O_2)} S_i F_i \tag{3.49}$$

and

$$N''_{H_2O}(f) = \sum_{i(O_2)} S_i F_i. \tag{3.50}$$

Here, S_i is the strength of absorption line i, and F_i is the shape factor of that line. The contributions of all absorption lines are summed together. $N''_D(f)$ is the dry continuum absorption due to Nitrogen and the Debye spectrum. We will not dwell on the details of these parameters, but the interested reader may consult the relevant ITU Recommendation, ITU-R P.676, for details [60]. Figure 3.8 depicts the specific attenuation in the troposphere from 1 to 350 GHz.

It can be seen that at GNSS frequencies ($\sim 1 - -2$ GHz), the specific attenuation is on the order of 10^{-3}–10^{-2} dB/km. For Earth-space paths, this typically results in negligible total attenuation. For a zenith path through the troposphere, the total gaseous attenuation can be expected to be 0.035 dB [27]. At low elevation angles, attenuation is more significant, with a typical value of 0.38 dB at an elevation angle of 5°.

3.4.4 Hydrometeor Attenuation

In Section 3.4.3, we considered the attenuation effects of water vapor – i.e. water in its gaseous state. However, typically there is also a significant amount of solid and liquid water in the atmosphere, such as rain, snow, ice, and clouds. Commonly, these non-gaseous particles are referred to as "hydrometeors." Among all atmospheric hydrometeors, rain is of greatest concern when designing a radiowave communication system. Ice absorbs less energy at radio frequencies than liquid water, and only rain droplets are large enough to cause significant attenuation. The water droplets which comprise clouds and fog have small diameters, which reduces the effect they have on incident signals at GNSS frequencies [45]. For these reasons, the remainder of this section will focus exclusively on the effects of rain. For information on attenuation by other hydrometeors, the reader is encouraged to consult the ITU-R Recommendation P.840-7 [61].

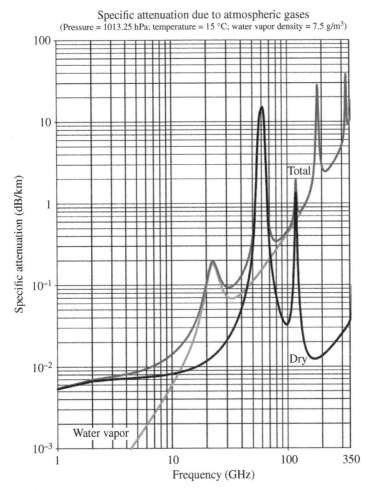

Figure 3.8 Specific attenuation due to tropospheric gases. Source: ITU-R P.676-11 [60]. Reproduced with permission.

The attenuation caused by rain varies with respect to the diameter of the rain drop. When the diameter is roughly the same size as the wavelength of the incident wave, attenuation is stronger [62]. Since rain droplet diameters are typically on the order of millimeters, attenuation tends to be stronger for millimeter wavelengths than centimeter wavelengths. To obtain the total attenuation A over a path through a rainstorm, one must consider the contribution of all raindrops along the path.

Rain drop diameter varies continuously over a certain range of sizes. Thus, it is necessary to consider the *drop size distribution* when calculating total attenuation. Integrating over this drop size distribution allows us to estimate the total attenuation [44, p. 156]:

$$A = 4.343 \, L \int_0^\infty \sigma_{ex}(D)N(D)\mathrm{d}D. \tag{3.51}$$

Here, A is the total attenuation in dB, L is the length of the path, $\sigma_{ex}(D)$ is the extinction cross-section (see Section 3.4.1) of the rain drop as a function of its diameter D, and $N(D)$ is the drop size distribution. If we set $L = 1$ km, the specific attenuation γ in dB/km is simply:

$$\gamma = 4.343 \int_0^\infty \sigma_{ex}(D)N(D)\mathrm{d}D. \tag{3.52}$$

This model of rain attenuation provides good physical intuition. However, in practice it is difficult to use because the size, shape, and orientation – and hence $\sigma_{ex}(D)$ and $N(D)$ – are very difficult to predict accurately. Instead, we must turn to empirical models to provide good estimates of rainfall attenuation.

Measurements have shown [63] that the specific attenuation of rain follows a power law relationship with the rain rate R of the form:

$$\gamma = kR^\alpha, \tag{3.53}$$

where k and α are empirically determined, and separate values are obtained for horizontal and vertical polarizations (denoted with subscripts h and v, respectively). Table 3.2 lists the values of k_h, k_v, α_h, and α_v at 1 and 2 GHz.

These values can then be used in Equation 3.53 to obtain the specific attenuation of horizontally and vertically polarized waves. To obtain values of k and α at intermediate frequencies, f and k must be interpolated logarithmically, and α must be interpolated linearly [44, p. 157].

Figure 3.9 shows the specific attenuation of the L1 GPS frequency $(f_{L1} = 1.57542 \, \text{GHz})$ as a function of rain rate. Values are shown for both horizontal and vertical polarizations.

For reference, a rain rate of less than 5 mm/h is considered light rain, while 100 mm/h represents a violent downpour, and rarely occurs for longer than a few moments at a time [45].

From Figure 3.9, it can be seen that even during torrential downpours, the attenuation of signals at GNSS frequencies due to rain is very low, typically

Table 3.2 Empirical values of rainfall power law coefficients at 1 and 2 GHz.

f (GHz)	k_h	k_v	α_h	α_v
1	0.000,038,7	0.000,035,2	0.912	0.880
2	0.000,154	0.000,138	0.963	0.923

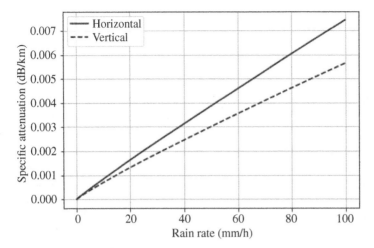

Figure 3.9 Specific attenuations in dB/km of the horizontal and vertical polarizations of the L1 GPS frequency as a function of rain rate.

< 0.01 dB/km. Therefore, for most Earth-space links, total rainfall attenuation is not a serious concern for typical GNSS applications [27].

3.5 Tropospheric Scintillations

The refractive index of the troposphere depends on temperature, pressure, and humidity, as discussed in Section 3.2. These quantities are not stationary or homogeneous throughout the troposphere, so it is commonly the case that the refractive index will change in both space and time. For example, turbulent air can cause rapid and irregular variations to the local refractive index. These local variations can cause incident electromagnetic waves to interfere with each other in a process known as scintillation.

For example, consider two incident waves with the same wavelength, λ, which are initially in phase with one another. Suppose the waves are traveling through a turbulent region of the troposphere, and each wave encounters a different index of refraction. Because of the differing refractive indices, the waves will travel through this region along paths of slightly different lengths. Due to this path length difference, they will develop a phase difference, ϕ. Now suppose the waves leave the turbulent region and once again travel at the same speed. For simplicity, we'll assume that the atmosphere is homogeneous along the rest of the path to the receiver, so they retain the phase difference they had as they exited the turbulent region.

The two waves may interfere constructively or destructively, depending on their relative phase. Interference is most destructive when the two waves are exactly one half wavelength out of phase with one another, or any integer multiple of wavelengths plus an additional half wavelength. That is, destructive interference is greatest when the difference in the path lengths is equal to $\lambda/2$, $3\lambda/2$, $5\lambda/2$, etc. On the other hand, interference is most constructive when the two path lengths are exactly an integer multiple of wavelengths apart: 0, λ, 2λ, 3λ, etc.

In the literature, it is common to refer to *Fresnel zones*. To understand the definition of a Fresnel zone, consider some *direct ray* signal with wavelength λ which travels along some straight path of length L. Now consider the path taken by some *indirect ray*, also with wavelength λ, which interferes with the direct ray. There are a number of (curved) paths that the indirect ray could take which have a length of $L + \lambda/2$, and would thus cause the two rays to destructively interfere. The set of all such paths defines the boundary of the first Fresnel zone. Similarly, the indirect ray could take a number of paths which have length $L + \lambda$, which would cause constructive interference with the direct ray. The set of all such paths defines the second Fresnel zone.

In general, the nth Fresnel zone is defined as the set of all paths for which the indirect ray's path length differs by $n\lambda/2$ from the path length of the direct ray. Notice that this means that all odd-numbered Fresnel zones refer to paths which cause maximum destructive interference between the two waves. Similarly, even-numbered Fresnel zones refer to paths that cause maximum constructive interference between the waves.

A brief word of warning: commonly, Fresnel zones are also discussed in the context of rays interfering due to reflection. In ground-to-ground links, reflection of grazing rays on the terrain induces a phase shift of about $180°$ (i.e. $\lambda/2$). This means that an indirect ray which travels a path length of $L + \lambda/2$ will incur an additional phase shift of $\lambda/2$ upon reflection and end up with a total phase difference of λ when it interferes with the direct ray and therefore constructively interfere. Fresnel zones are still defined in terms of the path length difference, which means that in the context of terrain reflection and interference, odd-numbered Fresnel zones cause constructive interference, while even-numbered Fresnel zones cause destructive interference. This is exactly the opposite of Fresnel zones caused solely by refraction! The reader should take care to consider which mechanism caused two rays to interfere, since this will determine which Fresnel zones are constructive and which are destructive.

We have just seen that a turbulent patch of the troposphere with a locally inhomogeneous index of refraction can cause incident signals to interfere.

Furthermore, the index of refraction in the turbulent region will change from one moment to the next, so the waves' mutual interference will also change over time. The changes in the turbulent region's refractive index are best modeled as a random process, which means that the amplitude fluctuations of the resultant wave are also random. For a carrier signal of the form $A(t)\sin(\omega t + \phi)$, we can define a term called the (tropospheric) scintillation intensity (measured in dB):

$$x(t) = 20\log_{10}\left(\frac{A(t)}{\bar{A}(t)}\right),$$ (3.54)

where $A(t)$ is the amplitude of the signal modeled as a random process and $\bar{A}(t)$ is the short-term mean amplitude. Experimental evidence has shown that $x(t)$ is normally distributed over short time spans [27, p. 522], which means that $A(t)$ has a lognormal distribution. Thus, the probability density of x (at some fixed time) can be expressed as:

$$p(x) = \frac{1}{\sigma_x\sqrt{2\pi}}\exp\left(-\frac{x^2}{2\sigma_x^2}\right),$$ (3.55)

where σ_x is the RMS standard deviation of x, in units of dB. Note that this definition only holds in the short-term, however. In the long term, changes in weather change the statistics of the scintillation intensity. This is reflected by experimental observations which show that $x(t)$ is not a stationary random process [64], so its standard deviation is not constant in time. Therefore, for long-term analysis of $x(t)$, it is necessary to also model σ_x as a random variable, and the probability density of x is conditional on σ_x:

$$p(x|\sigma_x) = \frac{1}{\sigma_x\sqrt{2\pi}}\exp\left(-\frac{x^2}{2\sigma_x^2}\right),$$ (3.56)

where σ_x has a mean value of σ_m and a standard deviation σ_σ. Its probability density can be modeled as:

$$p(\sigma_x) = \frac{1}{\sigma_\sigma\sigma_x\sqrt{2\pi}}\exp\left(-\frac{(\log\sigma_x - \log\sigma_m)^2}{2\sigma_\sigma^2}\right).$$ (3.57)

The CCIR has provided an expression which predicts σ_m, measured in dB, as a function of signal frequency f and elevation angle E [27, p. 523]:

$$\sigma_m = 0.025f^{\frac{7}{2}}(\csc E)^{-0.85},$$ (3.58)

where f is in GHz. Therefore, for the GPS L1 carrier frequency ($f_{L1} = 1.575,42$ GHz), σ_m is given by:

$$\sigma_m = 0.0326(\csc E)^{-0.85}.$$ (3.59)

For an elevation angle of 5°, this gives a mean RMS scintillation intensity of 0.259 dB [27, p. 523]. At larger elevation angles, the value is even smaller. At GNSS frequencies, tropospheric scintillations are not typically significant. However since σ_x is a random variable, the tropospheric scintillation intensity can become significant for a small percentage of the time, especially at low elevation angles. For example, approximately 10% of the time, σ_x is at least 0.9 dB at an elevation angle of 5° [27, p. 523].

4

Predictive Models of the Troposphere

4.1 Introduction

As discussed in Chapter 3, one of the most impactful tropospheric effects on global navigation satellite systems (GNSS) positioning is that of group delay. Because of its importance, several predictive models have been developed to estimate and eliminate tropospheric group delay error. Although highly accurate and sophisticated models of tropospheric propagation exist, they are often unsuitable for GNSS positioning since they are computationally expensive and time-consuming. For many GNSS applications, real-time positioning is often essential, so a group delay model must strike a balance between accuracy and computational latency. Two prominent group delay models are discussed in this chapter: the Saastamoinen model and the Hopfield model.

Section 4.2 focuses on the Saastamoinen model. A detailed derivation of the final closed-form expression for group delay is presented. This derivation requires a number of approximations, so an attempt has been made to motivate and justify their validity as often as possible. Following on from this discussion, Section 4.3 covers the Hopfield model which, in contrast to the Saastamoinen model, is an empirical model. In other words, several of the model parameters are inferred from a set of observed data, rather than being derived from first principles.

Finally in Section 4.4, this chapter closes with an overview of the U.S. Standard Atmosphere, which is a more sophisticated and comprehensive model of the atmosphere than the group delay models in the prior sections. The U.S. Standard Atmosphere provides altitude profiles of quantities such as pressure, temperature, and number densities of various molecular species. Such a model is useful in a broad range of contexts, including certain GNSS positioning applications which require highly precise positioning.

Tropospheric and Ionospheric Effects on Global Navigation Satellite Systems, First Edition.
Timothy H. Kindervatter and Fernando L. Teixeira.
© 2022 The Institute of Electrical and Electronics Engineers, Inc. Published 2022 by John Wiley & Sons, Inc.

4.2 Saastamoinen Model

In Section 3.2, we discussed the fact that the refractivity of the troposphere causes a delay in the travel of an electromagnetic wave. That delay can be expressed using Equation 3.19 (reproduced here for convenience):

$$\Delta^T = 10^{-6} \int_S \frac{77.6}{T(s)} \left(p(s) + 4810 \frac{e(s)}{T(s)} \right) ds, \tag{4.1}$$

where $T(s)$ is the temperature in Kelvins, and $p(s)$ and $e(s)$ are respectively the total pressure and water vapor partial pressure in millibars. Each quantity is a function of position along the signal's path. The tropospheric delay is of interest in GNSS applications because it induces observation error in GPS observables such as the pseudorange (see Section 2.2). If tropospheric delay can be accurately modeled and removed from the GPS observable, GPS positioning accuracy can be improved.

Unfortunately, it is infeasible to use Equation 4.1 to model tropospheric delay in the context of GNSS positioning. To do so would require data on $T(s)$, $p(s)$, and $e(s)$ for a number of points along the propagation path. It is infeasible to obtain empirical data for these quantities at every location of interest for GNSS positioning, even if only zenith paths are considered.[1] Furthermore, it is computationally expensive to use a theoretical model to simulate atmospheric properties, which makes that approach impractical for many GNSS positioning applications, in particular those which require positioning estimates in real time.

A solution to this problem was proposed by Saastamoinen [65, 66], who provided an empirical model given by the following equation:

$$\Delta^T \approx 0.002277 \left(p + \left(\frac{1255}{T} + 0.05 \right) e - 1.16 \tan^2 z \right) \sec z, \tag{4.2}$$

where z is the zenith angle, and T, p, and e are the temperature in Kelvins, total pressure in millibars, and water vapor partial pressure in millibars, respectively. What follows is a derivation of Equation 4.2, which has been adapted and expanded from a similar derivation in [66].

The following Figure depicts the problem, along with the notation that will be used:

Using the notation in Figure 4.1, we can write the tropospheric delay (in terms of refractive index n) as:

$$\Delta^T = \int_S (n-1)ds = \int_{r_1}^{r_2} (n-1) \sec z \, dr, \tag{4.3}$$

where the bounds of the integral are r_1, which represents the altitude of the GNSS receiver (henceforth, the *user altitude*), and r_2 is the base of the ionosphere

---

1 As discussed in Section 3.2, the zenith path delay can be used in conjunction with mapping functions to obtain an estimate of the slant path delay.

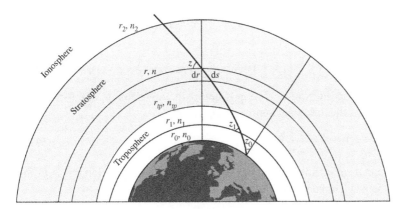

Figure 4.1 Refracted ray path of an electromagnetic wave traversing the troposphere.

(which occurs at an altitude of about 50 km). We will assume that $r_1 < r_{tp}$, where $r_{tp} = 10$ km is the altitude of the tropopause. In general, the zenith angle z is not constant at different points along the ray path. Via Snell's law, we know that at some altitude r where the index of refraction is n:

$$nr \sin z = n_1 r_1 \sin z_1. \tag{4.4}$$

We will need to derive an expression for $\sec z$ that allows us to integrate Equation 4.3. Beginning with Snell's law, and letting $y = nr/(n_1 r_1)$ for brevity, observe that via several trigonometric identities:

$$\sin^2 z = \frac{\sin^2 z_1}{y^2} = \frac{\tan^2 z_1 \cos^2 z_1}{y^2} = \frac{\sec^2 z_1 - 1}{y^2 \sec^2 z_1}. \tag{4.5}$$

Furthermore, noting that $\sin^2 z + \cos^2 z = 1$:

$$(1 - \cos^2 z) = \frac{\sec^2 z_1 - 1}{y^2 \sec^2 z_1} \Rightarrow \cos^2 z = 1 - \frac{\sec^2 z_1 - 1}{y^2 \sec^2 z_1}. \tag{4.6}$$

This can be rewritten slightly to obtain:

$$\cos^2 z = \frac{y^2 \sec^2 z_1 - \sec^2 z_1 + 1}{y^2 \sec^2 z_1}. \tag{4.7}$$

It follows that:

$$\sec^2 z = \frac{y^2 \sec^2 z_1}{y^2 \sec^2 z_1 - \sec^2 z_1 + 1} \Rightarrow \sec z = \frac{y \sec z_1}{\sqrt{y^2 \sec^2 z_1 - \sec^2 z_1 + 1}}. \tag{4.8}$$

Grouping similar terms, we get:

$$\sec z = \frac{y \sec z_1}{\sqrt{(y^2 - 1) \sec^2 z_1 + 1}}. \tag{4.9}$$

Using the fact that $1/\sqrt{x^2+1} \approx 1 - x^2/2$ to second order, we can write the following:

$$\sec z \approx y \sec z_1 \left(1 - \frac{1}{2}(y^2 - 1)\sec^2 z_1\right) = y \sec z_1 - \frac{1}{2}y(y^2 - 1)\sec^3 z_1. \quad (4.10)$$

Assuming that the index of refraction is approximately equal to one throughout the troposphere, the following approximation holds:

$$y \approx \frac{r}{r_1} = 1 + \frac{1}{r_1}(r - r_1). \quad (4.11)$$

Furthermore, if we substitute this approximation into the expression $y(y^2 - 1)$ and perform some algebraic manipulations, we obtain:

$$y(y^2 - 1) \approx \frac{2}{r_1}(r - r_1). \quad (4.12)$$

By substituting Equations 4.11 and 4.12 into Equation 4.10 and simplifying, we obtain the following approximation:

$$\sec z \approx \sec z_1 - \frac{1}{r_1}(\sec^3 z_1 - \sec z_1)(r - r_1). \quad (4.13)$$

Further, using the fact that $\tan^2 z_1 = \sec^2 z_1 - 1$, we can write:

$$\sec z \approx \sec z_1 - \frac{1}{r_1}\sec z_1 \tan^2 z_1 (r - r_1). \quad (4.14)$$

Thus, we can finally rewrite Equation 4.3 as:

$$\Delta^T \approx \sec z_1 \int_{r_1}^{r_2} (n - 1)dr - \frac{1}{r_1}\sec z_1 \tan^2 z_1 \int_{r_1}^{r_2} (n - 1)(r - r_1)dr. \quad (4.15)$$

Now we can proceed by solving each of these integrals separately.

4.2.1 First Integral

As discussed in Section 3.2, the molecular composition of the troposphere can be divided into two parts: dry air (composed of gases such as N_2 and O_2) and water vapor. It is convenient to express the total pressure in terms of the partial pressures of these components:

$$p = p_d + e. \quad (4.16)$$

Here, p is the total pressure in the troposphere, and p_d and e are the partial pressures of dry gases and water vapor, respectively. Both of the partial pressures can be modeled as ideal gases, so their densities can be expressed via the ideal gas law:

$$\rho_{dry} = \frac{p - e}{RT}, \quad \rho_{wet} = \frac{e}{R_w T}, \quad (4.17)$$

where R and R_w represent the gas constants of the dry and wet components, respectively, and T is temperature. Thus, we can express the total density of the troposphere as:

$$\rho = \frac{p}{RT} - \left(1 - \frac{R}{R_w}\right)\frac{e}{RT}. \tag{4.18}$$

The pressure at the user altitude, p_1, can be expressed as the total weight of the air in a column of unit cross section that extends from r_1 to the top of the atmosphere. This statement is expressed by the following integral:

$$mg = g \int_{r_1}^{\infty} \rho\, dr = p_1 \Rightarrow \int_{r_1}^{\infty} \rho\, dr = \frac{p_1}{g}. \tag{4.19}$$

At large altitudes, the atmosphere becomes very sparse and contributes little to the total weight of the column. We can simply use the bottom of the ionosphere, r_2, as the top of the column. This allows us to approximate pressure with the following integral:

$$\int_{r_1}^{r_2} \rho\, dr = \frac{1}{R}\int_{r_1}^{r_2} \frac{p}{T}dr - \frac{1}{R}\left(1 - \frac{R}{R_w}\right)\int_{r_1}^{r_2} \frac{e}{T}dr \approx \frac{p_1}{g}. \tag{4.20}$$

The refractivity of the troposphere can be written as:

$$n - 1 = \frac{(n_0 - 1)T_0}{p_0}\frac{p}{T} - c_w\frac{e}{T} + c'_w\frac{e}{T^2}, \tag{4.21}$$

where n is the refractive index of the troposphere, n_0, T_0, and p_0 are the refractive index of dry air, temperature, and pressure at the ground r_0. c_w and c'_w are constants. We may now integrate Equation 4.21 from r_1 to r_2 as follows:

$$\int_{r_1}^{r_2} (n - 1)dr = \frac{(n_0 - 1)T_0}{p_0}\int_{r_1}^{r_2} \frac{p}{T}dr - c_w\int_{r_1}^{r_2} \frac{e}{T}dr + c'_w\int_{r_1}^{r_2} \frac{e}{T^2}dr. \tag{4.22}$$

Now we see that if we rearrange Equation 4.20 like so:

$$\int_{r_1}^{r_2} \frac{p}{T}dr = \frac{p_1 R}{g} + \left(1 - \frac{R}{R_w}\right)\int_{r_1}^{r_2} \frac{e}{T}dr, \tag{4.23}$$

we can substitute it into Equation 4.22 to obtain:

$$\int_{r_1}^{r_2} (n - 1)dr = \frac{(n_0 - 1)T_0}{p_0}\left[\frac{p_1 R}{g} + \left(1 - \frac{R}{R_w}\right)\int_{r_1}^{r_2} \frac{e}{T}dr\right]$$
$$- c_w\int_{r_1}^{r_2} \frac{e}{T}dr \tag{4.24}$$
$$+ c'_w\int_{r_1}^{r_2} \frac{e}{T^2}dr.$$

Grouping like terms, we can simplify this to:

$$\int_{r_1}^{r_2} (n-1)\mathrm{d}r = \frac{R}{g}\frac{(n_0-1)T_0}{p_0}p_1$$
$$+ \left[\frac{(n_0-1)T_0}{p_0}\left(1-\frac{R}{R_w}\right) - c_w\right]\int_{r_1}^{r_2}\frac{e}{T}\mathrm{d}r \qquad (4.25)$$
$$+ c_w'\int_{r_1}^{r_2}\frac{e}{T^2}\mathrm{d}r.$$

In order to evaluate the two remaining integrals, we need to consider the profile of water vapor pressure. The exact character of this profile is highly variable over time and across different locations, but for average conditions in mid-latitudes, we can use the following approximation [66]:

$$e = e_1\left(\frac{T}{T_1}\right)^{\frac{-4g}{R\beta}}, \qquad (4.26)$$

where T_1 and e_1 are the temperature and water vapor pressure at the user altitude, respectively. The quantity β is the vertical gradient of temperature, that is:

$$\beta = \frac{\mathrm{d}T}{\mathrm{d}r}, \qquad (4.27)$$

also commonly referred to as the lapse rate. As we will discuss briefly later, β is approximately constant up to an altitude of about 10 km, at which point it abruptly drops to approximately zero and remains there up to an altitude of about 50 km. This will be important in our evaluation of the two water vapor pressure integrals.

We will begin by integrating the first of the two water vapor integrals. We can modify the integrand so that we are integrating with respect to temperature:

$$\int_{r_1}^{r_2}\frac{e}{T}\mathrm{d}r = \int_{r_1}^{r_2}\frac{e_1}{T}\left(\frac{T}{T_1}\right)^{\frac{-4g}{R\beta}}\frac{\mathrm{d}r}{\mathrm{d}T}\frac{\mathrm{d}T}{\mathrm{d}r}\mathrm{d}r = \int_{T_1}^{T_2}\frac{e_1}{T}\left(\frac{T}{T_1}\right)^{\frac{-4g}{R\beta}}\frac{1}{\beta}\mathrm{d}T, \qquad (4.28)$$

where we have used $\mathrm{d}r/\mathrm{d}T = 1/\beta$. As noted, β varies abruptly at the tropopause, so it is convenient to express this integral as the sum of two separate integrals:

$$\int_{T_1}^{T_2}\frac{e_1}{T}\left(\frac{T}{T_1}\right)^{\frac{-4g}{R\beta}}\frac{1}{\beta}\mathrm{d}T = \int_{T_1}^{T_{tp}}\frac{e_1}{T}\left(\frac{T}{T_1}\right)^{\frac{-4g}{R\beta}}\frac{1}{\beta}\mathrm{d}T + \int_{T_{tp}}^{T_2}\frac{e_1}{T}\left(\frac{T}{T_1}\right)^{\frac{-4g}{R\beta}}\frac{1}{\beta}\mathrm{d}T.$$
$$(4.29)$$

The evaluation of each integral proceeds identically, so we will ignore the bounds for now and reintroduce them later. By introducing $u = T/T_1$ and $\mathrm{d}u = \mathrm{d}T/T_1$ and letting $m = -g/(R\beta)$, the above integral write more compactly as

$$\frac{1}{\beta}\int\frac{e_1}{T}\left(\frac{T}{T_1}\right)^{\frac{-4g}{R\beta}}\frac{T_1}{T_1}\mathrm{d}T = \frac{e_1}{\beta}\int\left(\frac{T}{T_1}\right)^{4m}\frac{T_1}{T}\frac{\mathrm{d}T}{T_1} = \frac{e_1}{\beta}\int u^{4m-1}\mathrm{d}u. \quad (4.30)$$

Note that we can pull β out of the integral because it is constant within each region of interest. Evaluating this integral, we obtain:

$$\frac{e_1}{\beta} \int u^{4m-1} \, du = \frac{e_1}{\beta} \frac{1}{4m} u^{4m} = \frac{e_1}{\beta} \left(\frac{-R\beta}{4g} \right) \left(\frac{T}{T_1} \right)^{\frac{-4g}{R\beta}} = \frac{-e_1 R}{4g} \left(\frac{T}{T_1} \right)^{\frac{-4g}{R\beta}}.$$
$$(4.31)$$

Reintroducing the bounds, we see that the integral corresponding to the troposphere evaluates to:

$$\frac{-e_1 R}{4g} \left(\frac{T}{T_1} \right)^{\frac{-4g}{R\beta}} \Bigg|_{T_1}^{T_{tp}} = \frac{-e_1 R}{4g} \left(\frac{T_{tp}}{T_1} \right)^{\frac{-4g}{R\beta}} - \left(\frac{-e_1 R}{4g} \left(\frac{T_1}{T_1} \right)^{\frac{-4g}{R\beta}} \right)$$

$$= \frac{-e_1 R}{4g} \left(\frac{T_{tp}}{T_1} \right)^{\frac{-4g}{R\beta}} + \frac{e_1 R}{4g}.$$

Note that at T_{tp}, β abruptly becomes close to zero, so the exponent of the first term rapidly approaches $-\infty$ and therefore the whole first term approaches 0. Thus, we can ignore the first term entirely, leaving:

$$\frac{1}{\beta} \int_{T_1}^{T_{tp}} \frac{e_1}{T} \left(\frac{T}{T_1} \right)^{\frac{-4g}{R\beta}} dT = \frac{e_1 R}{4g}.$$
$$(4.32)$$

Since $\beta = 0$ above the tropopause, the same argument can be used to show that the stratospheric integral vanishes everywhere. That is:

$$\frac{-e_1 R}{4g} \left(\frac{T}{T_1} \right)^{\frac{-4g}{R\beta}} \Bigg|_{T_{tp}}^{T_2} = \frac{-e_1 R}{4g} \left(\frac{T_2}{T_1} \right)^{\frac{-4g}{R\beta}} - \left(\frac{-e_1 R}{4g} \left(\frac{T_{tp}}{T_1} \right)^{\frac{-4g}{R\beta}} \right) = 0,$$
$$(4.33)$$

and therefore:

$$\frac{1}{\beta} \int_{T_{tp}}^{T_2} \frac{e_1}{T} \left(\frac{T}{T_1} \right)^{\frac{-4g}{R\beta}} dT = 0.$$
$$(4.34)$$

Finally, using Equations 4.32 and 4.34 in to get the total integral shown in Equation 4.29, we find that:

$$\int_{r_1}^{r_2} \frac{e}{T} dr = \frac{e_1 R}{4g}.$$
$$(4.35)$$

A completely analogous derivation to the one just shown can be used to evaluate the second water vapor pressure integral. The end result is:

$$\int_{r_1}^{r_2} \frac{e}{T^2} dr = \frac{1}{\frac{4g}{R} + \beta} \left(\frac{e_1}{T_1} \right).$$
$$(4.36)$$

Finally, using Equations 4.35 and 4.36 in Equation 4.25, we can obtain the final expression for the first integral in Saastamoinen's tropospheric delay formula (i.e. Equation 4.15):

$$\int_{r_1}^{r_2} (n-1)dr = \frac{R}{g} \frac{(n_0 - 1)T_0}{p_0} p_1$$
$$+ \left[\frac{(n_0 - 1)T_0}{p_0} \left(1 - \frac{R}{R_w}\right) - c_w \right] \left(\frac{e_1 R}{4g} \right)$$
$$+ c_w' \left[\frac{1}{\frac{4g}{R} + \beta} \left(\frac{e_1}{T_1} \right) \right].$$

Rearranging slightly, we obtain the following expression:

$$\int_{r_1}^{r_2} (n-1)dr = \frac{R}{g} \frac{(n_0 - 1)T_0}{p_0} p_1$$
$$+ \left[\frac{1}{4} \frac{R}{g} \frac{(n_0 - 1)T_0}{p_0} \left(1 - \frac{R}{R_w}\right) - \frac{c_w}{4} \frac{R}{g} \right] \tag{4.37}$$
$$+ \left(\frac{c_w'}{\frac{4g}{R} + \beta} \frac{1}{T_1} \right) \right] e_1.$$

4.2.2 Second Integral

The second integral in Equation 4.15, that is:

$$\int_{r_1}^{r_2} (n-1)(r - r_1)dr, \tag{4.38}$$

is more complicated to evaluate than the first integral because it requires consideration of the vertical profile of the atmosphere. In particular, we need to find expressions for both $n - 1$ and $r - r_1$ in terms of the temperature and pressure profiles of the troposphere. This integral extends from the user altitude up to the base of the ionosphere. However, this region of the atmosphere can be split into two distinct regions: the troposphere and the stratosphere.[2] The troposphere is the lowest layer, extending from the ground up to an altitude of about 10 km, known as the tropopause. As noted before, at this height the temperature gradient becomes approximately zero and remains that way up to an altitude of about 50 km (which roughly coincides with the base of the ionosphere). Therefore, it is also convenient to split this Equation 4.57 into two separate integrals: one that

---

2 Note that the contribution of both the troposphere and stratosphere is considered when computing the total tropospheric delay. In the context of GNSS applications, the major sources of delay error are the ionosphere and the region below the ionosphere. Thus "tropospheric delay" is commonly used to simply mean "non-ionospheric delay" in GNSS literature [39]. Indeed, about 80% of the delay caused by the non-ionized atmosphere can be attributed to the troposphere [40].

represents the region of the atmosphere below the tropopause, and another that represents the region above the tropopause:

$$\int_{r_1}^{r_2} (n-1)(r-r_1)dr = \int_{r_1}^{r_{tp}} (n-1)(r-r_1)dr + \int_{r_{tp}}^{r_2} (n-1)(r-r_1)dr,$$

(4.39)

where r_{tp} is the altitude of the tropopause. We will begin by evaluating the integral below the tropopause.

From the ground up to an altitude of about 10 km, the temperature of the atmosphere decreases approximately linearly. That is, we can express the temperature T with respect to altitude r as follows:

$$T = T_1 + \beta(r - r_1),$$

(4.40)

where T_1 is the temperature at the user altitude. Since temperature and altitude are linearly related below 10 km, β can be assumed to be constant within that altitude range. In contrast, the pressure of the troposphere decreases exponentially with height. The relationship is given by:

$$p = p_1 \left(\frac{T}{T_1}\right)^{-\frac{g}{R\beta}},$$

(4.41)

where p_1 and T_1 are the pressure and temperature at the user altitude, g is the acceleration due to gravity, and R is the specific gas constant. For further details on the derivation of these temperature and pressure profiles, refer to [27, pp. 529–533].

We wish to determine the refractivity profile in terms of the temperature and pressure profiles. To do so, it is convenient to simplify Equation 4.25 by using the fact that water vapor only contributes about 10% of the total refractivity of the troposphere [66]. Therefore, the terms of Equations 4.21 and 4.25 corresponding to water vapor's contribution to refractivity (i.e. terms containing e) can be ignored to obtain the approximations:

$$n - 1 \approx \frac{(n_0 - 1)T_0}{p_0} \frac{p}{T},$$

(4.42)

$$\int_{r_1}^{r_{tp}} (n-1)dr \approx \frac{R}{g} \frac{(n_0 - 1)T_0}{p_0} p_1.$$

(4.43)

Substituting Equation 4.41 into Equation 4.42, we obtain:

$$n - 1 \approx \frac{(n_0 - 1)T_0}{p_0} \frac{p_1 \left(\frac{T}{T_1}\right)^{-\frac{g}{R\beta}}}{T}.$$

(4.44)

At the ground, i.e. when $r_1 = r_0$, we can substitute the values $T_1 = T_0$ and $p_1 = p_0$:

$$n - 1 \approx \frac{(n_1 - 1)T_1}{p_1} \frac{p_1 \left(\frac{T}{T_1}\right)^{-\frac{g}{R\beta}}}{T}.$$

(4.45)

After some simplifications, this can be rewritten as:

$$n - 1 \approx (n_1 - 1)\left(\frac{T}{T_1}\right)^{-\frac{g}{R\beta}-1}. \tag{4.46}$$

For simplicity, we will set $m' = -g/(R\beta) - 1$, which gives us:

$$n - 1 \approx (n_1 - 1)\left(\frac{T}{T_1}\right)^{m'}. \tag{4.47}$$

Now that we have an expression for $n - 1$, we must also find an expression for $r - r_1$. This can be obtained by simply rearranging Equation 4.40:

$$r - r_1 = \frac{T - T_1}{\beta} = \frac{T_1}{\beta}\left(\frac{T}{T_1} - 1\right). \tag{4.48}$$

We can now substitute Equations 4.47 and 4.48 into Equation 4.38 to obtain:

$$\int_{r_1}^{r_{tp}} (n_1 - 1)\left(\frac{T}{T_1}\right)^{m'} \frac{T_1}{\beta}\left(\frac{T}{T_1} - 1\right) dr. \tag{4.49}$$

Rewriting slightly, we obtain:

$$\frac{(n_1 - 1)T_1}{\beta} \int_{r_1}^{r_{tp}} \left(\frac{T}{T_1} - 1\right)\left(\frac{T}{T_1}\right)^{m'} \frac{dr}{dT}\frac{dT}{dr} dr. \tag{4.50}$$

Note that using Equation 4.27, we see that $dr/dT = 1/\beta$, so this simplifies to:

$$\frac{(n_1 - 1)T_1}{\beta^2} \int_{r_1}^{r_{tp}} \left(\frac{T}{T_1} - 1\right)\left(\frac{T}{T_1}\right)^{m'} dT. \tag{4.51}$$

This integral can be solved by parts. Let:

$$u = \left(\frac{T}{T_1} - 1\right), \tag{4.52}$$

and

$$dv = \left(\frac{T}{T_1}\right)^{m'} dT. \tag{4.53}$$

Then it follows that:

$$du = \frac{dT}{T_1}, \tag{4.54}$$

and

$$v = \frac{T^{m'+1}}{(m'+1)T^{m'}} = \frac{T}{m'+1}\left(\frac{T}{T_1}\right)^{m'}, \tag{4.55}$$

so that the integration by parts formula becomes:

$$\int \left(\frac{T}{T_1} - 1\right)\left(\frac{T}{T_1}\right)^{m'} dT = \left(\frac{T}{T_1} - 1\right)\frac{T}{m'+1}\left(\frac{T}{T_1}\right)^{m'}$$

$$-\int \frac{1}{m'+1}\left(\frac{T}{T_1}\right)^{m'+1} dT. \tag{4.56}$$

For now, we have chosen to ignore the bounds of the integral. They will be reintroduced later. Evaluating the integral on the right hand side and factoring out common terms results in:

$$\int \left(\frac{T}{T_1}-1\right)\left(\frac{T}{T_1}\right)^{m'} dT$$
$$= \frac{T}{m'+1}\left(\frac{T}{T_1}\right)^{m'}\left[\left(\frac{T}{T_1}-1\right)-\frac{1}{(m'+2)}\left(\frac{T}{T_1}\right)\right]. \tag{4.57}$$

With this result in mind, we see that evaluating Equation 4.51 results in the following:

$$\frac{(n_1-1)}{\beta^2}\left(\frac{T}{m'+1}\left(\frac{T}{T_1}\right)^{m'}\left[(T-T_1)-\frac{T}{(m'+2)}\right]\right)\Bigg|_{r_1}^{r_{tp}}. \tag{4.58}$$

Evaluating this expression at the bounds results in:

$$\frac{(n_1-1)}{\beta^2}\left[\left(\frac{T_{tp}}{m'+1}\left(\frac{T_{tp}}{T_1}\right)^{m'}\left[(T_{tp}-T_1)-\frac{T_{tp}}{(m'+2)}\right]\right)\right.$$
$$\left.+\left(\frac{T_1^2}{(m'+1)(m'+2)}\right)\right]. \tag{4.59}$$

If we group like terms, we end up with:

$$\frac{(n_1-1)}{\beta^2}\left[\frac{1}{(m'+1)(m'+2)}\left(T_1^2-T_{tp}^2\left(\frac{T_{tp}}{T_1}\right)^{m'}\right)\right.$$
$$\left.+\frac{T_{tp}}{m'+1}\left(\frac{T_{tp}}{T_1}\right)^{m'}(T_{tp}-T_1)\right]. \tag{4.60}$$

Note that at r_{tp}, $n = n_{tp}$ and $T = T_{tp}$ so Equation 4.47 becomes:

$$n_{tp}-1 \approx (n_1-1)\left(\frac{T_{tp}}{T_1}\right)^{m'} \Rightarrow \frac{n_{tp}-1}{n_1-1} \approx \left(\frac{T_{tp}}{T_1}\right)^{m'}. \tag{4.61}$$

Furthermore, using $m' = -g/(R\beta) - 1$ and after some algebra, Equation 4.60 becomes

$$\frac{(n_1-1)}{\beta^2}\left[\frac{R^2\beta^2}{g^2\left(1-\frac{R\beta}{g}\right)}\left(T_1^2(n_1-1)-T_{tp}^2(n_{tp}-1)\right)\frac{1}{n_1-1}\right.$$
$$\left.-\frac{R\beta^2}{g}T_{tp}\left(\frac{n_{tp}-1}{n_1-1}\right)\left(\frac{T_{tp}-T_1}{\beta}\right)\right]. \tag{4.62}$$

From Equation 4.40 evaluated at r_{tp}, we see that:

$$T_{tp} = T_1 + \beta(r_{tp} - r_1) \Rightarrow \frac{T_{tp} - T_1}{\beta} = r_{tp} - r_1. \tag{4.63}$$

If we substitute this into Equation 4.62 and distribute $(n_1 - 1)/\beta^2$, we finally obtain:

$$\frac{(n_1 - 1)T_1}{\beta^2} \int_{r_1}^{r_{tp}} \left(\frac{T}{T_1} - 1\right) \left(\frac{T}{T_1}\right)^{m'} dT$$

$$= \frac{R^2}{g^2(1 - \frac{R\beta}{g})} \left(T_1^2(n_1 - 1) - T_{tp}^2(n_{tp} - 1)\right) \tag{4.64}$$

$$- \frac{R}{g} T_{tp} \left(n_{tp} - 1\right) \left(r_{tp} - r_1\right).$$

Note that this equation holds for all values of β except $\beta = g/R$ (including $\beta = 0$).

Next, we will evaluate the second integral of Equation 4.39, which corresponds to the region of the atmosphere above the tropopause and below the ionosphere. Henceforth, we will refer to this as "the stratospheric integral." In this region, $\beta = 0$ and the pressure and refractivity vary as follows:

$$p = p_{tp} \exp\left(-\frac{g}{RT_{tp}}(r - r_{tp})\right), \tag{4.65}$$

$$n - 1 = (n_{tp} - 1)\exp\left(-\frac{g}{RT_{tp}}(r - r_{tp})\right), \tag{4.66}$$

where all parameters with the subscript tp correspond to the values of those parameters at the tropopause. Using Equation 4.66, the stratospheric integral can be rewritten as follows:

$$\int_{r_{tp}}^{r_2} (n - 1)(r - r_1)dr = \int_{r_{tp}}^{r_2} (n_{tp} - 1)e^{m(r - r_{tp})}(r - r_1)dr, \tag{4.67}$$

where $m = -g/(RT_{tp})$. The resulting expression can be rearranged slightly:

$$(n_{tp} - 1)\left[\int_{r_{tp}}^{r_2} r\, e^{m(r - r_{tp})}\, dr - r_1 \int_{r_{tp}}^{r_2} e^{m(r - r_{tp})}dr\right], \tag{4.68}$$

which, via integration by parts for the first integral and straightforward integration for the second, results in:

$$(n_{tp} - 1)\left[\frac{1}{m}(r - r_1)e^{m(r - r_{tp})} - \frac{1}{m^2}e^{m(r - r_{tp})}\right]\Bigg|_{r_{tp}}^{r_2}. \tag{4.69}$$

Evaluating this expression at the bounds gives:

$$(n_{tp} - 1)\left[\left(\frac{1}{m}(r_2 - r_1)e^{m(r_2 - r_{tp})} - \frac{1}{m^2}e^{m(r_2 - r_{tp})}\right) - \left(\frac{1}{m}(r_{tp} - r_1) - \frac{1}{m^2}\right)\right]. \tag{4.70}$$

This expression can be simplified by considering the typical values of the relevant quantities in the stratosphere. In particular, using the following values [66–68]:

$$g = 9.784 \frac{m}{s^2}, \quad R = 287 \frac{J}{K \cdot kg}, \quad T_{tp} = 222.15 \text{ K},$$

$$r_2 = 50 \text{ km}, \quad r_{tp} = 10 \text{ km}, \quad r_1 = 0 \text{ km},$$

we find that the first two terms in Equation 4.70 together contribute less than 1% to the total value of the expression. Therefore, we are justified in ignoring these terms, which simplifies the overall expression to:

$$(n_{tp} - 1)\left[-\left(\frac{1}{m}(r_{tp} - r_1) - \frac{1}{m^2}\right)\right]. \tag{4.71}$$

Then, substituting $m = -g/(RT_{tp})$ back in, we finally obtain the following approximate expression for the stratospheric integral:

$$\int_{r_{tp}}^{r_2} (n-1)(r-r_1) dr = \frac{RT_{tp}}{g}(n_{tp} - 1)(r_{tp} - r_1) + \frac{R^2 T_{tp}^2}{g^2}(n_{tp} - 1). \tag{4.72}$$

We are finally prepared to write the full expression for the total atmospheric integral given by Equation 4.39. Using Equations 4.64 and 4.72, we see that:

$$\int_{r_1}^{r_2} (n-1)(r-r_1) dr = \frac{R^2}{g^2(1 - \frac{R\beta}{g})} \left(T_1^2(n_1 - 1)) - T_{tp}^2(n_{tp} - 1) \right)$$

$$- \frac{RT_{tp}}{g}(n_{tp} - 1)(r_{tp} - r_1) \tag{4.73}$$

$$+ \frac{RT_{tp}}{g}(n_{tp} - 1)(r_{tp} - r_1) + \frac{R^2 T_{tp}^2}{g^2}(n_{tp} - 1).$$

Simplifying this expression, we find that the total atmospheric integral can be written as:

$$\int_{r_1}^{r_2} (n-1)(r-r_1) \, dr = \frac{R^2}{g^2}\left[\frac{T_1^2(n_1 - 1) - \frac{R\beta}{g} T_{tp}^2(n_{tp} - 1)}{1 - \frac{R\beta}{g}} \right]. \tag{4.74}$$

Using Equation 4.42, we see that:

$$n_1 - 1 \approx \frac{(n_0 - 1)T_0}{p_0} \frac{p_1}{T_1}, \tag{4.75}$$

and

$$n_{tp} - 1 \approx \frac{(n_0 - 1)T_0}{p_0} \frac{p_{tp}}{T_{tp}}. \tag{4.76}$$

Therefore, rearranging these equations in terms of $T_1(n_1 - 1)$ and $T_{tp}(n_{tp} - 1)$, respectively, Equation 4.74 can be rewritten as:

$$\int_{r_1}^{r_2} (n-1)(r-r_1)dr = \frac{R^2}{g^2} \frac{(n_0 - 1)T_0}{P_0} \left[\frac{P_1 T_1 - \frac{R\beta}{g} P^{tp} T^{tp}}{1 - \frac{R\beta}{g}} \right].$$
(4.77)

4.2.3 Putting Everything Together

We are finally in a position to write the full expression for Saastamoinen's tropospheric delay model. If we substitute Equations 4.37 and 4.77 into Equation 4.2, we obtain the following expression for the tropospheric delay:

$$\Delta^T \approx \frac{(n_0 - 1)RT_0}{P_0 g} P_1 \sec z$$

$$- \frac{R^2}{g^2} \frac{(n_0 - 1)T_0}{P_0} \left[\frac{P_1 T_1 - \frac{R\beta}{g} P^{tp} T^{tp}}{1 - \frac{R\beta}{g}} \right] \frac{1}{r_1} \sec z \tan^2 z$$

$$+ \left[\frac{1}{4} \frac{R}{g} \frac{(n_0 - 1)T_0}{P_0} \left(1 - \frac{R}{R_w} \right) - \frac{c_w}{4} \frac{R}{g} + \left(\frac{c'_w}{\frac{4g}{R} + \beta} \frac{1}{T_1} \right) \right] e_1 \sec z.$$
(4.78)

This formula can be simplified further by considering typical numerical values of the relevant parameters in the troposphere. We will use the following numerical values [66]:

$$(n_0 - 1)\frac{T_0}{P_0} = 77.62 \times 10^{-6} \frac{K}{mbar},$$

$$c_w = 12.92 \times 10^{-6} \frac{K}{mbar}, \quad c'_w = 371{,}900 \times 10^{-6} \frac{K^2}{mbar}.$$

In addition, the specific gas constants and acceleration due to gravity will be approximated by [66, 67]:

$$R = 287 \frac{J}{K \cdot kg}, \quad R_w = 461 \frac{J}{K \cdot kg}, \quad g = 9.784 \frac{m}{s^2}.$$

The temperature gradient typically takes values from about -5 to about -7 K/km in the troposphere. Selected values of β at various locations can be seen in Table 3 of Chapter 13 in [27, p. 532]. We will use the value:

$$\beta = -6.21 \frac{K}{km}.$$

Additionally, in [66], a table of values is given for the following quantity, which at sea level takes on a value of 1.16:

$$\frac{R}{r_1 g} \left[\frac{P_1 T_1 - \frac{R\beta}{g} P^{tp} T^{tp}}{1 - \frac{R\beta}{g}} \right] = 1.16.$$
(4.79)

Substituting these values into Equation 4.78, we finally obtain the model given at the beginning of this section (repeated here for convenience):

$$\Delta^T \approx 0.002277 \left(p + \left(\frac{1255}{T} + 0.05 \right) e - 1.16 \tan^2 z \right) \sec z. \tag{4.80}$$

4.3 Hopfield Model

In contrast to the Saastamoinen model, which was built up solely from first principles (along with a few simplifying assumptions), the Hopfield model is an empirical model. That is, where at least one of the parameters used to formulate the model is inferred from a set of observed data. Empirical models are especially useful for modeling the atmosphere because although it is possible to obtain solutions from first principles, the process is often highly involved and technical. In the case of the Hopfield model, two empirically determined parameters are used in conjunction with a few well-justified assumptions to obtain a model for tropospheric delay which agrees with measured values of tropospheric delay to within 0.08%.

Tropospheric delay depends on refractivity, so it is important to consider the vertical refractivity profile of the atmosphere in order to build a good model of tropospheric delay. As we have seen in Sections 3.2 and 4.2, the refractivity of the troposphere can be split into two components: dry and wet. In particular, recall that the equation for tropospheric refractivity is given by Equation 3.17 (reproduced here for convenience):

$$N = N_d + N_w = \frac{77.6}{T} \left(p + 4810 \frac{e}{T} \right). \tag{4.81}$$

Here, N_d is the dry component, N_w is the wet component, p is dry pressure in K/mbar, e is water vapor pressure in K/mbar, and T is temperature in Kelvins. The vertical profiles of both these two components drop off exponentially with height, but at different rates. For example, at an altitude of about 9–10 km, the dry pressure p is at about one third of its surface value, while e is approximately zero [39]. Therefore, the two components must be treated separately. We will begin by discussing the dry component.

The dry refractivity at height h is given by the equation below [39]:

$$N_d = N_{0_d} \left(\frac{T_0 - \beta h}{T_0} \right)^{m'}, \tag{4.82}$$

where N_{0_d} is the surface dry refractivity, T_0 is the surface temperature, $\beta = dT/dh$, and $m' = -g/(R\beta) - 1$ as seen before in the Saastamoinen model. The quantity T_0/β has units of length and can be interpreted as an equivalent height. In particular, it represents the height at which the temperature would fall to absolute zero if the lapse rate β remained constant [39]. We can rewrite Equation 4.82 as:

$$N_d = N_{0_d} \left(\frac{\frac{T_0}{\beta} - h}{\frac{T_0}{\beta}} \right)^{m'} . \tag{4.83}$$

This makes it more clear that the dry refractivity profile is of the form:

$$N_d = N_{0_d} \left(1 - \frac{h}{h_d} \right)^{m'} , \tag{4.84}$$

where h_d is the top of the dry layer of the troposphere and is analogous to T_0/β. For a given value of m', it is possible to determine the value of h_d using empirical data. Such a model would then be a suitable approximation of tropospheric refractivity. This is exactly the approach taken in the Hopfield model [39, 40].

The choice of m' is an important consideration. It depends on β, which itself varies in the troposphere. As noted before, typical values of β range between about -5 and -7 K/km. Figure 4.2 shows how m' varies with respect to β.

Some selected values of $m'(\beta)$ are $m'(-7) = 3.87$, $m'(-6) = 4.68$, and $m'(-5) = 5.81$. Note that as the exponent increases, it becomes more complicated to evaluate Equation 4.84. Since GNSS receivers must compute the tropospheric delay thousands of times per day or more, it is desirable for this calculation to be as simple as

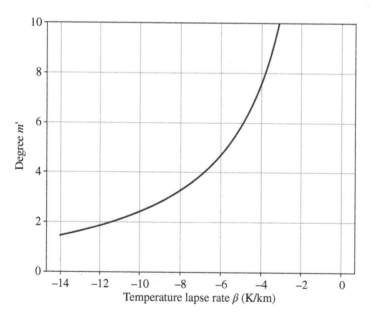

Figure 4.2 Degree of refractivity polynomial with respect to temperature lapse rate.

possible. The Hopfield model uses $m' = 4$ as a practical approximation which provides good agreement to empirical data while also allowing simple computation of Equation 4.84 [39]. Thus, the dry refractivity is modeled by:

$$N_d = \begin{cases} N_{0_d}(1 - \frac{h}{h_d})^4, & h \le h_d, \\ 0, & h > h_d. \end{cases} \tag{4.85}$$

Henceforth, we will refer to this model as a *quartic model*.

Hopfield similarly found that the vertical profile of the wet refractivity N_w could be well-modeled by a quartic expression, although with different parameters:

$$N_w = \begin{cases} N_{0_w}(1 - \frac{h}{h_w})^4, & h \le h_w, \\ 0, & h > h_w. \end{cases} \tag{4.86}$$

Here, h_w fulfills an analogous role as h_d did in the dry model: it represents the top of the wet layer of the troposphere. Since the total refractivity is the sum of the dry and wet components, the total refractivity is given by:

$$N = N_d + N_w = N_{0_d}\left(1 - \frac{h}{h_d}\right)^4 + N_{0_w}\left(1 - \frac{h}{h_w}\right)^4. \tag{4.87}$$

Recall also that the tropospheric delay is given by:

$$\Delta^T = 10^{-6} \int_S N \, ds, \tag{4.88}$$

where ds is the differential element along the electromagnetic wave's path S through the troposphere. Using Equation 4.87, for a zenith path this becomes:

$$\begin{aligned} \Delta^T &= 10^{-6} \int_0^{h_d} N_{0_d}\left(1 - \frac{h}{h_d}\right)^4 dh + 10^{-6} \int_0^{h_w} N_{0_w}\left(1 - \frac{h}{h_w}\right)^4 dh \\ &= \frac{10^{-6}}{5}\left(N_{0_d} h_d + N_{0_w} h_w\right), \end{aligned} \tag{4.89}$$

where N_{0_d} is only nonzero below h_d and N_{0_w} is only nonzero below h_w. To complete the model, values for each of the parameters in this equation must be determined. The surface refractivities N_{0_d} and N_{0_w} in a given location at a given time can be computed using Equation 4.81 by using measurements of the temperature, pressure, and water vapor pressure at the surface. That is, the tropospheric delay expression is given by:

$$\Delta^T = \frac{10^{-6}}{5}\left(77.6\frac{P}{T}h_d + 3.73 \times 10^5 \frac{e}{T^2}h_w\right). \tag{4.90}$$

The equivalent heights h_d and h_w can be computed empirically by performing a least-squares fit of data from meteorological balloons for example. In particular, measurements can be taken of vertical profiles of pressure, temperature, and relative humidity at various measurement stations. These measurements are then used to obtain a refractivity profile by using Equation 4.81, and the resulting profile is then numerically integrated to obtain an observed tropospheric delay value. Actual measurements were taken twice daily for one full year at a number of locations ranging from $-14°$ to $38°$ in latitude and $-170°$ to $166°$ in longitude [40]. The Hopfield model fits h_d to experimental data under the assumption that h_d varied at a given location as a function of surface temperature, that is:

$$h_d = h_{0_d} + a_d T, \tag{4.91}$$

where T is the surface temperature in degrees Celsius, h_{0_d} is the value of h_d when the surface temperature is 0 °C, and a_d is a constant. In this model, both h_{0_d} km and a_d km/°C are determined via least-squares. In particular, this was done by substituting Equation 4.91 into Equation 4.90 to obtain a theoretical expression for the tropospheric delay for given values of temperature and pressure (i.e. the values at the time of a given balloon flight). The theoretical expression was equated to the observed tropospheric delay values obtained from each balloon flight in a given one-year data set, and a least-squares fit was performed on this system of equations to obtain values for h_{0_d} and a_d for that data set. It was found that the difference between the theoretical and observed tropospheric delay values was very small – at most 0.08% across all data sets [40] – and that each data set produced consistent values of both h_{0_d} and a_d. The mean value of each of these parameters was taken across all data sets to obtain $h_{0_d} = 40.082$ km and $a_d = 0.148, 98$ km/°C. This gives us the empirically determined expression for h_d:

$$h_d = 40.082 + 0.14898T. \tag{4.92}$$

The same procedure was performed to obtain a fit of h_w, which was also assumed to vary as a function of surface temperature, i.e.

$$h_w = h_{0_w} + a_w T, \tag{4.93}$$

where T is temperature in degrees Celsius and h_{0_w} and a_w fulfill analogous roles to the dry parameters h_{0_d} and a_d. Values of h_{0_w} and a_w were obtained for each one-year data set by least-squares fit, but unlike their analogous dry parameters, the resulting values were not consistent across data sets. The local mean values of h_w varied from 8.6 to 11.5 km and the Hopfield model is not a fully satisfactory way to predict the wet component of tropospheric delay.

Despite this drawback, the Hopfield model still remains useful in many instances because the total tropospheric delay is dominated by the more stable, dry component. Approximately 90% of the total delay can be attributed to the dry term, so the error introduced by the high variance of the wet term can often be ignored in many applications.

4.4 U.S. Standard Atmosphere

Up to this point in this chapter, we have considered specialized models which are tailored to estimate a particular quantity, namely tropospheric group delay. In certain cases, a more general and more sophisticated model is required. Thus it is often desirable to have a model of the vertical profiles of atmospheric conditions which can be used for purposes such as aircraft performance considerations, altimeter calibrations, or calculation of GNSS signal paths (e.g. via ray tracing). The U.S. Standard Atmosphere is one such model which provides vertical distributions of temperature, pressure, and density from sea level up to an altitude of 1000 km. Several other reference atmospheric models also exist, such as the NASA Global Reference Atmospheric Model and the International Standard Atmosphere (ISA) published by the International Organization for Standardization as ISO Standard 2533:1975. The ISA and the U.S. Standard Atmosphere are both static models that follow very similar assumptions and methodology, and are indeed identical for altitudes up to 32 km. Due to its widespread use, we will focus here on the U.S. Standard Atmosphere. In particular, this section will provide a brief overview of the most recent iteration of the model available (published in 1976). A reader interested in a more detailed discussion is encouraged to consult the original technical document [69].

4.4.1 Model Assumptions

The U.S. Standard Atmosphere uses several simplifying assumptions in its formulation. These assumptions will be briefly discussed next.

Equilibrium Assumptions
The air is assumed to be dry, that is the contribution of water vapor to quantities such as pressure is not considered. Below an altitude of 86 km, atmospheric gases are assumed to be mixed homogeneously and are treated as an ideal gas. That is, their pressure at a given point in the atmosphere is given by:

$$P = \frac{\rho R T}{M} = \frac{N R T}{N_A},$$ (4.94)

where ρ and T are the total density and temperature, respectively, at that same point of the atmosphere. R is the ideal gas constant, M is the mean molecular weight, N is the total number density of molecules, and N_A is Avogadro's number. Alternatively, P can be given as a sum of partial pressures of each molecular species:

$$\boxed{P = \sum_i P_i = \sum_i n_i k T,}$$ (4.95)

where P_i is the partial pressure of the ith molecular species, n_i is the number density of the ith species, and k is Boltzmann's constant. Within the region of complete mixing, pressure is assumed to be in hydrostatic equilibrium and horizontally stratified. That is, the differential change in pressure with respect to altitude is:

$$dP = -g\rho \, dz, \tag{4.96}$$

where dP is the differential pressure, g is the (altitude-dependent) acceleration due to gravity, and dz is the differential of geometric (as opposed to geopotential) altitude. Note that, using Equations 4.94 and 4.96, we get:

$$\frac{dP}{P} = -\frac{gM}{RT} dz. \tag{4.97}$$

Also note that since $d\ln(P)/dP = 1/P$, we see that the differential change in the logarithm of pressure is equivalent to Equation 4.98:

$$d\ln(P) = \frac{dP}{P} = -\frac{gM}{RT} dz. \tag{4.98}$$

This model of pressure is sufficient below 86 km, but at higher altitudes, the assumption of hydrostatic equilibrium breaks down as molecular transport phenomena become important. Under these circumstances, the vertical flux of molecules must be considered. One way to do so is to use the *mass continuity equation* (or simply continuity equation), which is essentially just a way of stating the conservation of mass. In particular, the sum of all vertical molecular fluxes must be zero, since mass cannot be created nor destroyed. For a given molecular species i, this equation takes the following form:

$$n_i v_i + D_i \left(\frac{dn_i}{dz} + \frac{n_i(1 + \alpha_i)}{T} \frac{dT}{dz} + \frac{gn_iM_i}{RT} \right)$$
$$+ K \left(\frac{dn_i}{dz} + \frac{n_i}{T} \frac{dT}{dz} + \frac{gn_iM}{RT} \right) = 0, \tag{4.99}$$

where v_i is the vertical transport velocity of species i, D_i is the (altitude-dependent) molecular diffusion coefficient of species i diffusing through N_2, α_i is the thermal diffusion coefficient of species i, M_i is the molecular weight of species i, M is the mean molecular weight of the gas that species i is diffusing through, and K is the (altitude-dependent) eddy diffusion coefficient. Further discussion of these parameters can be found in the full technical document on the U.S. Standard Atmosphere [69]. However, one fact worth noting is that at altitudes below about 90 km, $K \gg D_i$ and $K \gg v_i$ for each species i in the atmosphere. Thus Equation 4.99 reduces to:

$$K \left(\frac{dn_i}{dz} + \frac{n_i}{T} \frac{dT}{dz} + \frac{gn_iM}{RT} \right) = 0. \tag{4.100}$$

Note that K can be divided out and the resulting equation can be rearranged to obtain:

$$\frac{\mathrm{d}n_i}{\mathrm{d}z} + \frac{n_i}{T}\frac{\mathrm{d}T}{\mathrm{d}z} = -\frac{gn_iM}{RT}, \tag{4.101}$$

which, after some rearrangement becomes:

$$\frac{\mathrm{d}n_i}{n_i} + \frac{\mathrm{d}T}{T} = -\frac{gM}{RT}\mathrm{d}z. \tag{4.102}$$

Also, note that from the definition of P_i shown in Equation 4.95, we can see that:

$$\mathrm{d}P_i = kT\,\mathrm{d}n_i + n_ik\,\mathrm{d}T \tag{4.103}$$

and so:

$$\frac{\mathrm{d}P_i}{P_i} = \frac{k(T\mathrm{d}n_i + n_i\mathrm{d}T)}{n_ikT} = \frac{\mathrm{d}n_i}{n_i} + \frac{\mathrm{d}T}{T}. \tag{4.104}$$

Therefore, we see that Equation 4.102 is equivalent to Equation 4.98, which tells us that the continuity equation reduces to our simplified model at lower altitudes.

Geopotential Altitude

Acceleration due to Earth's gravity varies with respect to latitude, longitude, and altitude. However, in equations where acceleration due to gravity, g, plays a role, it is usually desirable to be able to treat g as a constant value. For small changes in latitude, longitude, and altitude, g changes only negligibly, so it is acceptable to indeed treat it as a constant. However, over larger distances this approximation breaks down. Instead, it is convenient to define a gravity-adjusted measure of height, known as geopotential height.

In contrast to geometric height, which is the normal notion of altitude measured in meters, geopotential height is measured in terms of geopotential meters (m′). Geopotential meters are defined in terms of the work. Let W_1 be the amount of work required to move a unit mass ($m = 1$) one geometric meter through a region of space ($\Delta z = 1$ m) where the local acceleration due to gravity is uniformly $g_0 = 9.80665$ m/s^2 (the so-called "standard gravity" value, which is measured at the mean sea level). Then W_1 is given by:

$$W_1 = mg_0\Delta z = g_0. \tag{4.105}$$

Consider some other region where the local gravity g'. In this region, if $g' \neq g_0$, the work done to move a unit mass one geometric meter will not equal g_0:

$$W_2 = mg'\Delta z = g'. \tag{4.106}$$

We will define geopotential meters such that moving a unit mass through one geopotential meter ($\Delta H = 1$ m′) takes work exactly equal to g_0. That is:

$$W_2' = mg'\Delta H = g_0. \tag{4.107}$$

From this requirement, we see that one geopotential meter is given by:

$$\frac{W_2'}{W_1} = \frac{mg'\Delta H}{mg_0 \Delta z} = \frac{g_0}{g_0} \Rightarrow \Delta H = \frac{g_0}{g'}\Delta z. \tag{4.108}$$

This has the effect of making geopotential meters smaller than geometric meters if $g' > g_0$, larger than geometric meters if $g' < g_0$, and the same size as meters if $g' = g_0$. This is illustrated in Figure 4.3.

The so-called unit geopotential is given by $g_0' = 9.80665$ m/(s² m').

Since gravity is a conservative force, it can be expressed as the gradient of some potential function Φ:

$$\vec{F}_g = -\nabla\Phi, \tag{4.109}$$

where $\vec{F}_g$ is the force of gravity at some point. We can also define the gravity per unit mass (which is also a vector field) as:

$$\vec{g} = -\nabla\Phi. \tag{4.110}$$

The value of gravity at a given point is equal to the magnitude of this vector:

$$g = |\vec{g}| = |\nabla\Phi|. \tag{4.111}$$

A potential function defines equipotential surfaces along which the function remains constant. In the direction of a vector normal to an equipotential surface, the gradient can be expressed (in geometric meters) as:

$$\frac{d\Phi}{dz} = mg \Rightarrow d\Phi = mg \, dz. \tag{4.112}$$

Therefore:

$$\Phi = \int_0^{z_0} g \, dz. \tag{4.113}$$

The geopotential altitude H is defined as the height of any point above mean sea level (which is assumed to have zero potential), measured in geopotential meters.

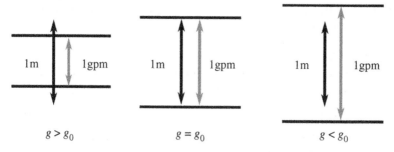

Figure 4.3 Illustration of how the size of geopotential meters (gpm) compares to geometric meters under various local gravity values.

H is given by the equation:

$$H = \frac{\Phi}{g_0'} = \frac{1}{g_0'} \int_0^{z_0} g \, dz.$$ (4.114)

Note that in order to transform geometric altitude into geopotential altitude, we need to know the dependence of g on geometric altitude. From the inverse square law, we know that on Earth's surface, $g_0 = GM_E/r_0^2$, where G is the gravitational constant, M_E is the mass of the Earth, and r_0 is the radius of the Earth. If we consider the ratio of the gravity g at some altitude z to the surface gravity, we get:

$$\frac{g}{g_0} = \frac{\dfrac{GM}{(r_0 + z)^2}}{\dfrac{GM}{r_0^2}} = \frac{r_0^2}{(r_0 + z)^2}.$$ (4.115)

Then, in order to evaluate the integral in 4.114, we can let $u = r_0 + z$ and $du = dz$ so that the integral becomes:

$$H = \frac{g_0}{g_0'} \int_{r_0}^{r_0 + z_0} \frac{r_0^2}{u^2} du = -\frac{g_0}{g_0'} \frac{r_0^2}{u} \Big|_{r_0}^{r_0 + z_0}.$$ (4.116)

Evaluating this expression at the endpoints of the integral gives:

$$H = -\frac{g_0}{g_0'} \left(\frac{r_0^2}{r_0 + z_0} - \frac{r_0^2}{r_0} \right) = \frac{g_0}{g_0'} \left(\frac{r_0 z_0}{r_0 + z_0} \right).$$ (4.117)

We also see that the differential of Equation 4.114 is given by[3]:

$$\boxed{g_0' \, dH = g \, dz.}$$ (4.118)

An example of where geopotential altitude is useful is in Equation 4.98. If Equation 4.98 were integrated in terms of geometric meters, g would not be constant with respect to altitude, and so could not be pulled out of the integral. To simplify calculations, it is preferable to substitute Equation 4.118 in to obtain:

$$d \ln(P) = \frac{dP}{P} = -\frac{g_0' M}{RT} dH.$$ (4.119)

Note that g_0' is constant with respect to geopotential altitude and can be pulled out of an integral, simplifying the calculation.

---

3 Note that this is equivalent to what we showed in Equation 4.114 (in the limit), since g_0/g' and g/g_0' are equivalent. The difference is that the first sets a reference point g_0 using geometric meters while the second uses a reference point g_0' using geopotential meters. In the first case, the local gravity g' is measured using geopotential meters, whereas in the second case, local gravity g is measured using geometric meters.

Mean Molecular Mass

As seen Equation 4.119 and similar equations in the preceding discussion, the mean molecular mass M must be known to carry out pressure computations. By definition, this mean mass is given by:

$$M = \frac{\sum_i n_i M_i}{\sum_i n_i},$$
(4.120)

where n_i is the number density of the ith species, and M_i is the ith species' molecular mass. Below an altitude of 86 km, constituent molecules of the atmosphere are assumed to be mixed homogeneously and transport processes such as diffusion are negligible. Therefore, the fraction of the total number density N which each species contributes remains the same throughout the entire region. Let F_i denote the fraction of the ith species at sea level, i.e.:

$$n_i = F_i N(z).$$
(4.121)

Table 4.1 shows values of F_i and M_i at sea level.

In the region below 86 km, the mean molecular mass is predicted to be:

$$M = M_0 = \frac{\sum_i F_i N(z) M_i}{\sum_i F_i N(z)},$$
(4.122)

Table 4.1 Molecular masses and fractional volumes of molecular species in the atmosphere at sea level.

Gas species	Molecular mass M_i (kg/kmol)	Fractional volume F_i (dimensionless)
N_2	28.013,4	0.780,84
O_2	31.998,8	0.209,476
Ar	39.948	0.009,34
CO_2	44.009,95	0.000,314
Ne	20.183	0.000,018,18
He	4.002,6	0.000,005,24
Kr	83.80	0.000,001,14
Xe	131.30	0.000,000,087
CH_4	16.043,03	0.000,002
H_2	2.015,94	0.000,000,5

Source: U.S. Standard Atmosphere [69]/National Oceanic and Atmospheric Administration/Public Domain.

and since $N(z)$ does not depend on the molecular species i, it can be factored out of both the numerator and denominator, we see that it cancels out and we obtain:

$$M = M_0 = \frac{\sum_i F_i M_i}{\sum_i F_i},$$ (4.123)

where $M_0 = 28.9655$ kg/kmol is the mean molecular mass at sea level. Note that this is a constant value, meaning that even though the total number density may vary with altitude, due to the homogeneous mixing below 86 km, the mean molecular mass remains at a constant value. However, just below 86 km, photochemical effects introduce a small amount of atomic oxygen that is not present at sea level. A different calculation is used to determine M at this height, which is outlined in Appendix A of [69]. This procedure results in a value $M_{86km} = 28.9522$ kg/kmol, which is about 0.04% less than the sea level value of M_0. To prevent a discontinuous jump in the profile of mean molecular mass, M is interpolated between M_0 at 80 km and $M_{86\ km}$ at 86 km in intervals of 0.5 km. The interpolated values can be expressed as a fraction M/M_0, and are shown in Table 4.2.

Table 4.2 Interpolated values of mean molecular mass M, given as a fraction of the mean molecular mass at sea level M_0.

Geometric height z (m)	Geopotential height H (m')	M/M_0 (dimensionless)
80,000	79,005.7	1.000,000
80,500	79,493.3	0.999,996
81,000	79,980.8	0.999,989
81,500	80,486.2	0.999,971
82,000	80,955.7	0.999,941
82,500	81,443.0	0.999,909
83,000	81,930.2	0.999,870
83,500	82,417.3	0.999,829
84,000	82,904.4	0.999,786
84,500	83,391.4	0.999,741
85,000	83,878.4	0.999,694
85,500	84,365.2	0.999,641
86,000	84,852.0	0.999,579

Source: U.S. Standard Atmosphere [69]/National Oceanic and Atmospheric Administration/Public Domain.

Molecular Scale Temperature

In a similar manner to mean molecular mass, three other quantities used in the U.S. Standard Atmosphere also exhibit slight discrepancies from their sea level values just below 86 km. These quantities, namely dynamic viscosity, kinematic viscosity, and thermal conductivity, cannot be corrected in the same manner as mean molecular mass. Instead a different approach is taken: each of these quantities requires a temperature value to be computed, so rather than correcting the quantities directly, the temperature value is corrected.

The corrected temperature, known as the molecular scale temperature, is given by:

$$T_M = T\frac{M_0}{M},\tag{4.124}$$

where M/M_0 is the same ratio given in Table 4.2. Note that with this definition, we can replace T in Equation 4.119 to obtain:

$$d\ln(P) = \frac{dP}{P} = -\frac{g_0'M_0}{RT_M}dH.\tag{4.125}$$

Note that this substitution accomplishes a similar goal to replacing geometric height z with geopotential height H in the sense that it makes the integration of pressure more tractable. In fact, if T_M is expressed as a piecewise-linear function of geopotential height, Equation 4.125 can be integrated analytically rather than numerically. To that end, T_M is represented in the U.S. Standard Atmosphere as follows:

$$\boxed{T_M = T_{M,b} + L_{M,b} \cdot (H - H_b).}\tag{4.126}$$

Here, line segments are indexed by b, which ranges from 0 to 7. Each line segment begins at geopotential altitude H_b and has a slope given by $L_{M,b}$, the lapse rate (i.e. the gradient with respect to altitude) of the molecular scale temperature. The function begins at $T_0 = 288.15\,\text{K}$, the temperature at sea level.

The resulting function, when plotted, is shown in Figure 4.4.

Kinetic Temperature

The molecular scale temperature is only defined for altitudes up to 86 km, so a different quantity is needed for greater altitudes. The so-called kinetic temperature T is the quantity shown on the right-hand side of Equation 4.124. Thus we see that by multiplying both sides of this equation by M/M_0 that below 86 km we have:

$$\boxed{T = T_M\frac{M}{M_0}.}\tag{4.127}$$

From 86 to 1000 km, T is defined differently. Just as T_M was defined as a piecewise function, so is T. Four different segments are used to define T; the first region

Figure 4.4 Piecewise-linear function representing the molecular scale temperature below 84.8520 km. Source: Adapted from U.S. Standard Atmosphere [69]/National Oceanic and Atmospheric Administration/Public Domain.

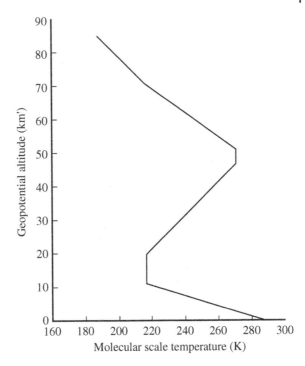

extends from $z_7 = 86$ km to $z_8 = 91$ km, the second region extends from z_8 to $z_9 = 110$ km, the third layer from z_9 to $z_{10} = 120$ km, and the fourth from z_{10} to $z_{12} = 1000$ km. Each segment of this piecewise function is based on a fit to observed data, e.g. from satellite measurements. We will discuss the definition of T for each segment in turn.

86–91 km: Throughout this region, the temperature remains constant, so

$$T = T_7 = 186.8673 \text{ K.} \tag{4.128}$$

This value is obtained by multiplying the value of T_M at 86 km (which equals 84.8520 km') by the ratio M/M_0 at that height. Using Equation 4.126 along with the values in Table 4.3, we obtain that $T_{M,7} = 186.946$ K. From Table 4.2, we see that $M/M_0 = 0.999579$. Multiplying this by $T_{M,7}$ produces the value given for T_7 above. Note that constant temperature implies that the lapse rate is zero throughout this region, i.e.

$$\frac{dT}{dz} = 0.0 \text{ K/km.} \tag{4.129}$$

91–110 km: In this region, the kinetic temperature is given by a segment of an ellipse, as defined by the following equation:

$$T = T_c + A\left[1 - \left(\frac{z - z_8}{a}\right)^2\right]^{\frac{1}{2}}, \tag{4.130}$$

where $T_c = 263.1905\,\mathrm{K}$, $A = -76.3232\,\mathrm{K}$, and $a = -19.9429\,\mathrm{km}$. The derivation of this equation is shown in Appendix B of [69]. In this region, the lapse rate is given by:

$$\frac{\mathrm{d}T}{\mathrm{d}z} = -\frac{A}{a}\left(\frac{z - z_8}{a}\right)\left[1 - \left(\frac{z - z_8}{a}\right)^2\right]^{-\frac{1}{2}}. \tag{4.131}$$

110–120 km: Here, T is defined by a linear function:

$$T = T_9 + L_9(z - z_9), \tag{4.132}$$

where $T_9 = 240\,\mathrm{K}$, which can be determined from the previous segments' equations for T, the lapse rate is $L_9 = 12.0\,\mathrm{K/km}$ and $z_9 = 110\,\mathrm{km}$. Note that in this region, the lapse rate is constant, i.e.

$$\frac{\mathrm{d}T}{\mathrm{d}z} = L_9 = 12.0\ \mathrm{K/km}. \tag{4.133}$$

120–1000 km: Above 120 km, the kinetic temperature is modeled to decrease according to the following exponential relationship:

$$T = T_\infty - (T_\infty - T_{10})\exp(-\lambda\xi(z)), \tag{4.134}$$

where

$$\lambda = \frac{L_{K,9}}{T_\infty - T_{10}} = 0.018,75 \tag{4.135}$$

and

$$\xi(z) = (z - z_{10})\left(\frac{r_0 + z_{10}}{r_0 + z}\right). \tag{4.136}$$

In these expressions, $T_\infty = 1000\,\mathrm{K}$ and r_0 is the Earth's radius. Using these same definitions, we can obtain the lapse rate:

$$\frac{\mathrm{d}T}{\mathrm{d}z} = \lambda(T_\infty - T_{10})\left(\frac{r_0 + z_{10}}{r_0 + z}\right)^2\exp(-\lambda\xi(z)). \tag{4.137}$$

Together, all of the segments of T combine to produce a temperature profile that looks like the one illustrated in Figure 4.5.

4.4.2 Computational Equations

As discussed briefly in Section 4.4.1, the U.S. Standard Atmosphere is divided into two different altitude regions; in particular, the region from 0 to 86 km, and the region from 86 to 1000 km. The equations used to compute various quantities differ depending on the region in question.

Figure 4.5 Kinetic temperature vs. geometric altitude from 0 to 500 km. Source: U.S. Standard Atmosphere [69]/ National Oceanic and Atmospheric Administration/Public Domain.

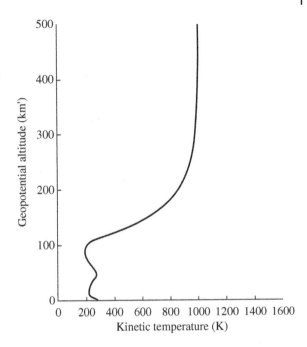

Pressure

In the region below 86 km, pressure is computed using Equation 4.98. Throughout Section 4.4.1, several simplifications were made to Equation 4.98 in order to make its integration analytically tractable. Ultimately, these simplifications resulted in Equation 4.125, which is reproduced here for convenience:

$$\mathrm{d}\ln(P) = \frac{\mathrm{d}P}{P} = -\frac{g_0' M_0}{R T_M}\mathrm{d}H. \tag{4.138}$$

Recalling that M_0 and R are constants, and that g_0' is constant with respect to geopotential altitude H, we see that integrating both sides results in:

$$\int \mathrm{d}\ln(P) = -\frac{g_0' M_0}{R} \int \frac{1}{T_M}\mathrm{d}H. \tag{4.139}$$

Below 86 km, T_M is a piecewise-linear function so each segment must be integrated separately. For a given segment of index b, Equation 4.126 can be substituted in for T_M, resulting in:

$$\int_{P_b}^{P} \mathrm{d}\ln(P) = -\frac{g_0' M_0}{R} \int_{H_b}^{H} \frac{1}{T_{M,b} + L_{M,b} \cdot (H - H_b)}\mathrm{d}H, \tag{4.140}$$

Table 4.3 Heights and lapse rates of each segment of the piecewise-linear function representing the molecular scale temperature between 0 and 84.8520 km'.

Segment index b	Geopotential height H (km')	Lapse rate $L_{M,b}$ (K/km')
0	0	−6.5
1	11	0.0
2	20	1.0
3	32	2.8
4	47	0.0
5	51	−2.8
6	71	−2.0
7	84.8520	

Source: U.S. Standard Atmosphere [69]/National Oceanic and Atmospheric Administration/Public Domain.

where $T_{M,b}$ and $L_{M,b}$ are given in Table 4.3. Note that the integral of the right-hand side depends on whether the lapse rate $L_{M,b}$ is nonzero. For segments in which $L_{M,b} = 0.0$, T_M simplifies to simply $T_M = T_{M,b}$, a constant value. Thus Equation 4.140 simplifies further to:

$$\int_{P_b}^{P} d\ln(P) = -\frac{g_0' M_0}{R T_{M,b}} \int_{H_b}^{H} dH.$$

(4.141)

If we carry out this integration, we see that:

$$\ln(P) - \ln(P_b) = -\frac{g_0' M_0}{R T_{M,b}}(H - H_b).$$

(4.142)

Simplifying further, this becomes:

$$P = P_b \exp\left(-\frac{g_0' M_0}{R T_{M,b}}(H - H_b)\right).$$

(4.143)

Returning to Equation 4.140 and instead assuming that $L_{M,b}$ is nonzero, performing the same integration gives:

$$\ln(P) - \ln(P_b) = -\frac{g_0' M_0}{R}\left[\frac{\ln(T_{M,b} + L_{M,b} \cdot (H - H_b))}{L_{M_b}} - \frac{\ln(T_{M,b})}{L_{M,b}}\right].$$

(4.144)

Distributing the negative sign on the right-hand side and simplifying, this results in:

$$\ln(P) - \ln(P_b) = \frac{g'_0 M_0}{R L_{M,b}} \left[\ln(T_{M,b}) - \ln(T_{M,b} + L_{M,b} \cdot (H - H_b)) \right]. \quad (4.145)$$

Applying several logarithm identities, we obtain:

$$\ln\left(\frac{P}{P_b}\right) = \ln\left(\left[\frac{T_{M,b}}{T_{M,b} + L_{M,b} \cdot (H - H_b)} \right]^{\frac{g'_0 M_0}{R L_{M,b}}} \right), \quad (4.146)$$

which may be further simplified to:

$$\boxed{ P = P_b \left[\frac{T_{M,b}}{T_{M,b} + L_{M,b} \cdot (H - H_b)} \right]^{\frac{g'_0 M_0}{R L_{M,b}}}. } \quad (4.147)$$

Thus together Equations 4.143 and 4.147 fully describe the computational model for pressure below 86 km.

As discussed in Section 4.4.1, the pressure relationship given by Equation 4.98 is only valid below 86 km because molecular transport phenomena can be ignored. From 86 to 1000 km, effects such as diffusion must be taken into account and instead Equation 4.95 must be used to determine pressure. Note that the number density of molecular species i does not remain constant with respect to altitude, so computational equations describing their vertical profiles must be provided with the model.

Number Density, Mass Density, and Mole Volume

Below 80 km, turbulent mixing causes the fractional concentrations of the molecular species in the atmosphere to remain constant. The number density of species i given in terms of its fractional concentration F_i of the total number density $N(z)$. Equation 4.121, which relates these quantities, is reproduced here:

$$n_i = F_i N(z). \quad (4.148)$$

Recall that the values of F_i are given in Table 4.1. From 80 to 86 km, these fractions must be corrected slightly by the value of the ratio M/M_0, whose values are given in Table 4.2.

In the region from about 86 to 120 km, due to molecular transport processes such as oxygen dissociation and diffusive separation, n_i must instead be given by more complex equations. In principle a system of equations, each of the form given by Equation 4.99, could be solved simultaneously to obtain the number densities of all major species above 86 km. However, this system of equations is coupled because the diffusion parameters D_i depend on the number densities of all other species.

To solve such a coupled system of equations would be computationally expensive, so instead the U.S. Standard Atmosphere makes a series of simplifications in order to obtain approximate values for each species' number density one at a time. We will not discuss this computational scheme in detail here, but in brief the idea is to compute molecular densities in the order: $n(N_2)$, $n(O)$, $n(O_2)$, $n(Ar)$, $n(He)$, and $n(H)$. The model starts with N_2 because its distribution is closest to that of static equilibrium. From there, each n_i is computed assuming the previous species as the background gas through which species i diffuses. For details on this process, the reader is encouraged to consult Section 1.3.2 of [69].

Above 120 km, it can be assumed that there is no longer any large-scale oxygen dissociation and the atmosphere is roughly in diffusive equilibrium. Because of these changes in atmospheric dynamics, the molecular diffusion calculations simplify and are no longer tightly coupled. This mean that computation of each n_i can be performed independently, which is much less computationally expensive and is comparable to the computation of pressure or density below 80 km.

The ideal approach to computing total number density in each of these altitude regions (below 86 km, between 86–120 km, and above 120 km) differs depending on the validity and ease of computing the parameters it depends on in each region. Thus, we can see that the equivalent expressions for N:

$$N = \frac{M_0}{M}\frac{N_A P}{T_M R} = \frac{N_A P}{TR} = \sum_i n_i, \tag{4.149}$$

are each useful in a different region. In particular, the expression containing T_M is valid below 86 km, while the two rightmost expressions are both useful above 86 km and either one may be used depending on whether T or $\sum n_i$ is more convenient or computationally efficient. Throughout this section, when multiple forms of a parameter are provided, the same rationale holds true. For example, the total mass density can be obtained from multiple expressions:

$$\rho = \frac{M_0 P}{T_M R} = \frac{MP}{TR} = \frac{\sum_i n_i M_i}{N_A}. \tag{4.150}$$

From this, the mole volume $V_m = M/\rho$ can also be determined from multiple expressions:

$$V_m = \frac{M T_M R}{M_0 P} = \frac{TR}{P} = \frac{N_A}{\sum_i n_i}. \tag{4.151}$$

4.4.3 Data Sources and Implementation

The analytical equations presented in the preceding sections are based on observed temperature and density profiles obtained from satellite, rocket, and radar measurements. In particular, below 86 km temperature data was used from eight rocket launch sites whose locations range from 6 °S to 71 °N in latitude and from 157 °W to 137 °E in longitude. At each location and at various altitudes, a mean monthly temperature value was obtained (or interpolated in the absence of available data for a given month). These monthly averages were themselves averaged to obtain an arithmetic mean temperature $\overline{T}$. Since temperature exhibits both annual and semi-annual variation, the variation in temperature over the course of a year was modeled using harmonic analysis. In particular, the regression equations are:

$$T = \overline{T} + A_1 \sin(ix + \phi_1) + A_2 \sin(2ix + \phi_2), \qquad (4.152)$$

for values of i ranging from 0 to 11. Here, $x = 360°/\text{period}$ and $i = 0$ corresponds to 15 January. The regression coefficients A_1 and A_2 are determined by best fit to the observed data. From these temperature curves, annual mean temperature values are then computed at each location and each altitude. The resulting temperature profile was then approximated using a piecewise-linear function, as discussed in Section 4.4.1 (under the Molecular Scale Temperature heading).

In a similar fashion, the temperature profile above 86 km was measured and a set of analytical equations was chosen to provide a good fit to the observed data. In particular, this equation was designed to fit the criteria below to make it useful and easy to work with:

1. The temperature profile must be a smooth function of geometric altitude with continuous first derivative.
2. The functions comprising the temperature profile must be flexible so that they may be adjusted to fit multiple data sets.

With these goals in mind, the equations presented in Section 4.4.1 (under the Kinetic Temperature heading) were chosen.

Above 86 km, molecular number densities must also be measured because effects such as oxygen dissociation and diffusion cause the density profiles of different molecular species to vary significantly. This is not necessary below 86 km because turbulent mixing causes the density profile of each species to remain approximately constant. At high altitudes, molecular density data is obtained via sources such as rocket-borne mass spectrometers, incoherent scatter radar measurements, and satellite drag data. This data, in conjunction with theoretical constraints such as the continuity equation, allow the molecular densities to be expressed using physically meaningful analytical equations. These equations are complex and are not discussed in this book, but may be found in the U.S. Standard Atmosphere 1976 reference document [69]. In addition, further examination of the data sources and model validation may be found in Part 2 of the same document.

The U.S. Standard Atmosphere has been implemented in a number of programming languages, including Fortran, Pascal(Delphi), C, C++, Fortran77, IDL, Python, QBasic Java, and JavaScript [70].

5

Physics of the Ionosphere

5.1 Introduction

There are many sources of error that degrade the performance of global navigation satellite system (GNSS) positioning, a number of which were discussed in Section 2.3. The most significant among these error sources are those due to the ionosphere [27]. For this reason, researchers have extensively studied ionospheric phenomena in an effort to better understand and predict their effects on GNSS performance.

Mitigation of these deleterious effects poses a serious engineering problem. In order to solve it, one must answer several questions. For example, what are the physical mechanisms that give rise to ionospheric error? How do we quantify their impact on GNSS positioning? What methods exist to reduce this impact? Presently, one of the most effective approaches to error mitigation is the use of computational models. So one may also ask: which computational models are available, and what is their relative efficacy in reducing GNSS error? The next few chapters will address all of these questions.

In order to fully understand ionospheric propagation effects, it is essential to first understand the nature of the ionosphere itself. In this chapter, we will present several physical derivations from first principles. Together, they comprise a basic framework for understanding the structure and behavior of the ionosphere. These foundational concepts will provide the reader with the background necessary to understand the subsequent chapters of this book.

The parameter of greatest interest in characterizing the ionosphere is electron density – i.e. the number of free electrons per unit volume. Gas molecules in the upper atmosphere are ionized when high energy photons or charged particles collide with them and knock one of their valence electrons loose. Nearly all of these highly charged particles originate from the Sun, so this chapter begins with a discussion of how these particles are generated and the path they take to reach the Earth.

Tropospheric and Ionospheric Effects on Global Navigation Satellite Systems, First Edition.
Timothy H. Kindervatter and Fernando L. Teixeira.

During the day, the change in electron density with respect to time is dominated by solar ionization. However, there are competing mechanisms which reduce electron density as well. Section 5.3 provides a brief overview of the physical and chemical processes that give rise to these competing effects. Electron density in the ionosphere also depends strongly on height. Sydney Chapman, one of the pioneers of ionospheric science, was the first to present a mathematical model of this relationship. Section 5.4 presents a derivation of his model.

Although Chapman's model was an important first step in our understanding of the ionosphere, it was far from complete. At large altitudes, the movement of free electrons significantly affects electron density – a fact that Chapman's model fails to account for. For example, electrons may be blown around by wind, or they may respond to the Earth's magnetic field. These so-called transport effects are responsible for a number of anomalous phenomena that are relevant to radiowave propagation, many of which are discussed in Section 6.4. Due to their importance, thorough derivations of a few major transport effects are presented in Section 5.5.

5.2 Solar-Terrestrial Relations

The Sun is by far the most prominent source of ionization in the Earth's atmosphere. Photons from across the electromagnetic spectrum are produced by the Sun, and those with sufficiently high energy are capable of knocking electrons loose from gas particles in the atmosphere, ionizing them. The Sun also ejects high energy particles (such as high-velocity electrons and protons), which are also capable of ionizing atmospheric gases in a process called *corpuscular ionization*.

In order to provide context for the ionization processes which form the cornerstone of this chapter, we will briefly discuss some of the important aspects of the solar-terrestrial environment. The regions of interest along the path between the Sun and the Earth are: the Sun, the interplanetary medium, the Earth's magnetosphere, and the Earth's atmosphere. Each of these regions plays a role in the production and transmission of electromagnetic waves and energetic particles from the Sun to the Earth. These topics are incredibly deep and complex in and of themselves, so only a cursory examination of them will be provided. The interested reader may consult [71] for further details.

5.2.1 The Sun

Nuclear Fusion
The Sun is classified as a main sequence star – middling in size, temperature, and luminosity as compared with other stars. It has a radius of 695,990 km, a mass of 1.989×10^{30} kg, and is made up almost entirely of Hydrogen (70%)

and Helium (28%). At its surface, it has a temperature of 5770 K, and acts as a blackbody radiator, appearing mostly yellow to the human eye. The total luminosity of the Sun is 3.846×10^{33} erg/s (1 erg/s = 10^{-7} W). The Earth orbits the Sun on a slightly elliptical path, its distance ranging from about 147.100×10^6 to 147.157×10^6 km [72].

Due to the enormous mass of the Sun, gravitational compression near the center is extreme, resulting in a core which is incredibly hot and densely packed. Models estimate the gravitational pressure of the core to be 2.477×10^{16} Pa, resulting in a density of 1.622×10^5 kg/m^3 and a temperature of 1.571×10^7 K [73]. Even at this temperature, the kinetic energy of the positively charged atomic nuclei in the core is insufficient to overcome the energy barrier produced by their Coulombic repulsion. However, due to the uncertainty principle, there is some nonzero probability for quantum tunneling to occur, allowing the nuclei to pass through this energy barrier [74]. If one nucleus tunnels to within a close enough distance of another, the strong nuclear force overpowers the Coulomb force and nuclear fusion occurs.

In the core of the Sun, atoms (mostly hydrogen) are stripped of their electrons entirely, leaving a roiling plasma of free protons and free electrons. Arguably the most significant type of fusion reaction that occurs in the Sun is the proton–proton chain reaction. This reaction begins when two Hydrogen nuclei (i.e. free protons) fuse to form a diproton (two protons, zero neutrons), a very unstable isotope of helium. The diproton immediately undergoes beta decay, releasing a positron and a neutrino, and producing Deuterium (one proton and one neutron) in the process. This is represented by the equation:

$$H^1 + H^1 \rightarrow D^2 + e^+ + \nu, \tag{5.1}$$

where H^1 is a Hydrogen atom, D^2 is Deuterium, e^+ is a positron, and ν is a neutrino. The newly formed Deuterium can also fuse with a Hydrogen atom to produce Helium-3 and a gamma ray:

$$D^2 + H^1 \rightarrow He^3 + \gamma. \tag{5.2}$$

Finally, two Helium-3 nuclei can fuse to produce a Helium-4 nucleus and two Hydrogen nuclei:

$$He^3 + He^3 \rightarrow He^4 + H^1 + H^1. \tag{5.3}$$

The released Hydrogen nuclei complete the cycle, providing more opportunities for the cycle to begin again with Equation 5.1. The whole proton–proton chain produces 26.732 MeV of energy, some of which is emitted as gamma rays, some as neutrinos and the rest becomes kinetic energy which creates the high temperatures of the core.

Whenever a positron is released by a reaction (such as in Equation 5.1), it will immediately annihilate with an electron, releasing two gamma rays in the process, each with the mass energy of an electron (which is equivalent to that of a positron): 511 keV.

$$e^+ + e^- = \gamma + \gamma. \tag{5.4}$$

As evidenced by Equations 5.2 and 5.4 (and also as the result of other reactions which were not discussed), gamma rays are constantly produced by nuclear fusion in the Sun's core. These rays will make their way to the surface of the sun, but along the way, the photons scatter off of other particles, losing some energy with each collision. The direction in which the photon scatters is random, and thus the photon will make a random walk from the core all the way to the surface of the Sun. Since the core of the Sun is so dense, the *mean free path*, or the distance it can travel before another collision occurs, is very small. Thus the photon is constantly being randomly redirected, massively lengthening the time it takes to reach the surface.

If the photon were able to travel in a straight line, it would reach the surface in 2.3 seconds, but due to the constant redirection, it actually takes thousands of years to complete its journey. During this time, the photon is continually losing energy via collisions, and since the energy of a photon is directly proportional to its frequency, it eventually changes from a gamma ray into a photon of lesser frequency. Depending on how long it takes to escape, the photon may reach the surface along any portion of the electromagnetic spectrum, though the majority of photons emitted by the Sun are in the visible spectrum.

The Sun's Magnetic Field

The Sun is not solid, like the Earth, but rather a cauldron of dynamic plasma. Plasma is a fluid, much like a gas or a liquid, so it is subject to the same hydrodynamic physics. The massive amount of heat produced by fusion in the Sun's core is carried away toward the surface by convective currents. The difference between plasma and other fluids, however, is that it is not electrically neutral. The convective flows of plasma are therefore continuously moving charge around, creating electric currents. These electric currents create extremely complex magnetic fields.

The exact processes by which magnetic fields arise in the Sun are the subject of ongoing research. It is known, however, that the magnetic behavior of the Sun is directly linked to its level of *activity*. Solar activity refers to the amount of electromagnetic radiation emitted by the Sun, as well as the frequency with which it ejects high energy particles. Solar activity is periodic, operating on an 11-year cycle during which the Sun becomes very active for a few years before calming down again. Astronomers have noticed that this level of activity is directly correlated with the number of sunspots (darker, cooler regions on the

Sun's surface) visible at any given time. The solar cycle and sunspot number will be discussed further in Section 6.4.

Sunspots appear in areas of high magnetic flux – in other words directly along a magnetic field line. The magnetic flux inhibits plasma convection, cooling the surrounding area and causing it to appear darker. During the quietest part of the 11-year cycle, the magnetic fields produced by the Sun are somewhat uniform. Near the equator, magnetic field lines are *closed*, meaning they loop up above the surface of the Sun before returning to the surface. Near the poles, the magnetic field lines are *open*, meaning they do not loop back to the surface, but rather extend outward into the solar system. At one of the Sun's poles the magnetic field lines point outward, while at the other pole they point inward. Thus, during low solar activity, the Sun's magnetic field approximates a dipole whose axis is aligned with the Sun's axis of rotation.

Over the course of 5.5 years, the magnetic fields become much more complex and erratic, following no clear pattern. The dipole nature of the magnetic field is lost, with both open and closed magnetic fields scattered across all latitudes. An illustration provided by NASA's Goddard Space Flight Center, [75] shown in Figure 5.1,

Over the remaining 5.5 years of the solar cycle, the pent-up magnetic energy is gradually released, and the magnetic field begins to align itself with the rotation of the Sun once again. A new dipole configuration is formed, but this time with the opposite polarity as before. Thus, the Sun's magnetic polarity makes one full cycle every 22 years [76].

During periods of high magnetic activity, the Sun's magnetic fields are constantly reorienting themselves. Typically field lines do not break nor merge with other field lines, but during high solar activity a phenomenon called magnetic reconnection occurs wherein field lines reconfigure themselves, releasing a large amount of energy in the process. This energy is transferred to charged particles in

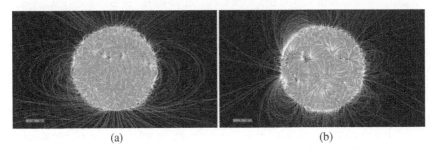

(a)   (b)

Figure 5.1 Relatively ordered magnetic field lines produced by the Sun in January 2011, shortly after the solar minimum of solar cycle 24 (a). Chaotic and disordered magnetic field lines in July 2014, during the solar maximum of solar cycle 24 (b). Source: Garner 2016 [75]/Reproduced with permission of NASA.

the form of kinetic energy, accelerating them to near the speed of light. At the time of this writing, the exact conditions under which magnetic reconnection occurs in the Sun are still unknown, but they seem to be directly correlated with the occurrence of solar flares and coronal mass ejections (CMEs) [77–80].

Solar flares are large explosions, which occur in proximity to sunspots. They produce bright flashes of electromagnetic radiation, emitting photons across the electromagnetic (EM) spectrum. This includes a large amount of ionizing radiation in the UV, X-ray, and gamma ray spectra. These high energy photons cause abnormally high levels of ionization in the ionosphere, which can be problematic for radio frequency communication links such as GNSS.

When charged particles are accelerated near the Sun's surface, they will follow magnetic field lines. If these field lines are closed, the particles will simply "leap" above the surface of the Sun before returning to the surface. These events are known as prominences or filaments. On the other hand, if the magnetic field lines are open, the particles will not return to the Sun and will instead shoot out into the solar system. This type of event is called a CME. If charged particles from a CME arrive at the Earth, they will bombard the atmosphere, causing corpuscular ionization and briefly complicating radio communications. This is known as a geomagnetic storm, and can last several hours.

5.2.2 The Interplanetary Medium

The entire Sun is a giant ball of plasma which becomes less dense farther away from the core, so it does not have a surface in the same way that the Earth does. Instead, the surface of the Sun is defined by the photosphere, which is the boundary that causes the Sun to appear as a yellow disc from Earth. Above the photosphere lies the Sun's "atmosphere," the lowest layer of which is the chromosphere extending up to a height of about 5000 km. The chromosphere is a less homogeneous region in which long, thin tendrils of luminous gas known as *spicules* and *fibrils* create a "hairy" appearance.

Above the chromosphere is the Corona, a region extending millions of kilometers above the photosphere, comprised of much hotter, but much less dense plasma than the photosphere. This is why the Corona is only visible to the naked eye during a total solar eclipse. Portions of the Corona which lie within closed magnetic field lines trap most charged particles and keep them relatively close to the Sun. Conversely, areas with open magnetic field lines, called Coronal holes, allow a large number charged particles to escape out into the solar system. Coronal holes have a lower density and temperature than other potions of the corona, causing them to appear darker.

Near the surface of the Sun, charged particles move slowly, but they may be carried along magnetic field lines to higher altitudes. In the Corona, the tremendously high temperature creates pressure gradient forces much stronger than the effect of the Sun's gravity, accelerating them to supersonic speeds. Even if the magnetic field lines are closed, particles with sufficiently high velocity can escape and flow out into the solar system. In coronal holes, where the magnetic field lines are open, the magnetic force on the particles does not act to redirect them toward the Sun, so they escape with an even higher velocity.

This outflow of charged particles from the Sun is known as solar wind. Solar wind is comprised of mostly protons and electrons, each with a typical density of $5 \times 10^6 \, \text{m}^{-3}$ [81]. Since positive and negative charges are equally abundant, the solar wind as a whole is electrically neutral. Solar wind is not to be confused with CMEs – CMEs are intermittent bursts which eject particles at near-relativistic speeds and last for only a few hours. Solar wind is a continuous stream of particles which travel much slower, at about 200–1000 km/s, depending on the level of solar activity.

As solar wind extends away from the Sun, it travels along open magnetic field lines out into the solar system. Plasma is a very good conductor, which has the effect of "freezing" the magnetic field lines and the plasma together, forcing them to follow the same trajectory through the solar system. This magnetic field, known as the interplanetary magnetic field (IMF), would extend radially outward from the Sun if it were stationary. However, due to the Sun's rotation, these radial lines are dragged along in a spiral pattern, much like a jet of water as it is sprayed from

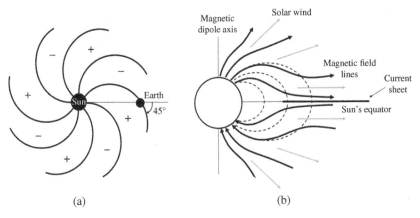

(a)   (b)

Figure 5.2 (a) Spiral pattern of the IMF viewed from above the Sun's equatorial plane. (b) Formation of current sheets viewed from within the Sun's equatorial plane.

a rotating sprinkler. Thus the IMF arrives at the Earth at an angle of 45°, which has been confirmed by experimental observation [72]. The shape of the IMF is illustrated in Figure 5.2a.

When viewed from above the equatorial plane, as in Figure 5.2a, the IMF is divided into "sectors" of alternating magnetic polarity. The reason for this is illustrated by Figure 5.2b, which displays the Sun's magnetic field from within the equatorial plane (i.e. orthogonal to the perspective of Figure 5.2a). During low solar activity, the magnetic field lines near the equator form a dipole, meaning they point away from the Sun in one (either northern or southern) hemisphere and toward the Sun in the other. Solar wind at the equator distorts this ideal dipole, causing field lines directly above and below the equator to become nearly parallel. This forms a neutral current sheet at the equator, above which the magnetic field has one polarity and below which it has the opposite polarity. This current sheet occurs in a wavy pattern, like a ballerina's tutu [72]. Thus, as the Earth orbits the Sun, it periodically passes above and below the current sheet, giving the impression of alternating sectors of magnetic polarity. The number of sectors is always even, an equal number having positive and negative polarity.

5.2.3 Earth's Magnetic Field

Earth's has its own magnetic field, which we call the geomagnetic field. The geomagnetic field arises in Earth's core, which is comprised of two layers, each consisting mainly of a nickel–iron alloy. Despite the similar chemical composition, the inner core is solid while the outer core is liquid. This is because at high enough pressures, the melting point of iron increases dramatically [82]. The solid inner core has a radius of 1220 km, and at the inner core boundary (ICB), the temperature is about 6000 K, the pressure is 330 GPa, and the density is 1.33×10^4 kg/m^3. The molten outer core sits on top of the inner core and extends to a radius of 3490 km, above which lies a layer called the mantle. At the core–mantle boundary (CMB), the temperature is about 4300 K, the pressure is 136 GPa, and the density is 1.07×10^4 kg/m^3 [83, 84].

The geomagnetic field arises from a phenomenon called the geodynamo. Hot molten metal near the ICB will become less dense and rise toward the CMB, where it transfers heat to the mantle. After losing some heat, the metal will become denser and sink again. This circulation of fluid creates a convection current. Since this molten nickel–iron alloy is conductive, its movement induces a magnetic field. Due to the Coriolis force, the movement of the metal tends to form columns which roughly align with the Earth's axis of rotation [85]. Thus the geomagnetic field may be approximated as a dipole whose poles are near the geographic poles of the Earth.

Due to the dynamic nature of the geodynamo, the poles of the geomagnetic field are not constant and typically wander over time, though not very far. The axis of the geomagnetic dipole is inclined with respect to the Earth's axis of rotation by about 11°. If the dipole is imagined as a bar magnet, the south end of the magnet actually lies at the north magnetic pole, and vice versa. In other words, magnetic field lines point out of the southern hemisphere and loop around, pointing down into the northern hemisphere. This can be seen in Figure 5.3. There is evidence of the geomagnetic field spontaneously reversing its polarity; this appears to occur randomly, though very infrequently. The last reversal occurred 780,000 years ago and on average reversals only occur every 200,000–300,000 years [86], so it is not an important consideration for engineering purposes.

A brief note: in the literature, the terms "north magnetic pole" and "south magnetic pole" are used frequently. In these terms, north and south refer to the *hemisphere* that the magnetic pole lies in. That is, the north magnetic pole is in the northern hemisphere and the south magnetic pole is in the southern hemisphere. The reader should take care not to confuse this with the usage of north and south in the context of a bar magnet. The bar magnet terminology will not be used in this book going forward, so "north" and "south" will always refer to the hemisphere containing the magnetic pole.

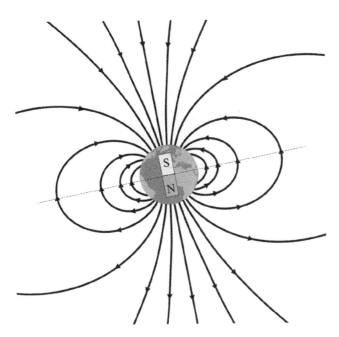

Figure 5.3 Dipole nature of the geomagnetic field.

At large scales, magnetic field lines are often described using *L-shells*. A given L-shell corresponds to the set of magnetic field lines which cross Earth's magnetic equator at a certain height. This height, measured in Earth radii, is called the L-value. For example, the L-shell corresponding to an L-value of 3 encompasses all magnetic field lines which cross the magnetic equator 3 Earth radii from the center of the Earth. Figure 5.4 depicts several L-shells. Note that the L-value need not be integer-valued; any real number may be used to define an L-shell at the corresponding number of Earth radii. Integer L-values were chosen for Figure 5.4 merely for clarity.

A given L-shell reaches its maximum altitude at the magnetic equator. At other magnetic latitudes, the same L-shell will lie at a lower altitude. This altitude can be determined via the expression

$$r = L \cos^2 \phi_m, \tag{5.5}$$

where r is the distance, measured in Earth radii, between the center of the Earth and the point on the L-shell described by the L-value L at a given magnetic latitude ϕ_m. As an example, consider a point on the $L = 3$ shell. If this point lies at a magnetic latitude of $\phi_m = 45°$, its height above the center of the earth will be $r = 3\cos^2(45) = 1.5$ Earth radii.

Locally, the magnetic field vector can be described using the coordinate system shown in Figure 5.5.

The coordinates X, Y, and Z refer to geographic north, geographic east, and down, respectively. The field vector F points along the magnetic meridian, a line connecting the north and south magnetic poles. F always points from the south magnetic pole to the north magnetic pole. H, the horizontal component of F, lies along the magnetic meridian, which makes an angle D, called the declination angle, with respect to geographic north. F dips below the horizontal by an angle I, called the inclination angle or dip angle, measured with respect to

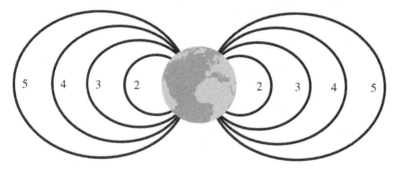

Figure 5.4 Selected L-shells of Earth's magnetic field.

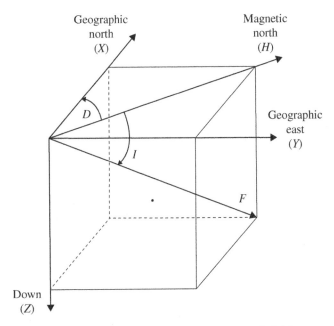

Figure 5.5 Reference frame of the local geomagnetic field.

H. Some relationships between these parameters are as follows:

$$F = \sqrt{X^2 + Y^2 + Z^2}, \quad H = \sqrt{X^2 + Y^2}, \quad F = \sqrt{H^2 + Z^2},$$
$$X = H\cos D, \qquad Y = H\sin D, \qquad X = Y\tan D, \tag{5.6}$$
$$H = F\cos I, \qquad Z = F\sin I, \qquad Z = H\tan I.$$

The seven parameters – F, X, Y, Z, H, D, and I – are called the magnetic elements, and any three of them may be used to specify F [81]. F has SI units of teslas (T), and the magnitude of F ranges from about 25,000 to 65,000 nT at the surface of the Earth. Other units used to report magnetic field intensity are gauss $(G = 10^{-4}\,T)$ and gammas $(1\gamma = 1\,nT)$.

The inclination angle can range from $-90°$ (up) to $90°$ (down). At the south magnetic pole, F points straight up. If one were to travel north along a magnetic field line, they would observe F rotating downward, reaching $0°$ at the magnetic equator. Past the magnetic equator, F continues to rotate downward, finally pointing straight down at the north magnetic pole.

The north magnetic pole is slightly displaced from the geographic north pole, and the declination angle measures the eastward deviation of magnetic north with respect to geographic north. Since a compass always points toward magnetic north, the declination angle may be thought of as the angle between the compass arrow and the direction of geographic north.

Parameters of the geomagnetic field at a given point on Earth may be modeled using the International Geomagnetic Reference Field (IGRF). Figures 5.6–5.8 show the level curves of the modeled magnetic field intensity F, inclination I, and declination D for the year 2015. The level curves for each parameter are given special names: lines of constant intensity are called *isodynamic* lines, lines of constant inclination are called *isoclinic* lines, and lines of constant declination are called *isogonic* lines.

An online mapping tool based on the IGRF was used to generate Figures 5.6–5.8. In addition to these parameters, maps of the horizontal H and vertical Z components of the field intensity may be generated. This mapping tool is available at [87], and the mathematical formulation of the current IGRF model may be found in [88].

5.2.4 The Magnetosphere

Near the surface of the Earth, the dipole approximation of the geomagnetic field remains accurate. At very large altitudes, this approximation is no longer valid because incoming solar wind significantly distorts the shape of the geomagnetic field. About 10 Earth radii (~65,000 km) upstream of the solar wind, high energy charged particles collide with the geomagnetic field creating a magnetohydrodynamic shock wave, also known as a bow shock. This has the effect of compressing the geomagnetic field on the dayside of the Earth, while on the night side, a long tail extends for over 200 Earth radii. This effect is analogous to the wake created by a stone in a stream of water.

Most of the charged particles in the solar wind are deflected away from the Earth along the bow shock in a region called the magnetosheath. Within the magnetosheath is the boundary of the geomagnetic field, called the magnetopause. Below the magnetopause is a region called the magnetosphere. A fraction of charged particles from the solar wind manage to make it into the magnetosphere, either via turbulent exchange on the dayside magnetosphere, or at the polar cusps, which may be observed in Figure 5.9.

Van Allen Belts

Within the magnetosphere, there are two torus shaped rings of energetic charged particles trapped by Earth's magnetic field. These rings are called the Van Allen radiation belts. The inner belt, which extends from an altitude of about 1000 km to about 2 Earth radii, consists mainly of electrons with energies of hundreds of keV and protons with energies up to 100 MeV. The electrons originate mostly

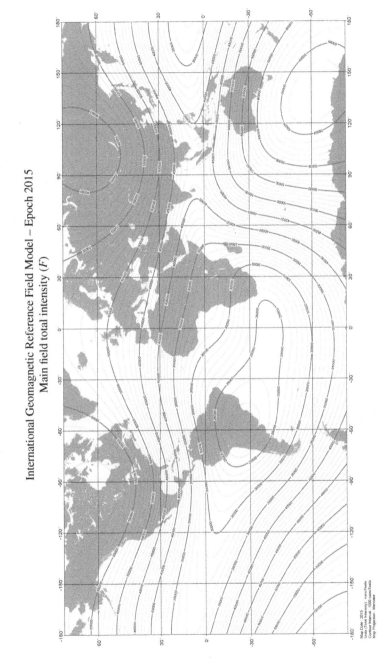

International Geomagnetic Reference Field Model – Epoch 2015
Main field total intensity (F)

Figure 5.6 IGRF isodynamic lines for 2015.

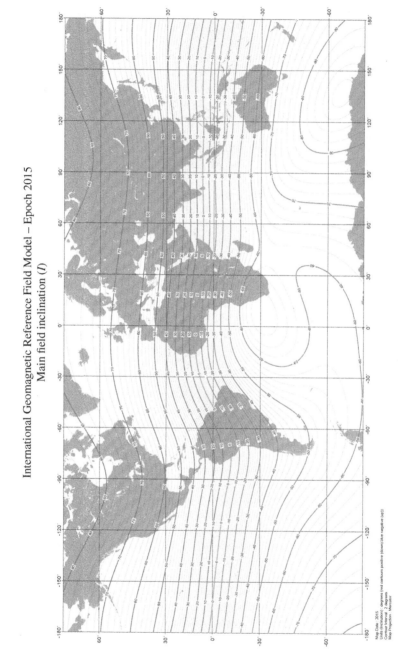

International Geomagnetic Reference Field Model – Epoch 2015
Main field inclination (*I*)

Figure 5.7 IGRF isoclinic lines for 2015.

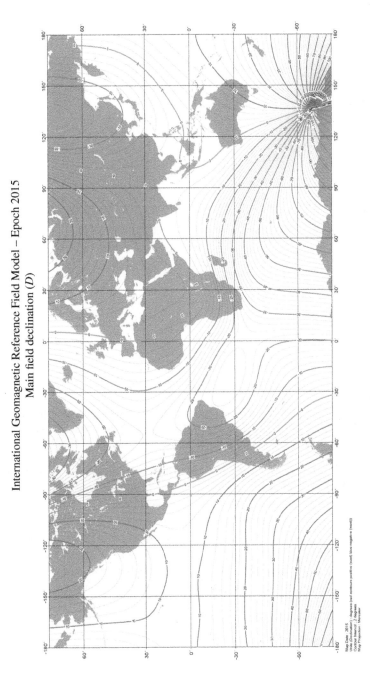

Figure 5.8 IGRF isogonic lines for 2015.

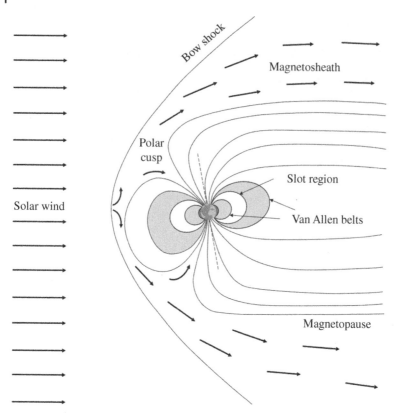

Figure 5.9 Earth's magnetosphere, compressed on the dayside by solar wind. Source: Figure 2.12 of [81]. Reproduced with permission of the Institution of Engineering and Technology (IET).

from the solar wind, while low energy protons diffuse up from the upper reaches of the ionosphere. High energy protons are thought to be created when cosmic gamma rays interact with atomic nuclei in the atmosphere, ejecting free neutrons up into the inner Van Allen belt. These neutrons have a half-life of about 12 minutes, after which they beta decay into a proton, electron and neutrino [89]. The charged particles are then trapped by the geomagnetic field and become part of the Van Allen belt. The outer Van Allen belt reaches from about 4–7 Earth radii, and is comprised mostly of MeV electrons from the solar wind.

There is a gap between the inner and outer Van Allen belts called the slot region, in which charged particle density is significantly lessened. This gap is caused by the interaction of charged particles with very low frequency (VLF) "whistler-mode" waves originating from lightning in the atmosphere [90]. In September 2012, a third distinct belt of charged particles was discovered in the range of ~3 to 3.5 Earth radii, comprised of MeV electrons. It only existed temporarily – for about four weeks – before becoming disrupted and disappearing once again [91].

Particles trapped in the Van Allen belts travel in circular patterns along the Earth's magnetic field lines. The magnetic force on a particle is equivalent to a centripetal force. In the plane perpendicular to the local magnetic field lines, this force is given by the equation $F = qvB = mv^2/r$, where B is the local strength of the magnetic field, q is the charge of the particle, and m is its mass. The particle rotates with a radius of gyration r which, via rearranging the previous equation, can be expressed as $r = mv/qB$.

As the particle approaches Earth along a field line, B increases, so r decreases and the radial component of its velocity increases. A particle's kinetic energy is split between the radial velocity component $v_\perp$, and the component along a magnetic field line $v_\parallel$. The particle energy remains constant, so as $v_\perp$ increases, $v_\parallel$ must decrease. Thus the particle will move more and more slowly along the magnetic field line until it stops entirely, and then "mirrors," reversing its direction and traveling back along the field line.

In this way, particles remain trapped, traveling back and forth between the northern and southern hemispheres. In addition to this north-south motion, protons will slowly drift westward while electrons will drift eastward. This differential motion of charge gives rise to a so-called ring current. This current induces its own magnetic field which opposes the geomagnetic field. This phenomenon is typically observed as a short term variation in the geomagnetic field. Further discussion of magnetic variations and disturbances is presented in Section 6.4.

The height at which a charged particle reverses direction is called the mirror height, and if the mirror height is less than about 100 km, the particle can be absorbed, causing corpuscular ionization of neutral molecules [81]. Ionization due to charged particles typically occurs at high latitudes, causing auroras. Appropriately, these latitudes are called auroral latitudes, and a number of additional effects, which can affect radio wave propagation arise in these regions.

The lowest portion of the magnetosphere is called the plasmasphere. This region extends from 1000 km above the Earth's surface to an upper boundary called the plasmapause, which typically lies at about 5 or 6 Earth radii. This region, which includes the inner Van Allen belt, is characterized by a cold plasma of positive ions – mostly Hydrogen ions (H^+), but also some Helium ions (He^+). Due to the large number of H^+ ions, this region is often referred to as the protonosphere. Ion density at 1000 km is on the order of 10^{10} ions/m^3, and decays gradually out to the plasmapause, where there is an abrupt drop of one or two orders of magnitude [81].

Auroral Ovals and Polar Caps

Charged particles in the magnetosphere are guided by Earth's magnetic field lines, and most of them become trapped in the Van Allen belts. Those with sufficiently high energy are able to penetrate into dense regions of the atmosphere at high altitudes, causing ionization and creating auroras. The regions in which auroral events occur are known as the auroral ovals, rings about 2000–3000 km in diameter centered on the magnetic poles [92]. The regions inside the auroral ovals are known as the polar caps.

Solar particle events, which are associated with solar flares and CMEs, send large numbers of high energy particles (primarily protons) at high speeds toward the Earth. These particles can penetrate all the way to the D layer of the ionosphere and cause abnormally high ionization, which can severely inhibit radio communications at these latitudes. This type of event is called polar cap absorption (PCA), and can last anywhere from a few hours to a few days.

Complex interactions between the solar wind, the IMF and the geomagnetic field cause a charge separation at the magnetopause, setting up an electric field known as the *magnetospheric electric convection field*, which points from dawn to dusk across the polar cap [92]. Further discussion of the origin of this electric field may be found in [93]. Due to the high conductivity of the E layer of the ionosphere (see Section 5.5.4), currents are induced by the electric field, flowing from day to night across the polar cap. At auroral latitudes these currents are especially strong, and are referred to as the auroral electrojet. These currents follow magnetic field lines and typically flow within the auroral ovals, but can extend to higher or lower altitudes during solar particle events.

Significant features of the auroral electrojet may be seen in Figure 5.10. The directed contour lines indicate the direction of current flow, while the arrow pointing straight across the polar cap represents the electric convection field. At each of the compass directions, a time in hours is listed using magnetic local time (MLT) – i.e. the position of the Sun, or solar direction, relative to the magnetic pole. Magnetic noon points toward the sun, so in this figure, the solar direction is at the top of the diagram.

The outer boundary of the figure represents a magnetic latitude of 50°, and each + mark represents a latitude increase of 10°. In other words, the + marks indicate magnetic latitudes of 60°, 70°, and 80°. The ⊚ symbol represents the magnetic pole. At low latitudes between 18 and 00 MLT, a trough of ionization exists, limiting current flow. This trough exists because ion production and transport processes

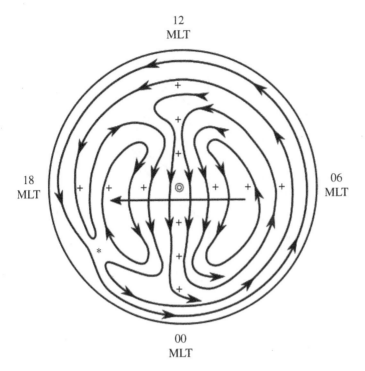

12
MLT

18
MLT

06
MLT

00
MLT

Figure 5.10 Features of the auroral ovals and polar cap. Source: Adapted from Figure 16 of [92]. Reproduced with permission of the Institution of Engineering and Technology (IET).

are insufficient to maintain a strong F2 layer at this time of day [92]. The center of the trough region is denoted by the * symbol in the diagram.

5.2.5 Earth's Atmosphere

Earth's atmosphere can be divided into layers in a number of ways, depending on certain properties of interest. Each layer is denoted with the suffix "sphere" and the upper boundary of a layer is denoted with the suffix "pause." Figure 5.11 illustrates various ways in which the atmosphere is commonly divided.

Based on temperature, the atmosphere can be divided into the troposphere, stratosphere, mesosphere, and thermosphere. Temperature is a function of height, and in the lower atmosphere it alternates between decreasing and increasing with altitude. The tropopause, stratopause, and mesopause denote the maxima and minima of the temperature function. Above the mesopause, in the thermosphere,

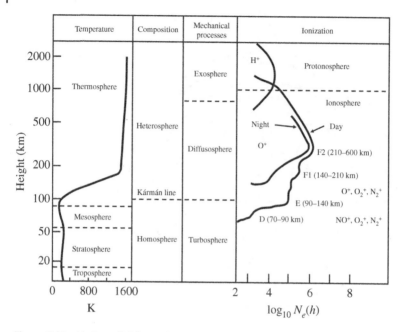

Figure 5.11 Various divisions of the atmosphere. Source: Adapted from Figure 10.6 of [45]. Reproduced with permission of Wiley.

temperature increases dramatically. These high temperatures play an important role in the production of thermospheric winds, which give rise to a number of complicated transport phenomena in the ionosphere.

In the turbosphere, the atmosphere is highly turbulent, which keeps the chemical composition of the atmosphere relatively uniform. Due to this chemical homogeneity, this region of the atmosphere is called the homosphere. Since the effects in the turbosphere give rise to this homogeneity, the turbopause and the homopause are roughly equivalent. It should be noted that this is also the height of the Kármán Line, the line above which the atmosphere becomes too thin to support aerodynamic flight. This boundary, at about 100 km, is used by the Fédération Aéronautique Internationale (FAI) as the official boundary between Earth's atmosphere and space [94]. This distinction is fairly arbitrary, and the United States has never officially recognized a specific altitude as the boundary of space, but the Kármán Line serves as a useful rule of thumb.

Above the turbopause is a region called the diffusosphere, in which diffusive separation becomes important and the atmosphere is no longer homogeneous. Hence, this region is also called the heterosphere. At very large altitudes, the atmosphere is very sparse – so sparse, in fact, that molecules rarely collide and thus cease to behave like a gas. In this region, called the exosphere, molecules with sufficient upward velocity can escape the atmosphere entirely.

The lowest layers of the ionosphere lie within the turbosphere, where the turbulent mixing facilitates recombination of electrons and ions, which limits the degree of ionization at these heights. The majority of the ionosphere lies within the diffusosphere, where the heterogeneity of the atmosphere reduces recombination and allows much higher levels of ionization. We will also see later that diffusion plays an important role in complex transport effects in the upper layers of the ionosphere. The sparseness of the exosphere contributes to the decline in ionization at extreme altitudes, eventually resulting in an upper boundary on the ionosphere.

The ionosphere is a region of the atmosphere which begins at a height of about 70 km and extends to at least 600 km above the surface of the Earth. However the upper boundary, also known as the ionopause, is not well defined. There is some debate as to where to designate its altitude, but in the literature the ionopause is commonly designated at 1000 km. In the ionosphere, gas molecules are ionized by incident solar radiation, freeing a large number of electrons. This high density of free electrons turns the ionosphere into a plasma.

Later in this chapter, we will derive the electron density as a function of height, and we will see that the ionosphere may itself be subdivided into discrete layers. During the day, when the electron density is highest, there exist four distinct layers – labeled D, E, F1, and F2 in increasing order of altitude. At night, the electron density decreases, so the D and E layers tend to disappear while the two F layers merge.

The plasma nature of the ionosphere changes the way that electromagnetic signals propagate. In Chapter 7, we will discuss several ionospheric propagation effects that are of critical importance to consider when designing a GNSS system. For example, signals propagate more slowly through the ionosphere, resulting in a time delay. If this time delay is not accounted for, the GNSS receiver will produce an incorrect positioning calculation.

The effects of the ionosphere can be grouped into two major categories: effects dependent upon the total electron content (TEC) along the propagation path, and effects arising from ionospheric irregularities which are stochastic in nature. Effects belonging to the former category are easier to model, and include phenomena such as Faraday rotation and group delay. Conversely, due to their inherent randomness, ionospheric irregularities are more difficult to model. Ionospheric scintillations are the most important irregularity to consider at GNSS frequencies, since they can cause significant signal fading.

The remainder of this chapter will discuss the structure and nature of the ionosphere. This background will leave the reader poised for Chapter 7 wherein we will discuss how these factors impact electromagnetic wave propagation. These topics will be approached from the context of GPS signal propagation – i.e. a one-way traverse of the ionosphere at frequencies between 1.0 and 1.5 GHz.

5.3 Physics of Ionization

5.3.1 Neutral Atmosphere

The majority of this chapter concerns the physics of the ionosphere. As its name implies, the ionosphere is characterized by its abundance of ionized gas. Before considering the effects of ionized gas, it is important to first understand the physics of neutral gases. In this section, we will briefly discuss a few important quantities and concepts which will be important for derivations later in this chapter.

The Barometric Law

The following derivation is adapted from [45]. Consider a cylindrical volume of the atmosphere with cross section A and height dh. The bottom of this cylinder is at a height h and is subject to a pressure p. The top of the cylinder is therefore at a height $h + dh$, and since pressure decreases with height, the pressure at the top of the cylinder will be $p - dp$. Figure 5.12 illustrates this.

Recall that pressure is related to force by the expression $p = F/A$. The force exerted by the fluid is merely its weight mg. Thus the pressure of a static fluid of volume V and height h is given by:

$$p = \frac{F}{A} = \frac{mgh}{V} = \rho gh, \tag{5.7}$$

Figure 5.12 Cylindrical volume of air.

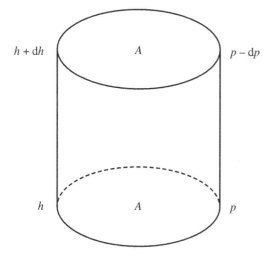

$h + dh$ A $p - dp$

h A p

where $\rho = m/V$ is the density of the fluid and g is the acceleration due to gravity. The density may also be expressed as:

$$\rho = nm_s, \tag{5.8}$$

where n is the number of molecules per unit volume (also known as the number density) and m_s is the mean molecular mass. Further, the atmosphere obeys the ideal gas law:

$$p = nk_BT, \tag{5.9}$$

where $k_B = 1.38 \times 10^{-23}$ J/K is the Boltzmann constant, and T is the absolute temperature.

The pressure exerts forces on the cylinder. We will designate a coordinate system in which upward forces are positive and downward forces are negative. The bottom of the cylinder is subject to an upward force Ap and the top is subject to a downward force $-A(p - dp)$. We will assume that the pressure at a given height is symmetric in all directions, so the no net horizontal force is applied. The net force on the cylinder due to gas pressure is thus given by the sum of these two vertical forces:

$$dF_b = Ap - A(p - dp) = A\,dp. \tag{5.10}$$

This net upward differential force dF_b is a buoyant force. We will also assume that there is no vertical wind, meaning the total force on the air molecules must be in equilibrium. For this to be the case, the buoyant force must equal the downward force of gravity. Therefore the differential force due to gravity, dF_g is given by:

$$dF_g = -A\,dp = -A\rho g\,dh = -Anm_s g\,dh. \tag{5.11}$$

Note that dF_g is negative because it points downward, and Equations 5.7 and 5.8 have been used. Equating the buoyant force and the gravitational force gives:

$$dp = -nm_s g\,dh. \tag{5.12}$$

Now we may use the ideal gas law (Equation 5.9) to replace n. Making this substitution and moving p to the left hand side gives:

$$\frac{dp}{p} = -\frac{m_s g}{k_B T}\,dh. \tag{5.13}$$

Integrating both sides from height h_0 to height h gives:

$$\ln(p) - \ln(p_0) = -\int_{h_0}^{h} \frac{m_s g}{k_B T}\,dh, \tag{5.14}$$

where p_0 is the pressure at h_0. Since $\ln(p) - \ln(p_0) = \ln(p/p_0)$, we can rewrite Equation 5.14 as:

$$p = p_0 \exp\left(-\int_{h_0}^{h} \frac{m_s g}{k_B T}\,dh\right), \tag{5.15}$$

In general, m_s, g, and T are all functions of height. However, for small differences in height these parameters are nearly constant. Under this assumption, we may define a parameter called the scale height:

$$H = \frac{k_B T}{m_s g}.$$
(5.16)

Therefore Equation 5.15 becomes:

$$p = p_0 \exp\left(-\int_{h_0}^{h} \frac{dh}{H}\right),$$
(5.17)

which may be evaluated to obtain:

$$p = p_0 \exp\left(-\frac{h - h_0}{H}\right).$$
(5.18)

Pressure Gradient Force

Whenever there exists a pressure difference across a surface, this implies a net difference in the force on the surface, as per the equation $p = F/A$. Returning to the notation used in Figure 5.12, we see that the cylinder of air has area A and height dh. The mass of the air inside the cylinder is given by:

$$m = \rho A\, dh,$$
(5.19)

where $\rho = m/V$ is the density of the air in the cylinder volume V. The pressure difference $-dp$ between the bottom and the top of the cylinder gives the net force $F = -A\, dp$. Substituting Equation 5.19 into $F = ma$ and equating this force with $F = A\, dp$, we can write:

$$\rho A a\, dh = -A\, dp \Rightarrow a = -\frac{1}{\rho}\frac{dp}{dh}.$$
(5.20)

Multiplying both sides by m gives:

$$F = -\frac{1}{n}\frac{dp}{dh},$$
(5.21)

where n is the number density. Note that this expression only takes into account the vertical pressure difference because the horizontal forces on the rounded face of the cylinder cancel out. For some arbitrary volume, this symmetry cannot be assumed, so there may be net horizontal force as well. To generalize the force due to a pressure differential, we can write:

$$\vec{F} = -\frac{1}{n}\nabla p,$$
(5.22)

where the gradient term ∇p encompasses pressure derivatives in any direction in 3D space.

5.3.2 Ionization

Ionization is a phenomenon in which an electrically neutral atom either gains or loses an electron, resulting in a net charge. In the upper atmosphere, high energy photons from the Sun collide with gas molecules, stripping them of electrons and thus ionizing them. Photo-ionization, or ionization due to photon collisions, can be represented mathematically by the equation:

$$a + \frac{hc}{\lambda} \rightarrow a^+ + e^-, \tag{5.23}$$

where a is some arbitrary neutral atom, hc/λ is the energy of an incident photon of wavelength λ, a^+ is a positive ion, and e^- is an electron. This process can only occur if the photon energy exceeds the ionization energy E_{ion}. To gain a sense for how ionization energy is determined, the following semi-classical[1] derivation is provided.

Consider the energy of an electron in the nth quantized orbital shell ($n = 1, 2, 3,$ ...) of a Bohr atom with Z protons in its nucleus. The total energy of the electron is given by the sum of its kinetic and potential energies:

$$E = \frac{m_e v^2}{2} - \frac{kZe^2}{r}, \tag{5.24}$$

where $m_e = 9.109 \times 10^{-31}$ kg is the mass of an electron, $e = 1.602 \times 10^{-19}$ C is the elementary charge, v is the velocity of the electron, $k = 1/(4\pi\varepsilon_0) = 8.988 \times 10^9$ N m^2/C^2 is Coulomb's constant, and r is the radius of the electron orbit. Note that an electron has charge $-e$ and the nucleus of the atom has charge $+Ze$. The only unknowns in this equation are v and r. We can solve for v by considering the centripetal force on the electron. The source of the centripetal force is the Coulomb force between it and the nucleus:

$$\frac{m_e v^2}{r} = \frac{kZe^2}{r^2}. \tag{5.25}$$

We may rearrange this equation in terms of v to obtain

$$v^2 = \frac{kZe^2}{m_e r}. \tag{5.26}$$

---

1 This semi-classical approach is not a perfect model of the system – we now know that the Bohr model of the atom is incorrect and that electrons do not actually orbit the nucleus. Instead, electron orbitals must be described by a wave function which describes the probability of observing an electron in a given location around the nucleus. This model requires the use of quantum mechanics [95]. Bohr's model works best for hydrogen-like atoms (atoms or ions with only one electron in their outermost valence shell). The above derivation breaks down for atoms with more valence electrons. Additionally, inter-atomic bond energies alter the ionization energy of molecules as compared with atoms. Details on this phenomenon may be found in [96]. Despite these shortcomings, the semi-classical approach is simple and provides a good conceptual understanding of ionization energy.

Now we can rewrite Equation 5.24, substituting v^2 to obtain

$$E = \frac{kZe^2}{2r} - \frac{kZe^2}{r} = -\frac{kZe^2}{2r}. \tag{5.27}$$

At this point, all terms in the energy equation are known quantities except for r. Recall that the angular momentum of an electron is quantized, and is given by

$$L = rm_e v = n\hbar, \tag{5.28}$$

where $\hbar = 1.055 \times 10^{-34}$ J s/rad is the reduced Planck constant. Solving this equation for v and substituting it into Equation 5.26 lets us write:

$$\frac{n^2 \hbar^2}{r^2 m_e^2} = \frac{kZe^2}{m_e r}. \tag{5.29}$$

We may solve this equation for r:

$$r = \frac{n^2 \hbar^2}{kZe^2 m_e}. \tag{5.30}$$

Finally, substituting r into Equation 5.27 gives:

$$E_n = -\frac{k^2 Z^2 e^4 m_e}{2n^2 \hbar^2}, \tag{5.31}$$

where E_n indicates the energy of the nth bound state of the electron, which is quantized because n can only assume integer values. All of the constants may be computed to obtain a numerical value. In this section, all constants were given in terms of Joules, but bound state energies are very small and typically expressed in electron-volts (eV). The conversion between Joule and electron-volt is $1 \text{ eV} = 1.602 \times 10^{-19}$ J. When all constants are computed and converted to eV, the numerical value is found to be 13.6 eV. Therefore, Equation 5.31 may be simplified to:

$$E_n = -13.6 \frac{Z^2}{n^2}, \tag{5.32}$$

where E_n is measured in eV.

The bound state energy of an electron is the amount of energy that must be imparted to the electron to knock it loose from its orbital – in other words the bound state energy is precisely the ionization energy. The point when the photon energy equals the ionization energy is known as the ionization limit. Since the photon energy is proportional to its wavelength, the ionization limit may also be expressed in terms of a wavelength. If the wavelength of an incident photon is shorter than the ionization limit, the atom or molecule in question will be ionized. The ionization limits of several ionizing agents abundant in the ionosphere are given in Table 5.1.

Table 5.1 Ionization limits in nm of molecular species present in the ionosphere.

Molecular species	Ionization limit (nm)
N_2	79.6
O	91.1
H	91.1
O_2	102.7
NO	134

Source: Data from Rishbeth [92].

Due to absorption, ionizing photons will penetrate the atmosphere to different heights depending on their frequencies. In general, higher frequency (i.e. higher energy) photons penetrate deeper into the ionosphere. The depth to which a photon will penetrate the ionosphere is also dependent upon a quantity known as the *absorption cross section*, measured in m^2. This quantity is a measure of the probability that absorption will occur for a photon traveling through a given medium. The absorption cross section is given by the equation:

$$\sigma = \frac{2\pi\kappa}{\lambda_0 n},$$ (5.33)

where n is here the number of absorbing particles per unit volume in m^{-3}, λ_0 is the vacuum wavelength of the photon, and κ is a quantity known as the extinction coefficient (for details on κ, see Appendix B.3.1). The extinction coefficient is different for every atom and molecule, so the absorption cross section is dependent upon the molecular composition of the medium.

The rate of production of ion-electron pairs, measured in ions per unit volume per unit time ($m^{-3}\,s^{-1}$) is dependent upon the electron cross section as well. This production rate for a given atomic species is:

$$\boxed{q = \zeta\sigma nS.}$$ (5.34)

ζ is the *ionization efficiency*, or the number of ions produced per unit time per unit power, measured in $W^{-1}\,s^{-1}$. ζ is dependent upon the particular chemical species being ionized. S is the incident solar flux density in W/m^2, which is directly related to the number of incident photons. Whenever an atom is ionized, a photon is absorbed and so S decreases, changing the production rate going forward. The intensity therefore varies as a function of height. This concept will be treated more rigorously in Section 5.4. In an ionized medium, electrons are unbound and may move freely. However, the ionosphere as a whole is still electrically neutral.

5.3.3 Recombination and Attachment

The reverse of ionization is called recombination; free electrons and ions can recombine to form a neutral atom. This phenomenon occurs constantly in the ionosphere according to one of two processes. In the first process, a positive ion a^+ and a free electron e^- encounter one another. This produces a neutral atom a and some excess energy E, as shown by the equation [72]:

$$a^+ + e^- \rightarrow a + E. \tag{5.35}$$

In the second process, a positive ion first interacts with a neutral diatomic molecule b_2. In the ionosphere, O_2 and N_2 are abundant diatomic molecules which mediate this reaction. One of the atoms in the diatomic molecule is supplanted by the ion in a single replacement reaction, resulting in a positive molecular ion ba^+ and a leftover neutral atom b. The following equation summarizes this:

$$a^+ + b_2 = ba^+ + b. \tag{5.36}$$

The molecular ion then recombines with a free electron, and the molecule dissociates in a process known as dissociative recombination. The result is two neutral atoms a and b. Dissociative recombination is given by the equation:

$$ba^+ + e^- = b + a. \tag{5.37}$$

The probability of recombination is related to both the number of positive ions per unit volume N_i and the number of free electrons per unit volume N_i. The recombination rate for a given ionic species is:

$$l_r = \alpha N_e N_i, \tag{5.38}$$

where α is the recombination coefficient, which is dependent on the specific ions involved.

In addition to recombination, free electrons can be removed from the ionosphere if an electron attaches itself to a neutral atom to produce a negative ion a^- and some excess energy E. This process is represented by the equation:

$$e^- + a = a^- + E. \tag{5.39}$$

The probability of attachment is dependent on the number of electrons per unit volume N_e and the number of neutral atoms per unit volume n. The attachment rate for a given atomic species is:

$$l_c = \beta N_e, \tag{5.40}$$

where $\beta = kn$, in which k is the attachment coefficient, related to the ability for the atomic species in question to attract an electron. In the ionosphere, ions and electrons are always created in pairs, and other sources of ionization are negligible, so we may assume $N_e = N_i$. The total electron loss due to both recombination and attachment at intermediate heights of the ionosphere is given by [81]:

$$L = \frac{\beta \alpha N_e^2}{\beta + \alpha N_e}. \tag{5.41}$$

When $\beta \gg \alpha N_e$, as in the E and F1 layers, Equation 5.41 reduces to 5.38. When $\beta \ll \alpha N_e$, as in the F2 layer, Equation 5.41 reduces to 5.40.

5.3.4 Photochemical Processes in the Ionosphere

It is worth discussing ion-electron production and loss in the context of Earth's atmosphere. The chemical composition of the atmosphere is shown in Table 5.2 [97].

This table has been cut off at 5 ppm, but a number of trace compounds including Methane, molecular Hydrogen, Ozone and others exist in trace amounts. Separate from this table is water vapor, which typically accounts for up to 4% of the total atmosphere. The remaining fraction of the atmosphere is divided up among the compounds in Table 5.2. For example, if water vapor accounts for 4% of the total atmosphere, nitrogen accounts for 78.11% of the remaining 96%, or 74.98% of the total atmosphere.

Table 5.2 Chemical composition of the dry atmosphere.

Chemical species	Symbol	Concentration
Nitrogen	N_2	78.11%
Oxygen	O_2	20.95%
Argon	Ar	0.93%
Carbon dioxide	CO_2	0.04%
Neon	Ne	18.18 ppm
Helium	He	5.24 ppm

Source: Data from [97].

By far the most abundant chemical species are molecular Nitrogen and molecular Oxygen. These molecules may be ionized via the following reactions:

$$N_2 + \frac{hc}{\lambda} \rightarrow N_2^+ + e^-, \tag{5.42}$$

$$O_2 + \frac{hc}{\lambda} \rightarrow O_2^+ + e^-. \tag{5.43}$$

These positive ions may recombine with free electrons via dissociative recombination. This results in two neutral atoms, as shown in the following reactions:

$$N_2^+ + e^- \rightarrow N + N, \tag{5.44}$$

$$O_2^+ + e^- \rightarrow O + O. \tag{5.45}$$

Atomic oxygen can be ionized itself, or it may participate in a replacement reaction with N_2^+ to form a nitric oxide ion and a neutral nitrogen atom:

$$O + \frac{hc}{\lambda} \rightarrow O^+ + e^-, \tag{5.46}$$

$$N_2^+ + O \rightarrow NO^+ + N. \tag{5.47}$$

Atomic oxygen ions can take part in single replacement reactions with either N_2 or O_2, as follows:

$$O^+ + N_2 \rightarrow NO^+ + N, \tag{5.48}$$

$$O^+ + O_2 \rightarrow O_2^+ + O. \tag{5.49}$$

Finally, nitric oxide ions can dissociatively recombine:

$$NO^+ + e^- \rightarrow N + O. \tag{5.50}$$

A number of physical and chemical processes inform the relative number of each ionic species at a given altitude. An outline of important effects is provided:

- Although N_2 is abundant in the atmosphere, N_2^+ ions do not exist in large quantities since they react very quickly via Equations 5.44 and 5.47.
- At lower altitudes, where molecular number density is high, N_2 and O_2 molecules are prevalent. O^+ ions react quickly with these molecules according to Equations 5.48 and 5.49, leaving few O^+ ions.
- NO^+ is abundant at low altitudes. Direct ionization of NO creates NO^+, but NO is not abundant in the atmosphere. The prevalence of NO^+ is due to reactions 5.47 and 5.48.
- At higher altitudes, N_2 and O_2 are less abundant, so reactions 5.48 and 5.49 occur less frequently. When they do occur, the resulting NO^+ and O_2^+ ions quickly recombine with the abundant free electrons, via reactions 5.45 and 5.50. Because of this, NO^+ and O_2^+ are not abundant at higher altitudes.

- Molecular ions can recombine quickly and readily with electrons via dissociative recombination. This mechanism is not available for atomic ions; instead atomic ions must radiate excess energy as light during recombination – a process known as radiative recombination. Radiative recombination is inefficient and occurs much less frequently than dissociative recombination.
- Ions may also recombine via collisions. To satisfy conservation of momentum and energy as well as quantum mechanical principles, atomic ions must be involved in three-body collisions to recombine, whereas molecular ions only need to be involved in two-body collisions [92].
- At low altitudes, both two-body and three-body collisions occur with some regularity due to the high density of the atmosphere. At high altitudes, three-body collisions become much less common. Thus O^+ will combine less frequently ($\sim 10^{-5}$ times less [81, p. 63]) than N_2^+, O_2^+, and NO^+ at higher altitudes, leaving O^+ as the most abundant ion.

In summary, O_2^+ and NO^+ are the most prevalent ions at low altitudes, with some N_2^+ ions present as well. As altitude increases, these molecular ions decrease in number, while O^+ ions increase. Molecular ions dominate up to about 200 km, at which point O^+ takes over as the dominant ion. Information in this section has been adapted from [92].

5.4 Chapman's Theory of Ionospheric Layer Formation

Qualitatively, the nature of ionospheric ionization is straightforward to determine. At high altitudes, there is a large amount of solar flux but the atmosphere is very sparse, so there are few molecules to be ionized. As a result, ion production is relatively small. As photons travel farther down into the atmosphere, many of them are absorbed, decreasing the intensity of the solar flux. At some point the intensity will be so small that ionization is negligible, despite the high density of ionizable molecules. Between these two extremes, ion production is greater due to the combination of sufficient solar flux intensity and molecular number density. There is a point at which the combination of these parameters provides maximum ion production. Ion production increases monotonically from low values at low altitudes up to this maximum value, then decreases monotonically to low values again at high altitudes [45].

Less straightforward is a quantitative description of this process. It was Sydney Chapman who first formalized ionospheric ionization using a quantitative model in 1931. He discovered that the variation in electron density with respect to height creates regions of the ionosphere which act differently from one another. Although the density of free electrons varies continuously with height, ionospheric regions can be roughly divided into discrete layers, each layer displaying distinct properties. These layers will be explored further in Section 6.3.

A succinct derivation of Chapman's model will be provided next, which has been adapted from [45]. The following simplifying assumptions will be made in the derivation [72, 92]:

1. The atmosphere consists of a single, exponentially distributed ionizable gas.
2. The atmosphere is plane stratified and not subject to diffusion, turbulence, or horizontal variations.
3. The temperature of the atmosphere is constant, and therefore the scale height $H = k_b T / m_s g$ is also constant.
4. Electron production is only caused by photo-ionization via absorption of monochromatic light.

The derivation consists of six steps (a)–(f) described next.

(a) *Calculation of the solar flux as a function of height*: The solar flux may be incident on the ionosphere at some arbitrary angle χ with respect to zenith. Consider a cylinder of atmospheric gas tilted at this angle, as shown in Figure 5.13. The top of the cylinder, at height $h + dh$, receives a solar flux density $S + dS$. The bottom of the cylinder receives a flux density of only S, meaning the total loss of flux in the volume is given by $A\, dS$. This loss is related to the absorption

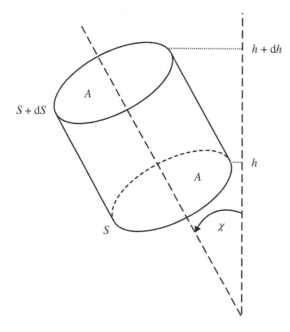

Figure 5.13 Slanted cylindrical volume of air.

cross section σ of the molecules by the expression:

$$A \, dS = \sigma n S \, dV, \tag{5.51}$$

where n is the number density of molecules, and dV is the differential volume. dV may also be written as:

$$dV = A \sec \chi \, dh. \tag{5.52}$$

Substituting dV into Equation 5.51 and moving S to the left hand side gives:

$$\frac{dS}{S} = \sigma n \sec \chi \, dh. \tag{5.53}$$

Noting that n and σ are both dependent on height, we can integrate both sides from ∞ to h to obtain:

$$\ln S(h) - \ln S(\infty) = \sec \chi \int_{\infty}^{h} \sigma(h')n(h')dh'. \tag{5.54}$$

The negative of the integral on the right hand side represents a unitless parameter known as the optical depth, which we may denote τ:

$$\tau(h) = - \int_{\infty}^{h} \sigma(h')n(h')dh'. \tag{5.55}$$

Using a logarithmic identity on the left hand side and substituting in τ on the right side gives:

$$\ln \left(\frac{S(h)}{S(\infty)} \right) = -\tau \sec \chi. \tag{5.56}$$

Let $S(\infty) = S_{\infty}$ represent the flux at a height large enough that absorption is negligible. The flux as a function of height may now be written as:

$$S(h) = S_{\infty} \, e^{-\tau \sec \chi}. \tag{5.57}$$

(b) *Calculation of the solar flux derivative with respect to height*: Equation 5.57 may be differentiated with respect to height to determine the change in S over a small height dh as

$$\frac{dS}{dh} = -S_{\infty} \sec \chi \, e^{-\tau \sec \chi} \frac{d\tau}{dh}. \tag{5.58}$$

From Equation 5.55, we know that τ is an integral with respect to h, therefore:

$$\frac{d\tau}{dh} = -\sigma n. \tag{5.59}$$

Thus we may write Equation 5.58 as:

$$\frac{dS}{dh} = \sigma n S_{\infty} \sec \chi \, e^{-\tau \sec \chi}. \tag{5.60}$$

(c) *Derivation of the ion production per unit volume as a function of height*:
The power absorbed by the volume is equal to the decrease in solar flux:

$$dP = A dS = A \frac{dS}{dh} dh. \tag{5.61}$$

The number of ions produced in the volume is directly proportional to the absorbed power. This relationship is given by the expression:

$$N_i = \zeta \, dP = \zeta A \frac{dS}{dh} dh, \tag{5.62}$$

where ζ is the ionization efficiency. The number of ions produced in the volume is also equal to the production rate q times the differential volume. Using Equation 5.52, we see that:

$$N_i = q \, dV = q A \sec \chi \, dh. \tag{5.63}$$

From 5.62 and 5.63 it follows that:

$$q \sec \chi = \zeta \frac{dS}{dh}. \tag{5.64}$$

Finally, using Equation 5.60 to substitute in for dS/dh, we arrive at

$$q(\chi, h) = \zeta \sigma n S_\infty \, e^{-\tau \sec \chi}. \tag{5.65}$$

(d) *Derivation of the optical depth as a function of height via barometric law*:
For simplicity, the reference height H will be chosen as the height for which optical depth equals unity. We may assume that σ and ζ have little variation over a region of interest. We may also assume that $N(h)$ varies in a similar fashion to Equation 5.18, that is:

$$n(h) = n_0 \exp\left(-\frac{h - h_0}{H}\right), \tag{5.66}$$

where n_0 is the number density of molecules at height h_0. Recall that $H = k_B T / m_s g$ is the scale height. Substituting this equation into the expression for optical depth given by Equation 5.55, we may write:

$$\tau(h) = -\sigma n_0 \int_\infty^h \exp\left(-\frac{h' - h_0}{H}\right) dh = \sigma n_0 H \exp\left(-\frac{h - h_0}{H}\right). \tag{5.67}$$

Evaluating this equation at h_0 gives $\tau(h_0) = \sigma n_0 H = \tau_0$, the optical depth at height. Given this definition, the expression for $\tau(h)$ can be written as:

$$\tau(h) = \tau_0 \exp\left(-\frac{h - h_0}{H}\right). \tag{5.68}$$

At this point, we may arbitrarily choose our reference height h_0 such that $\tau(h_0) = \sigma n(h_0) H = 1$. This height has physical significance. From

Equation 5.57 we see that when the Sun is at zenith ($\chi = 0$), the flux density at the reference height, $S(h_0)$, has decreased to $1/e$, or about 37% of the initial flux density S_∞. Shortly, we will also show that the ion production rate when the Sun is at zenith, i.e. $q(\chi = 0, h)$ is maximized at the reference height.

We will also introduce a few more simplifying notations. Let

$$z = \frac{h - h_0}{H}. \tag{5.69}$$

Since we have defined τ such that $\tau_0 = 1$, this allows us to write $\tau = e^{-z}$ and $n = n_0\, e^{-z}$. Again noting that $\tau_0 = 1$, we may write $n_0 = \frac{1}{\sigma H}$.

Utilizing all of these simplifications, we may rewrite Equation 5.65 as a function of χ and z:

$$q(\chi, z) = \zeta \sigma n_0\, e^{-z} S_\infty \exp(-e^{-z} \sec \chi) = \frac{\zeta S_\infty}{H} \exp(-z - e^{-z} \sec \chi). \tag{5.70}$$

(e) *Determination of the height of maximum ion production as the reference height:* We will now prove that q is maximized at $\chi = 0$ and $h = h_0$. First we will find the maximum of q with respect to χ. We must set the partial derivative of $q(\chi, z)$ equal to zero.

$$\frac{\partial q}{\partial \chi} = -\frac{\zeta S_\infty}{H} e^{-z} \exp(-z - e^{-z} \sec \chi) \sec \chi \tan \chi = 0. \tag{5.71}$$

This function is a product of three functions of χ: $\exp(-z - e^{-z} \sec \chi)$, $\sec \chi$, and $\tan \chi$. If any of these three functions equals zero, then the entire function will equal zero. Because neither $\exp(-z - e^{-z} \sec \chi)$ nor $\sec \chi$ have any real roots, the only solution to Equation 5.71 is when $\tan \chi = 0$, or when $\chi = \pm m\pi$ (with $m = 0, 1, 2, \ldots$). Note again that χ represents the angle with respect to zenith of incident solar flux. The ion production rate is only physically meaningful for the interval $\chi \in [-\pi/2, \pi/2]$, which represents angles for which the Sun is above the horizon. Therefore we see that the only solution to Equation 5.71 within our interval is $\chi = 0$, when the Sun is at zenith. In this case, $\sec \chi = 1$ and we have:

$$q(0, z) = \frac{\zeta S_\infty}{H} \exp(-z - e^{-z}). \tag{5.72}$$

To find the maximum of this function with respect to z, we set the partial derivative with respect to z equal to zero:

$$\frac{\partial q(0, z)}{\partial z} = \frac{\zeta S_\infty}{H} (-1 + e^{-z}) \exp(-z - e^{-z}) = 0. \tag{5.73}$$

Since $\exp(-z - e^{-z})$ has no real roots, so the only solution to Equation 5.73 is when $-1 + e^{-z} = 0$, or when $z = -\ln(1) = 0$. From Equation 5.69 if $z = 0$ then $h = h_0$, thus proving that the reference height is the altitude for which ionization is maximized. Evaluating q at both $\chi = 0$ and $h = h_0$ gives the maximum ion production rate:

$$q(\chi = 0, z = 0) = q_0 = \frac{\zeta S_\infty}{H} e^{-1} \Rightarrow q_0 e = \frac{\zeta S_\infty}{H}. \tag{5.74}$$

Substituting this into Equation 5.70 gives:

$$q(\chi, z) = q_0 \exp(1 - z - e^{-z} \sec \chi). \tag{5.75}$$

(f) *Determination of the electron density as a function of height*: The final step in the derivation consists of determining the electron density profile as a function of height, $N_e(\chi, z)$. To do this, we must account not only for the ion production rate $q(\chi, z)$ but also the recombination rate $l_r(\chi, z)$. For now, we will neglect attachment. A separate electron density function dependent on attachment instead of recombination will be considered shortly. From assumption 2, we ignore transport phenomena such as diffusion, so the parameter d in Equation 5.136 may be neglected. Thus, the rate of change of electron density is simply the difference of the production and recombination rates:

$$\frac{dN_e}{dt} = q(\chi, z) - l_r(\chi, z). \tag{5.76}$$

Recall that the recombination rate is given by the equation:

$$l_r = \alpha N_e N_i, \tag{5.77}$$

where α is the recombination coefficient, N_i is the number density of positive ions, and N_e is the number density of electrons. Ionization always produces electrons and positive ions in pairs, so using assumption 1 above, which says that electron production is solely due to photoionization, we have $N_e = N_i$. Therefore, Equation 5.77 can be simplified to:

$$l_r = \alpha N_e^2. \tag{5.78}$$

The electron density only changes significantly over long timescales, so we may assume equilibrium $(dN_e/dt \approx 0)$ in the production and recombination rates over short timescales, i.e.

$$q(\chi, z) \approx l_r(\chi, z). \tag{5.79}$$

Using Equations 5.78 and 5.79, we obtain:

$$N_e = \sqrt{\frac{q(\chi, z)}{\alpha}}. \tag{5.80}$$

Substituting in Equation 5.75, we can now write N_e as:

$$N_e(\chi, z) = \sqrt{\frac{q_0}{\alpha}} \exp\left[\frac{1}{2}(1 - z - e^{-z}\sec \chi)\right]. \tag{5.81}$$

A similar electron density profile may be determined if attachment is considered instead of recombination. For clarity, we will denote this electron density $\tilde{N}_e$. The rate of change of this electron density is given by the difference of the production rate $q(\chi, z)$ and the attachment rate $l_c(\chi, z)$. The recombination rate may be neglected, and effects such as diffusion may be ignored, as before. In this case,

$$\frac{d\tilde{N}_e}{dt} = q(\chi, z) - l_c(\chi, z). \tag{5.82}$$

Recall that $l_c = \beta\tilde{N}_e$ where β is dependent on the neutral molecular number density n. $\tilde{N}_e$ is the electron number density. Since $n = n_0 e^{-z}$, it follows that

$$l_c(\chi, z) = \beta_0 e^{-z}\tilde{N}_e, \tag{5.83}$$

where $\beta_0 = kn_0$ (k is the attachment coefficient). Once again we may assume equilibrium over short timescales, which means $q(\chi, z) \approx l_c(\chi, z)$. Therefore we can write:

$$\tilde{N}_e = \frac{q(\chi, z)}{\beta_0}e^z \tag{5.84}$$

and using Equation 5.75, we finally obtain:

$$\tilde{N}_e(\chi, z) = \frac{q_0}{\beta_0}\exp(1 - e^{-z}\sec \chi). \tag{5.85}$$

Figures 5.14–5.16 show the electron densities as functions of z and χ, as given by Equations 5.81 and 5.85. Figure 5.14 shows the electron densities for both recombination and attachment for $\chi = 0°$. Figures 5.15 and 5.16 show the recombination and attachment electron densities respectively, each for various values of χ. In each of these plots, the vertical axis has been normalized, which can be done by introducing a scaling factor N_s equal to the coefficient leading the exponential term in each case. In other words, in Equation 5.81, $N_s = \sqrt{q_0/\alpha}$ and for Equation 5.85, $N_s = q_0/\beta_0$. The electron densities may be normalized by dividing by their respective scaling factors.

Chapman theory is a fairly accurate model of layer formation at lower altitudes, including the D, E and F1 layers of the ionosphere. The Chapman model does not take into account transport effects, which become non-negligible at F2 layer altitudes. Therefore the F2 layer is not a Chapman layer and cannot be described by the equations derived in this section. However, more sophisticated models have

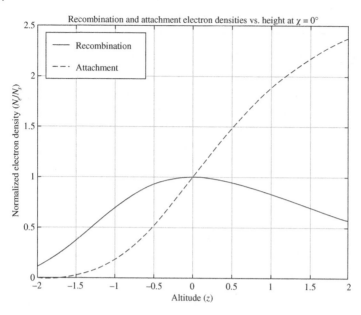

Figure 5.14 Electron densities for recombination and attachment for $\chi = 0°$.

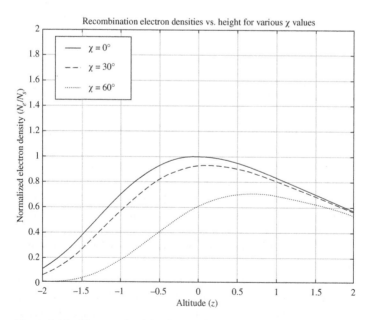

Figure 5.15 Electron densities for recombination at various χ values.

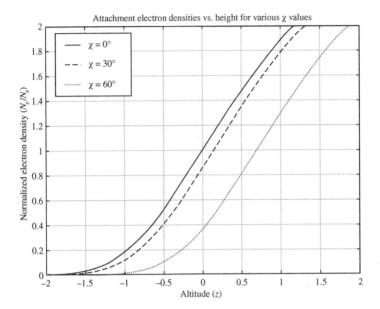

Figure 5.16 Electron densities for attachment at various χ values.

been successful in predicting F2 layer behavior. One example is the International Reference Ionosphere, which is discussed in detail in Section 8.4.

5.5 Plasma Transport

In the D, E, and F1 layers, only the electron production parameter q and loss parameter L are needed to model the local electron density. In the F2 layer, it is important to also consider the movement of electrons and ions. This movement is known as plasma transport. Electrons and ions in the ionosphere are subject to a number of transport processes including diffusion, neutral winds, and vertical electromagnetic drift. The net loss in electrons or ions per unit volume due to transport is given by:

$$d = -\nabla \cdot N\vec{v}, \tag{5.86}$$

where N denotes number density and $\vec{v}$ denotes velocity. When considering electrons $N = N_e$ and $\vec{v} = \vec{v}_e$. Similarly, when considering ions, $N = N_i$ and $\vec{v} = \vec{v}_i$.

The velocities for both electrons and ions may be found via their equations of motion. At large scales, the accelerations on these particles are negligible, so the forces on them must be in equilibrium. In the ionosphere, the ions of interest have charge $+e$, so we may write the equations of motion for ions and electrons respectively as [98, p. 129]:

$$m_i \frac{d\vec{v}_i}{dt} = -m_i \vec{g} - \frac{1}{N_i} \nabla p + e(\vec{E} + \vec{v}_i \times \vec{B})$$

$$- m_i v_{in} \left(\vec{v}_i - \vec{U} \right) - m_e v_{ei} \left(\vec{v}_i - \vec{v}_e \right) = 0,$$

$$m_i \frac{d\vec{v}_e}{dt} = -m_e \vec{g} - \frac{1}{N_e} \nabla p - e(\vec{E} + \vec{v}_e \times \vec{B})$$

$$- m_e v_{en} \left(\vec{v}_e - \vec{U} \right) - m_e v_{ei} \left(\vec{v}_e - \vec{v}_i \right) = 0,$$

$$(5.87)$$

where m is mass, $\vec{g}$ is the acceleration due to gravity, p is pressure, e is the elementary charge, $\vec{E}$ is an electric field, $\vec{B}$ is a magnetic field, and $\vec{U}$ is the velocity of neutral air. In each of these equations, the five terms on the right hand side represent, respectively, the gravitational force, the force due to a pressure gradient, the Lorentz force, collisions with neutral particles, and collisions between electrons and ions.

The parameters v_{in}, v_{en}, and v_{ei} represent the *collision frequencies* between ions and neutral particles, electrons and neutral particles, and electrons and ions, respectively. These terms are given by [99, p. 382]:

$$v_{en} = 5 \times 10^{-1} \, nT^{\frac{1}{2}},$$

$$v_{in} = 3.35 \times 10^{-21} \, \frac{n}{M^{\frac{1}{2}}},$$

$$v_{ei} = \left[34 + 8.46 \log_{10} \left(\frac{T^{\frac{3}{2}}}{N_e^{\frac{1}{2}}} \right) \right] N_i T^{-\frac{3}{2}},$$

$$(5.88)$$

where n is the number density of neutral molecules, T is temperature, M is the mean molecular mass, and N_e and N_i are the number densities of ions and electrons, respectively.

5.5.1 Diffusion

Diffusion with a Vertical Magnetic Field

The effects of diffusion can be determined using the equations of motion. Only vertical diffusion will be considered, because the effects of horizontal diffusion are negligible in the ionosphere. Therefore, only the vertical components v_h of the velocities in Equation 5.87 are important. Additionally, the pressure gradient term reduces to a derivative only with respect to h. With these simplifications, we may rearrange our equations of motion to get:

$$\frac{d}{dh}\left(N_i k_B T_i\right) = -N_i m_i g + N_i e(\vec{E} + \hat{h}v_{i,h} \times \vec{B})_h - N_i m_i v_{in}\left(v_{i,h} - U_h\right)$$
$$- N_i m_e v_{ei}\left(v_{i,h} - v_{e,h}\right),$$

$$\frac{d}{dh}\left(N_e k_B T_e\right) = -N_e m_e g - N_e e(\vec{E} + \hat{h}v_{e,h} \times \vec{B})_h - N_e m_e v_{en}\left(v_{e,h} - U_h\right)$$
$$- N_e m_e v_{ei}\left(v_{e,h} - v_{i,h}\right), \tag{5.89}$$

where k_B is Boltzmann's constant. From here on, we will make the following simplifications:

- $m_i \gg m_e$, the mass of an electron is negligible compared with the mass of an ion.
- $N = N_e = N_i$, electrons and ions have the same number density since they are created in equal numbers.
- $v_h = v_{e,h} = v_{i,h}$, the vertical velocities of both electrons and ions are equal. We will call this velocity v_h the plasma drift velocity.
- $m_i v_{in} \gg m_e v_{en}$, collisions between neutral molecules and electrons are negligible compared with collisions between neutral molecules and ions.

We will also make the following assumptions:

- $T = T_e = T_i$, electrons and ions are in thermal equilibrium.
- Temperature is independent of height.
- $\vec{U} = 0$, neutral air is at rest.
- The geomagnetic field is vertical, so $\hat{h}v_h \times \vec{B} = 0$.

Using all of this information, we can write Equation 5.89 as a scalar equation, solely in terms of vertical components:

$$\frac{d}{dh}\left(Nk_B T\right) = -Nm_i g + NeE_h - Nm_i v_{in}v_{i,h},$$
$$\frac{d}{dh}\left(Nk_B T\right) = -Nm_e g - NeE_h - Nm_e v_{en}v_{e,h}. \tag{5.90}$$

Adding these two equations gives:

$$2\frac{d}{dh}\left(Nk_B T\right) = -N(m_i + m_e)g - N(m_i v_{in} + m_e v_{en})v_h. \tag{5.91}$$

From our assumptions, we know that m_e and $m_e v_{en}$ are negligible compared with m_i and $m_i v_{in}$, respectively. Therefore the former terms may be dropped, which results in the equation:

$$2\frac{d}{dh}\left(Nk_B T\right) = -Nm_i g - Nm_i v_{in}v_h. \tag{5.92}$$

Rearranging this equation in terms of v_h gives:

$$v_h = -\frac{1}{m_i v_{in}}\left[\frac{2k_B T}{N}\frac{dN}{dh} + m_i g\right]. \tag{5.93}$$

If we factor out $2k_BT$ from every term, we get the coefficient $D = 2k_BT/m_i\nu_{in}$ in front. D is known as the ambipolar diffusion coefficient. If we define $H_{pl} = 2k_BT/m_ig$ as the plasma scale height, we see that:

$$\upsilon_h = -D\left[\frac{1}{N}\frac{dN}{dh} + \frac{1}{H_{pl}}\right]. \tag{5.94}$$

Note that the plasma scale height H_{pl} is twice that of a neutral gas. This is because the ions and electrons diffuse together in a process known as ambipolar diffusion [81].

Diffusion with an Inclined Magnetic Field

Recall that in this derivation, we assumed the geomagnetic field to be completely vertical. In reality this is not the case, so $\hat{h}\upsilon_h \times \vec{B}$ will not be zero. We will now observe the effect this has on Equation 5.94. This derivation has been adapted from [100].

Consider a system with Cartesian coordinates (x, y, h) in which the xh-plane is parallel to the geomagnetic field. Thus, the magnetic field is given by $\vec{B} = \hat{x}B_x + \hat{h}B_h$ and the velocity is given by $\vec{\upsilon} = \hat{x}\upsilon_x + \hat{y}\upsilon_y + \hat{h}\upsilon_h$. We may now return to Equations 5.87. Recall that the pressure gradient and gravity only act vertically (i.e. $\nabla_x p = \nabla_y p = 0$ and $\vec{g} = \hat{h}g$). $\vec{E} = \hat{h}E_h$ is also assumed to be completely vertical. We may also make the same simplifications we made previously, so $\vec{\upsilon}_e = \vec{\upsilon}_i = \hat{h}\upsilon_h$ and $\vec{U} = 0$. Using all this information, we obtain the following three equations for ions:

$$e\upsilon_{i,y}B_h - m_i\nu_{in}\upsilon_{i,x} = 0, \tag{5.95}$$

$$-e(\upsilon_{i,x}B_h - \upsilon_{i,h}B_x) - m_i\nu_{in}\upsilon_{i,y} = 0, \tag{5.96}$$

$$-m_ig - \frac{1}{N_i}\frac{d}{dh}\left(N_ik_BT_i\right) + e(E_h - \upsilon_{i,y}B_x) - m_i\nu_{in}\upsilon_{i,h} = 0. \tag{5.97}$$

Three analogous equations for electrons may be obtained as well. The cyclotron frequency (also known as the gyro frequency or Larmor frequency) of a charged particle is given by $\omega_c = qB/m$ where q is the magnitude of the particle's charge (see Section B.7.1). With this definition, we may write:

$$\omega_{i,x} = \frac{eB_x}{m_i}, \quad \omega_{i,h} = \frac{eB_h}{m_i}, \quad \omega_{e,x} = -\frac{eB_x}{m_e}, \quad \omega_{e,h} = -\frac{eB_h}{m_e}, \tag{5.98}$$

which lets us rewrite Equations 5.95 and 5.96 as:

$$\upsilon_{i,x} = \frac{\omega_{i,h}\upsilon_{i,y}}{\nu_{in}}, \tag{5.99}$$

$$\upsilon_{i,y} = \frac{\omega_{i,x}\upsilon_{i,h} - \omega_{i,x}\upsilon_{i,x}}{\nu_{in}}. \tag{5.100}$$

Solving these two equations simultaneously, we obtain:

$$v_{i,y} = \frac{v_{in}\omega_{i,x}}{v_{in}^2 + \omega_{i,h}^2} v_{i,h}. \tag{5.101}$$

Substituting Equation 5.101 into Equation 5.97, we get:

$$-m_i g - \frac{1}{N_i}\frac{d}{dh}\left(N_i k_B T_i\right) + eE_h - e\left(\frac{v_{in}\omega_{i,x}}{v_{in}^2 + \omega_{i,h}^2}v_{i,h}\right)B_x - m_i v_{in} v_{i,h} = 0. \tag{5.102}$$

This may be rewritten as:

$$-m_i g - \frac{1}{N_i}\frac{d}{dh}\left(N_i k_B T_i\right) + eE_h - m_i v_{in}\left(1 + \frac{\omega_{i,x}^2}{v_{in}^2 + \omega_{i,h}^2}\right)v_{i,h} = 0. \tag{5.103}$$

The analogous equation for electrons may also be derived in the same fashion, and is found to be:

$$-m_e g - \frac{1}{N_e}\frac{d}{dh}\left(N_e k_B T_e\right) - eE_h - m_e v_{en}\left(1 + \frac{\omega_{e,x}^2}{v_{en}^2 + \omega_{e,h}^2}\right)v_{e,h} = 0. \tag{5.104}$$

Our earlier simplifications that $N = N_i = N_e$, $T = T_i = T_e$, and $v_h = v_{i,h} = v_{e,h}$ still apply. Using this information and summing Equations 5.103 and 5.104, we obtain:

$$-(m_i + m_e)g - \frac{2}{N}\frac{d}{dh}\left(N k_B T\right) - m_i v_{in}\left(1 + \frac{\omega_{i,x}^2}{v_{in}^2 + \omega_{i,h}^2}\right)v_h$$
$$- m_e v_{en}\left(1 + \frac{\omega_{e,x}^2}{v_{en}^2 + \omega_{e,h}^2}\right)v_h = 0. \tag{5.105}$$

Again referring to our earlier simplifications, $m_i \gg m_e$ and $m_i v_{in} \gg m_e v_{en}$. Additionally, $\omega_i \gg v_{in}$ and $\omega_e \gg v_{en}$ in the F2 layer. The smaller terms are negligible and may be ignored, leaving:

$$-m_i g - \frac{2k_B T}{N}\frac{dN}{dh} - m_i v_{in}\left(1 + \frac{\omega_{i,x}^2}{\omega_{i,h}^2}\right)v_h = 0. \tag{5.106}$$

Note that $\omega_{i,x}/\omega_{i,h}$ reduces to B_x/B_h, which may be related to the inclination of the magnetic field, I, (also known as the magnetic dip) by:

$$\frac{B_h}{B_x} = \tan I = 2\tan\Phi. \tag{5.107}$$

Here, I has also been expressed in terms of the dipole latitude Φ, defined by the equation

$$\sin\Phi = \sin\phi\sin\phi_0 + \cos\phi\cos\phi_0\cos(\lambda - \lambda_0), \tag{5.108}$$

where the geographic coordinates (ϕ, λ) represent the latitude and longitude at a point P. $\phi_0 = 80.5°$ N and $\lambda_0 = 72.8°$ W are the coordinates of the geomagnetic north pole, as defined by the IGRF-12 model [88].

Using Equation 5.107, we may now rewrite 5.106 as:

$$-m_i g - \frac{2k_B T}{N} \frac{dN}{dh} - m_i v_{in} \left(1 + \cot^2 I\right) v_h = 0. \tag{5.109}$$

Rearranging this equation and recalling that $1 + \cot^2 I = \csc^2 I$, we may write:

$$v_h = -\frac{2k_B T}{m_i v_{in}} \left[\frac{1}{N} \frac{dN}{dh} + \frac{m_i g}{2k_B T} \right] \frac{1}{1 + \cot^2 I} = -\frac{2k_B T}{m_i v_{in}} \sin^2 I \left[\frac{1}{N} \frac{dN}{dh} + \frac{m_i g}{2k_B T} \right]. \tag{5.110}$$

Finally, using our definitions from earlier, we see that:

$$v_h = -D \sin^2 I \left[\frac{1}{N} \frac{dN}{dh} + \frac{1}{H_{pl}} \right], \tag{5.111}$$

where D is the ambipolar diffusion constant and H_{pl} is the plasma scale height. Note that as $B_x \to 0$ (i.e. the magnetic field is completely vertical), $\sin^2 I \to 1$ and Equation 5.111 reduces to 5.94.

5.5.2 Neutral Winds

Another transport process significant to the movement of ionization in the ionosphere is the presence of neutral winds with velocity $\vec{U}$. The majority of the ionosphere lies within the thermosphere, which is a region of the atmosphere extending from about 90 km to as high as 1000 km above Earth's surface. In the thermosphere, absorption of ultraviolet and X-ray radiation from the Sun causes heating during the day. At night, the absence of sunlight allows the thermosphere to cool significantly. The temperature difference between the day side and the night side of Earth is around 200–300 K [92]. This difference in temperature creates a pressure differential that causes wind to blow from the day side (high temperature, high pressure) to the night side (low temperature, low pressure).

Winds of neutral particles are subject to an equation of motion similar to those of ions and electrons (as shown in Equation 5.87). The equation of motion is given by:

$$m_n \frac{d\vec{U}}{dt} = -m_n \vec{g} - \frac{1}{n} \nabla p - 2\vec{\Omega} \times \vec{U} - v_{in} \left(\vec{U} - \vec{v_i} \right) - \frac{\mu}{n} \nabla^2 \vec{U}. \tag{5.112}$$

Here, m_n, $\vec{U}$, and n are the mass, velocity and number density of neutral particles, respectively. $\vec{g}$ is acceleration due to gravity, p is pressure, $\vec{\Omega}$ is the angular velocity of Earth's rotation, v_{in} is the collision frequency of ions with neutral particles, $\vec{v_i}$ is the velocity of ions, and μ is the viscosity of the air. In this equation of motion, the terms on the right hand side are, in order of appearance: gravity, differential pressure force, Coriolis force, ion drag due to ion-neutral collisions, and fluid shear force.

The vertical pressure nearly exactly balances gravity, removing them from the equation of motion. If we divide through by m_n, we see that the coefficient leading the shear force term becomes $v = \mu/\rho$, where ρ is density. This coefficient is called the kinematic viscosity of the fluid. At intermediate altitudes, n is large, so the kinematic viscosity is small, allowing greater flow due to shear stress. This allows the formation of sporadic E layers in the ionosphere, as discussed in Section 6.3. At higher altitudes, n is small so the kinematic viscosity is large, smoothing out wind shears. This is why sporadic layers cannot form at high altitudes in the ionosphere.

The equation of motion earlier is derived from the Navier–Stokes equations, which are simply a statement of the conservation of momentum. In addition to momentum, conservation of mass must be satisfied for any fluid flow. Conservation of mass can be expressed via the continuity equation:

$$\frac{\partial n}{\partial t} = -\nabla \cdot n\vec{U}. \tag{5.113}$$

The vertical winds represented by the vertical component of the divergence are an important transport process in the ionosphere, especially at F2 layer altitudes.

Via collisions, neutral molecules transfer momentum to ions, and due to the Coulomb force, these ions move electrons along with them. It should be noted that there is a maximum distance to which an ion can attract an electron. This length, known as the Debye length, is limited due to shielding of the ion by the cloud of electrons associated with it. At a point one Debye length away from the center of the ion, the electric field due to the ion becomes negligible. The Debye length of a quasineutral cold plasma – such as that of the ionosphere – is given by:

$$\lambda_D = \sqrt{\frac{\epsilon_0 k_B T_e}{e^2 N_e}}, \tag{5.114}$$

where $\epsilon_0 = 8.85 \times 10^{-12} \text{ m}^{-3} \text{ kg}^{-1} \text{ s}^4 \text{ A}^2$ is the permittivity of free space, T_e is the electron temperature, and N_e is the electron number density. As an example, in the F2 layer, $N_e = 10^{11}$ electrons/m^3, $T_e = 1000 \text{ K}$, so $\lambda_D \approx 7 \text{ mm}$ [81, p. 25].

If the collision frequency is large compared with the cyclotron frequency, the plasma moves with the neutral wind. Conversely, if the collision frequency is much smaller than the cyclotron frequency, the neutral wind is not strong enough to move charged particles in a direction perpendicular to the geomagnetic field $\vec{B}$. The charged particles will "slide" along the magnetic field lines, like rings on a bar. At mid-latitudes, due to the dip angle I of the magnetic field, poleward winds will cause plasma to descend, while equatorward wind will lift the plasma to a higher

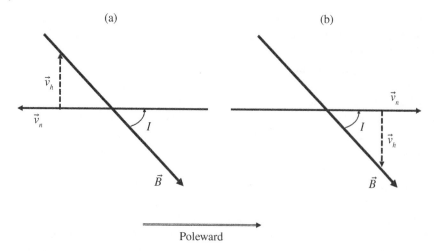

Figure 5.17 Plasma motion due to horizontal neutral winds. (a) Equatorward wind. (b) Poleward wind.

altitude [81]. The vertical component of the plasma drift velocity is given by

$$\vec{v}_h = \frac{(\vec{U} \cdot \vec{B})\vec{B}}{B_0^2},$$ (5.115)

where B_0 is the magnitude of $\vec{B}$. Figure 5.17 illustrates this vector relation.

5.5.3 Electromagnetic Drift

A third transport process in the ionosphere is electromagnetic drift. In the presence of an electric field $\vec{E}$, positive ions will move parallel to the field and electrons will move anti-parallel. If $\vec{E}$ is parallel to the geomagnetic field $\vec{B}$, the charged particles will move with velocity $\vec{v}$ parallel (or anti-parallel) to the magnetic field, so $\vec{v} \times \vec{B} = 0$. Therefore there is no magnetic force on the particles, and their motion is defined solely by $\vec{E}$. However, if $\vec{E}$ is inclined at some angle with respect to $\vec{B}$, then $\vec{v} \times \vec{B} \neq 0$; in other words, the charge carrier is subject to a magnetic force. This process, known as electromagnetic drift, is especially important in the F2 layer because, unlike neutral winds, it allows plasma to move in a direction perpendicular to the geomagnetic field.

Consider the case where $\vec{E} \perp \vec{B}$. Motion of charged particles in the plane perpendicular to $\vec{B}$ is dependent upon the ratio of the neutral collision frequency to the cyclotron frequency, i.e. ν/ω. Note that this ratio is different for ions and electrons. The angle of travel within the plane is measured with respect to the axis which is orthogonal to both $\vec{E}$ and $\vec{B}$, and is given by:

$$\theta = \arctan(\frac{v}{\omega}). \tag{5.116}$$

Motion along these trajectories is cycloidal in nature. Figure 5.18 shows several sample idealized trajectories for various ratios of v/ω. In these trajectories, all forces other than the Lorentz force have been ignored, and particles are assumed to collide with neutral particles at regular intervals of $1/v$, bringing them to rest after each collision. In this figure, a ratio of $\omega_i/\omega_e = 10$ is used so that the gyrations of both electrons and ions are visible. In reality this ratio is $\mathcal{O}\left(10^4\right)$. The numbers in parentheses represent the altitudes at which the conditions occur.

In the case where $v \ll \omega$, ions and electrons both move in the same direction, given by $\vec{E} \times \vec{B}$. The vertical component of the electromagnetic drift velocity in this case is:

$$\boxed{\vec{v}_h = \frac{\vec{E} \times \vec{B}}{B_0^2}.} \tag{5.117}$$

Figure 5.19 illustrates this.

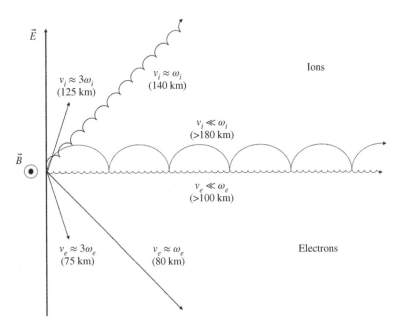

Figure 5.18 Cycloid trajectories of ions and electrons in plane perpendicular to geomagnetic field. Source: Adapted from Figure 31 of [98]. Reproduced with permission of Academic Press.

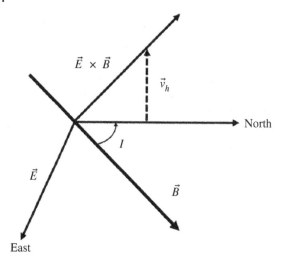

Figure 5.19 Electromagnetic drift due to a west-to-east electric field.

5.5.4 Combined Effects of Neutral Wind and Electromagnetic Drift

We can summarize the effects of neutral winds and electromagnetic drift using equations of motion. The following information is adapted from [98]. We may ignore the effects of gravity, pressure-gradient forces, and electron-ion collisions. Thus Equation 5.87 may be written as:

$$e(\vec{E} + \vec{v}_i \times \vec{B}) - m_i \nu_{in} \left(\vec{v}_i - \vec{U} \right) = 0,$$
$$-e(\vec{E} + \vec{v}_e \times \vec{B}) - m_e \nu_{en} \left(\vec{v}_e - \vec{U} \right) = 0. \tag{5.118}$$

Consider a system with coordinates (x, y, z), but with $z \neq h$ in general. We will assume a magnetic field solely in the z-direction $\vec{B} = \hat{z}B$, while the wind velocity $\vec{U}$, the electric field $\vec{E}$ and the charge carrier velocity $\vec{v}$ have components in all three directions. Thus, in general we may write three scalar equations as follows:

$$\pm eE_x \pm ev_y B - m\nu v_x + m\nu U_x = 0,$$
$$\pm eE_y \mp ev_x B - m\nu v_y + m\nu U_y = 0, \tag{5.119}$$
$$\pm eE_z - m\nu v_z + m\nu U_z = 0.$$

In these and all following expressions, the top sign refers to ions and the bottom sign refers to electrons. Here, v is the collision frequency with neutral particles (ν_{in} for ions and ν_{en} for electrons). As before, m and v are respectively the mass and velocity of the charge carrier in question, and are each given the subscripts i for ions or e for electrons. If we define

$$\vec{F} = \pm e\vec{E} + m\nu\vec{U}, \tag{5.120}$$

we may rewrite the three expressions in Equation 5.119 as follows:

$$F_x = mvv_x \mp ev_yB,$$ (5.121)

$$F_y = mvv_y \pm ev_xB,$$ (5.122)

$$F_z = mvv_z.$$ (5.123)

We wish to solve these equations in terms of $\vec{v}$. v_z is simple to solve for, and is given by:

$$v_z = \frac{1}{mv}F_z.$$ (5.124)

To find v_x and v_y, we must solve Equations 5.121 and 5.122 simultaneously. First, we will sum these two equations, which gives us:

$$F_x + F_y = (mv \pm eB)v_x + (mv \mp eB)v_y.$$ (5.125)

Similarly, differencing these equations gives:

$$F_x - F_y = (mv \mp eB)v_x - (mv \pm eB)v_y.$$ (5.126)

Next we will solve for v_x. To do this, we must first solve Equation 5.125 in terms of v_y, which gives:

$$v_y = \frac{F_x + F_y - (mv \pm eB)v_x}{mv \mp eB}.$$

Substituting this expression into Equation 5.126 yields:

$$F_x - F_y = (mv \mp eB)v_x - (mv \pm eB)\left(\frac{F_x + F_y - (mv \pm eB)v_x}{mv \mp eB}\right).$$

If we multiply both sides by $mv \mp eB$ and rearrange slightly, we obtain:

$$(mv \mp eB)(F_x - F_y) + (mv \pm eB)(F_x + F_y) = \left[(mv \mp eB)^2 + (mv \pm eB)^2\right]v_x.$$

Expanding out each side of the equation allows us to cancel some terms, leaving:

$$mvF_x \pm eBF_y = \left[(mv)^2 + (eB)^2\right]v_x.$$

Solving for v_x, factoring out a $1/vm^2$, and recalling that $\omega = eB/m$, we finally obtain:

$$v_x = \frac{1}{mv}\left[\frac{v^2}{v^2 + \omega^2}F_x \pm \frac{\omega v}{v^2 + \omega^2}F_y\right].$$ (5.127)

If we instead solve Equation 5.125 in terms of v_x, we see that:

$$v_x = \frac{F_x + F_y - (mv \mp eB)v_y}{mv \pm eB}.$$

Substituting this into Equation 5.126 gives:

$$F_x - F_y = (mv \mp eB)\left(\frac{F_x + F_y - (mv \mp eB)v_y}{mv \pm eB}\right) - (mv \pm eB)v_y. \qquad (5.128)$$

Performing similar steps as we did when solving for v_x, we see that the expression for v_y becomes:

$$v_y = \frac{1}{mv}\left[\mp\frac{\omega v}{v^2 + \omega^2}F_x + \frac{v^2}{v^2 + \omega^2}F_y\right]. \qquad (5.129)$$

We may write Equations 5.124, 5.127, and 5.129 in matrix form:

$$\begin{bmatrix} v_x \\ v_y \\ v_z \end{bmatrix} = \begin{bmatrix} k_1 & \pm k_2 & 0 \\ \mp k_2 & k_1 & 0 \\ 0 & 0 & k_0 \end{bmatrix} \begin{bmatrix} F_x \\ F_y \\ F_z \end{bmatrix}. \qquad (5.130)$$

Here, each k is the ratio of a velocity to a force, and therefore measures mobility per unit charge. The mobility components are summarized as follows, including the commonly used names for each component. The middle column represents the direction for which a given mobility component is important, with respect to the directions of $\vec{B}$ and $\vec{E}$.

Component name	Direction	Mobility parameter equation
Direct, longitudinal:	$\left(\parallel \vec{B},\ \parallel \vec{E}\right)$	$k_0 = \dfrac{1}{mv} = \dfrac{1}{Be}\dfrac{\omega}{v}$
Transverse, Pedersen:	$\left(\perp \vec{B},\ \parallel \vec{E}\right)$	$k_1 = \dfrac{1}{mv}\dfrac{v^2}{v^2 + \omega^2} = \dfrac{1}{Be}\dfrac{v\omega}{v^2 + \omega^2}$
Hall:	$\left(\perp \vec{B},\ \perp \vec{E}\right)$	$k_2 = \dfrac{1}{mv}\dfrac{\omega v}{v^2 + \omega^2} = \dfrac{1}{Be}\dfrac{\omega^2}{v^2 + \omega^2}$

These results agree with our previous conclusions regarding neutral winds and electromagnetic drift. Recall our definition of $\vec{F}$, from Equation 5.120. We see that this force takes into account the contribution of both neutral winds and an electric field. Forces due to winds in the same direction as $\vec{B}$, or due to the component of the electric field parallel to $\vec{B}$, will only move a charge carrier parallel to the magnetic field. Therefore, the charge carrier will experience no magnetic force. In other words, a charge carrier moving in the h-direction is only affected by direct mobility, so its motion will remain in the h-direction. This is illustrated by the fact that $v_z = k_0 F_z$, as seen in Equation 5.130.

The components of the wind and electric field which are in the xy-plane (i.e. the plane perpendicular to $\vec{B}$) are more complicated. Motion of charge carriers due to winds in the direction of the electric field, or due to the electric field itself, are mediated by Pedersen mobility, k_1. Motion due to winds perpendicular to the electric field or due to perpendicular electromagnetic drift are mediated by Hall mobility, k_2. In general, motion will be dependent upon all of these effects and will be neither parallel nor perpendicular to $\vec{E}$. Because of this, the trajectory is dependent upon both Pedersen and Hall mobilities. The relative contribution of each is dependent upon v and ω. All of this is reflected in Equation 5.130, which shows that the components of the trajectory, v_x and v_y, are each dependent on a weighted sum of F_x and F_y. The weights, k_1 and k_2 are determined by the values of v and ω.

It is helpful to consider different cases under specific conditions. For motion in the plane perpendicular to $\vec{B}$, the direction of motion is dependent on the ratio v/ω. Note that v decreases exponentially with altitude, while ω is nearly constant, so v/ω decreases with altitude. This ratio also differs for electrons and ions. Different cases of v/ω are considered, and the heights at which these ratios occur for ions and electrons are presented [92].

- *Electric field parallel to magnetic field* $\left(\vec{E} \parallel \vec{B}\right)$: At all heights, an electric field parallel to $\vec{B}$ causes a drift velocity of $\pm eE_0/mv$. Ions and electrons move parallel to $\vec{B}$ in opposite directions, producing a current. This current is primarily carried by electrons, for which k_0 is greater than for ions.

  At low altitudes, v is large due to the high number density n of neutral molecules, so the mobility k_0 is small and the velocities of ions and electrons are both small. Thus little to no current is produced at low altitudes (i.e. conductivity is low). Conversely, at higher altitudes n decreases, so v also decreases. Thus the mobility and therefore the drift velocities are higher, producing larger currents (i.e. high conductivity).

- *Wind parallel to magnetic field* $\left(\vec{U} \parallel \vec{B}\right)$: At all heights, wind parallel to $\vec{B}$ causes a drift of $mvU/mv = U$ which is also parallel to the magnetic field. Note that this speed is independent of m and v and charge q, so ions and electrons both move at the same speed in the same direction, and no net current is induced.

- *Electric field perpendicular to magnetic field* $\left(\vec{E} \perp \vec{B}\right)$: $v \gg \omega$ (up to 100 km for ions, up to 60 km for electrons): ω is negligible compared with v, so $k_1 = k_0 \gg k_2$. Therefore, motion is identical to the case when $\vec{E} \parallel \vec{B}$. Ions and electrons move in opposite directions, but because of the large collision frequency, they move slowly and conductivity is low at this altitude.

  - $v \approx \omega$ (~125 km for ions, ~75 km for electrons): An electric field produces drift inclined to itself. The angle of this trajectory is given by $\arctan(v/\omega) \approx 45°$.

  - $v \ll \omega$ (above 150 km for ions, above 90 km for electrons): v is negligible compared with ω, so $k_1 \ll k_2 \approx 1/eB$. The drift velocity is $eE_0/eB_0 = E_0/B_0$

is due to electromagnetic drift. This drift, also known as Hall drift, may be written vectorially as $\vec{E} \times \vec{B}/B^2$ for both ions and electrons. $v \ll \omega$ for both ions and electrons above 150 km. In other words, at this height all charge carriers move in the same direction, which can be seen in Figure 5.18. Since the drift velocity is independent of m, v, and q, ions and electrons move at the same speed. Thus, since all charge carriers move in the same direction at the same speed, no net currents are formed at this altitude.

- *Wind perpendicular to magnetic field* $\left(\vec{U} \perp \vec{B} \right)$: $v \gg \omega$ (up to 100 km for ions, up to 60 km for electrons): ω is negligible compared with v, so $k_1 = k_0 \gg k_2$. Therefore, motion is identical to the case when $\vec{E} \parallel \vec{B}$. Ions and electrons move in the same direction at the same speed, so no net current is induced.
  - $v \approx \omega$ (~125 km for ions, ~75 km for electrons): Wind produces drift inclined to itself. The angle of this trajectory is given by $\arctan(v/\omega) \approx 45°$.
  - $v \ll \omega$ (above 150 km for ions, above 90 km for electrons): v is negligible compared with ω, so $k_1 \ll k_2 \approx 1/eB$. The drift velocity is mvU/eB_0, and since v is very small, the drift velocity is almost zero. In other words, wind does not move ions or electrons across the magnetic field at this altitude. Since there is no movement, no current is produced.

Electric currents typically flow between altitudes of 60 and 150 km, though currents are small below 100 km because electron density N_e is small. The heights of peak conductivity correspond to the E and lower F1 layers. In these regions, ions and electrons travel in different directions due to the ratio v/ω (see Figure 5.18), creating strong currents called electrojets. Above about 150 km, currents become small again because ions and electrons begin to move in the same direction.

As we have seen in this section, conductivity is an important factor in ionospheric propagation. Conductivity, denoted σ, can be expressed via the point Ohm's law, which represents the relationship between the current density induced by an electric field. At an atomic scale, current density can be associated to the relative velocity of charge carriers (see Section B.4). In the ionosphere, the relationship between these parameters is given by:

$$\vec{J} = N_e e(\vec{v}_i - \vec{v}_e) = \overline{\overline{\sigma}}\vec{E}, \tag{5.131}$$

where $\vec{J}$ is current density and conductivity $\overline{\overline{\sigma}}$ is in general a 3×3 tensor due to the anisotropy of the ionosphere. Following our previous discussion, we can write $\overline{\overline{\sigma}}$ more explicitly as

$$\overline{\overline{\sigma}} = \begin{bmatrix} \sigma_1 & -\sigma_2 & 0 \\ \sigma_2 & \sigma_1 & 0 \\ 0 & 0 & \sigma_0 \end{bmatrix}, \tag{5.132}$$

where the components σ_0, σ_1, and σ_2 are all positive quantities, defined as follows:

Direct, longitudinal: $\sigma_0 = Ne^2\left(k_{0e} + k_{0i}\right),$

Transverse, Pedersen: $\sigma_1 = Ne^2\left(k_{1e} + k_{1i}\right),$

Hall: $\sigma_2 = Ne^2\left(k_{2e} - k_{2i}\right).$

Note that despite the opposite signs of k_{2e} and k_{2i}, σ_2 is always positive because $k_{2e} \gg k_{2i}$ at all heights in the ionosphere [98].

In general, the current $\overline{\overline{\sigma}}\vec{E}$ in the ionosphere can have a vertical component. However, the conductive region of the ionosphere is limited in vertical extent, and charged particles will begin to accumulate near the boundaries of the conductive region, since they are unable to flow into the adjacent regions of low conductivity. This induces an electric polarization that counteracts the vertical component of the electric field $\vec{E}$, eventually reducing the vertical flow of current to zero. Therefore, the flow of current will only be horizontal, and we can reduce $\overline{\overline{\sigma}}$ to a matrix of the form:

$$\overline{\overline{\sigma}}' = \begin{bmatrix} \sigma_{xx} & \sigma_{xy} & 0 \\ -\sigma_{yx} & \sigma_{yy} & 0 \\ 0 & 0 & 0 \end{bmatrix}. \tag{5.133}$$

Here, x refers to the local direction of magnetic south and y to the local direction of magnetic east. The components of this matrix are dependent on the local magnetic dip angle I – in other words, they vary with geomagnetic latitude. The components of the matrix in Equation 5.133 may be written explicitly as follows:

$$\sigma_{xx} = \frac{\sigma_0 \sigma_1}{\sigma_0 \sin^2 I + \sigma_1 \cos^2 I},$$

$$\sigma_{xy} = \frac{\sigma_0 \sigma_2 \sin I}{\sigma_0 \sin^2 I + \sigma_1 \cos^2 I}, \tag{5.134}$$

$$\sigma_{yy} = \sigma_1 + \frac{\sigma_2^2 \cos^2 I}{\sigma_0 \sin^2 I + \sigma_1 \cos^2 I}.$$

Typically, $\sigma_0 \gg \sigma_1$ and $\sigma_0 \gg \sigma_2$, so at most geomagnetic latitudes, $\sigma_{xx} \approx \sigma_1/\sin^2 I$, $\sigma_{xy} \approx \sigma_2/\sin I$ and $\sigma_{yy} \approx \sigma_1$. However, this changes near the magnetic equator, where $I = 0$. There, the conductivities reduce to

$$\sigma_{xx} = \sigma_0,$$

$$\sigma_{xy} = 0, \tag{5.135}$$

$$\sigma_{yy} = \frac{\sigma_1^2 + \sigma_2^2}{\sigma_1} = \sigma_3.$$

Near the magnetic equator, the east-west conductivity σ_{yy} is equivalent to σ_3, which is referred to as Cowling conductivity. This conductivity is very large, which gives rise to a powerful eastward-directed current known as the equatorial electrojet. The extent of the equatorial electrojet is limited to a few degrees on either side of the magnetic equator, where $\sigma_0 \sin^2 I \ll \sigma_1 \cos^2 I$. Later, in Section 6.4, we will see that the equatorial electrojet is partly responsible for an anomaly in the ionization of the equatorial F2 region. For further details on ionosphere conductivity, the interested reader may refer to [98, pp. 136–142].

5.5.5 Continuity Equation

The rate of change in the electron density of the ionosphere with respect to time is related to the rate of electron production q, the rate of electron recombination l_r, the rate of electron attachment l_c, and the plasma transport term d [72]. This rate of change can be expressed as:

$$\frac{dN}{dt} = q - L + d, \tag{5.136}$$

where $N = N_e$ for electrons and $N = N_i$ for ions. Note that the loss term L is negative because it represents removal of electrons and ions from the ionosphere. Substituting in Equations 5.41 and 5.86 for q, L, and d respectively results in:

$$\frac{dN}{dt} = \zeta \sigma n S - \frac{\beta \alpha N^2}{\beta + \alpha N} - \nabla \cdot (N\vec{v}), \tag{5.137}$$

where ζ, α, and β are the effective coefficients for ionization, recombination and attachment, respectively. These coefficients take into account the contribution of all chemical species involved. $\vec{v}$ represents the drift velocity of electrons ($\vec{v} = \vec{v}_e$) or ions ($\vec{v} = \vec{v}_i$) due to diffusion, neutral winds and electromagnetic drift. As discussed previously, the contribution of the horizontal components to $\nabla \cdot (N\vec{v})$ are negligible, so we may reduce Equation 5.137 to:

$$\boxed{\frac{dN}{dt} = \zeta \sigma n S - \frac{\beta \alpha N^2}{\beta + \alpha N} - \frac{d(Nv_h)}{dh},} \tag{5.138}$$

where v_h is the vertical component of the drift velocity.

Below about 200 km, in the D, E, and F1 layers, the transport term d is small, so the photochemical terms q and L dominate. The time constant of recombination reactions is short, meaning loss occurs quickly and frequently. Any ion-electron pairs which are produced will rapidly recombine, meaning $q \approx L$, so dN_e/dt is small. Transport effects can be added as a perturbation if necessary [98]. In the F2 layer, d becomes comparable with q and L, creating a balance between photochemical and transport phenomena. At even greater heights, d dominates the continuity equation, and transport phenomena become the most important consideration.

6

Experimental Observation of the Ionosphere

6.1 Introduction

In Chapter 5, we presented a mathematical framework of the ionosphere's basic structure and behavior. It is essential to complement this theoretical approach with empirical observation. A number of measurement techniques have been developed to collect data on various ionospheric parameters. In many cases, this data validates the theoretical predictions discussed in Chapter 5. However, it is also common for experiments to produce surprising results that reveal the shortcomings of the prevailing theory.

There are a few major parameters that can be used to characterize the ionosphere; namely: electron density, electron temperature, ion temperature, and ion composition. These parameters are dependent on a number of factors, including altitude, latitude, time of day, season, and solar activity. There is no single measurement technique that can provide global data on all parameters at all times. So instead, several complementary techniques are used, each providing their own strengths. In Section 6.2, we will discuss a few of the most prominent ionospheric measurement techniques.

In the remainder of this chapter, we will discuss some insights that ionospheric observation has provided. For example, in Section 6.3, we will discuss the layered structure of the ionosphere in detail. Each layer has its own distinct properties, meaning the ionosphere behaves very differently at different altitudes. The layer naming convention presented in Section 6.3 will be adopted throughout the remainder of this book, so it is strongly recommended that the reader familiarize themselves with the concepts in this section.

Section 6.4 contains a collection topics pertaining to the variability of the ionosphere. The degree to which a given point in the ionosphere is ionized is directly dependent on the amount of solar radiation that reaches that point. In general,

Tropospheric and Ionospheric Effects on Global Navigation Satellite Systems, First Edition.
Timothy H. Kindervatter and Fernando L. Teixeira.
© 2022 The Institute of Electrical and Electronics Engineers, Inc. Published 2022 by John Wiley & Sons, Inc.

electron density in the ionosphere follows predictable patterns – e.g. greater during the day than at night, greater in the summer than the winter, and greater at equatorial latitudes than polar latitudes.

If solar radiation were the *only* factor that determined ionization, electron density in the ionosphere would follow these trends very cleanly. Of course, solar radiation is not the only factor that affects electron density. As we discussed in Section 5.5, plasma transport effects can have a significant effect, especially in the F2 layer. Section 6.4.1 discusses a few of the surprising exceptions to the expected electron density trends, also known as *ionospheric anomalies*.

Ionization also follows a mostly predictable pattern with respect to solar activity. Every 11 years, the sun cycles between high and low solar activity, and electron density in the ionosphere largely follows suit. However, occasionally the sun will suddenly and violently erupt in events such as solar flares and coronal mass ejections (CMEs). If one of these eruptions is pointed toward Earth, it can cause dramatic spikes in ionization, which can cause serious problems for satellite communications and GNSS. These topics will all be touched upon in Section 6.4.2.

As we saw in Section 5.5, the Earth's magnetic field is a significant factor in plasma transport. Further, we will see in Section 7.2 that the presence of the geomagnetic field induces an inherent anisotropy in the ionosphere that changes how electromagnetic (EM) waves propagate. It's clear that the state of Earth's magnetic field is highly relevant to ionospheric study. Unfortunately, its behavior is not simple. Globe-spanning current systems, which are driven by various mechanisms, cause fluctuations in the geomagnetic field. Section 6.4.3 discusses these different mechanisms, as well as the standard indices used to quantify their effect.

This chapter finally concludes with Section 6.4.4, which discusses the effect of irregularities in the density of the ionosphere. For example, localized disturbances such as thunderstorms can create small-scale oscillations, or fluid dynamics can introduce bubbles of differing electron density. These disturbances can affect the way GNSS signals propagate. For example, a phenomenon called spread F creates a patchy, inhomogeneous structure in the ionosphere that degrades the integrity of a signal in a phenomenon called ionospheric scintillation. Scintillation will be discussed more thoroughly in Section 7.4.

6.2 Ionospheric Measurement Techniques

Nearly one century ago, Sir Edward Appleton demonstrated the existence of the ionosphere by transmitting radio signals into the atmosphere and observing the interference patterns in their reflections. Since then, researchers have developed a number of theoretical models of the ionosphere's behavior and composition.

These include Chapman's, which was discussed in Section 5.4, as well as the models that will be discussed in Chapter 8. To validate such models, researchers also needed methods for experimentally confirming these theoretical predictions which required data on ionospheric parameters. To that end, a number of measurement techniques were developed. A few of the most widely-used ones will be presented in this section.

Ionospheric measurement methods can be classified into two major groups: remote sensing and in situ. Remote sensing techniques utilize radio signals to probe the ionosphere from afar, relying on reflection, refraction, or scattering of the waves to infer ionospheric characteristics. In contrast, in situ measurements are performed by probing the ionosphere directly. To accomplish this, instruments are installed on rockets which are then flown through the region of interest.

6.2.1 Ionosondes

Ionospheric sounding was the first method by which the ionosphere was measured, and is still one of the most widely used techniques today. Ionospheric sounders, or *ionosondes*, are frequency-swept pulsed radars which operate by transmitting signals toward the ionosphere. Signals of certain frequencies will be reflected due to the high concentration of free electrons. The ionosonde receives these reflected signals and records the time delay between transmission and reception. The frequency of transmission is gradually increased from pulse to pulse, allowing the ionosonde to generate a plot of time delay vs. frequency known as an *ionogram*.

The measured time delay is equivalent to the two-way travel time of the signal. Since radio waves propagate through the atmosphere at approximately the speed of light, their travel time can be used to deduce the height at which they reflect. This height is directly dependent on the number of free electrons, so ionograms provide a representation of the electron density profile of the ionosphere. Since electron density is arguably the most important characteristic of the ionosphere, ionograms have become one of the most important tools for ionospheric study. A worldwide network of ionosondes has collected data across all latitudes for more than 50 years, granting scientists and engineers a nuanced view of the structure and dynamics of the ionosphere.

Ionospheric sounding can be accomplished from either below or above the ionosphere, and both methods must be used in tandem to achieve a full picture. Most ionosondes are "bottomside" sounders, which are located on the ground. These include both vertical sounders, which transmit directly upward, and oblique sounders, which transmit upward at an angle. "Topside" sounders are space-based instruments (typically installed on satellites or aboard rockets) which fly above the ionosphere and transmit signals downward. Topside sounding is essential for characterizing the uppermost portion of the ionosphere, since there is a certain height at which signals from a bottomside sounder will escape the ionosphere entirely, producing no reflection for the ground-based ionosonde to receive. A few of the best known topside sounders include the Alouette 1 and 2 satellites, the International Satellites for Ionospheric Study (ISIS 1 and 2), and the Ionospheric Sounding Satellite (ISS-b) [101–103]. Although topside sounding is of critical importance for ionospheric observation, it follows the same basic principle as bottomside sounding. Therefore, in this section, we will only consider the case of ground-based (i.e. bottomside) sounders.

For simplicity, throughout this section, a lossless, isotropic ionosphere is assumed. These complicating factors may be reintroduced by using the Appleton–Hartree equation. The reader may refer to Section 7.2 for a discussion on this topic. In order interpret the information in an ionogram, we must consider what it means for a signal to be reflected by the ionosphere. Strictly speaking, signals are not actually reflected, but rather refracted so strongly that they return to Earth. This can be seen in Figure 6.1. If we model the ionosphere as a horizontally stratified medium, Snell's law describes how a signal will refract as it propagates.

$$n(h) \sin \theta(h) = \sin \theta_I. \tag{6.1}$$

Here, $n(h)$ is the index of refraction of the ionosphere as a function of height, $\theta(h)$ is the angle of the ray with respect to vertical, and θ_I is the angle of incidence at the bottom of the ionosphere. For simplicity, it has been assumed that the index of

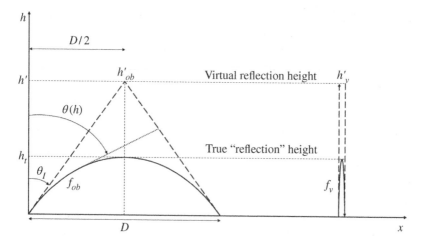

Figure 6.1 True trajectory vs. virtual trajectory for an obliquely incident wave (left) and a vertically incident wave (right). Source: Adapted from Figure 11.13 of [45]. Reproduced with permission of Wiley.

refraction below the ionosphere is 1. The signal will be refracted in an arc, reaching some maximum height before angling back toward the Earth. The top of this trajectory is considered the signal's point of "reflection." At some point, the propagating ray will be refracted so much that it travels perpendicular to the vertical, i.e. $\theta = 90°$, which reduces Equation 6.1 to

$$n(h_t) = \sin \theta_I, \tag{6.2}$$

where h_t is the true height of the signal's "reflection." From Appendix B.4.2, we see that the index of refraction is related to the plasma frequency of a conducting medium by the expression:

$$n^2 = 1 - \frac{\omega_N^2}{\omega^2}, \tag{6.3}$$

where we have denoted the plasma frequency as ω_N and the frequency of the transmitted signal as ω. The plasma frequency of the ionosphere is dependent on electron density N which, as we saw previously, is dependent on height. Therefore, the ionospheric plasma frequency itself is also a function of height. Using this in conjunction with Equation 6.2, we can write

$$1 - \frac{\omega_N^2}{\omega^2} = \sin^2\theta_I. \tag{6.4}$$

Rearranging slightly, we can obtain $1 - \sin^2\theta_I$ on the left hand side, which reduces to $\cos^2\theta_I$. Frequencies can also be represented in Hz by canceling a factor of 2π. Thus we see that

$$\cos^2\theta_I = \frac{f_N^2}{f^2} \Rightarrow f = f_N \sec \theta_I. \tag{6.5}$$

This equation tells us that a signal propagating at frequency f, launched at an initial incidence angle θ_I will reflect at some height h_t where f_N takes on a value such that Equation 6.5 is satisfied.

In the case of vertical sounding, $\theta_I = 0$, so $\sec \theta_I = 1$. Therefore:

$$f_v = f_N \sec 0 = f_N, \tag{6.6}$$

which indicates that the signal will reflect at a height where its frequency of propagation (denoted f_v here to emphasize the case of vertical sounding) is equal to the local plasma frequency. In the more general case of oblique sounding, θ_I is arbitrary, so we have the equation:

$$f_{ob} = f_N \sec \theta_I. \tag{6.7}$$

Using Equation 6.6, we can rewrite this as

$$f_{ob} = f_v \sec \theta_I \tag{6.8}$$

This relation is known as the secant law.

Additionally, when $\theta_I = 0$ we see from Equation 6.1 that $n(h_t) = 0$. In other words, the signal will reach its maximum propagation height when the index of refraction is zero. Recalling from Section B.4.2 that $\omega_N^2 = Ne^2/m_e\varepsilon_0$, we can write f_N as:

$$f_N = \frac{\omega_N}{2\pi} = \frac{1}{2\pi}\sqrt{\frac{Ne^2}{m_e\varepsilon_0}} = 8.98\sqrt{N}. \tag{6.9}$$

In this expression we have used $e = 1.602 \times 10^{-19}$ C, $m_e = 9.11 \times 10^{-31}$ kg, and $\varepsilon_0 = 8.85 \times 10^{-12}$ m^{-3} kg^{-1} s^4 A^2. We see that the value of the plasma frequency is dependent on the electron density N. Therefore, layers of the ionosphere with less ionization will reflect lower frequencies, and layers with higher ionization will reflect higher frequencies. If the frequency of the wave exceeds the largest plasma frequency of any layer of the ionosphere, it will escape into space and not be reflected at all.

The maximum electron density of the ionosphere is on the order of 10^{11}–10^{12} e$^-$/m^3, occurring at a height of 300–500 km above the surface of the Earth. Therefore, according to Equation 6.9, in order to pass through the ionosphere signals must have a frequency on the order of 10^6–10^7 Hz. This corresponds to the upper end of the high frequency (HF) band, so we can be assured that any signals in the very high frequency (VHF) band or higher will pass through the ionosphere. This is one of the reasons that L-band (ultra high frequency [UHF]) frequencies were chosen for GNSS signals.

Modeling this refraction problem is fairly cumbersome, so it is more convenient to instead consider the equivalent case where the transmitted signal travels along a straight line as though in a vacuum, then reflects at some virtual reflection point directly above the peak of the true trajectory. Upon reflection, the signal travels along a straight line back to Earth, arriving in the same location as the true trajectory. This scenario is shown in Figure 6.1. In this case, we can simply express $\sec\theta_I$ in terms of the geometry of the equivalent problem as follows:

$$\sec\theta_I = \sqrt{\left(\frac{D}{2h'}\right)^2 + 1}, \tag{6.10}$$

where D is the horizontal distance traversed by a reflected signal and h' is its virtual reflection height.

Breit and Tuve's Theorem

We must justify our assertion that the simplified virtual reflection problem is, in fact, equivalent to our original problem. In our original problem, we assumed a horizontally stratified ionosphere, so the signal will experience the same index of refraction on the way down as it did on the way up. Thus the true trajectory will be a symmetric arc, so it is valid to assume that the virtual reflection point is directly above the vertex of the arc.

However, we must also show that the travel time of the signal is identical in both cases. The following derivation has been adapted from [45]. First we will consider the travel time along the virtual trajectory. Referring to the geometry of Figure 6.1, we see that the distance traversed by a signal from the ground to the virtual reflection point is given by the hypotenuse of a triangle with legs $D/2$ and h'_{ob}. The angle between the vertical and the virtual trajectory is θ_I, so we can write the length of the hypotenuse as $(D/2) \csc \theta_I$. The signal traverses an identical path on the way down, so the total distance traversed is $D \csc \theta_I$. For the virtual trajectory, the signal is assumed to be traveling in a vacuum, so it travels at velocity c. Therefore, the total travel time of the signal is given by

$$\Delta t' = \frac{D}{c} \csc \theta_I. \tag{6.11}$$

The travel time of along the true trajectory is a bit more complicated. The information in the signal is carried at the group velocity $v_g < c$, which is dependent on the group index of refraction n_g. From Section A.2, we have:

$$n_g = \frac{c}{v_g} = c \frac{dk}{d\omega}. \tag{6.12}$$

We may rewrite k as $n\omega/c$, and substituting in n using Equation 6.3 we see that:

$$n_g = c \frac{d}{d\omega} \left(\frac{\omega}{c} \sqrt{1 - \frac{\omega_N^2}{\omega^2}} \right). \tag{6.13}$$

Using the product rule, we can evaluate this derivative, and after some algebraic manipulation, we get:

$$n_g = c \frac{1}{c \sqrt{1 - \frac{\omega_N^2}{\omega^2}}} = \frac{1}{n}. \tag{6.14}$$

As $n \to 0$, $n_g \to \infty$, and $v_g \to 0$. In other words, as the signal approaches the top of its trajectory, its group velocity decreases, finally reaching zero at the reflection point before increasing again in the opposite direction.

The plasma frequency ω_N is a function of height, so the group index and thus the group velocity are as well. The time for the signal to propagate some differential distance ds along the true trajectory is:

$$dt = \frac{ds}{v_g(h)} = \frac{dx}{v_g(h) \sin \theta(h)}, \tag{6.15}$$

where x indicates the horizontal direction and $\theta(h)$ is the angle made with respect to vertical at height h. Using Equations 6.1, 6.12 and 6.14, we see that

$$dt = \frac{dx}{cn \sin \theta(h)} = \frac{dx}{c \sin \theta_I}. \tag{6.16}$$

To find the total travel time of the signal, we must integrate along the whole path:

$$\Delta t = \frac{\csc\theta_I}{c}\int_0^D dx = \frac{D}{c}\csc\theta_I. \tag{6.17}$$

We see that this result is equivalent to the expression we found in Equation 6.11, so we conclude that

$$\Delta t = \Delta t', \tag{6.18}$$

confirming that the signal's travel time is identical in both cases. This result is known as Breit and Tuve's theorem.

Martyn's Theorem

It is also possible to prove that the virtual reflection height is identical no matter the angle of incidence of the signal. Consider some signal traveling at an oblique angle. Using Equation 6.3, we can say $n_{ob}^2 = 1 - f_N^2/f_{ob}^2$. Similarly, a signal traveling vertically has $n_v^2 = 1 - f_N^2/f_v^2$. Via the secant law, we can write

$$n_v^2 = 1 - \frac{f_N^2}{f_{ob}^2\cos^2\theta_I} \Rightarrow f_N^2 = f_{ob}^2\cos^2\theta_I\left(1 - n_v^2\right). \tag{6.19}$$

If we plug this result into our equation for n_{ob}^2, we get

$$n_{ob}^2 = 1 - \frac{f_{ob}^2\cos^2\theta_I\left(1 - n_v^2\right)}{f_{ob}^2} = \sin^2\theta_I + n_v^2\cos^2\theta_I. \tag{6.20}$$

For the oblique signal, Snell's law gives $n_{ob}^2\sin^2\theta(h) = \sin^2\theta_I$, assuming $n = 1$ below the ionosphere. Thus, we can write

$$n_{ob}^2 = n_{ob}^2\sin^2\theta(h) + n_v^2\cos^2\theta_I. \tag{6.21}$$

After rearranging and simplifying, we finally see that

$$n_{ob}^2\cos^2\theta(h) = n_v^2\cos^2\theta_I. \tag{6.22}$$

The travel time along the oblique path to the maximum height h_t is given by:

$$\frac{\Delta t_{ob}}{2} = \int_{S_0}^{S_{h,t}} \frac{ds}{v_{g,ob}} = \int_{S_0}^{S_{h,t}} \frac{ds}{cn_{ob}} = \int_0^{h_t} \frac{dh}{cn_{ob}\cos\theta(h)}, \tag{6.23}$$

where the bounds S_0 and $S_{h,t}$ are the positions along the oblique path S at heights 0 and h_t, respectively. Using Equation 6.22, we can write:

$$\frac{\Delta t_{ob}}{2} = \int_0^{h_t} \frac{dh}{cn_v\cos\theta_I} = \sec\theta_I\int_0^{h_t} \frac{dh}{c/n_{g,v}} = \sec\theta_I\int_0^{h_t} \frac{dh}{v_{g,v}} = \sec\theta_I\frac{\Delta t_v}{2}. \tag{6.24}$$

Using Breit and Tuve's theorem, we know that the travel time along a real path Δt is the same as the travel time along a virtual path $\Delta t'$. Thus,

$$\frac{\Delta t'_{ob}}{2} = \sec \theta_I \frac{\Delta t'_v}{2}. \tag{6.25}$$

Rearranging this expression slightly and multiplying both sides by c, we get

$$c\frac{\Delta t'_{ob}}{2} \cos \theta_I = c\frac{\Delta t'_v}{2}. \tag{6.26}$$

Since a signal traveling along a virtual path is assumed to propagate at the speed of light, the quantity $c\Delta t'/2$ is equivalent to the one-way virtual path length. For a vertical path, the one-way path length is simply the virtual reflection height, so the right hand side reduces to h'_v:

$$c\frac{\Delta t'_{ob}}{2} \cos \theta_I = h'_v. \tag{6.27}$$

An oblique path's length is greater than the virtual reflection height, but they may be related via trigonometry. Figure 6.1 shows the one-way virtual path as a dashed line, the hypotenuse of a right triangle with legs $D/2$ and h'_{ob}. From the geometry of the figure, it can be seen that

$$h'_{ob} = c\frac{\Delta t'_{ob}}{2} \cos \theta_I, \tag{6.28}$$

and we finally conclude that

$$\boxed{h'_{ob} = h'_v.} \tag{6.29}$$

In other words, the virtual reflection height is identical regardless of the transmitted signal's angle of incidence. This result is known as Martyn's theorem.

Interpreting Ionograms

Via Breit and Tuve's theorem, the time delay measured on the ionogram is identical for both the real path and the virtual path. The time delay axis of an ionogram can be scaled by $c/2$ to obtain a plot of frequency vs. virtual height. It should be noted that since signals actually travel at a velocity slightly less than c through the atmosphere, the true height of reflection h_t will be lower than h'.

From the information we have presented in this section, it is possible to predict the shape of an ionogram. An ionosonde will transmit a signal, which will return with some time delay that can be scaled to a virtual reflection height h'. As the ionosonde sweeps upward in frequency, the reflection height will increase, corresponding to the larger electron density at greater altitudes.

In certain regions of the ionosphere, the electron density changes slowly with height, which can be interpreted on the h vs. N curve as a region with a steep slope.

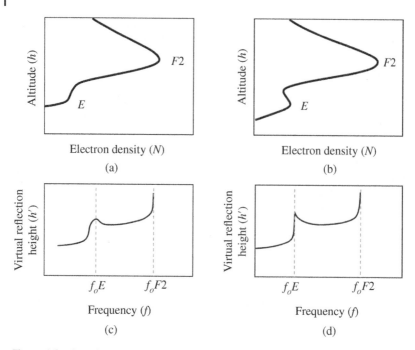

Figure 6.2 Sample electron density profiles (a, b) and corresponding idealized ionograms (c, d). Only E and F2 layers are shown. Source: Adapted from Figure 11.7 of [45]. Reproduced with permission of Wiley.

Consider a signal with a reflection height in this region. As it approaches its reflection height, it will slow down considerably, but due to the slow change of N with respect to height, it will need to travel a relatively large distance before reaching the reflection height. In other words, this signal will spend a large amount of time in the region where it is slowed down the most, causing an especially large time delay.

The large time delay creates a "hump" in an ionogram. If the electron density profile becomes steep without becoming completely vertical, the ionogram will exhibit a rounded hump. This is shown in Figure 6.2a,c, in which the electron density of the E layer becomes steep, but not completely vertical. On the other hand, if the electron density profile reaches a local maximum, there will be some point at which its slope is completely vertical. This creates a sharp cusp in the ionogram, which can be seen in Figure 6.2b,d.

Regardless of the shape of the ionogram's hump, the frequency which produces the maximum time delay within a given layer is called the critical frequency of that region.[1] The critical frequency of a given region is denoted f_o followed by the layer it corresponds to. For example, the critical frequency of the E layer is denoted f_oE. At frequencies higher than f_oF2, signals escape the ionosphere entirely, so the reflection height is infinite.

The peak electron density for a given layer occurs at the critical frequency. Recalling that any point on an ionogram represents the point where $f = f_N$, we see via Equation 6.9 that the peak electron density N_m is given by

$$N_m = \frac{f_o^2}{80.6}. \tag{6.30}$$

The peak electron density of a given layer is denoted in a similar fashion to the critical frequency. For example, the peak density in the E layer is denoted N_mE. Recall that the virtual height h' represented on an ionogram does not represent the true height of reflection h_t. Therefore an ionogram cannot be directly used to determine the true height of a region's peak electron density, h_m. The virtual height is related to true height by the equation

$$h' = \int_0^{h_t} n_g dh = \int_0^f n_g \frac{dh}{df_N} df_N + h(0), \tag{6.31}$$

where $h(0)$ is the height at the bottom of the ionosphere, for which f_N is assumed to equal zero. This expression can be used to find the electron density profile, but this process is not straightforward, so it will not be presented in this book. The interested reader may find a derivation in [98, pp. 55–58].

Maximum Usable Frequency

Suppose one wishes to establish a skywave link between two locations on the ground separated by a fixed horizontal distance D (curvature of the Earth may be neglected). The link will transmit signals obliquely at some frequency f_{ob}. Reception at D is not possible for all values of f_{ob}; in fact there is a maximum frequency beyond which a transmitted signal will be unable to arrive at this distance. This frequency, called the maximum usable frequency (MUF) is an important design consideration for skywave links.

---

1 In this section, we have ignored the anisotropy of the ionosphere. In reality, this anisotropy will give rise to two waves: an ordinary wave and an extraordinary wave (see Section B.6). Each wave will have its own separate ionogram curve, with separate critical frequencies. The ordinary wave is identical to the case we have described in this section. The extraordinary wave's ionogram is shifted slightly to the right; i.e. the extraordinary wave reflects at slightly higher frequencies. Ordinary wave critical frequencies are denoted f_o and extraordinary wave critical frequencies are denoted f_x. For more information, the reader may consult [45, 98].

It is possible to determine whether propagation at distance D is possible for a given f_{ob} through the combined information provided by an ionogram of the local ionospheric conditions, and a plot known as a transmission curve. Transmission curves will be briefly discussed before we return to the topic of MUF.

Using Equations 6.8 and 6.10, we see that

$$\frac{f_{ob}}{f_v} = \sec \theta_I = \sqrt{\left(\frac{D}{2h'}\right)^2 + 1}, \tag{6.32}$$

which, when rearranged in terms of h', becomes

$$h' = \frac{D}{2\sqrt{\left(\frac{f_{ob}}{f_v}\right)^2 - 1}}. \tag{6.33}$$

This equation can be used to plot the virtual height h' that corresponds to a vertical frequency f_v, for some choice of fixed D and fixed f_{ob}. This type of plot is called a transmission curve, and is based solely on the geometry of the propagation path. In other words, information on local ionospheric conditions is not included. Figure 6.3 shows a family of transmission curves for a fixed distance D and various propagation frequencies f_{ob}.

Using Breit and Tuve's theorem and Martyn's theorem, we showed that the virtual reflection height of a signal is independent of the signal's angle of incidence. Therefore ionograms and transmission curves can be easily superimposed, since both of their vertical axes display virtual reflection height h'. Figure 6.4 shows an example of such a superposition.

If the ionogram intersects a transmission curve (defined by some fixed D and fixed f_{ob}), it indicates that a skywave link over the distance D may be established by transmitting that frequency f_{ob}. At a certain frequency, the transmission curve will no longer intersect the ionogram at all, meaning no link may be established on that frequency over a distance D. This can be seen in Figure 6.4, where point M shows that the highest frequency that may be used to transmit over D is approximately $f_{ob} = 13$ MHz. This highest frequency is appropriately named the MUF. For further information on skywave links, ionograms, transmission curves, and MUF, the reader is encouraged to consult Chapter 11 of [45].

Each layer of the ionosphere has its own MUF, and the MUF varies for different values of D. Therefore, in order to be completely unambiguous, when measured MUF values are reported, they are followed by the distance (measured in km) and the name of the layer that they correspond to. For example, the MUF for the $F1$ layer at a distance of 2000 km is denoted MUF(2000)$F1$. If no layer is listed, it is implied that the data refers to the F2 layer. In Section 8.2.2, we will see that one of the most commonly used GNSS error mitigation models, NeQuick, utilizes MUF as an important input.

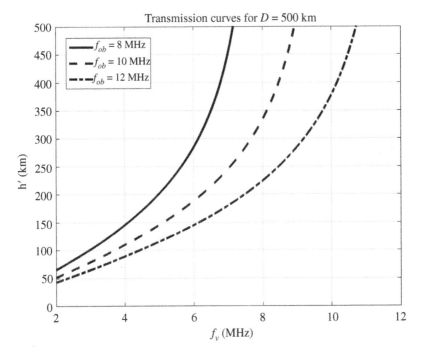

Figure 6.3 Family of transmission curves for fixed D and various fixed values of f_{ob}.

6.2.2 Incoherent Scatter Radar

Another powerful method for monitoring the ionosphere is incoherent scatter radar (ISR). This technique involves transmitting radar pulses toward the ionosphere and looking for the backscattered radiation that bounces off of the numerous free electrons. Much like ionosondes, these radars receive reflected signals and use the time delay between transmission and reception to determine the height of reflection. However, unlike ionosondes, ISRs typically operate at frequencies above the largest plasma frequency of the ionosphere. Therefore, the majority of the transmitted signal will escape into space, but a small fraction of its power will be reflected back by electrons along its path. Since ISRs do not rely on the transmitted signal returning to Earth, it is possible for them to make measurements beyond the F2 peak. ISRs are therefore a useful tool for observing the topside ionosphere – a region that bottomside sounders are incapable of measuring.

This extended capability comes at a cost. Since the backscattered radiation has only a small percentage of the original signal's power, high power transmitters and very large, high gain antennas are required to implement an ISR. For example, the Millstone Hill ISR at the MIT Haystack Observatory utilizes a transmitter with

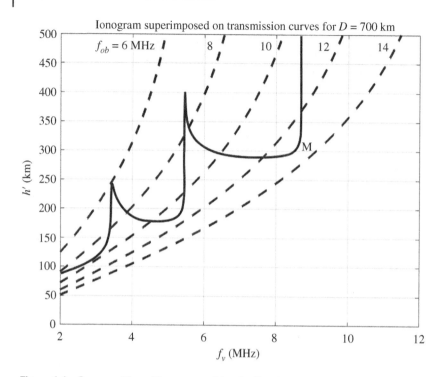

Figure 6.4 Superposition of ionogram with a family of transmission curves. Intersection points indicate the possibility of transmitting a signal over distance D.

an average power output of 150 kW and a peak transmit power of 2.5 MW. Two separate parabolic antennas, each of which operate at 440 MHz, are used: one fully steerable antenna which is 46 m in diameter with a gain of 42.5 dBi, and one fixed zenith-pointing antenna with a 68 m diameter and gain of 45.5 dBi. When used in conjunction, these antennas are capable of measuring the characteristics of the ionosphere over 1000 miles away from the observatory. More information on the MIT Haystack Observatory and the Millstone Hill ISR can be found on the Haystack Observatory website [104].

Table 6.1 lists several of the currently operating ISR facilities across the world. Excluding the EISCAT system, all of these radar facilities are solely monostatic – that is, their transmitter and receiver antennas are co-located. The EISCAT system also operates monostatic radars at several of its facilities. For example, the station in Svalbard, Norway has a monostatic radar operating at 500 MHz, and the station in Tromsø, Norway has one that operates at 928.5 MHz. In addition, the Tromsø station has a tristatic radar that operates at 224 MHz. The

Table 6.1 Locations and transmission characteristics of several currently operating incoherent scatter radar facilities.

Facility	Latitude	Longitude	Peak transmit power (MW)	Transmit frequency (MHz)
Jicamarca, Peru	11.9 °S	76.9 °W	5–6	50
Millstone, Hill, MA, USA	42.6 °N	71.5 °W	2.5	440
Kharkiv, Ukraine	48.5 °N	36.0 °E	2.5	150
Irkutsk, Russia	52.2 °N	104.5 °E	2.5	150
Sondrestrom, Greenland	67.0 °N	51.0 °W	3.6	1290
EISCAT Tromsø, Norway	69.6 °N	19.2 °E	2.2	928.5
			3.0	224
EISCAT Svalbard, Norway	78.2 °N	16.0 °E	1	500
AMISR Poker Flat, AK, USA	65.1 °N	147.5 °W	1.5–2	450
AMISR Resolute Bay, NU, Canada	74.7 °N	95.0 °W	1.5–2	450
ALTAIR Roi-Namur, Marshall Islands	9.4 °N	167.5 °E	4	422
QJISR Qujing, China	25.6 °N	103.8 °E	2	500

Source: Adapted from Table 14.1 of [107]. Reproduced with permission of Cambridge University Press.

tristatic radar is capable of receiving signals both at the transmit site in Tromsø as well as two additional receiver stations – one in Kiruna, Sweden and another in Sodankylä, Finland [105]. The remote receivers in the tristatic system were originally designed to receive at 928.5 MHz as well, but due to interference with the 930 MHz communications band this functionality was discontinued in 2012 [106]. Another collection of ISRs, dubbed AMISR, is comprised of three separate monostatic radar arrays: one in Poker Flat, Alaska, and two in Resolute Bay, Nunavut, Canada – one of which faces north toward the Arctic Circle and the other faces south toward North America.

ISRs are a powerful tool for ionospheric radio science, but due to their immense size and cost, they cannot be implemented as widely or as easily as ionosondes. Nonetheless, they remain a staple of ionospheric observation, so this section will discuss the basics of their operation and the information that they provide.

Theory of ISR Operation

In the early twentieth century, J.J. Thomson proposed that individual electrons are capable of scattering electromagnetic waves. The radar cross section (RCS) of an electron, denoted σ_e, is given by the expression

$$\sigma_e = 4\pi(r_e \sin \psi)^2, \tag{6.34}$$

where $r_e = 2.818 \times 10^{-15}$ m is the classical electron radius and ψ is the angle between the observer and the incident wave's direction of propagation. The RCS may be thought of as the efficiency with which an object reflects signals in the direction of the observer. Individual electrons will scatter incoming signals in different directions. These scattered signals will in general have a random distribution of phases with respect to each other. In other words, the radiation is *incoherent*, hence the name ISR. It should be noted that the presence of ions introduces a degree of coherence to the returned signal, so some sources prefer to call the technique Thomson scatter radar instead. Both names are used in the literature, but the name ISR is more common, so that terminology will be retained throughout this book.

Since the phases from the scattered signals are random, the time averaged power at the radar receiver will equal the number of electrons that scattered a signal times the power received from a single electron. For some region of the ionosphere with electron density N_e, it can be shown (as in [108]) that, assuming random thermal motion and no magnetic field, the RCS σ within a unit volume is given by

$$\sigma = N_e \sigma_e \tag{6.35}$$

given a radar's output power, it is possible to predict the received power by using the radar equation:

$$P_r = \frac{P_t G_t G_r \lambda^2 \sigma L}{(4\pi)^3 R^4}, \tag{6.36}$$

where P_r is the received power, P_t is the transmitted power, G_t and G_r are the gains of the transmitting and receiving antennas, respectively, λ is the wavelength of the radar wave, R is the range between the antenna and the target, and L is a loss factor ≤ 1, which takes into account losses due to factors such as atmospheric absorption or polarization mismatch between the wave and the antenna.

When an ISR is used in practice, P_t, λ, G_t, and G_r are all known a priori. P_r is measured upon reception, R can be determined from the time delay of the signal, and L can be estimated using empirical models for atmospheric loss. Therefore, the only unknown remaining in the radar equation is σ, which can be solved for. Then, since σ_e is known given the angle of incidence of the received wave, Equation 6.35 can be used to obtain the electron density at range R. In this way, an electron density profile of the ionosphere can be obtained. This technique is suitable for altitudes up to about 1000 km, allowing measurement of the entire ionosphere.

Measuring Temperature, Velocity, and Ion Composition Using ISRs

In addition to electron density, ISRs are capable of measuring a number of other characteristics of the ionospheric plasma. These include the temperature of the electrons and ions, the mean drift velocity of the plasma, and the densities of different ion species. All of this is made possible by studying the shape of the received power spectrum. Figure 6.5 shows the typical shape of such a spectrum.

In this figure, many features can be seen. The dashed line denotes the ISR transmit frequency. The offset of the spectrum's center frequency from this line is the Doppler shift f_d. From this shift, the mean velocity of the plasma along the line of sight can be deduced. In the figure shown, the received spectrum is shifted slightly downward in frequency, indicating that the plasma has a net velocity away from the receiver. The half power width can also be seen in Figure 6.5. This information can be used to determine the plasma temperature. However, this is not straightforward, so a bit of explanation is warranted. What follows is a brief history of how this relationship was discovered.

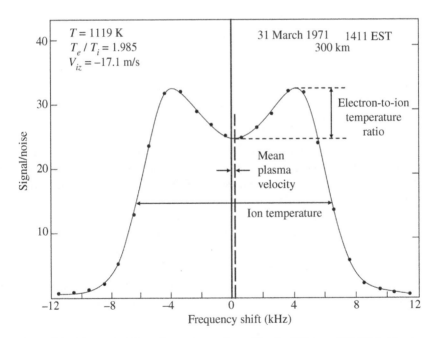

Figure 6.5 Theoretical fit to data points collected by the Millstone Hill ISR in March 1971. From the spectrum's features, it is possible to determine information about the electron and ion temperatures of the plasma as well as its mean velocity along the line of sight. These values are shown in the upper left. Source: Adapted from Figure 14.5 of [107]. Reproduced with permission of Cambridge University Press.

The free electrons in the ionosphere are in constant random thermal motion. Therefore, a number of electrons will be moving along the line of sight of an ISR, some toward and some away. These moving electrons induce varying Doppler shifts in the scattered waves, meaning that the power spectrum of the received signal should be spread out in frequency. Following this line of reasoning, it was originally thought that a radar signal incident on some finite volume of the ionosphere would receive a power spectrum with a Gaussian shape, whose half-power width is proportional to the Doppler shift of an electron approaching the radar with mean thermal speed, Δf_e, where

$$\Delta f_e = \frac{1}{\lambda} \sqrt{\frac{8kT_e}{m_e}}. \tag{6.37}$$

In this equation, k is Boltzmann's constant, T_e is the electron temperature, m_e is the mass of an electron, and λ is the wavelength of the transmitted signal. When K.L. Bowles successfully measured backscattered radar signals from the ionosphere in 1958, he discovered that the received power spectrum was indeed spread out around the transmitted frequency, but the bandwidth was much narrower than predicted by Equation 6.37. Subsequent theoretical studies concluded that positive ions were the cause of this discrepancy. It was found that the scattering from a plasma is dependent on the ratio of the radar wavelength to the Debye length of the plasma. Recall from Equation 5.114 that the Debye length λ_D of a quasineutral cold plasma (like the ionosphere) is dependent on the electron density and electron temperature. Typical Debye lengths in the ionosphere are 1 cm or less at altitudes below 1000 km [108].

Organized motion in a plasma cannot persist on scales smaller than λ_D, so for wavelengths much smaller than the Debye length, scattering is due to individual electrons and the half-power width of the received spectrum is related to Equation 6.37, as originally expected. However, ISR wavelengths are typically much larger than 1 cm. The wavelength of the highest frequency (i.e. smallest wavelength) ISR in operation today is about 23 cm. For wavelengths larger than the Debye length, scattering is due not only to the motion of free electrons but rather density fluctuations caused by oscillations in the plasma. There are three principal components to these oscillations: 1) ion-acoustic waves governed by the motion of positively charged ions, 2) electron waves induced at the plasma frequency, and 3) electron waves induced at the electron gyrofrequency. Figure 6.6 shows each of these lines in an ISR spectrum. The equation for this power spectrum is very complex and is not presented here, but can be found in sources such as [108].

Note that although there appears to be only a single ion line, if we zoom into a narrower frequency range, the double-humped shape from Figure 6.5 emerges. With the understanding that these frequency lines correspond to resonant modes in plasma oscillations, the origin of the strange shape becomes clearer. There are

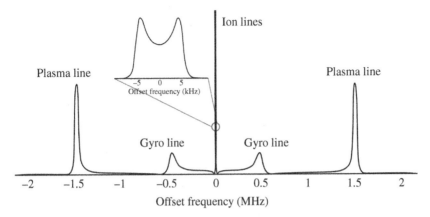

Figure 6.6 The ion, plasma, and gyro lines of an ISR spectrum. Source: Adapted from Figure 1 of [109]. Reproduced with permission of Springer Nature.

actually two ion lines, each of which has a single peak. If these peaks were separated sufficiently in frequency, they would be easily distinguishable, just like the plasma lines and gyro lines in Figure 6.6. However, since these peaks are so close together, they overlap and create the distinctive double-humped shape.

The ion lines are very tall, meaning that the majority of the received power lies in the narrow bandwidth they span. The half-power width of this peak is a good approximation of the thermal motion of the plasma, and is on the order of Δf_i, the Doppler shift of an ion approaching the radar with mean thermal speed. Δf_i has a very similar form to Equation 6.37:

$$\Delta f_i = \frac{1}{\lambda} \sqrt{\frac{8kT_i}{m_i}}, \tag{6.38}$$

where T_i is the temperature of the ions and m_i is the mass of the dominant positive ion.

The relationship between the radar wavelength and the Debye length is commonly encapsulated in a parameter α, given by

$$\alpha = \frac{4\pi \lambda_D}{\lambda}. \tag{6.39}$$

It can be seen that for radar wavelengths λ much smaller than the Debye length, $\alpha > 1$. Conversely, wavelengths much larger than the Debye length give $\alpha < 1$. Figure 6.7 shows various ion lines from ISR power spectra where $T_e = T_i$ and the values of α differ. It can be seen that when α is large (small radar wavelength), a Gaussian shape is present, whereas when α is small (large radar wavelength), the shape begins to develop "wings." Note that in this figure, the frequency scale

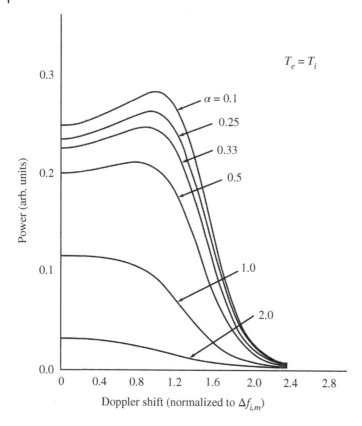

Figure 6.7 Ion lines of various ISR power spectra with $T_e = T_i$ and differing values of α. The frequency scale has been normalized to the Doppler shift of an ion approaching at the mean thermal speed of the ions. Source: Adapted from Figure 2 of [108]. Reproduced with permission of the Institute of Electrical and Electronics Engineers (IEEE).

has been normalized to the Doppler shift corresponding to an ion approaching the observer at the mean thermal speed of the ions. We will denote this quantity $\Delta f_{i,m}$.

Figure 6.7 assumes that the ion and electron temperatures of the ionosphere are equal. However, this is not always the case; ions are many times more massive than electrons, so they are not accelerated to high velocities (i.e. high temperatures) as readily. Therefore, in general the ratio between the electron and ion temperatures, T_e/T_i, will not equal one. This ratio is reflected in the height of the ion lines' wings above the valley between them. The higher the electron temperature relative to the ion temperature, the more pronounced the wings will be. Figure 6.8 displays the ion lines of several spectra of O^+ ions where $\alpha = 1$ and the ratio T_e/T_i varies. The frequency scale is normalized to in the same manner as that of Figure 6.7.

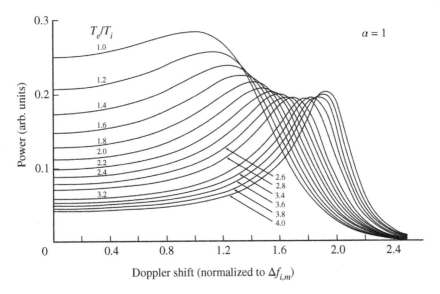

Figure 6.8 Ion lines of ISR power spectra of O^+ for $\alpha = 1$ and various ratios of electron temperature to ion temperature. The frequency scale has been normalized to the Doppler shift of an ion approaching at the mean thermal speed of the ions. Source: Adapted from Figure 3 of [108]. Reproduced with permission of the Institute of Electrical and Electronics Engineers (IEEE).

Lighter ions can achieve high velocities more easily than heavy ions. Faster motion corresponds to a larger Doppler shift, so lighter ions will have spectra spread very widely around their center frequency, while the spectra of heavier ions will be narrower. The spectra of three positive ions commonly found in the ionosphere – H^+, He^+, and O^+ – can be seen in Figure 6.9. Note that the frequency scale is normalized to Δf_e, rather than Δf_i in this figure.

The information presented in this section is just a basic introduction to ISR theory. Each of the topics thus far may be expanded upon further, and additional considerations must be made under certain circumstances. For example, at lower altitudes collisions with neutral molecules cannot be ignored, as they significantly impact the returned power spectrum. ISR measurements can also be altered if the radar signal propagates nearly perpendicular to Earth's magnetic field. This entire section is based on the implicit assumption that the velocity distribution of the plasma's thermal motion follows a Maxwellian distribution. However, in regions where there are powerful magnetic fields, such as at high latitudes, the velocity distribution can become non-Maxwellian, significantly altering the shape of the received spectrum. Such effects have been studied in, for example,

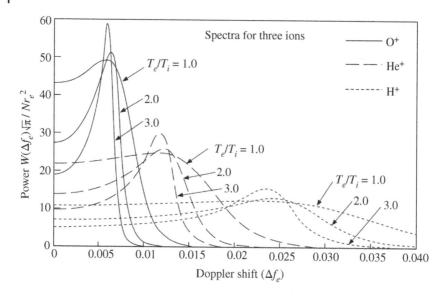

Figure 6.9 Ion lines of ISR power spectra for Hydrogen, Helium and Oxygen ions, each with differing temperature ratios. The frequency scale has been normalized to the value of Δf_e. Source: [107]. Reproduced with permission of Cambridge University Press.

[110]. For further discussion on ISR theory, the reader may refer to [108, 111], Chapter 14 of [107], and a number of further sources cited therein.

6.2.3 In Situ Measurements

Remote sensing methods such as ionospheric sounding and ISR are well-understood, widely used, and reliable methods for measuring the ionosphere. However, these techniques have one major weakness: they cannot be used to obtain a clear picture of the lowest regions of the ionosphere. At altitudes below about 100 km, neutral molecules are more abundant, and thus there are more electron-neutral collisions than at higher altitudes. It will be shown in Chapter 8 that absorption of an electromagnetic signal's energy is directly dependent upon the electron collision rate. This effect is also frequency dependent, causing much stronger absorption in low frequency signals, such as those used by ionosondes and ISRs. The severe attenuation of these signals makes remote sensing techniques an impractical choice for measurement of the lower ionosphere. Instead, measurements must be taken in situ by an apparatus installed aboard a rocket or satellite. In this section, two commonly used plasma measurement devices will be introduced: the Langmuir probe and the retarding potential analyzer (RPA).

Langmuir Probes

Put simply, a Langmuir probe is merely a conductor placed in a plasma. The probe is given a bias voltage V_B with respect to a reference electrode, and V_B may be swept to produce a current of varying magnitude. Through analysis of the current–voltage (henceforth I–V) relationship, it is possible to deduce some parameters of the plasma, such as the electron and ion densities, electron temperature, and plasma potential. What follows is a simple introduction to the theory of Langmuir probe operation.

When a conductor is placed in a plasma, it will be bombarded by charged particles due to their thermal motion. In a typical plasma like the ionosphere, electrons will have higher velocities than ions. This is because the mass of the electrons is much smaller than that of any ion, so the electrons' velocities will be much higher at the same kinetic energy (i.e. same temperature). Typically, the electron temperature of the ionosphere is greater than the ion temperature, but even when they are comparable the electron velocities will always be higher. Therefore, since the electron and ion densities of the ionosphere are roughly equal, after being immersed in the plasma, the probe will encounter electrons more quickly than ions and will begin to develop a negative charge. Over time, the increased negative charge will begin to repel electrons and attract ions, lessening the further increase of negative charge. Eventually, the probe accumulates enough charge to achieve equilibrium. Note that since the ionosphere is a quasineutral[2] plasma, it will not have the same potential as the negatively charged probe. The probe will always float to some potential V_f, which is lower than the surrounding plasma potential V_p. One might expect that this potential difference would induce current flow, but this is not the case because the negatively charged probe will attract a sheath of positive ions around it, which electrically screens its effect on the surrounding plasma. The characteristic shielding thickness of the sheath is given by the Debye length λ_D, which has been discussed in previous sections. Note that if V_f were instead greater than V_p, a sheath would still form, but in this case having negative charge.

Until now, we have assumed that the voltage of the probe was allowed to float freely until it reached an equilibrium value. However, the probe can also be held at a chosen bias potential V_b by a power supply. If V_b differs from V_f, it will induce a positive or negative current depending on whether V_b is higher or lower than V_f. Thus, by sweeping V_b from low to high, it is possible to obtain an I–V curve. An idealized I–V curve obtained by a Langmuir probe is shown in Figure 6.10.

Note that the reference potential of 0 V is dependent upon the specific conditions of the experiment. The values of V_b, V_f, and V_p with respect to zero are not important; rather it is their values relative to each other which matter.

---

2 Although charged particles are abundant, their charges cancel out when viewed in aggregate, causing the plasma to appear neutral.

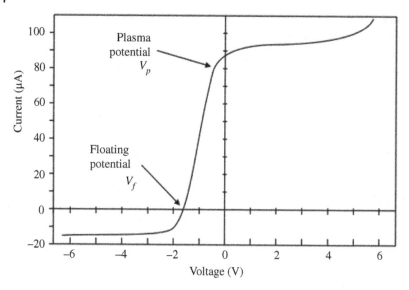

Figure 6.10 Sample idealized *I–V* curve of a Langmuir probe. The horizontal axis shows the bias potential applied to the probe.

The total current at the probe is the sum of the electron current and the ion current. It is simplest to analyze each of these constituent currents individually. First, let's consider the values of each current when $V_b = V_p$. In this case no sheath exists since there is no potential difference between the probe and the surrounding plasma. Thus the electron current is related to the number of electrons which encounter the collecting area of the probe, and the same is true of the ion current and the number of encountered ions. Assuming Maxwellian velocity distributions for both electrons and ions, the electron and ion currents are respectively given by

$$I_{es} = \frac{1}{4}A_p N_e \sqrt{\frac{8kT_e}{\pi m_e}}, \tag{6.40}$$

and

$$I_{is} = \frac{1}{4}A_p N_i \sqrt{\frac{8kT_i}{\pi m_i}}, \tag{6.41}$$

respectively. Note that these are the magnitudes of the currents only, and that their signs will be opposite one another. The convention is to denote electron current as positive (current flowing out into the plasma) while ion current is negative (current flowing in from the plasma). In these equations, A_p is the collecting area of the probe, k is Boltzmann's constant, N is number density, T is temperature, and m is mass, with subscripts e and i referring to electrons and ions respectively. As V_b is increased or decreased from V_p, the situation changes.

First we will consider the change in the electron current, I_e, as V_b is changed. As V_b is decreased from V_p, electrons begin to be repelled by the probe, as discussed earlier. Only the fraction of electrons with sufficient kinetic energy to overcome the repulsive force are able to be collected, decreasing the electron current. The current decreases exponentially as V_b decreases, eventually reaching zero when $V_b \ll V_p$. An electron current of zero indicates that all electrons are repelled from the probe.

If V_b is increased from V_p, electrons are instead accelerated toward the probe. The electron current is limited by the collection area of the probe, so one might expect I_e to saturate beyond this point. In practice, however, a slight increase in current is observed as V_b is increased beyond V_p before it levels off. This is because the formation of the sheath around the probe actually increases its effective collection area, slightly increasing the received current. This process is complex and depends on the geometry of the probe in question. Some discussion of this phenomenon is presented in [107, 112].

In total, the electron current as a function of V_b can be represented as a piecewise function as follows:

$$I_e(V_b) = \begin{cases} I_{es} \exp\left(\dfrac{-e(V_p - V_b)}{kT_e}\right), & V_b \leq V_p, \\ I_{es}, & V_b > V_p, \end{cases}$$

where I_{es} is the electron saturation current. The slight increase beyond V_p has been ignored for the sake of simplicity.

Largely, the ion current I_p follows a similar but mirrored trend when compared with the electron current, as one might expect. As V_b is increased past V_p, the probe begins to repel ions, reducing the ion current exponentially. When $V_b \gg V_p$, all ions are repelled and $I_p = 0$. As V_b is decreased from V_p, the ion current saturates quickly. Unlike the analogous situation for electron current, the sheath does not increase the effective collection area of the probe. If it were the case in the ionosphere that $T_e \approx T_i$, the ion current would saturate to the value given by Equation 6.41. However, in the ionosphere $T_e \gg T_i$, and in this case Equation 6.41 is no longer valid. Instead, the ion current saturates to a value known as the Bohm current, given by

$$I_{Bohm} = 0.605 A_p N_i e \sqrt{\frac{kT_e}{m_i}}. \tag{6.42}$$

Curiously, this current is dependent on the electron temperature rather than the ion temperature. The reasons for this are beyond the scope of this book, but the reader may refer to [112, 113] for further information. In total, the ion current can be expressed piecewise as follows:

$$I_i(V_b) = \begin{cases} -I_{Bohm} \exp\left(\dfrac{e(V_p - V_b)}{kT_i}\right), & V_b \geq V_p, \\ -I_{Bohm}, & V_b < V_p. \end{cases}$$

Summing I_e and I_i gives the total I–V relationship as it is depicted in Figure 6.10 (small spurious currents I_{sp} may also arise, for example due to photoelectrons ejected by solar photons incident upon the probe). The electron current is typically at least one order of magnitude greater than the ion current. Therefore, I_i and I_{sp} may be considered negligible compared with I_e. If this assumption is made in the electron retarding region ($V_b < V_p$), the exponential I–V relationship may be used to determine the value of T_e. The plasma potential V_p can be determined by finding the inflection point in the I–V curve, i.e. when $\frac{d^2 I_e}{dV_b^2} = 0$. The electron density can then be found by measuring the current at this point and using Equation 6.40. More detail into these calculations is presented in Chapter 14 of [107].

Note that if the ion current were deduced at V_p, Equation 6.42 could instead be used to obtain the ion density. Since the electron and ion densities in the ionosphere are essentially equal, either method will produce the same answer. Also note that since Equation 6.42 is dependent on the electron temperature, there is no easy way to measure the ion temperature using a basic Langmuir probe.

Retarding Potential Analyzers

While Langmuir probes are useful tools for in situ ionospheric measurements, they have a number of limitations. For example, incident solar photons can eject electrons via the photoelectric effect, inducing spurious currents which make the total current measurement noisy. Additionally, it is difficult to determine certain parameters, such as ion temperature, for the reasons touched upon previously. To address these problems, as well as a few others, a measurement device more sophisticated than a Langmuir probe has been developed. This device is known as a retarding potential analyzer (RPA), and a brief overview of its operation as well as benefits it provides over basic Langmuir probes will be given in this section.

The basic principle of an RPA is fundamentally the same as that of a Langmuir probe: charged particles impinging upon a conductor may be used to deduce plasma parameters. However, the major difference between the two is that an RPA is designed to include several electrostatic grids which may be used to screen out particles of a certain charge or energy. The structure of a typical RPA may be seen in Figure 6.11.

An RPA consists of an aperture, three to four electrostatic grids, and a collector plate seated inside and insulated from a conductive frame. Each grid can be electrically biased to hold some potential in order to screen out certain particles.

Figure 6.11 Schematic of a typical RPA with four grids between the aperture and the collector.

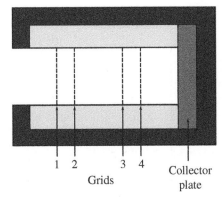

1 2   3 4   Collector
Grids   plate

Typically, RPAs are used to measure ion currents, so the grid potentials are chosen such that electrons are screened out while letting ions pass through. The first grid is either set at the floating potential of the plasma or is grounded in order to smooth any perturbations in the plasma entering the aperture. The second grid is set at a large negative potential in order to repel all incoming electrons, since they are not of interest. The third grid is swept from zero volts up to some large positive voltage. This progressively screens out ions of higher and higher energies that, as we will see shortly, can be used to obtain a rough picture of ion temperature and composition. The fourth grid is set at a negative potential as a secondary electron repulsion grid. This grid is sometimes omitted, but can reduce noise at the collector by screening out electrons ejected from within the RPA due to particle bombardment. Ions then collide with a conductive collector at the back of the RPA, inducing an ion current which can be measured in much the same way as with a typical Langmuir probe.

The key feature of an RPA is the third, voltage swept grid. As the voltage increases, ions will need larger kinetic energies to overcome the repulsion. By analyzing the resultant $I–V$ curve, one can determine the kinetic energy distribution of the collected ions, and thus estimate the ion temperature of the surrounding plasma. The $I–V$ curve of an RPA will also display clear drop-offs at several points. To understand this, first consider a plasma with no thermal motion. In this case, the velocity of every ion when it hits the collector will simply be the spacecraft's velocity with respect to the plasma. Thus, the kinetic energy for all ions of the same mass will be the same. As the voltage of the RPA is swept upwards, there are certain points at which lighter ion species will no longer have enough energy to overcome the repulsive force. When this happens, there will be a sharp drop-off in the $I–V$ curve, indicating a sudden decrease in the number of ions reaching the detector. In this way, it is possible to determine the different ion species present in the plasma.

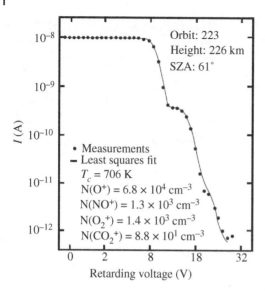

Figure 6.12 *I–V* curve and derived plasma parameters from Venus' atmosphere, as reported by the RPA aboard the Pioneer Venus Orbiter. Source: [107]. Reproduced with permission of Cambridge University Press.

In this hypothetical case, an RPA would be able to measure the ion composition of a plasma with very high precision. Unfortunately, this utility is confounded somewhat by the thermal motion of the ions. This has the effect of smoothing out the steep, well defined drop-offs in ion energy. Despite this, the energy transitions are typically distinct enough to be observed in an RPA *I–V* curve, as seen in Figure 6.12.

6.3 Morphology of the Ionosphere

The ionosphere is typically divided into three distinct layers called the D, E, and F layers. This nomenclature was introduced by Edward V. Appleton in the 1920s. Appleton was the first to experimentally confirm the existence of the ionosphere by demonstrating that radio waves could be reflected by the upper atmosphere. He labeled the first layer he discovered the "E layer," to denote the fact that electric fields were reflected. Subsequently, Appleton discovered two additional layers: one above the E layer which he denoted with the letter F, and another below the E layer which he denoted with the letter D.

It has since been discovered that the F layer can be divided further into two regions – the F1 and F2 layers – with differing properties. The F2 layer is higher in altitude. Following this same convention, the E layer is sometimes divided into E1 and E2 regions, though this is uncommon in the literature. The lowest portion of the ionosphere is sometimes referred to as the C layer, though this layer is unimportant for most applications so it is either subsumed into the D layer or ignored completely.

As discussed in the preceding Chapter 5, electron density varies continuously with height. The boundaries of these layers are defined primarily by differences in chemical composition and physical processes, which vary significantly with altitude. These factors are highly dynamic, varying greatly with latitude, time of day, season, and solar activity. Therefore the boundaries of each region may only be roughly defined. In the literature, reported values for the boundaries of a given layer often vary as much as 10–20 km. The boundaries reported in this book are those given in [72]. Complicating factors at equatorial, auroral and polar latitudes can change the behavior of the layers. These effects will be discussed in Section 6.4, but this section will focus on layer formation at mid-latitudes.

6.3.1 C Layer

The C layer is the lowest layer of the ionosphere, ranging from approximately 50–70 km above the surface of the Earth. Only very high energy photons – cosmic gamma rays with MeV or higher energies – are able to penetrate to this depth [92]. These photons are capable of ionizing all chemical species in the atmosphere. The Sun only emits gamma rays occasionally, during solar flare events, causing sudden ionospheric disturbances (SIDs) in the ionosphere. SIDs typically only last about a half hour [92], but can cause significant problems for radiowave propagation.

Typically, gamma ray flux comes from *cosmic gamma rays*, which originate in distant supernovae, neutron stars, black holes, quasars, and other exotic astrophysical phenomena. Because of the large distances between the Earth and these objects, this flux is small – $\mathcal{O}\left(10^{-2}\right)$ erg/(cm^2 s) (note that 1 erg/s = 10^{-7} W) [114]. For comparison, the total solar flux over all wavelengths at the top of the atmosphere is $\mathcal{O}\left(10^{6}\right)$ erg/(cm^2 s). C layer ionization can still happen at night, but the relatively minuscule flux of photons is unable to ionize the atmosphere to a large degree. Thus the C layer is insignificant in most cases.

6.3.2 D Layer

The D layer is the region from about 70–90 km above the Earth's surface. Photons with wavelength 1 nm or shorter can penetrate to this depth. This means that in addition to gamma rays, hard X-rays ($\lambda = 0.1–1$ nm) which are capable of ionizing N_2 and O_2 also reach the D layer. The flux of solar X-rays is dependent upon solar activity, which includes the natural 11-year cycle of solar irradiance (discussed in Section 5.2.1), as well as solar flares. Table 6.2 [115, p. 85] shows some sample hard X-ray fluxes in erg/(cm^2 s) under varying solar activity conditions.

Table 6.2 Orders of magnitude of hard X-ray flux values measured in erg/(cm² s) for various wavelengths and solar activity conditions.

	$\lambda = 0.2$ nm	$\lambda = 0.4$ nm	$\lambda = 0.6$ nm
Completely quiet	10^{-8}	10^{-7}	10^{-6}
Quiet	10^{-7}	10^{-6}	10^{-5}
Slightly disturbed	10^{-6}	10^{-5}	10^{-4}
Disturbed	10^{-5}	10^{-4}	10^{-3}
Flares, class 2	10^{-4}	10^{-3}	10^{-2}
Flares, class 3	10^{-3}	10^{-2}	10^{-1}

Source: Friedman [115]/Annual Reviews.

Since X-ray flux is small at most times during the solar cycle, the primary source of ionization in the D layer is actually due to a spectral line of hydrogen which happens to reach the D layer through a window in the absorption cross section of O_2 [81]. This spectral line, the Lyman-α line, has a wavelength of 121.6 nm and is capable of ionizing nitric oxide (NO) in the atmosphere. Although NO is relatively less abundant than O_2 and N_2, the large Lyman-α flux – 5.1 erg/(cm² s) [115, p. 72] – is capable of producing a significant number of NO^+ ions.

At D layer heights, NO^+ and O_2^+ are the most prevalent ions, while O^+ and N_2^+ exist in lesser quantities. Water vapor is present at altitudes below about 85 km, and can interact with NO^+ to form larger molecular ions such as H_3O^+, $H_5O_2^+$, and $H_7O_4^+$.

Recombination and attachment are both important in the D layer. The high density of the region results in a large number of collisions. In addition to two-body collisions which allow molecular ions such as N_2^+, O_2^+, and NO^+ to recombine, three-body collisions may also occur at this altitude, allowing atomic ions such as O_2^+ to recombine as well (see Section 5.3.4). At night, electrons may attach to neutral molecules to form negative ions, and the negative ions can subsequently recombine with positive ions. For these reasons, electron loss is relatively high in the D layer.

During the day, loss offsets production to some extent, causing the D layer to have a lower electron density than the E or F layers – approximately 10^8–10^9 electrons/m³, varying somewhat with season and solar activity [72]. At night, in the absence of solar radiation, recombination and attachment cause the D layer to disappear almost completely, leaving only a small amount of ionization which is induced by cosmic rays.

In the D layer, the temperature is fairly constant and electrical conductivity is low due to the high collision frequency. This means that dynamic processes such as plasma transport, thermospheric winds, and atmospheric tides are not important to consider. Thus ionization is dependent almost entirely on ion-electron production q and loss L in this region. At high latitudes, a phenomenon known as polar cap absorption occurs; this will be discussed briefly in Section 5.2.4.

6.3.3 E Layer

The E layer extends from about 90–140 km above Earth's surface. Soft X-rays (λ = 1–10 nm) and extreme ultraviolet (EUV) radiation of wavelength 91.2–102.6 nm can penetrate to this depth and produce ionization at all times. Note that wavelengths in the 10–91.1 nm spectrum are also theoretically able to penetrate to this depth, but in practice they are absorbed strongly be the F1 and F2 layers before reaching the E layer. Included in the EUV spectrum is the Lyman-β line at 102.6 nm capable of ionizing O_2. NO^+ and O_2^+ ions are the most prevalent, while O^+ and N_2^+ have secondary abundance [72].

Near solar maximum – the peak of the 11-year solar cycle – the total flux of X-rays below 10 nm is 1 erg/(cm^2 s). At solar minimum, this flux is smaller, only about 0.1 erg/(cm^2 s). However, the maximum-to-minimum flux ratio is much greater for hard X-rays (shorter wavelengths) than soft X-rays (longer wavelengths). In the 0.2–0.8 nm band, this ratio is a factor of several hundred; in the 0.8–2 nm band the ratio is about 45; in the 4.4–6 nm band, the ratio is about 7 [115, p. 78]. In other words, soft X-ray flux varies relatively little over the solar cycle. Soft X-ray flux increases during solar flare events, but this increase is also relatively small compared with the increase in hard X-ray and gamma ray flux. EUV flux from the strongest solar emission lines in the 91.1–102.6 nm range totals about 0.141 erg/(cm^2 s) [115, p. 72].

In the E layer, the neutral molecule density n is still fairly large, much larger than electron density N_e. Referring to Equation 5.41, we see that β, which depends on n, will therefore be much greater than αN_e. Since $\beta \gg \alpha N_e$, Equation 5.41 reduces to $L = \alpha N_e^2$, describing a recombination-type or α-type Chapman layer. Thus the electron density profile in this region is closely modeled by Equation 5.81. Electron density is higher in this region than in the D layer, at approximately 10^{11} electrons/m^3. At night, ionization decreases greatly due to recombination, down to approximately 5×10^9 electrons/m^3, leaving the E layer only weakly ionized.

The E layer can sometimes be divided into two subregions – the E1 layer an altitude of about 100 km, and the E2 layer above it. An ionosphere observation station in Haringhata, Calcutta, has observed regular formation of an E2 region in the local ionosphere. This layer typically forms in the afternoon (c. 1300 hours local time) at about 150–160 km in altitude. The E2 layer first forms as a "ridge" in the F1 layer profile, but descends over the course of several hours, acting as a distinct layer before finally merging with the E1 layer [116]. The E2 layer is rarely mentioned in the literature, and its properties are still the subject of research, though the cause of formation appears to be solely due to UV radiation [72, p. 42].

Diffusion is unimportant in the E layer, but the effects of neutral winds and electromagnetic drift are significant. At the altitudes spanned by the E layer, the collision frequency is much smaller than the cyclotron frequency for electrons, so they are subject to a strong Hall drift, moving them perpendicular to both $\vec{E}$ and $\vec{B}$ (see Figure 5.18). Conversely, the collision and cyclotron frequencies of ions are still comparable, so their trajectories are subject to both Pedersen and Hall drifts, meaning they are deflected at some angle $\theta = \arctan(v_i/\omega_i)$. This separation in charge movement produces a current in the E region. This phenomenon, called the ionospheric dynamo, produces strong electrojets at both equatorial and auroral latitudes. These electrojets will be discussed in Section 6.4.

6.3.4 Sporadic E Layer

Occasionally a dense patch of ionization can be detected at altitudes of approximately 100–120 km. This layer is known as a sporadic E layer, denoted Es. The formation of Es seems to be unrelated to the formation of the standard E layer and does not follow any discernible periodic pattern, hence the name sporadic E. A sporadic E layer sometimes appears as a thick sheet which completely obscure higher layers, and at other times it appears patchy and partially transparent. Es can be spread horizontally over very large regions or concentrated to a small area, and it can even appear strongly at night.

At mid-latitudes, sporadic E layers form due to wind shear. As discussed in Section 5.5.4, the motion of charged particles is dependent on both neutral winds and electromagnetic drift. At this altitude, the collision frequency and cyclotron frequencies of ions are comparable, so their trajectory is deflected by some angle with respect to the wind. For a westward wind, this trajectory drives ions downward, while an eastward wind drives them upwards. For electrons, the collision frequency is much smaller than the cyclotron frequency, so electrons must move along the magnetic field lines.

Consider a region in which an eastward wind is blowing at one altitude, and a westward wind is blowing at a slightly higher altitude, as shown in Figure 6.13[3] [98]. Ions are forced upward by the eastward (lower) wind and downward by the westward (upper) wind. This causes the ions to cluster at an intermediate altitude.

---

3 Note that in this figure, $\vartheta = \arctan(\omega_i/v_i)$. The argument of the arctan function here is the reciprocal of the argument of the arctan function in Equation 5.116. This is because the angle is measured with respect to the wind velocity vector, as opposed to a line orthogonal to $\vec{E}$, as in Equation 5.116. In other words, angle ϑ is the complement of θ shown in that equation.

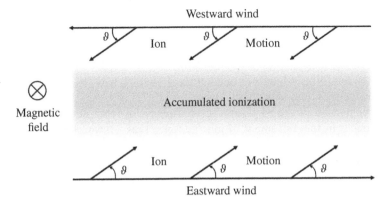

Figure 6.13 Redistribution of ionization due to wind shear, resulting in a sporadic E layer.

This dense patch of ionization corresponds to the sporadic E layer. In this formation process, electrons move in such a way that the effect of the charge separation is neutralized, so Es is electrically neutral.

In addition to wind-shear at mid-latitudes, Es layers form via different processes at both equatorial and auroral latitudes. At the equator, complex interactions between the equatorial electrojet and the geomagnetic field give rise to patchy sporadic E layers [98]. At auroral latitudes, intense localized bombardment from high energy particles can create sporadic E layers.

The mean lifetime of ions in the ionosphere is only a few minutes. This is much shorter than the observed lifetime of sporadic E layers, which typically last several hours. This means that sporadic E layers should consist primarily of longer-lived metal ions that have likely been ablated from meteors. Indeed, mass spectrometry results have indicated the presence of metallic positive ions in sporadic E layers [117].

6.3.5 F1 Layer

The F1 layer reaches from about 140–210 km in altitude. Ionization in this region is due to EUV radiation of wavelength 10–91.1 nm band. As in the E layer, O_2^+ and NO^+ ions dominate, with secondary prevalence of O^+ and N_2^+. Flux of EUV radiation from 14 to 91.1 nm has been found to be about 1.9 erg/(cm^2 s) [118, p. 794]. This varies somewhat with solar activity, but these variations are small compared with those in X-ray flux. Thus there is always an appreciable amount of flux in the EUV spectrum. Significant spectral lines in this band include the Helium II 30.4 nm line and the Helium I 58.4 nm line.

As in the E layer, $\beta \gg \alpha N_e$, so loss is described by $L = \alpha N_e^2$. Thus the F1 layer is also a recombination-type Chapman layer, and can be modeled by Equation 5.81. Electron density in the F1 layer is higher than in both the D and E layers, ranging from 2×10^{11} to 10^{12} electrons/m^3, depending on solar activity. The ionization in this region does not fully disappear at night, instead merging with the F2 layer to create a single region simply called the F layer.

In the lower portion of the F1 layer, the ionospheric dynamo effect persists due to differences in the movement of ions and electrons. However, at higher altitudes, the collision frequency of ions becomes negligible compared with their cyclotron frequency, and they begin to move in the same direction as electrons (see Figure 5.18). At this height, the conductivity of the ionosphere once again reduces and the electrojet currents can no longer flow. The effects of diffusion are still negligible in this region.

6.3.6 F2 Layer

The F2 layer extends from 210 to 600 km in altitude. At this altitude, O$^+$ becomes the dominant ion for reasons discussed in Section 5.3.4. Ionization is caused by EUV radiation, as in the F1 layer, and wavelengths below 91.1 nm are particularly important for ionizing atomic oxygen.

The ionosphere is sparse at this altitude, so ions are unable to recombine readily. Thus, in this region $\beta \ll \alpha N_e$ so Equation 5.41 reduces to $L = \beta N_e$. Since β is dependent on the density of neutral molecules, L is small at this height, allowing for large electron densities. Indeed, the F2 layer has the highest electron density of any region of the ionosphere, reaching as high as 8×10^{12} electrons/m^3. The F2 layer remains strongly ionized at night due to transport effects which cause vertical drift of ionization downward from the perpetually ionized protonosphere. Since the F2 layer is cooler at night, it descends slightly and combines with the F1 layer to form a single region simply called the F layer.

Recall that $\beta = kn$, where n is the density of neutral particles. Thus $\beta = k'n[N_2] + k''n[O_2]$, where k' and k'' are the attachment coefficients of N_2 and O_2, respectively. Since the densities of O_2 and N_2 decrease rapidly with height above the F1 layer, β decreases more quickly with altitude than production q does. Therefore, the ratio q/β increases. This seems to indicate runaway ion production as altitude increases, but this is not we observe. The missing piece of the puzzle is the effect of diffusion.

At F2 layer heights, diffusion becomes an important consideration. Vertical drift due to diffusion works together with photochemical loss to offset production at higher altitudes. In the F2 layer, the transport term d becomes comparable to the production term q and the loss term L. There is a point when production and loss are in balance; this point is called the F2 peak (denoted h_mF2). At this height, we have the following relations, where the subscript m indicates the value at the F2 peak:

$$\beta_m \approx \frac{D_m}{H^2},$$
$$N_m \approx \frac{q_m}{\beta_m} = \frac{S_\infty n[O]}{k'n[N_2] + k''[O_2]}, \tag{6.43}$$

where D_m is the ambipolar diffusion constant, H is the scale height, S_∞ is the solar flux at the top of the atmosphere, and $N_m = N_e = N_i$ is the density of ions and electrons. We see that N_m therefore depends on the ratio of neutral atoms to neutral molecules at h_mF2.

Via the continuity equation (see Section 5.5.5), when q, L and d are in balance, the change in electron density with respect to time becomes zero. Thus, at h_mF2, electron density remains nearly constant. Above the F2 peak, electron loss due to diffusion becomes greater than production, causing the electron density to decline with altitude.

The point at which $\beta_m \approx D_m/H^2$ varies with temperature, since H is dependent on temperature. Therefore, h_mF2 also changes with temperature. Wind and electromagnetic drift can also affect h_mF2. The interested reader may consult [92, p. 303] and [98, pp. 178, 179] for a brief discussion on the variation of h_mF2.

6.3.7 Topside Ionosphere

Far above the F2 peak, at altitudes higher than about 700 km, the number of O^+ ions begins to decline with altitude. In their place, H^+ and He^+ ions become abundant, with H^+ being the most prevalent. H^+ ions are simply free protons, so this region is often called the protonosphere. Electron density in this region is lower than in the F2 region, at about 10^{10} electrons/m^3. This is expected due to the strong influence of diffusion at this altitude. The protonosphere and the F2 layer continually exchange ions via the reaction:

$$O^+ + H \rightleftharpoons H^+ + O. \tag{6.44}$$

During the day, upward diffusion causes the forward reaction, in which ionization is transferred from the F2 layer to the protonosphere. At night, the reverse reaction takes place as downward diffusion supplies the F2 layer with ionization from the protonosphere. In this way, the F2 layer can maintain its ionization throughout the night by drawing upon the reservoir of ionization in the protonosphere.

6.4 Variability of the Ionosphere

6.4.1 F2 Layer Anomalies

Because the degree to which the ionosphere is ionized depends on the incident solar flux, the electron density is highly variable, changing with time of day, season, location on earth, and solar activity. The D, E and F1 layers of the ionosphere are relatively simple, and they generally adhere to the Chapman layer formulation presented in Section 5.4. The F2 layer is far more complicated and cannot be considered a Chapman layer, as it is subject to non-negligible transport phenomena. The complicated interactions in the F2 layer are the subject of ongoing research, but several well-documented anomalies have been observed. In each of the following sections, the behavior predicted by Chapman's theory of the ionosphere is presented, followed by discussion of anomalies which deviate from this expected behavior. Although most of these anomalies occur in the F2 layer, some effects may also perturb the lower layers, especially away from mid-latitudes.

The vast majority of ionizing flux comes from the sun, so electron density is much higher during the day than at night. Within one diurnal cycle, maximum ionization typically occurs around 1200 hours, and minimum ionization at about 0300 hours. In addition to diurnal variation, there is a difference in the average amount of ionization during the summer as compared with the winter. Solar flux is typically higher during the summer. For example, one study found that the total ionization during the summer was 14% higher on average than in the winter [119]. The diurnal and seasonal cycles are sufficient to explain daily and yearly variations in ionization in the D, E, and F1 layers. In the F2 layer, several anomalies in these trends have been observed.

Early Morning and Late Afternoon Anomalies

In 1962, researchers at the University of Sydney analyzed satellite measurements of the F region of the ionosphere. They found that in addition to the expected ionization maximum at 1200 hours, a second maximum appeared year-round at 1800 hours. In the winter, a third maximum appeared at 0500 hours [119]. These anomalies, respectively called the late afternoon anomaly and the early morning anomaly, indicate the impact of transport effects, temperature changes, and traveling ionospheric disturbances (TIDs) which produce large-scale movement of ionization.

Winter or Seasonal Anomaly

In the winter, ionization in the F2 layer during the day is actually higher than in the summer; this is known as the daytime winter anomaly (DWA). It has been proposed that this anomaly is due to a change in the atomic-to-molecular ratio of neutral species in the ionosphere. During the day, heating in the summer hemisphere causes convection currents that carry lighter elements toward the winter hemisphere, thus changing the ratio of O to N_2 in both hemispheres [120, 121]. As seen in Equation 6.43, this ratio is equivalent to q/β, and when this quantity is larger, electron density is greater. As O atoms are blown away from the summer hemisphere, q/β decreases, causing ionization to decrease as well. Conversely, in the winter hemisphere, the influx of O atoms increases q/β, thus increasing ionization. In the literature, the DWA is often referred to simply as the winter anomaly or sometimes as the seasonal anomaly. The DWA is more pronounced in the northern hemisphere than in the southern hemisphere, which is related to the asymmetry in the Earth's magnetic field about the geographic equator. The DWA is also more pronounced during solar maximum than solar minimum [122].

Figure 6.14 [119] shows the diurnal variation of several measurables that can be used to determine ionization at a given height. Figure 6.14a shows the fading rate of a satellite signal vs. time. The fading rate can be directly related to the amount of ionization (see [119] for details). Thus, the DWA can be easily observed in this figure. Figure 6.14b shows the F2 critical frequency of the ordinary ray (f_oF2) vs. time, which is related to the degree of ionization, as we discussed in Section 6.2. Finally, Figure 6.14(c) shows the "thickness index" vs. time. The thickness index of the ionosphere is simply the ratio of the fading rate to $(f_oF2)^2$. The early morning and late afternoon anomalies are both evident in this figure.

Nighttime Winter Anomaly

Normally during the winter, ionization drops dramatically at night, causing the average daily ionization to be lower than in the summer. However, at geomagnetic mid-latitudes during low solar activity, a nighttime winter anomaly (NWA) has also been observed, wherein ionization at night is higher during the winter than in the summer. The NWA appears to occur only in the northern hemisphere due to an asymmetry in the geomagnetic field around Earth's geographic equator. The DWA is not observed in years when the NWA occurs, and vice versa, so the phenomena appear to be decoupled. Plots of electron density vs. time may be found in [123], where the NWA effect may be clearly seen.

Annual Anomaly

An annual anomaly also exists, and appears to occur in phase with the elliptical orbit of the Earth around the sun. The global F2 electron density – i.e. the sum of the daytime F2 layer electron densities in the northern and southern hemispheres – is greater at perihelion in December than at aphelion in June. Solar flux is indeed higher at perihelion due to the Earth's closer proximity to the Sun. However, the asymmetry in Earth–Sun distance only predicts 7% higher ionization in

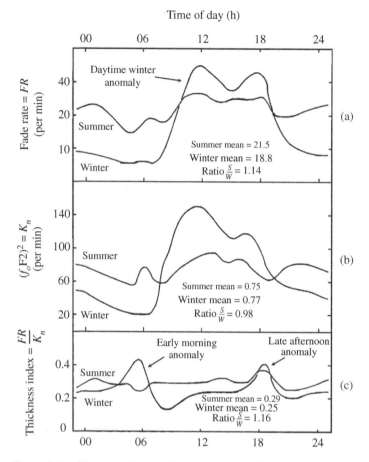

Figure 6.14 Diurnal variation of (a) fading rate, (b) F2 ordinary ray critical frequency, and (c) thickness index. Source: Adapted from Figure 3 of [119]. Reproduced with permission of Wiley.

December as compared with June, while experimental evidence indicates a variation of as much as 30%. Attempts to explain this discrepancy have been made, for example in [124], but these attempts were unsuccessful. At the time of this writing, the processes governing the discrepancy in the annual anomaly ionization remain unexplained.

Semi-Annual Anomaly

In addition to the seasonal (winter) and annual anomalies, a semi-annual anomaly also exists, in which daytime F2 layer electron densities are greater during the equinoxes than during the solstices. This anomaly is primarily observed at low latitudes, but also at mid latitudes in the Southern Hemisphere [125]. The semi-annual anomaly is prevalent in the southern hemisphere, but not the Northern Hemisphere for reasons again related to the asymmetry of the magnetic field about the geographic equator. An explanation for this anomaly has been proposed in [126]: during the solstices, the Northern and Southern Hemispheres are heated unevenly, inducing global thermospheric winds. These winds cause turbulent mixing that increases the number of N_2 molecules at F2 layer heights, thus decreasing the ratio of O to N_2. From Equation 6.43, we see that this also decreases q/β, which in turn decreases the level of ionization. Conversely during the equinoxes, both hemispheres are heated evenly so there is less turbulent mixing. Fewer N_2 molecules reach the F2 layer, so the ratio of O to N_2 remains high, and therefore q/β is large and ionization does not decrease. The interested reader may consult [122, 127] for further discussion and references on the seasonal, semi-annual and annual anomalies.

Equatorial Anomaly

The Earth's magnetic field is responsible for dynamic phenomena which can significantly change the level of ionization. The effects which occur vary greatly with geomagnetic latitude (for brevity, throughout the remainder of this section, geomagnetic latitude will be shortened to latitude). At mid-latitudes, ionization generally follows regular and predictable variations with time and solar activity. However, at lower and higher latitudes, a number of anomalies have been observed. Generally, the Earth can be divided into several zones [128, pp. 140–141]:

- *Equatorial zone*: 30° on either side of the geomagnetic equator.
- *Mid-latitude Zone*: Latitudes of 30 − 60° in both hemispheres.
- *Auroral zone*: Latitudes from 60 − 75° in both hemispheres.
- *Polar cap*: Region extending from a latitude of 75° to the geomagnetic pole in each hemisphere.

Curiously, there exists a "trough" of unexpectedly low F2 layer ionization on either side of the magnetic equator, extending to latitudes of ±16°. Further, at ±16–18°, there are peaks of higher ionization than anticipated. This phenomenon is known as the equatorial anomaly or the Appleton anomaly. This effect can be observed in Figure 6.15 (data from [98]).

The equatorial anomaly is a complex process which can only be explained by considering the myriad transport processes at work in the F2 layer. Diurnal heating of the Earth induces tidal winds, and at E layer altitudes these neutral winds work in tandem with the geomagnetic field to move electrons and ions in different directions. This phenomenon was discussed in Section 5.5.4. Differential movement of charge causes the ionosphere to be conductive, and this conductivity is

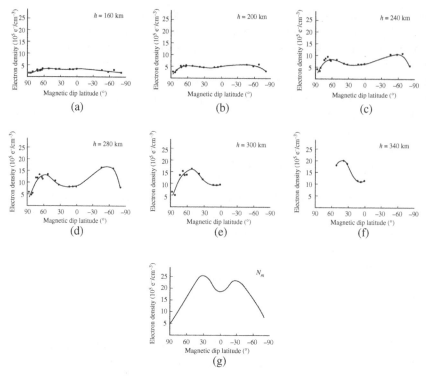

Figure 6.15 Variation in electron density with respect to magnetic dip angle at various heights. Figures (a)–(f) display empirical data taken at noon on magnetically quiet days in September 1957. Figure (g) depicts the theoretical peak electron density of the ionosphere vs. magnetic dip latitude, which occurs at the F2 layer peak height, $h_m F2$. Source: Data from Rishbeth and Garriott [98].

especially large near the magnetic equator. On the day side of the ionosphere, this gives rise to a large, eastward-directed current called the equatorial electrojet.

The equatorial electrojet sets up an eastward-pointing electric field $\vec{E}$. This electric field, in conjunction with Earth's magnetic field $\vec{B}$, causes an upward drift of plasma, with velocity $\vec{E} \times \vec{B}/B_0^2$, as described in Section 5.5.3. Charged particles continue to move upward in this fashion until pressure gradient forces become appreciable and slow their ascent to a halt. At this point, the particles will descend under the force of gravity, following Earth's magnetic field lines as they do. This process can be visualized in Figure 6.16. The field lines carry the particles to higher, tropical latitudes, where they remain [127, 129]. The combination of these mechanisms is referred to as the equatorial fountain, and accounts for the observed redistribution of ionization at low magnetic latitudes. Quantitative models of the equatorial fountain have been developed, for example, in [129, 130].

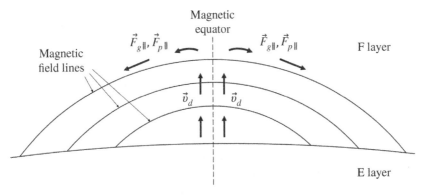

Figure 6.16 The equatorial fountain effect. Charged particles drift upward with velocity $\vec{v}_d$, then diffuse downward along magnetic field lines. $\vec{F}_{g\parallel}$ and $\vec{F}_{p\parallel}$ are the components of gravity and pressure gradient forces parallel to the magnetic field lines.

6.4.2 Solar Activity

As we have seen, ionization in the ionosphere is caused primarily by incident solar radiation. Solar flux follows a predictable, periodic pattern that repeats every 11 years. One of the most commonly used indicators of solar activity is the flux of 10.7 cm ($f = 2.8\,\text{GHz}$) radio waves, denoted Φ. This quantity is commonly expressed in units known as solar flux units or SFU. One SFU is equivalent to $10^{-22}\,\text{W}/(\text{m}^2\,\text{Hz})$ or $10^{-19}\,\text{erg/cm}^2$. Φ typically ranges from about 70 SFU during low solar activity to about 250 SFU during high solar activity [45]. It has been observed that solar flux is directly correlated with the number of sunspots visible on the Sun at a given time. Sunspots tend to group together, though individual spots are also commonly observed. The number of sunspots is measured using the sunspot number R, given by the equation

$$R = k(10g + s), \tag{6.45}$$

where g is the number of sunspot groups, s is the number of individual sunspots, and k is a correction factor (typically close to 1) which accounts for the observation equipment. At times it can be difficult to properly count sunspots; for example, it is not always obvious whether a large collection of spots should be considered one group or two. Despite this imprecision, R is still a widely used index of solar activity, since we have over 100 years' worth of measured data for the parameter. In Figure 6.17, the correlation between solar flux and sunspot number can be clearly observed.

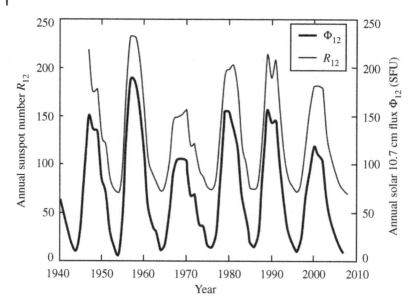

Figure 6.17 Comparison of annual sunspot number to total annual 10.7 cm solar flux.

Commonly in ionospheric work, the 12-month running mean sunspot number R_{12} is referenced, which is given by the equation:

$$R_{12} = \frac{1}{12}\left(\sum_{k=n-5}^{n+5} R_k + \frac{1}{2}\left(R_{n+6} + R_{n-6}\right)\right), \tag{6.46}$$

where R_k is the mean value of R over a single month k, and R_{12} is the smoothed index for the month $k = n$. Similarly, the 10.7 cm flux Φ can be averaged over a 12 month period to obtain the parameter Φ_{12}. Φ_{12} can be expressed in terms of R_{12} as follows [72]:

$$\Phi_{12} = 63.7 + 0.728R_{12} + 8.9 \times 10^{-4}R_{12}^2. \tag{6.47}$$

Measured and predicted values of R_{12} and Φ_{12} are available from several organizations, such as the Sunspot Index Data Centre (SIDC) in Brussels, Belgium [72].

Solar Flares

A solar flare is a sudden flash of electromagnetic radiation which occurs in close proximity to a sunspot. The number of flares per solar rotation N_F is statistically related to the mean sunspot number $\overline{R}$:

$$N_F = C(\overline{R} - 10). \tag{6.48}$$

Here, C is a proportionality factor. Sample values for C are 1.98 for the solar cycle which peaked in 1937, and 1.47 for the solar cycle which peaked in 1947 [81].

It is easiest to observe solar flares in red $H\alpha$ light ($\lambda = 656.3\,$nm), though they are commonly observed in wavelengths across the electromagnetic spectrum, from radio waves to X-rays. Flares typically only last a few minutes to a few hours. A two character identifier is used to classify solar flares by their size and brightness. Flares are categorized by size according to the following scale [81]:

Subflare: ≤ 2.0 square degrees,
Importance 1: 2.1–5.1 square degrees,
Importance 2: 5.2–12.4 square degrees,
Importance 3: 12.5–24.7 square degrees,
Importance 4: ≥ 24.8 square degrees.

For reference, the Sun and the Moon each span about 0.2 square degrees when viewed from the surface of the Earth. The character used to indicate flare size is the number of its importance, and subflares are denoted S. The brightness of a solar flare is categorized as using one of three characters: f = faint, n = normal, or b = bright. Under this classification, the least dramatic flare event is a faint subflare, denoted Sf, and the most dramatic flare event is a bright importance 4 flare, denoted 4b. Flares in the X-ray spectrum of 0.1–0.8 nm are given their own separate classification based on their flux density (Table 6.3):

Note that each class represents a factor of 10 increase in flux over the previous one. The letter designating the flare class is followed by a number representing the flux within that order of magnitude. For example an M2.4 flare represents a flux of $2.4 \times 10^{-5}\,$W/m^2. Class X flares have no upper bound. The largest solar flare ever recorded, which occurred in November 2003, saturated the X-ray detectors designed for flare measurement. By extrapolating data from subsequent ionospheric observation, the flare was eventually classified as an X45 [131].

Solar flares are of importance to ionospheric studies because they are capable of inducing a number of related phenomena, collectively termed SIDs. SIDs occur due to the increased ionization produced by the increased flux during a solar flare. Brief descriptions of SIDs relevant to ionospheric propagation are presented [98]:

Table 6.3 Classification of X-ray flares, 0.1–0.8 nm.

Class	Peak flux ϕ (W/m^2)
A	$\phi < 10^{-7}$
B	$10^{-7} \leq \phi < 10^{-6}$
C	$10^{-6} \leq \phi < 10^{-5}$
M	$10^{-5} \leq \phi < 10^{-4}$
X	$10^{-4} \leq \phi$

Short wave fadeout (SWF): Increased ionization in the D layer causes increased absorption of HF signals, particularly those at the lower end of the band. HF signals reflected from the E and F layers must traverse the D layer at least twice, and are strongly attenuated each time they pass through.

Sudden cosmic noise absorption (SCNA): Like SWF, SCNA is an absorption phenomenon, but is typically observed in the 20–30 MHz range. At these frequencies, signals are capable of escaping the ionosphere, so they only experience absorption along a one-way path, in contrast to SWF.

Sudden frequency deviation (SFD): The reflection height of a given signal suddenly lowers, and the index of refraction in the region changes rapidly. Together these effects change the effective path length of the signal, resulting in a Doppler shift [132].

Sudden phase anomaly (SPA): VLF waves are reflected at a lower height than normal, which results in a phase advance.

Sudden enhancement of atmospherics (SEA): Atmospherics, also known as "sferics" are VLF electromagnetic waves induced by lightning in the lower atmosphere. During a solar flare, the field strength of atmospherics is increased [133].

Solar flare effect (SFE): Also known as a magnetic crochet. Due to the increased level of ionization, the conductivity of the ionosphere is increased, allowing greater current flow. This causes a disturbance in the Earth's magnetic field.

Geomagnetic Storms

During periods of high activity, charged particles are commonly ejected from the Sun. One dramatic example of such a discharge of particles is a coronal mass ejection (CME). CMEs occur due for the same reasons as solar flares, though the two events do not always occur simultaneously. Although CMEs eject particles at relativistic speeds, these particles have mass and thus cannot reach Earth as quickly as EM radiation from solar flares. Photons from a solar flare will reach Earth after about eight minutes, while particles from a CME can take hours or days to reach Earth.

The Earth is constantly subjected to a stream of charged particles from the Sun called the solar wind, but during a CME the flux of particles is dramatically increased. The sudden increase in solar wind pressure creates a shockwave throughout the magnetosphere. Via complex interactions between the solar wind, the interplanetary magnetic field, and the geomagnetic field, strong electric fields are created in the magnetosphere, increasing the flow of electric current. A number of charged particles also penetrate into the ionosphere, where they ionize gas molecules and create auroras. The details of these interactions are beyond the scope of this book, but the interested reader may consult [71] for further information.

When a geomagnetic storm disturbs the ionosphere sufficiently, skywave links and ionospheric communication will be negatively impacted. Extremely powerful storms are even capable of disrupting the operation of electrical equipment on the surface of the Earth. One famous example is the Carrington event, the most severe geomagnetic storm to hit Earth in recorded history. The storm, which occurred in September 1859, produced auroras as far south as Central America and the Caribbean [134]. Campers in the Rocky Mountains were reportedly awakened shortly after midnight by the auroral light overhead, and were so convinced that it was daybreak that they began preparing breakfast! Unfortunately, the storm was responsible for more than brilliant auroras. Extreme disturbances in Earth's magnetic field induced currents in telegraph networks around the world, sending incomprehensible messages and administering shocks to telegraph operators.

A Carrington-class event would wreak havoc on the society of the twenty-first century, which is increasingly reliant on electrical networks for power distribution and communication. A small taste of the damage a CME could inflict on modern society was felt during the geomagnetic storm of 1989. Though not as powerful as the Carrington event, the storm induced ground currents so powerful that they overloaded the Québec power grid, leaving the entire province without power for 12 hours. It also knocked satellites out of their proper orbits for several hours and disrupted wireless communications [135].

Numerous engineers and scientists work on safeguards that will help mitigate damage in case a CME hits Earth in the future. For example, NASA's Solar Shield project is designed to forecast the way currents would flow in the US power grid during a geomagnetic storm. This will help to identify the parts of the grid that are at the highest risk, and would allow engineers to disconnect them temporarily while the storm passes [136].

6.4.3 Magnetic Variation

Electric currents in the ionosphere produce magnetic fields which superimpose on the field produced by the geodynamo. The magnetic fields produced by these currents are observed as disturbances in the value of the geomagnetic field. Various current sources, which will be discussed shortly, can give rise to magnetic disturbances. Some of these disturbances vary smoothly and periodically, while others vary suddenly and sharply. Days on which disturbances vary smoothly are called quiet or q days, while days where disturbances are less predictable are called disturbed or d days.

For reasons discussed in Section 5.5.4, currents primarily flow in the E layer of the ionosphere. A number of distinct phenomena drive these currents, which together form a complex global current system called the ionospheric dynamo. Through extensive study of the ionosphere, researchers have been able to isolate and classify dynamo currents according to the forces that drive them. In the remainder of this section, we will discuss the major current systems in the ionosphere, as well as key parameters used in their measurement and analysis.

Sq Current

The most significant ionospheric current is the solar quiet or *Sq* current. The *Sq* current is caused by atmospheric tides driven by solar radiation. As discussed in Section 5.5.2, differential solar heating of the atmosphere gives rise to large scale wind patterns. In a similar fashion to ocean tides, these wind patterns follow global-scale periodic oscillations called atmospheric tides. Fourier analysis of atmospheric tides has revealed large amplitude oscillations of diurnal (24 hours) and semi-diurnal (12 hours) periods [92]. Oscillations with periods of eight hours and six hours have also been observed, though with reduced amplitudes relative to the diurnal and semi-diurnal tides.

Atmospheric tides can also be induced by the gravitational pull of large objects in the solar system. Indeed, the gravitational pull of the Sun contributes to the tides we have just discussed. However, it has been found that this contribution is minor compared with that of solar heating: 99% of the diurnal solar tide is due to radiation, while only 1% is due to gravity. Similarly, the relative contributions of radiation and gravity to the semi-diurnal tide are 90% and 10%, respectively [137, pp. 189, 190].

In Section 5.5.4 it was shown how wind can induce currents in the E layer via separation of ions and electrons. In this way, solar atmospheric tides induce massive, globe-spanning vortices of current whose magnitudes are on the order of hundreds of kiloampere, collectively referred to as the *Sq* current. The shape of the *Sq* current pattern can be seen in Figure 6.18. In this figure, it is easily observed that currents are nearly parallel close to the magnetic (i.e. dip) equator. This indicates the existence of the equatorial electrojet referred to in Section 6.4.1.

The magnetic fields produced by the *Sq* current superimpose on the geomagnetic field, causing deviations from its undisturbed value. These variations are on the order of tens to hundreds of gammas (recall that $1 \gamma = 1$ nT), while the base value of the geomagnetic field varies from 25,000 to 65,000 gammas over the surface of the Earth. Over the course of one day, the *Sq* current pattern remains fixed relative to the Sun, which means that a measurement station on the surface of the Earth will rotate beneath it and observe a periodic variation in magnetic field intensity. Figure 7 in [137] shows the diurnal variation of the horizontal (H) and vertical (Z) components of the magnetic field at various latitudes, averaged over 6 selected quiet days. Similarly, Figure 8 from the same source shows the diurnal variation in the declination angle (D) of the magnetic field, as observed from the same measurement stations and averaged over the same 6 quiet days as the data in Figure 7.

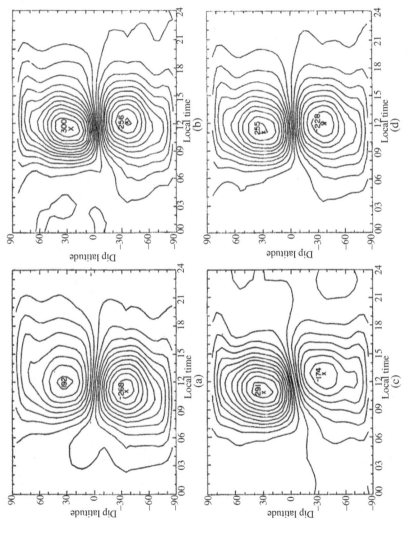

Figure 6.18 *Sq* currents during (a) the D season, (b) the E season, and (c) the J season. Part (d) Shows the yearly average. The numbers listed in the center of each set of isolines are measured in kA. The difference in magnitude between two consecutive lines is 25 kA, and the magnitude at midnight is zero. Source: [99]. Reproduced with permission of Academic Press.

The *Sq* current pattern varies seasonally, inducing stronger currents in the summer hemisphere than the winter hemisphere. Figure 6.18 shows the induced currents for months near the December solstice (D season), near the equinoxes (E season), and near the June solstice (J season), as well as a yearly average. Data in this figure was taken during the International Geophysical Year (IGY), a project lasting from 1 July 1957 to 31 December 1958.

The *Sq* current also varies over the course of a solar cycle, and is about 50% stronger during solar maximum than solar minimum [81, p. 47].

L Current

As discussed previously, gravitational influences of large celestial objects are capable of inducing atmospheric tides. The gravitational force of the Moon creates significant atmospheric tides, in a similar manner to the way it creates ocean tides. Consider a simplified model of the effects of the Moon's gravity on the Earth and Earth's atmosphere. In this model, Earth and its atmosphere will be considered spherical and homogeneous. Only five points will be considered: Earth's center of mass, the points of the atmosphere nearest and farthest from the moon (denoted N and F in Figure 6.19), and two points 90° away from the near and far points (denoted P_1 and P_2).

First we will analyze the gravitational forces on each of these five points from a frame of reference outside of the Earth–Moon system. In this frame of reference, point N will experience a large gravitational force, Earth's center of mass will experience a smaller force, and point F will experience an even smaller force. Points P_1 and P_2 will experience forces of the same magnitude, each with components both parallel and perpendicular to the Earth–Moon line.

If the situation is instead observed from an Earth-centered frame of reference, we must vectorially subtract the force experienced at Earth's center of mass from each point. When this is done, the forces at points N and F point radially outward with the same magnitude, while the forces at points P_1 and P_2 point radially inward with the same magnitude. This can be seen in Figure 6.19b. Thus it is easy to see that Earth's spherical atmosphere will be stretched along the Earth–Moon line and compressed along the line perpendicular to it, creating an ellipsoid, as seen in Figure 6.19c.

A measurement station on the surface of the Earth will rotate beneath the atmosphere, crossing the Earth–Moon line twice per lunar day. The distinction of a "lunar day" must be made because the Moon moves along with the rotation of the Earth, thus slightly changing the direction of the Earth–Moon line over time. If the measurement station begins below point N, it will take approximately 24 hours and 50 minutes for the station to arrive once again at point N [138].

Air movement is greater in the atmospheric "bulges" along the Earth–Moon line, creating a system of currents collectively known as the L current. The L

Figure 6.19 Lunar tidal forces on Earth and Earth's atmosphere from (a) a frame of reference outside the Earth–Moon system and (b) an Earth-centered Frame of reference. Part (c) shows the (exaggerated) net effect of lunar tidal forces on the shape of Earth's atmosphere.

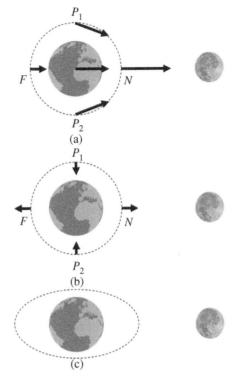

current is much smaller in magnitude than the *Sq* current, only achieving magnitudes of a few kiloamperes. The magnetic variation induced is commensurately smaller as well. This variation is lunar semi-diurnal, having a period of about 12.4 hours. On the sunlit side of the Earth, the conductivity of the ionosphere is greater than the night side due to the high level of ionization. Thus, the strength of the *L* current will be asymmetric whenever the Earth–Moon line is roughly aligned with the Earth–Sun line, since larger currents will be induced in the day-side bulge than the night-side bulge. In this way, the oscillation of the *L* current has a component with a period of one *lunation* in addition to the lunar semi-diurnal component previously mentioned. If the Moon begins along the Earth–Sun line, one lunation is the time it takes for the Moon to return to the Earth–Sun line. If the Earth stood still, this would take 27.3 solar days (also known as a sidereal month). However, since the Earth revolves around the Sun, this revolution takes slightly longer – about 29.5 days on average (known as a lunation, a lunar month, or a synodic month). This is completely analogous to the distinction between solar days and lunar days.

Similar to the *Sq* current, the *L* current varies seasonally. Stronger currents are observed in the summer hemisphere than in the winter hemisphere. The same seasonal distinctions – the December season (D), Equinoctial season (E) and June season (J) – are used for the L current. Plots of the L current pattern for each of these three seasons, as well as a yearly average, may be seen in Figure 6.20.

DP1 and DP2 Currents

The continuous stream of charged particles from the solar wind creates a current at the boundary of the magnetosphere, causing small variations in the geomagnetic field. The contribution of these currents to overall magnetic field variation is small, typically on the order of a few gammas [137]. Motion of charged particles in the Van Allen belts also affects the Earth's magnetic field. Notably, the ring current decreases the horizontal component of the magnetic field. Magnetic disturbances due to the ring current are denoted D_{st}, though they are typically only significant during geomagnetic storms.

Much more powerful currents form at auroral latitudes. The conductivity at these latitudes is high due to corpuscular ionization by the bombardment of charged particles from the solar wind, and complex interactions between the magnetosphere and ionosphere set up an electric convection field across the polar cap, driving the motion of charged particles. The auroral electrojet referred to in Section 5.2.4 is the most prominent example of such a current and has a magnitude on the order of MA (mega amperes).

Magnetic disturbances caused by the auroral electrojet are often called polar substorms or geomagnetic bays. The strongest variations are on the order of several hundred gammas, and typically last only a few hours [93]. If the disturbance lasts more than a few hours, it is referred to as the longitudinal disturbance and is denoted DS. Disturbances from the auroral electrojet can extend beyond auroral latitudes as well. The representative current system for these disturbances is referred to as $DP1$ (D = disturbance, P = polar). A simplified depiction of the $DP1$ current system, which assumes equal current flow in both the east and west directions, is shown in Figure 6.21. This "classical" formulation of the $DP1$ current is equivalent to DS. Note that this figure is roughly equivalent to that of the auroral electrojet shown in Figure 5.10.

Further discussion of the DP1 current system, as well as more sophisticated representations of the current pattern under more complex conditions, may be found in [93].

Using data from the IGY, it was discovered that the solar daily variation Sq of the geomagnetic field had an additional component at polar latitudes on quiet days [139]. This component was termed S_q^p. Upon first inspection, this current pattern appears similar to DS, but there are two key differences. Firstly, activation of the auroral electrojet is not observed. Secondly, current in sub-auroral latitudes connects directly to current flowing over the polar cap, and so cannot be considered return current from the auroral electrojet.

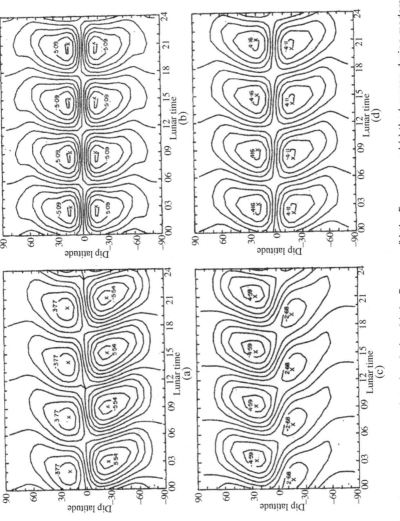

Figure 6.20 Mean *L* currents over one lunation during (a) the D season, (b) the E season, and (c) the J season during moderate solar activity. (d) Shows the yearly average. The numbers listed in the center of each set of isolines are measured in kA. The difference in magnitude between two consecutive lines is 1 kA. Source: Matsushita [99]. Reproduced with permission of Academic Press.

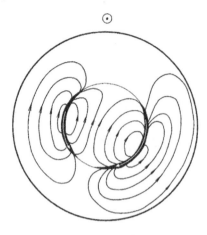

Figure 6.21 *DP*1 current viewed from above the magnetic pole, assuming equal current flow in westward and eastward directions. The symbol at the top of the diagram indicates the solar direction. Source: Obayashi and Nishida [93]. Reproduced with permission of Springer Nature.

Later, it was observed that a number of geomagnetic fluctuations with period one hour were observed to closely resemble the shape of the S_q^p pattern [140]. These fluctuations were denoted D_p^e, where the superscript e emphasizes the fact that these fluctuations are enhanced at the equator. It was concluded that the S_q^p disturbance was only one component of a larger system. This larger system is commonly denoted *DP*2.

The representative current system for *DP*2 consists of two polar vortices, one on the morning side and one on the night side [141]. In addition, *DP*2 extends all the way to equatorial regions, where current flows zonally (i.e. eastward or westward). The pattern of the *DP*2 current system can be seen in Figure 6.22b, and the subset S_q^p can be seen in Figure 6.22a. It should be noted that these two diagrams are scaled differently – (a) only extends to a geomagnetic latitude of 60°, while (b) extends all the way to the magnetic equator.

In summary, geomagnetic disturbances in polar regions can be classified into two main categories: *DP*1 and *DP*2. *DP*1 includes disturbances caused by the auroral electrojet, and "classical" polar magnetic disturbances including polar substorms, bays, and *DS*. When *DP*2 lasts only a few hours, it is observed as D_p^e, whereas when it persists for several days, it is observed as S_q^p [142].

*DP*1 and *DP*2 can be contrasted in several ways [93]:

1. *DP*1 originates from activation of the auroral electrojet, while *DP*2 does not.
2. *DP*1 varies by location, while *DP*2 appears coherently across the world.
3. Variations superimposed on *Sq* are *DP*1 type on disturbed days and *DP*2 type on quiet days.

The magnetic fluctuation, in gammas, induced at various latitudes by *DP*1 and *DP*2 can be seen in Figure 6.23. The horizontal axis lists time measured in hours, and the selected hours are given near midnight Magnetic Local Time (MLT). The

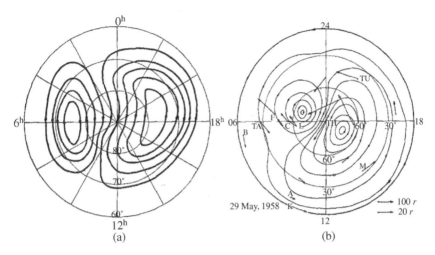

Figure 6.22 Equivalent current systems for (a) S_q^p and (b) *DP2* disturbances. Times at the cardinal directions are listed in MLT. Source: Obayashi and Nishida [93]. Reproduced with permission of Springer Nature.

vertical axis lists five different latitude zones where measurements were taken. Below the name of each latitude zone, a parenthetical note is included, which indicates the current system responsible for the magnetic fluctuations in that zone. It can easily be seen that *DP1* occurs strongly for a short period near midnight at auroral latitudes, while *DP2* is prominent at the poles and equator.

The interested reader can find detailed information on the *DP1* and *DP2* current systems in [93, 141–144].

Magnetic Indices

In order to quantify the level of disturbance in the Earth's magnetic field, the scientific community has adopted a set of standard indices. One such index, the K index, is an integer that indicates the level of magnetic disturbance. K is assigned an integer value from 0 to 9, with 0 representing minimal magnetic disturbance and 9 representing maximal disturbance. A K integer is designated to each three hour universal time (UT) interval (0000–0300, 0300–0600, …, 2100–2400) every day. Thus one full day is represented by a sequence of eight integers. For example, the sequence of K values for a quiet day might look like: 0, 1, 1, 2, 3, 3, 1, 0.

The K integer is determined based on another parameter: the amplitude range R. R is a measure of the amplitude variation in the H, Z, and D components of the local magnetic field. R is simply the highest amplitude value of a given component minus the lowest value, measured in gammas. The component with the largest R, also known as the most disturbed element, is used to determine which

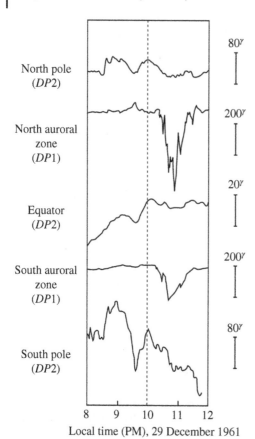

North pole
(*DP*2)

80^γ

North auroral
zone
(*DP*1)

200^γ

Equator
(*DP*2)

20^γ

South auroral
zone
(*DP*1)

200^γ

South pole
(*DP*2)

80^γ

8   9   10   11   12
Local time (PM), 29 December 1961

Figure 6.23 Magnetic variation, measured in gammas, at various latitudes due to *DP*1 and *DP*2. *DP*1 is seen in the auroral zones, while *DP*2 is seen at the poles and the equator. Source: Adapted from Figure 6 of [93]. Reproduced with permission of Springer Nature.

integer value to assign to K. Larger values of R correspond to larger values of K. The mapping between R and K is not linear, but quasi-logarithmic in nature.

The mapping used at a given observatory is permanent, but each observatory uses its own mapping. For example, $K = 9$ corresponds to $R = 1500$ nT in Godhaven, Denmark, while $K = 9$ corresponds to $R = 300$ nT in Honolulu, Hawaii. This is done so that observatories at different latitudes report roughly the same K values on disturbed days [81].

When relating R to K, the daily solar variation Sq, the lunar daily variation L, solar flare effects, and the after-effects of a disturbance must be removed first [81]. R values are then mapped to one of the ten possible integer values of K on a quasi-logarithmic scale. Table 6.4, which has been adapted from [81], gives an example for a mid-latitude station (Niemegk: 52° 04′N, 12° 40′E).

This table indicates that any R measured to be smaller than 5 nT is classified as $K = 0$. Similarly, any R between 5 and 10 nT is classified as $K = 1$, any R between 10 and 20 nT is classified as $K = 2$, and so on [145].

Table 6.4 K scale representing the relationship between K and R (nT) at a selected mid-latitude station.

K	0	1	2	3	4	5	6	7	8	9
R	5	10	20	40	70	120	200	330	500	

Source: Davies [81]. Reproduced with permission of the Institution of Engineering and Technology (IET).

Table 6.5 Conversion from K_p scale to a_p scale.

K_p	0_0	0_+	1_-	1_0	1_+	2_-	2_0	2_+	3_-	3_0	3_+	4_-	4_0	4_+
a_p	0	2	3	4	5	6	7	9	12	15	18	22	27	32

K_p	5_-	5_0	5_+	6_-	6_0	6_+	7_-	7_0	7_+	8_-	8_0	8_+	9_-	9_0
a_p	39	48	56	67	80	94	111	132	154	179	207	236	300	400

Source: Davies [81]. Reproduced with permission of the Institution of Engineering and Technology (IET).

The so-called planetary 3h index K_p further divides the K scale into 28 distinct increments [146]. Each K integer is separated into three values – that is to say, the scale is instead incremented by thirds: 0.00, 0.33, 0.67, 1.00, 1.33, 1.67, ..., 8.67, 9.00. For historical reasons, this scale is commonly mapped to the sequence 0_0, 0_+, 1_-, 1_0, 1_+, 2_-, ..., 9_-, 9_0 in the literature.

It is useful to define a parameter which represents the average daily level of magnetic disturbance. However, due to the nonlinear nature of the K (or equivalently, the K_p) scale, averaging R values will not accomplish this. In order to define a useful parameter, we must first convert to a linear scale. The new linear index a_p is called the equivalent 3h range and is related to K_p as shown in Table 6.5.

It can be clearly seen that a_p is scaled to range from 0 to 400. a_p approximately represents the amplitude variation of the most disturbed element among H, Z, and D. At 50° geomagnetic latitude, a_p is measured in units of 2.0 nT [72]. The number of nT per a_p is different depending on the latitude of the observatory, in order to retain the consistency of the scale across different stations, as described earlier. The average of the eight 3h a_p values over the course of one day is defined as the "daily equivalent planetary amplitude," denoted A_p. As an example, A_p for the quiet day with K_p index sequence 0_0, 1_-, 1_0, 2_0, 3_+, 3_0, $1+$, 0_+ is given by:

$$A_p = \frac{1}{8}(0 + 3 + 4 + 7 + 18 + 15 + 5 + 2) = 6.75, \tag{6.49}$$

where each K_p index has been replaced with its corresponding a_p value, and then averaged. K_p and A_p values are regularly published by the GFZ Helmholtz Center in Potsdam, Germany [147].

Quiet and disturbed days are ranked using three numbers, each given equal weight. These numbers are (i) the sum of the day's eight K_p values, (ii) the sum of the squares of the eight K_p values, and (iii) the greatest K_p value. The average of these three numbers is taken every day, and every day of a month is ranked in order of magnetic activity. The ten quietest and five most disturbed days each month are selected and used as points of reference when analyzing magnetic activity.

Additional indices are used for specific applications. For example, the auroral electrojet index AE is calculated from the fluctuation in the H component of the magnetic field at auroral latitudes. The maximum (upper) value of H is denoted AU and the minimum (lower) value is denoted AL. Thus $AE = AU - AL$ represents the variation in H, i.e. ΔH at any given time. Further, the average of AU and AL is denoted AO, i.e. $AO = (AU + AL)/2$ [81].

During geomagnetic storms, a storm variation index $D = D_{st} + DS$ can be defined, where D_{st} is the variation due to the ring current, and DS is the variation due to auroral electrojet activity. D_{st}, which measures the fluctuation in H, begins at some value, then decreases rapidly to some minimum value before gradually returning to its pre-storm value. This can be seen in Figure 6.24.

Geomagnetic storms can begin gradually, or can be triggered by an abrupt change in solar wind pressure known as a storm sudden commencement (SSC). During an SSC, a shock wave is sent throughout the magnetosphere, compressing

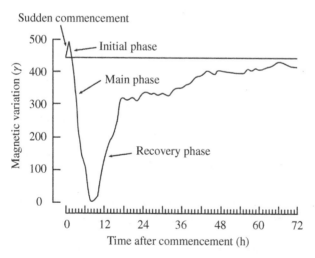

Figure 6.24 Magnetic field variation during an SSC, measured in gammas. The horizontal axis displays the time after the storm's commencement, measured in hours. Source: Adapted from Figure 2.14 of [81]. Reproduced with permission of the Institution of Engineering and Technology (IET).

Earth's magnetic field. After this *initial phase*, the magnetic field begins to rapidly decrease. The *main phase* of the storm lasts from the point when D_{st} decreases below the pre-storm value and ends when it reaches its lowest value. Finally, the *recovery phase* is the phase during which the magnetic field slowly returns to its pre-storm value. D_{st} and AE values are published regularly by the Data Analysis Center for Geomagnetism and Space Magnetism Faculty of Science in Kyoto, Japan [148].

6.4.4 Ionospheric Irregularities

A few miscellaneous ionospheric effects remain to be discussed. This section will only provide a brief overview of these phenomena, though references to supplementary literature will be provided and the interested reader is encouraged to seek additional information from these sources.

Traveling Ionospheric Disturbances

On occasion, an ionogram will display a "kink" that travels downward in frequency over time. This phenomenon, known as a TID is caused by the propagation of low frequency pressure waves in the atmosphere. These waves are referred to as atmospheric gravity waves (AGWs), and arise when the atmosphere is perturbed. Initially, buoyancy and gravity are in equilibrium, but if this balance is upset by a perturbation, it causes an oscillation as the two competing forces attempt to restore equilibrium. A more detailed discussion of the production of TIDs by AGWs may be found in [149].

Perturbations which give rise to AGWs (and thus TIDs) are produced by a number of sources. Generally, TIDs are designated as large scale or medium scale, depending on their wavelength, period, and velocity of propagation. Medium scale TIDs have wavelengths of 300 km or smaller, periods of 10–60 minutes, and travel at 50–300 m/s [72]. These can arise from, for example, air flow over a mountain range, shear forces near the jet stream, seismic activity, or even acoustic waves generated by thunder or nuclear detonations [150]. Large scale TIDs have wavelengths of 1000–4000 km, periods of one to three hours, and propagate faster than 300 m/s. These can arise as planetary or Rossby waves (global-scale inertial waves caused by the Coriolis force) or due to auroral activity [92].

TIDs typically propagate at hundreds of kilometers per hour, and can travel thousands of kilometers before dying down. They have been observed to cause oscillations in electron density at altitudes of about 150–600 km, causing deviations of about 10–20% from the baseline density [98]. For further information on the characteristics, detection, and consequences of TIDs, see the sources cited throughout this discussion, as well as [151–156].

Spread F

Ionograms of the F layer sometimes appear "fuzzy" and spread out, rather than comprising sharp, well defined traces. This occurs because the ionosphere at this altitude can become diffuse and patchy, causing radio waves to reflect randomly with frequency. Topside sounding of the ionosphere has revealed that spread F irregularities span distances of hundreds of meters to many kilometers, displaying electron density variations of a few percent [157]. The behavior of the spread F phenomenon varies considerably with geomagnetic latitude, season, and solar activity.

Spread F is primarily observed at auroral and equatorial latitudes [158], though it has also been observed less frequently at mid-latitudes [159]. At high latitudes, spread F is positively correlated with geomagnetic activity, while at equatorial and mid-latitudes, the correlation is negative. Near the magnetic poles, spread F can occur during both day and night, most commonly around the equinoxes. Elsewhere, the phenomenon is mainly observed at night, mostly near the equinoxes at equatorial latitudes and during the winter at mid-latitudes [72].

During solar minimum, spread F is observed more commonly than during solar maximum. It has been proposed that spread F does not actually occur less frequently during solar maximum, but rather that it occurs at larger heights, above the F2 peak [160]. Thus, any waves which propagate to this height will escape the ionosphere and cannot be used to create an ionogram for these altitudes, making spread F difficult to observe via ionospheric sounding.

Spread F is found to occur most commonly at equatorial latitudes, and the physical mechanisms involved in its formation have been extensively studied. Complex post-sunset transport effects in these regions create depletion regions or "bubbles" of reduced plasma density, giving the ionosphere a patchy, inhomogeneous structure. This structure arises due to a fluid-dynamical phenomenon known as Rayleigh–Taylor instability, which occurs at the interface of two fluids with different densities. If the two fluids were able to remain in perfect equilibrium, their interface would simply be a flat surface. However, perturbations (for example, AGWs in the ionosphere) will disrupt this equilibrium and cause the fluids to push against one another.

Consider a scenario in which a heavier (high-density) fluid is resting in equilibrium on top of a lighter (low-density) fluid. Initially, their interface will be planar. Now consider a slight perturbation which causes a sinusoidal spatial variation in the interface between the fluids, as seen in Figure 6.25.

The region in which the light fluid pushes into the heavy fluid (near $x = 0$), the pressure is increased on the heavy fluid, causing it to flow into the neighboring regions. Similarly, in the two regions where the heavy fluid pushes into the light fluid, the pressure increases on the light fluid, causing it to flow into the middle. This creates a positive feedback loop in which any initial imbalance is

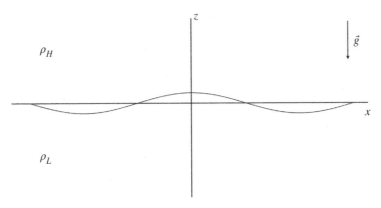

Figure 6.25 Perturbed interface between a heavy fluid of density ρ_H and light fluid of density ρ_L. Acceleration due to gravity $\vec{g}$ points in the $-z$-direction.

Figure 6.26 Time evolution of Rayleigh–Taylor instability in two fluids of differing density. The top fluid is higher density and the bottom fluid is lower density. Source: Adapted from Figure 3 of [161]. Reproduced with permission of Elsevier.

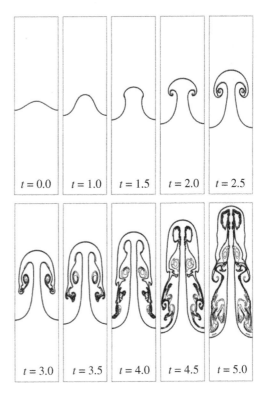

$t = 0.0$ $t = 1.0$ $t = 1.5$ $t = 2.0$ $t = 2.5$

$t = 3.0$ $t = 3.5$ $t = 4.0$ $t = 4.5$ $t = 5.0$

magnified, causing the amplitude of the initial perturbation to grow exponentially. This runaway growth of any perturbations means the system is unstable, hence the name Rayleigh–Taylor instability.

At first, Rayleigh–Taylor instabilities grow linearly, but after a short time nonlinear effects become noticeable. Complicated fluid dynamic interactions create vortices and turbulent mixing around the ascending light fluid, creating a mushroom-shaped plume. The reader is likely familiar with this mushroom shape, which can be seen in some volcanic eruptions and detonations of nuclear weapons. Figure 6.26 shows a numerical simulation of Rayleigh–Taylor instability, adapted from [161].

Much work has been done in applying Rayleigh–Taylor stability to the case of the ionosphere. The formation of bubbles of varying electron density has been observed, especially at equatorial and auroral latitudes. These irregularities affect the propagation of electromagnetic waves, causing fluctuations in their amplitude, phase, and angle of arrival. These fluctuations, known as ionospheric scintillations, are of great importance to consider when designing a satellite communication system such as GPS. Due to their importance, they will be covered in detail in Chapter 7.

The details of Rayleigh–Taylor instability and the mechanisms by which it arises in the ionosphere are beyond the scope of this book. For a deeper discussion of the general theory of Rayleigh–Taylor instability, the interested reader may reference [162, 163]. For analysis of how Rayleigh–Taylor instabilities give rise to the spread F phenomenon in the equatorial ionosphere, see [164–169].

7

Ionospheric Propagation

7.1 Introduction

With the background provided in Chapters 5 and 6, we are now prepared to explore the propagation effects caused by the interaction of electromagnetic waves with the ionospheric plasma. Ionospheric propagation is much more complex than tropospheric propagation, which is reflected by the expression for the ionosphere's refractive index, also known as the Appleton–Hartree equation. In this equation, it can be seen that many factors influence the propagation of electromagnetic waves through the ionosphere, such as oscillation and collision of electrons as well as the anisotropy introduced by the Earth's magnetic field.

In order to appreciate the Appleton–Hartree equation, as well as the implications it has for global navigation satellite systems (GNSS) signal propagation, a detailed derivation is presented in Section 7.2. As we will see, the fully general form of the Appleton–Hartree equation is cumbersome and challenging to work with. Therefore, the end of Section 7.2 is dedicated to a few important simplifications which are valid at GNSS frequencies. In Section 7.3, we move on to discuss the propagation effects induced by the so-called background ionosphere – that is, the effects introduced when the ionospheric plasma is in a steady state. These effects can all be characterized in terms of a single parameter of interest: the total electron content (TEC) along a wave's propagation path.

In general, electromagnetic waves experience several effects while traversing the ionosphere, including refraction, absorption, and dispersion. However, at GNSS frequencies these effects are of relatively little importance. The most important propagation effects to consider in the context of GNSS are group delay, phase advance, and Faraday rotation. Each of these effects is discussed in order to appreciate the reasons for their relative importance, but greater attention is given to those which most significantly impact GNSS operation.

Tropospheric and Ionospheric Effects on Global Navigation Satellite Systems, First Edition.
Timothy H. Kindervatter and Fernando L. Teixeira.
© 2022 The Institute of Electrical and Electronics Engineers, Inc. Published 2022 by John Wiley & Sons, Inc.

In contrast to Section 7.3, Section 7.4 is concerned with ionospheric propagation effects that arise due to instabilities in the ionospheric plasma. These instabilities give rise to bubbles of electron density that differ from the surrounding regions. These bubbles can cause electromagnetic waves to interfere, causing random fluctuations in their amplitude and phase. Just like tropospheric scintillations, if a signal is thereby sufficiently attenuated, it can cause the receiver to lose lock with a satellite. This results in an undesirable effect known as a cycle slip, which is touched upon in the discussion of the carrier phase observable in Section 2.2.

7.2 Magnetoionic Propagation

As an electromagnetic wave propagates through the ionosphere, its behavior is strongly affected by the physical properties of the medium. Firstly, the high degree of ionization causes the ionosphere to act like a plasma, which is a conducting medium, and secondly, the presence of the Earth's magnetic field creates an inherent anisotropy in the ionosphere. The net effect of these qualities is encapsulated in the index of refraction of the medium. In this section, we present a derivation of the Appleton–Hartree equation, which expresses the index of refraction of the ionosphere.

The index of refraction of the ionosphere is dependent on the motion of electrons within it. The equation of motion for an electron in the presence of an electromagnetic wave propagating through the ionosphere is given by the equation:

$$m_e \left(\frac{\mathrm{d}^2 \vec{r}}{\mathrm{d}t^2} + v \frac{\mathrm{d}\vec{r}}{\mathrm{d}t} \right) = -e \left(\vec{E} + \frac{\mathrm{d}\vec{r}}{\mathrm{d}t} \times \vec{B} \right). \tag{7.1}$$

Here, m_e is the mass of an electron, e is the elementary charge, $\vec{E}$ is the electric field, $\vec{B}$ is the magnetic field, and $\vec{r}$ is the displacement of the electron. v represents the electron collision frequency, a measure of how likely an electron is to collide with another particle in the ionosphere.

On the left hand side, we have terms that represent the inertial force of the electron and the force due to collisions, which may be thought of as a frictional force. On the right hand side are the force due to the time-varying electric field component of the electromagnetic wave and the force due to the Earth's magnetic field (which is assumed to be static since it is essentially constant over short timescales). Typically, the force due to the magnetic field would also include a time-varying term associated with the magnetic field component of the electromagnetic wave. However, since this time-varying component is very small relative to the static component, its contribution is negligible and will be neglected for the purposes of this derivation.

The reader is encouraged to refer to Section B.7 for an in-depth discussion of propagation in media similar to the ionosphere. Though the derivation in this section will be very similar to that in the appendix, there are a few key differences to note. Since the ionosphere is a plasma, most of its electrons are free electrons and are therefore not subject to the restoring force $m_e\omega_0^2\vec{r}$ seen in Equation B.25; in addition, the electron collision frequency ν is equivalent to the damping ratio Γ for the case of the ionosphere.

We will define our coordinate system as shown in Figure 7.1, in which the wave propagates in the $+z$ direction, at some angle θ_B with respect to the Earth's magnetic field vector $\vec{B}$, which lies in the yz-plane. Note that we could have defined our coordinate system such that the y-axis is reversed, in which case θ_B would be an obtuse angle. Regardless of the orientation of y, the x-axis must be defined such that the coordinate system is right-handed.

The magnetic field vector has components along the y- and z-axes. We can define a component parallel to the direction of propagation (i.e. along the z-axis), called the longitudinal component

$$B_\parallel = B_0 \cos\theta_B, \tag{7.2}$$

where B_0 is the magnitude of $\vec{B}$, as well as a component perpendicular to the direction of propagation (i.e. along the y-axis) called the transverse component:

$$B_\perp = B_0 \sin\theta_B. \tag{7.3}$$

Using these definitions in Equation 7.1, we can now split the vector equation into three separate scalar equations. Beginning with the x-component, we see that the

Figure 7.1 Coordinate system for propagation with respect to Earth's magnetic field.

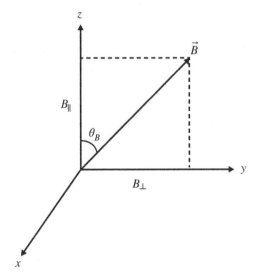

equation of motion in this direction becomes

$$m_e \left(\frac{d^2x}{dt^2} + v\frac{dx}{dt} \right) = -e \left(E_x + B_\| \frac{dy}{dt} - B_\perp \frac{dz}{dt} \right). \tag{7.4}$$

The propagating wave is sinusoidal and can be represented in phasor form: $E(\vec{r}, t) = \text{Re}\{\underline{E}(\vec{r})e^{j\omega t}\}$, where $\underline{E}$ is the complex amplitude and ω is the carrier frequency of the wave in radians/s. Since the differential equation is linear, each component of the position $\vec{r}$ will also be sinusoidal, and can be similarly represented as phasors of the same form. All complex amplitudes in this section will be represented using the underbar notation. Thus, plugging these phasors into the differential equation, we see that

$$m_e \left(-\omega^2 \underline{x} + j\omega v\underline{x} \right) = -e \left(\underline{E}_x + j\omega B_\| \underline{y} - j\omega B_\perp \underline{z} \right). \tag{7.5}$$

A time-varying polarization vector $\vec{P} = \text{Re}\{\underline{P} \, e^{j\omega t}\}$ is induced by the electromagnetic wave (see Section B.2), given by the equation $\vec{P} = -Ne\vec{r}$, where N is the number density of free electrons. Thus, we can write each component of the displacement vector in terms of its corresponding polarization component:

$$\underline{x} = -\frac{\underline{P}_x}{Ne}, \quad \underline{y} = -\frac{\underline{P}_y}{Ne}, \quad \underline{z} = -\frac{\underline{P}_z}{Ne}. \tag{7.6}$$

We can substitute these values into Equation 7.4 to obtain

$$m_e \left(\omega^2 \frac{\underline{P}_x}{Ne} - j\omega v\frac{\underline{P}_x}{Ne} \right) = -e \left(\underline{E}_x - j\omega B_\| \frac{\underline{P}_y}{Ne} + j\omega B_\perp \frac{\underline{P}_z}{Ne} \right). \tag{7.7}$$

If we multiply both sides of this equation by ε_0/e, we get

$$\frac{\varepsilon_0 m_e}{Ne^2} \omega^2 \underline{P}_x \left(1 - j\frac{v}{\omega} \right) = -\varepsilon_0 \underline{E}_x + j\omega \frac{\varepsilon_0}{Ne} B_\| \underline{P}_y - j\omega \frac{\varepsilon_0}{Ne} B_\perp \underline{P}_z. \tag{7.8}$$

We can modify the last two terms on the right hand side as follows:

$$\frac{\varepsilon_0 m_e}{Ne^2} \omega^2 \underline{P}_x \left(1 - j\frac{v}{\omega} \right) = -\varepsilon_0 \underline{E}_x + j\omega \frac{\varepsilon_0 m_e}{Ne^2} \frac{eB_\|}{m_e} \underline{P}_y - j\omega \frac{\varepsilon_0 m_e}{Ne^2} \frac{eB_\perp}{m_e} \underline{P}_z, \tag{7.9}$$

which will allow us to make use of some well-known quantities. We will denote the plasma frequency as ω_p. In the ionosphere, we have

$$\omega_p^2 = \frac{Ne^2}{\varepsilon_0 m_e}. \tag{7.10}$$

Another useful quantity is the cyclotron frequency, which is given by

$$\omega_c = -\frac{eB}{m_e}. \tag{7.11}$$

Both of these quantities are measured in units of radians/s. We can split the cyclotron frequency into longitudinal and transverse components as follows:

$$\omega_{\parallel} = -\frac{eB_{\parallel}}{m_e}, \quad \omega_{\perp} = -\frac{eB_{\perp}}{m_e}. \tag{7.12}$$

Using these definitions, we can define the unitless quantities

$$X = \frac{\omega_p^2}{\omega^2}, \quad Y_{\parallel} = \frac{\omega_{\parallel}}{\omega}, \quad Y_{\perp} = \frac{\omega_{\perp}}{\omega}, \quad Z = \frac{v}{\omega}. \tag{7.13}$$

Substituting these values into Equation 7.9 and multiplying both sides by X, we get

$$\underline{P}_x(1 - jZ) = -\epsilon_0 X \underline{E}_x - jY_{\parallel}\underline{P}_y + jY_{\perp}\underline{P}_z. \tag{7.14}$$

We can rearrange this equation slightly to obtain

$$\epsilon_0 X \underline{E}_x = -\underline{P}_x(1 - jZ) - jY_{\parallel}\underline{P}_y + jY_{\perp}\underline{P}_z. \tag{7.15}$$

Finally, we have an equation that expresses E_x in terms of the components of $\vec{P}$. Recall that this derivation was only for the x-direction. If we return to Equation 7.1 and perform similar derivations in the y- and z-directions, we obtain the analogous equations

$$\epsilon_0 X \underline{E}_y = -\underline{P}_y(1 - jZ) + jY_{\parallel}\underline{P}_x \tag{7.16}$$

for the y-direction and

$$\epsilon_0 X \underline{E}_z = -\underline{P}_z(1 - jZ) - jY_{\perp}\underline{P}_x \tag{7.17}$$

for the z-direction. We now wish to use Equations 7.15–7.17 to solve for the index of refraction. We currently have six unknowns – $\underline{E}_x, \underline{E}_y, \underline{E}_z, \underline{P}_x, \underline{P}_y,$ and $\underline{P}_z$ – but only three equations. Therefore, we must find several more independent equations to help solve the system.

From Maxwell's equations, we have (for a source free medium):

$$\nabla \times \vec{H} = \frac{\partial \vec{D}}{\partial t} = \frac{\partial}{\partial t}(\epsilon_0 \vec{E} + \vec{P}). \tag{7.18}$$

For an electromagnetic field traveling in the z-direction, all components of $\vec{H}$ have dependence $e^{-j(kz-\omega t)}$. Thus, the three scalar equations corresponding to Equation 7.18 are

$$jk\underline{H}_y = j\omega(\epsilon_0 \underline{E}_x + \underline{P}_x), \tag{7.19}$$

$$jk\underline{H}_x = -j\omega(\epsilon_0 \underline{E}_y + \underline{P}_y), \tag{7.20}$$

$$0 = j\omega(\epsilon_0 \underline{E}_z + \underline{P}_z). \tag{7.21}$$

Similarly, we also have from Maxwell's equations

$$\nabla \times \vec{E} = -\mu_0 \frac{\partial \vec{H}}{\partial t},$$ (7.22)

which corresponds to the three scalar equations

$$jk\underline{E}_y = -j\omega\mu_0\underline{H}_x,$$ (7.23)

$$jk\underline{E}_x = j\omega\mu_0\underline{H}_y,$$ (7.24)

$$0 = -j\omega\mu_0\underline{H}_z.$$ (7.25)

If we solve Equation 7.24 for H_y and plug it into Equation 7.19, we get

$$k^2 = \omega^2\mu_0\varepsilon_0 \left(1 + \frac{\underline{P}_x}{\varepsilon_0\underline{E}_x}\right).$$ (7.26)

We know that $\mu_0\varepsilon_0 = 1/c^2$ and $n = kc/\omega$, so we can rewrite this as

$$n^2 = 1 + \frac{\underline{P}_x}{\varepsilon_0\underline{E}_x}.$$ (7.27)

Similarly, if we solve Equation 7.23 for H_x and plug it into Equation 7.20, we can rearrange to obtain

$$n^2 = 1 + \frac{\underline{P}_y}{\varepsilon_0\underline{E}_y}.$$ (7.28)

If we set Equations 7.27 and 7.28 equal and simplify, we see that

$$\frac{\underline{P}_x}{\underline{E}_x} = \frac{\underline{P}_y}{\underline{E}_y}.$$ (7.29)

Finally, we may rearrange this equation to define a quantity R:

$$R := \frac{\underline{P}_y}{\underline{P}_x} = \frac{\underline{E}_y}{\underline{E}_x},$$ (7.30)

where R is the polarization ratio of the wave.

We finally have enough information to solve our system of equations. First, we see that from Equation 7.21, $\underline{E}_z = -\underline{P}_z/\varepsilon_0$. Substituting this into Equation 7.17, we can rewrite it as

$$X\underline{P}_z = P_z(1 - jZ) + jY_\perp P_x.$$ (7.31)

If we now solve this equation for $\underline{P}_z$ and plug it into Equation 7.15, we obtain

$$\varepsilon_0 X\underline{E}_x = -\underline{P}_x(1 - jZ) - jY_\parallel\underline{P}_y + \frac{Y_\perp^2\underline{P}_x}{(1 - X - jZ)}.$$ (7.32)

Using Equation 7.30, we can now write this equation solely in terms of $\underline{E}_y$ and $\underline{P}_y$,

$$\varepsilon_0 X \frac{\underline{E}_y}{R} = -\frac{\underline{P}_y}{R}(1 - jZ) - jY_\parallel \underline{P}_y + \frac{Y_\perp^2 \underline{P}_y}{R(1 - X - jZ)}. \tag{7.33}$$

If we now use Equation 7.30 in Equation 7.16 to eliminate P_x, we see that

$$\varepsilon_0 X \underline{E}_y = -\underline{P}_y (1 - jZ) + jY_\parallel \frac{\underline{P}_y}{R}. \tag{7.34}$$

By dividing both sides of this equation by R, we get

$$\varepsilon_0 X \frac{\underline{E}_y}{R} = -\frac{\underline{P}_y}{R}(1 - jZ) + jY_\parallel \frac{\underline{P}_y}{R^2}. \tag{7.35}$$

Note that the left hand sides of Equations 7.33 and 7.35 are identical, so we may equate their right hand sides. Simplifying, we see that

$$-jY_\parallel R^2 + \frac{Y_\perp^2 R}{(1 - X - jZ)} = jY_\parallel. \tag{7.36}$$

Solving this quadratic equation gives

$$\boxed{R = \frac{-j}{Y_\parallel}\left[\frac{Y_\perp^2}{2(1 - X - jZ)} \mp \sqrt{\frac{Y_\perp^4}{4(1 - X - jZ)^2} + Y_\parallel^2}\right].} \tag{7.37}$$

What this equation tells us is that there are only two specific polarizations, called characteristic polarizations or *eigenpolarizations*, that satisfy both Equations 7.33 and 7.35 simultaneously. The physical interpretation of this fact is that there only exist two polarizations that are able to retain their polarization state as they propagate through the ionosphere. All other polarizations will be distorted as they travel. This concept will be explored further in Section 7.3.5.

When Equation 7.37 is satisfied, Equations 7.33 and 7.35 both produce the same ratio $\underline{P}_y/\underline{E}_y$, given by

$$\frac{\underline{P}_y}{\underline{E}_y} = -\frac{\varepsilon_0 X}{1 - jZ - jY_\parallel R^{-1}}. \tag{7.38}$$

We can use this ratio in Equation 7.28 to obtain

$$n^2 = 1 - \frac{X}{1 - jZ - jY_\parallel R^{-1}}. \tag{7.39}$$

Substituting in for R finally gives us the Appleton–Hartree equation:

$$\boxed{n_\pm^2 = 1 - \frac{X}{1 - jZ - \dfrac{Y_\perp^2}{2(1 - X - jZ)} \pm \sqrt{\dfrac{Y_\perp^4}{4(1 - X - jZ)^2} + Y_\parallel^2}}.} \tag{7.40}$$

In this equation, the top sign corresponds to the top sign of Equation 7.37 and represents propagation of an ordinary wave (o-wave) through the ionosphere. Similarly, the bottom sign corresponds to the bottom sign of Equation 7.37 and represents propagation of an extraordinary wave (x-wave). The reader may consult Section B.6 for information on ordinary and extraordinary waves.

7.2.1 Simplifications of the Appleton–Hartree Equation

The Appleton–Hartree equation is clearly very complicated, which makes it difficult to work with in its most general form. Therefore, it is desirable to determine some reasonable approximations of the formula to simplify our analysis of ionospheric phenomena. The first approximation we will make is that of a "collisionless ionosphere." If the carrier frequency ω of a propagating wave is much higher than the average collision frequency v of electrons in the ionosphere, then $Z = v/\omega$ will be negligibly small. We can therefore approximate $Z \approx 0$, which allows us to simplify the Appleton–Hartree equation to

$$n_{\pm}^2 = 1 - \frac{2X(1-X)}{2(1-X) - Y_{\perp}^2 \pm \sqrt{Y_{\perp}^4 + 4(1-X)^2 Y_{\parallel}^2}}. \tag{7.41}$$

Note that this removes the imaginary part of n. Since the imaginary part is responsible for absorption, we can say that this approximation describes an essentially lossless ionosphere. The average electron collision frequency of the ionosphere is studied in [170] under various conditions, and is found to have values $\mathcal{O}\left(10^6\right)$ Hz or less. Compare this to GNSS signal frequencies, which are $\mathcal{O}\left(10^9\right)$ Hz. We see that $v \ll \omega$ and therefore the approximation is justified at GNSS frequencies.

In a collisionless ionosphere, the eigenpolarizations are given by

$$R = \frac{-j}{Y_{\parallel}} \left[\frac{Y_{\perp}^2}{2(1-X)} \mp \sqrt{\frac{Y_{\perp}^4}{4(1-X)^2} + Y_{\parallel}^2} \right], \tag{7.42}$$

where the plus sign corresponds to the ordinary wave and the minus sign corresponds to the extraordinary wave. Recall that $Y_{\parallel} = Y \cos\theta_B$ and $Y_{\perp} = Y \sin\theta_B$, where θ_B is the angle of propagation with respect to the Earth's magnetic field. In other words, the eigenpolarizations change depending on the direction of propagation.

Consider the case of longitudinal propagation, where $\theta_B = 0$. We see that $Y_{\parallel} = Y$ and $Y_{\perp} = 0$. The top sign, which corresponds to the ordinary wave, therefore has an eigenpolarization with ratio $+j$, which describes a left-hand circularly polarized (LHCP) wave (see Section A.3). Similarly, the extraordinary wave, which corresponds to the bottom sign, has an eigenpolarization with ratio $-j$, which describes a right-hand circularly polarized (RHCP) wave. In other words, for a wave traveling completely parallel to the Earth's magnetic field the ordinary wave will only retain its polarization if it is LHCP. Similarly, the extraordinary wave under the same circumstances will only retain its polarization if it is RHCP.

Note that when $\theta_B = \pi$, $Y_\parallel = -Y$ and $Y_\perp = 0$. In this case, the ordinary wave will have $R = -j$ and the extraordinary wave will have $R = +j$. These eigenpolarizations are still circular, but they are oppositely sensed as compared with the case where $\theta_B = 0$. In other words, when a wave is propagating anti-parallel to Earth's magnetic field, the ordinary wave only retains its polarization if it is RHCP, and the extraordinary wave will only retain its polarization if it is LHCP.

Now consider the case of transverse propagation, where $\theta_B = \pi/2$. In this case, $Y_\parallel = 0$ and $Y_\perp = Y$. We see that for the ordinary wave, the terms inside the brackets of Equation 7.42 cancel out, leaving $R = 0$. For the extraordinary wave, the term inside the brackets is nonzero, but as $Y_\parallel \downarrow 0$, $R \to +\infty$ and as $Y_\parallel \uparrow 0$, $R \to -\infty$. These eigenpolarizations correspond to linearly polarized waves; in particular, $R = 0$ is horizontally polarized, and $R = \pm\infty$ is vertically polarized. Thus, for a wave traveling completely perpendicular to the Earth's magnetic field, the ordinary wave's polarization will only remain unchanged if it is vertically polarized, and the extraordinary wave's polarization will only remain unchanged if it is horizontally polarized.

Note that for the case where $\theta_B = -\pi/2$, $Y_\parallel = 0$ and $Y_\perp = -Y$. This does not change anything for the ordinary wave, since all $Y_\perp$ terms are raised to even powers, so the negative sign makes no difference. Thus it is still true that $R = 0$, i.e. the ordinary wave is horizontally polarized. For the extraordinary wave, as $Y_\parallel \downarrow 0$, $R \to -\infty$ and as $Y_\parallel \uparrow 0$, $R \to +\infty$. However, since both $R = +\infty$ and $R = -\infty$ correspond to vertical polarization, the extraordinary wave is also unchanged from the case where $\theta_B = \pi/2$.

In general, for some arbitrary θ_B, the eigenpolarizations will be elliptical. If we consider an o-wave, sweeping θ_B from 0 to π, the eigenpolarization will begin left-hand circular ($R = +j$). As θ_B increases, it will become left-hand elliptical before finally becoming linear ($R = 0$) at $\theta_B = \pi/2$. As θ_B further increases, it will once again become elliptical, but will now be right-handed. Finally, once $\theta_B = \pi$, the wave will become right-hand circular ($R = -j$). An x-wave will follow the opposite pattern, beginning right-hand circular at $\theta_B = 0$, becoming right-hand elliptical, smoothly developing through a linear polarization ($R = \pm\infty$) at $\theta_B = \pi/2$, then becoming left-hand elliptical before finally becoming left-hand circular at $\theta_B = \pi$. Note that the polarizations of the o- and the x-waves follow the same pattern if swept toward $-\pi$ rather than toward π.

The ellipticity of eigenpolarizations for propagation oblique to Earth's magnetic field is dependent on the relative values of X, $Y_\parallel$, and $Y_\perp$. For certain relationships between these quantities, the elliptical eigenpolarizations can appear to be nearly circular or nearly linear. First consider the case where $Y_\perp^2 \ll |Y_\parallel| |1 - X|$.

Observing Equation 7.42, we see that if we consider $Y_\perp^2$ to be negligible, then R reduces to the longitudinal case. We call this *quasi-longitudinal* (Q_L) propagation, and under this approximation propagating waves appear to be essentially circularly polarized. If we extend this approximation to Equation 7.41, we can simplify the expression greatly:

$$n_\pm^2 = 1 - \frac{X}{1 \pm \left| Y_\parallel \right|}. \tag{7.43}$$

Recall that $Y_\parallel$ and $Y_\perp$ are dependent on θ_B, so the condition for Q_L propagation only holds for some range of angles. For example, for $\omega_p = 3 \times 10^7$ rad/s and $\omega_c = 8 \times 10^6$ rad/s (typical values for the ionosphere), a 10 MHz wave satisfies the Q_L approximation for angles as large as 50° [45]. This range increases further at higher frequencies. Table 7.1 shows the range of valid angles for the Q_L approximation at various frequencies.

From this table, it is clear that at GPS frequencies, the Q_L approximation applies for nearly all angles of propagation. Figure 7.2 shows the development of R with respect to θ_B for the Q_L case.

Because of this, GNSS signals are designed to propagate as circularly polarized waves. In practice, the RHCP polarization is chosen for GPS. Since it is impossible for a signal to be perfectly circularly polarized, a small amount of the signal's power is contained in the LHCP component. It can be seen in Figure 7.2 that an RHCP signal at GPS frequencies travels as an x-wave for $\left| \theta_B \right| < \pi/2$ and as an o-wave for $\left| \theta_B \right| > \pi/2$. The Earth's magnetic field vector is defined such that it points from south to north, so any GPS signal traveling north will propagate as an x-wave and any signal traveling south will propagate as an o-wave. This is shown in Figure 7.3.

For a more in-depth study of the behavior of polarization in anisotropic media, the reader is encouraged to refer to [171].

Converse to the Q_L approximation, consider the case where $Y_\perp^2 \gg \left| Y_\parallel \right| \left| 1 - X \right|$. This allows us to assume that $Y_\parallel$ is negligible, which causes Equation 7.42 to reduce to the transverse case, so this approximation is called the *quasi-transverse*

Table 7.1 The transition angle θ_t at which the Q_L approximation is no longer valid for various frequencies.

Frequency (GHz)	θ_t (°)
0.2	89.76
0.5	89.90
1.0	89.95
3.0	89.98

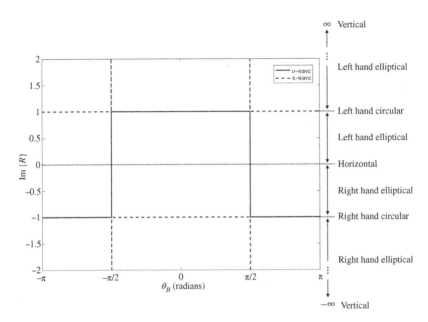

Figure 7.2 Polarization ratio vs. angle of propagation with respect to Earth's magnetic field.

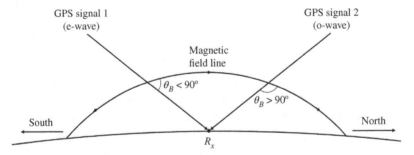

Figure 7.3 Difference in mode of propagation for northbound vs. southbound RHCP GPS signals.

(Q_T) approximation, in which all propagating signals appear to be essentially linearly polarized. Under this approximation, Equation 7.41 reduces to

$$n^2 = 1 - \frac{X}{1 - \dfrac{Y_\perp^2}{2(1-X)} \pm \dfrac{Y_\perp^2}{2(1-X)}}. \tag{7.44}$$

However, this approximation is not as useful as the Q_L approximation, since at most frequencies of interest for radio communications, the Q_L approximation holds for a much larger range of angles. It should be noted that in general, the Q_L and Q_T approximations also apply if collisions are considered. The expressions for n^2 in these cases can be obtained by simplifying Equation 7.40 for $Y_\perp = 0$ and $Y_\parallel = 0$, respectively.

If both electron collisions and the anisotropy of the ionosphere are ignored, the Appleton–Hartree equation simplifies greatly, simply becoming

$$n^2 = 1 - X, \tag{7.45}$$

which corresponds to an isotropic medium wherein any wave polarization would remain unchanged.

7.3 Propagation Effects of the Background Ionosphere

Electromagnetic waves traveling through the ionosphere are subject to a number of propagation effects, many of which arise directly from the presence of free electrons along its propagation path. In this section, we will define a parameter known as the TEC, and we will see how this parameter can be used to quantify the effects of propagation phenomena such as refraction, dispersion, and Faraday rotation.

Ionospheric propagation effects are dependent not only on TEC but also on frequency. Table 7.2 (from [172]) summarizes the most important propagation effects at various radio frequencies. Note that GPS signals propagate at approximately 1 GHz, so the values denoted in that column are of particular interest. The values in this table are based on a high TEC value, characteristic of daytime electron density values at low latitudes during high solar activity. This is the worst-case scenario for propagation in an undisturbed ionosphere, so the values listed in this table represent the maximum value one can typically expect from each source of error. It can be seen that for GPS frequencies, which lie between 1 and 1.6 GHz, the only significant ionospheric effects among those listed in the table will typically be group delay and Faraday rotation. All other effects are negligibly small at GPS frequencies and may be safely neglected in most situations.

7.3.1 Total Electron Content

As noted before, a number of important ionospheric propagation effects depend on a parameter known as the TEC, which represents the number of free electrons present along a signal's propagation path. As discussed in Chapter 4, electron density varies significantly with height, so in order to determine the TEC over a propagation path S, we must perform an integration:

$$\text{TEC} = \int_S N_e(s)\,ds, \tag{7.46}$$

Table 7.2 Estimated maximum ionospheric effects for a one-way traversal at an elevation angle of 30°. A TEC of 10^{18} electrons/m² is assumed.

Effect	Frequency dependence	0.1 GHz	0.25 GHz	0.5 GHz	1 GHz	3 GHz	10 GHz
Faraday rotation	$1/f^2$	30 rotations	4.8 rotations	1.2 rotations	108°	12°	1.1°
Group delay	$1/f^2$	25 μs	4 μs	1 μs	0.25 μs	0.028 μs	0.0025 μs
Refraction	$1/f^2$	<1°	<0.16°	<2.4′	<0.6′	<4.2′	<0.36′
Variation in the direction of arrival (RMS)	$1/f^2$	20′	3.2′	48′	12′	1.32′	0.12′
Absorption (auroral/polar cap)	$1/f^2$	5 dB	0.8 dB	0.2 dB	0.05 dB	6×10^{-3} dB	5×10^{-4} dB
Absorption (mid-latitude)	$1/f^2$	<1 dB	<0.16 dB	<0.04 dB	<0.01 dB	<0.001 dB	<1×10^{-4} dB
Dispersion	$1/f^3$	0.4 ps/Hz	0.026 ps/Hz	0.0032 ps/Hz	0.0004 ps/Hz	1.5×10^{-5} ps/Hz	4×10^{-7} ps/Hz

Source: ITU Recommendation P.618-12 [172]. Reproduced with permission.

where $N_e(s)$ is the electron density in electrons/m^3 as a function of position s. The TEC represents a column of electrons with unit cross-sectional area and a length $L = \int_S ds$ (measured in m). The TEC is often measured in TECU (total electron content units), where 1 TECU = 10^{16} electrons/m^2.

The TEC along an oblique path is called slant TEC or STEC, while propagation along the zenith path relative to a given receiver is called the vertical TEC or VTEC. For modeling purposes, VTEC is typically quoted [173], typically ranging from 10^{16} to 10^{18} electrons/m^2. The slant total electron content (STEC) is related to the vertical total electron content (VTEC) by the angle between the propagation path and the vertical, i.e. the zenith angle χ'. The expression relating the two quantities is:

$$\text{STEC} = \sec \chi' \ \text{VTEC}. \tag{7.47}$$

The quantity $\sec(\chi')$ is often denoted F and referred to as the slant factor.

In order to estimate χ', the ionosphere can be modeled by assuming the entire electron content is concentrated into an infinitesimally thin spherical shell at a height h_I, which represents the centroid of the electron density distribution and typically lies at about 300–400 km in altitude. This model is known as the single-layer ionosphere model. A propagating signal will intersect the spherical layer at some point along its propagation path; this point is referred to as the ionospheric piercing point (IPP). In this model, the zenith angle χ' is measured at the IPP, and is given by the equation

$$\chi' = \arcsin\left(\frac{a}{a+h_I} \sin \chi\right), \tag{7.48}$$

where $a \approx 6400$ km is the radius of the Earth and χ is the zenith angle at the receiver. χ can be calculated by using the known satellite position (provided by satellite ephemeris data) and approximated receiver coordinates. The single-layer ionosphere model is shown in Figure 7.4. *Note*: In this figure, E is the elevation angle, ψ is the geocentric angle between the receiver and the IPP, and φ is the geocentric angle between the receiver and the transmitter. These parameters will be referenced in subsequent sections. In the literature a point called the sub-ionospheric point (SIP), which represents the projection of the IPP onto the Earth, is commonly referenced. The sub-ionospheric point is labeled SIP in Figure 7.4.

As we learned, the electron density of the ionosphere, and thus the TEC, is highly variable, changing with time of day, season, location on earth, and solar activity. TEC is highest at latitudes within $\pm 15°$ to $\pm 20°$ on either side of Earth's magnetic equator [11]. The day to day variability has a standard deviation of about ± 20 to $\pm 25\%$ of the monthly average [174]. Ionospheric irregularities can also disturb the electron density, causing short-term changes in TEC. Because of these variations, it can be very difficult to predict the precise value of the TEC

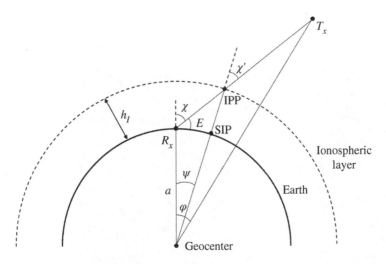

Figure 7.4 Geometry of the single-layer ionosphere model for computing χ'.

during transmission of a given signal. Several prominent modeling techniques for estimating TEC will be discussed later in this text.

Various models exist for quantifying TEC based on the principles we have discussed in this section. One such model, called the Global Ionospheric Scintillation Model (GISM), is capable of producing global maps of TEC estimates. GISM runs another model, known as NeQuick, as a subroutine to accomplish this (both GISM and NeQuick will be discussed in some detail later in this chapter). Figures 7.5 and 7.6 show global maps of TEC calculated by GISM. Both figures show the TEC on 1 January 2000 and assume a 10.7 cm solar flux Φ of 150 SFU. Figure 7.5 calculates the TEC at noon (12:00) UTC, while Figure 7.6 calculates the TEC twelve hours later, at midnight (24:00) UTC. It can be clearly observed in these figures that the highest TEC values occur at equatorial latitudes. It is also evident that the location of peak TEC migrates over the course of the day. These observations are sensible since solar flux is largest at the equator, and varies diurnally due to the rotation of the Earth.

The TEC between a given satellite and receiver does not remain constant over time. As a satellite orbits the Earth, the length of the propagation path and the elevation angle with respect to the receiver will change. Thus, even if the electron density of the ionosphere remained constant, there would be variation in the TEC solely due to the motion of the satellite. However, we know that the electron density *does* change, experiencing variation diurnally and seasonally, as well as over the course of the solar cycle. In addition to these expected periodic variations, there are numerous anomalies and irregularities which cause short term fluctuations in electron density. For example, solar events such as coronal mass ejections (CMEs)

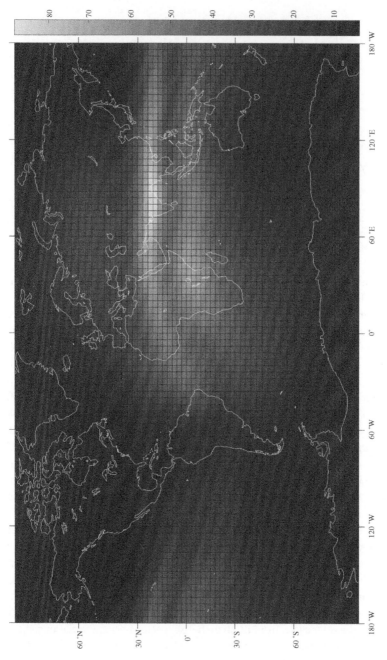

Figure 7.5 Global map of TEC modeled by GISM for 1 January 2000 at 12:00 UTC, assuming Φ = 150 SFU.

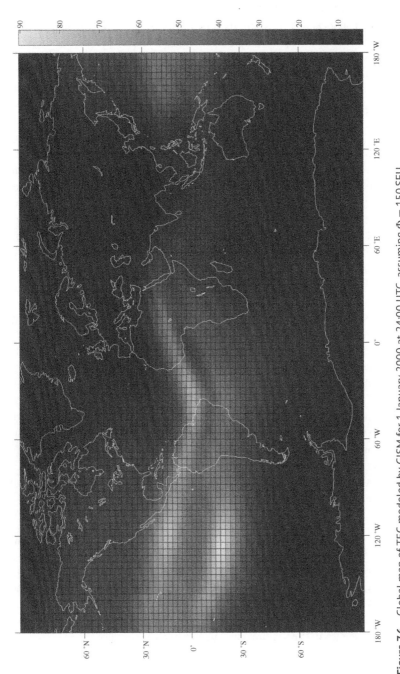

Figure 7.6 Global map of TEC modeled by GISM for 1 January 2000 at 24:00 UTC, assuming $\Phi = 150$ SFU.

and solar flares can cause sudden increases in ionization, and atmospheric gravity waves can cause localized fluctuations called traveling ionospheric disturbances (TIDs). A number of ionospheric anomalies were covered in Section 6.4, and special attention will be given to ionospheric scintillations in Section 7.4 due to their significant impact on GPS performance.

7.3.2 Ionospheric Refraction

GPS observables, such as pseudorange, are formulated theoretically assuming GPS signals travel at the speed of light along a straight path from the satellite to a receiver. However, due to the variable index of refraction of the atmosphere the wave will be refracted as it propagates, causing it to travel along a non-straight, deviated trajectory. The indices of refraction of the troposphere and ionosphere are governed by different physical phenomena. The former is discussed in Section 3.2, while the latter is described by the Appleton–Hartree equation, which we discussed earlier in this chapter. This section will discuss ionospheric refraction exclusively.

As seen from the Appleton–Hartree equation, the index of refraction of the ionosphere is directly proportional to the number density of free electrons N_e. As discussed in Sections 5.4 and 6.3, N_e varies continuously and monotonically with altitude up to a certain peak value, then decreases monotonically as altitude further increases. Thus the index of refraction behaves similarly, bending the signal in the manner shown in Figure 7.7.

Due to this bending, the true propagation path of a signal is slightly longer than the straight path. The total effective length S of a propagation path can be found by integrating the index of refraction over the path. Due to the anisotropy of the ionosphere, the wave fronts will not always be parallel to the direction of power flow. This phenomenon, called spatial walk-off, is discussed in some detail in Section B.6.5. Thus, a factor $\cos \alpha$ must be included in order to account for this effect, where α is the angle between the wave vector $\vec{k}$ (indicating the direction normal to the wavefronts) and the Poynting vector $\vec{S}$ (indicating the direction of power flow).

$$S = \int_S n(s) \cos \alpha \, ds. \tag{7.49}$$

Here, $n(s)$ is the index of refraction as a function of position. For frequencies in the very high frequency (VHF) band and higher, $\alpha \approx 0$ so that $\cos \alpha \approx 1$ and the spatial walk-off effect can be ignored. GPS signals fall within this frequency range, so we are justified in henceforth neglecting $\cos \alpha$. From this equation, it follows that the path length in the vacuum case, where $n = 1$ is simply:

$$S_0 = \int_{S_0} ds. \tag{7.50}$$

The path length difference between these two cases is simply

$$\Delta S = S - S_0 = \int_S n(s)ds - \int_{S_0} ds, \tag{7.51}$$

which can be split into two terms as follows:

$$\Delta S = S - S_0 = \int_S [n(s) - 1]\, ds + \left(\int_S ds - \int_{S_0} ds \right), \tag{7.52}$$

where the last term is the length difference between the actual and straight paths. A GPS receiver will interpret the measurement from a signal traveling along the refracted path as an apparent range S, which is longer than the true range S_0. It is possible to determine both of these quantities solely in terms of the index of refraction and geometric parameters. The apparent range from a satellite to a ground-based receiver, as given in [175], is:

$$S = \int_0^{h_t} \frac{n^2(a+h)dh}{\sqrt{n^2(a+h)^2 - n_0^2(a+h_0)^2\cos^2\theta}}$$
$$+ \int_{h_t}^{h_0} \frac{(a+h)dh}{\sqrt{n^2(a+h)^2 - n_0^2(a+h_0)^2\cos^2\theta}}. \tag{7.53}$$

The first term in this equation represents the contribution of the troposphere, and the second term represents that of the ionosphere. Here, h_t represents the height of the boundary between the lower atmosphere and the ionosphere (measured above sea level), and h_0 is the altitude of the satellite (measured above sea level). n is the refractive index at h, n_0 is the refractive index at h_0, a is the radius of Earth, and θ is the apparent depression angle (as measured from the satellite's frame of reference) at h_0.

The true range from the satellite to the receiver is given by

$$S_0 = a\frac{\sin\varphi}{\cos\alpha}, \tag{7.54}$$

where α is the true depression angle at h_0, given by

$$\alpha = \arctan\left(\frac{a+h_0}{a} \csc\varphi - \cot\varphi \right), \tag{7.55}$$

and φ (which can be seen in Figure 7.4) is the geocentric angle between the satellite and the receiver, given by:

$$\varphi = n_0(a+h_0)\cos\theta \int_0^{h_0} \frac{dh}{(a+h)\sqrt{n^2(a+h)^2 - n_0^2(a+h_0)^2\cos^2\theta}}. \tag{7.56}$$

When a GPS signal reaches a receiver, it will arrive at a slightly different angle than if it had traveled along a straight path. This is illustrated in Figure 7.7, where

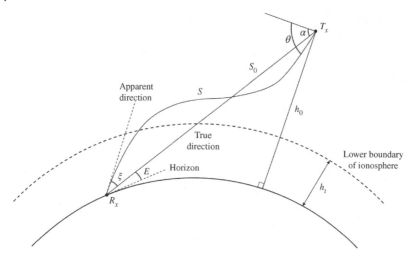

Figure 7.7 Comparison of the refracted ray path with the true range. The deviated path S is greatly exaggerated for emphasis.

the elevation angle E describes the direction that the signal is expected to arrive. However, due to refraction the signal will arrive from a different direction, and the angular error ξ describes the angle between the true direction to the satellite and the apparent direction, from whence the signal arrived. The angular error, as given in [81], is:

$$\xi \approx \frac{(S_0 + a\sin E)a\cos E}{h_I(2a + h_I) + a^2\sin^2 E}\frac{\Delta S}{S_0} \approx \frac{(S_0 + a\sin E)\cos E}{2h_I + a\sin^2 E}\frac{\Delta S}{S_0}. \tag{7.57}$$

Here, S_0 refers again the true range between the satellite and receiver, ΔS is the path length difference between the direct and refracted paths, and h_I is the height to the mean ionospheric height, as described previously in the single layer model, typically between 300 and 450 km.

As a general trend, refraction decreases as the frequency increases; at VHF and higher frequencies refraction effects are negligibly small and the actual propagation path does not deviate significantly from a straight path except under unusual atmospheric conditions and less common scenarios such as very low elevation angles. For GPS applications, it is often acceptable to assume $S \approx S_0$. In this case, Equation 7.52 simplifies to

$$\Delta S \approx \int_{S_0} [n(s) - 1]\, ds. \tag{7.58}$$

In Section 7.3.3, we will see that, for most GPS applications, we can assume $\Delta S \approx 40.3\,\text{TEC}/f^2$, where f is the frequency of the signal. Note that the dependence of ΔS on frequency makes ξ sensitive to frequency as well.

7.3.3 Group Delay and Phase Advance

As an electromagnetic wave travels through the ionosphere, it is refracted and will not travel at a speed exactly equal to c. If the pseudorange and carrier phase observables of a GPS system are formulated assuming a propagation velocity of c, the measured value of each observable will differ slightly from its theoretical value. The pseudorange observable is associated with the information encoded on the carrier wave (i.e. the modulation signal), which propagates at the group velocity v_{gr}. Conversely, the carrier phase observable is associated with the phase of the carrier wave, which propagates at the phase velocity v_{ph}. For information on the group and phase velocities of a wave, the reader may refer to Section A.2.

The ionosphere is dispersive, meaning the phase velocity and group velocity of the wave will assume different values (see Section B.3.2 for more information on dispersive media). Because of this, the error induced in the pseudorange measurement will not be the same as the error induced in the carrier phase measurement. In this section, we will derive the expression for each error term and compare them.[1]

The phase velocity and the group velocity of a wave each have their own refractive index associated with them. These relationships are given by

$$v_{ph} = \frac{c}{n_{ph}} \tag{7.59}$$

and

$$v_{gr} = \frac{c}{n_{gr}}, \tag{7.60}$$

respectively, where n_{ph} is the phase index (typically just called the index of refraction as we did before) and n_{gr} is the group index. In a dispersive medium, these values will be different from one another. By differentiating v_{ph} with respect to wavelength λ, we get:

$$\frac{dv_{ph}}{d\lambda} = -\frac{c}{n_{ph}^2} \frac{dn_{ph}}{d\lambda}. \tag{7.61}$$

A slightly modified version of Equation A.41 gives

$$v_{gr} = v_{ph} - \lambda \frac{dv_{ph}}{d\lambda}. \tag{7.62}$$

Plugging Equations 7.59–7.61 into Equation 7.62 gives

$$\frac{c}{n_{gr}} = \frac{c}{n_{ph}} + \lambda \frac{c}{n_{ph}^2} \frac{dn_{ph}}{d\lambda}, \tag{7.63}$$

---

1 This derivation has been adapted from [19].

or equivalently,

$$\frac{1}{n_{gr}} = \frac{1}{n_{ph}} \left(1 + \lambda \frac{1}{n_{ph}} \frac{dn_{ph}}{d\lambda} \right). \tag{7.64}$$

By inverting both sides of Equation 7.64 and using the fact that $(1 + \delta)^{-1} \approx (1 - \delta)$ if $\delta \approx 0$, we obtain

$$n_{gr} = n_{ph} \left(1 - \lambda \frac{1}{n_{ph}} \frac{dn_{ph}}{d\lambda} \right) \tag{7.65}$$

and hence

$$n_{gr} = n_{ph} - \lambda \frac{dn_{ph}}{d\lambda}. \tag{7.66}$$

Since $\lambda = c/f$, it follows that

$$\frac{d\lambda}{df} = -\frac{c}{f^2} = -\frac{\lambda f}{f^2} = -\frac{\lambda}{f}. \tag{7.67}$$

Rearranging, we see that

$$\frac{d\lambda}{\lambda} = -\frac{df}{f}, \tag{7.68}$$

and Equation 7.66 becomes

$$n_{gr} = n_{ph} + f \frac{dn_{ph}}{df}. \tag{7.69}$$

Using a power series expansion, we can express n_{ph} as follows:

$$n_{ph} = 1 + \frac{c_2}{f^2} + \frac{c_3}{f^3} + \frac{c_4}{f^4} + \dots, \tag{7.70}$$

where the coefficients c_i are independent of frequency, but do depend on N_e, the number density of free electrons in the ionosphere. Higher order terms may be neglected and by truncating the expansion to second order, we obtain

$$n_{ph} = 1 + \frac{c_2}{f^2}. \tag{7.71}$$

Referring to the Appleton–Hartree equation under the assumption of a collisionless, isotropic ionosphere we get

$$n_{ph}^2 = 1 - X = 1 - \frac{\omega_p^2}{\omega} = 1 - \frac{1}{(2\pi f)^2} \frac{N_e e^2}{\varepsilon_0 m_e}. \tag{7.72}$$

Here, we have expressed the plasma frequency in terms of the number density of electrons N_e, the elementary charge $e = 1.602 \times 10^{-19}$ C, the permittivity of free space $\varepsilon_0 = 8.85 \times 10^{-12}$ F/m, and the mass of an electron 9.11×10^{-31} kg. Substituting in all numerical values where possible and simplifying, we see that

$$n_{ph}^2 = 1 - 80.6 \frac{N_e}{f^2}. \tag{7.73}$$

Taking the square root of both sides and using the first two terms of a Taylor series expansion ($\sqrt{1-x} \approx 1 - \frac{1}{2}x$), we get

$$n_{ph} = 1 - 40.3 \frac{N_e}{f^2}. \tag{7.74}$$

This approximation is justified because maximum values of N_e are $\mathcal{O}(10^{11})$–$\mathcal{O}(10^{12})$ electrons/m^3, while GPS signals propagate at about 1 GHz, or $\mathcal{O}(10^9)$. Therefore the numerical value of the factor $80.6 N_e / f^2$ will be small for any N_e experienced in the ionosphere. Comparing Equations 7.71 and 7.74, we see that

$$c_2 = -40.3 N_e.$$

Differentiating Equation 7.71 expression with respect to f gives

$$dn_{ph} = -2 \frac{c_2}{f^3} df. \tag{7.75}$$

If we now substitute Equations 7.71 and 7.75 into Equation 7.69, we get

$$n_{gr} = 1 + \frac{c_2}{f^2} - 2f \frac{c_2}{f^3} = 1 - \frac{c_2}{f^2} = 1 + 40.3 \frac{N_e}{f^2}. \tag{7.76}$$

Since N_e is always positive, we see from Equations 7.74 and 7.76 that $n_{gr} > 1$ and $n_{ph} < 1$ and thus $v_{gr} < c$ and $v_{ph} > c$.[2] This tells us that the information encoded onto a signal will experience a delay as it travels through the ionosphere. This is called a *group delay*. Conversely, the wavefronts of the carrier wave will travel more quickly through the ionosphere, resulting in a *phase advance*. Because of this, the measured pseudorange of a GPS signal will be larger than the theoretical value, and the measured carrier phase will be smaller than the theoretical value.

First Order Ionospheric Error

Group delay and phase advance can be quantified by determining the difference in path length between the refracted path and the vacuum path. As seen in Section 7.3.2, the path length difference is given by

$$I = \Delta S = \int_S n \, ds - \int_{S_0} ds. \tag{7.77}$$

---

2 Note that this does not violate special relativity because the phase of a signal carries no information.

The path length difference is equivalent to the ionospheric error term, so we have used the notation I for the path length difference to remain consistent with the notation used in Chapter 1. Substituting $n = n_{ph}$ into this equation gives

$$I_{1,ph} = \int_S \left(1 - 40.3 \frac{N_e}{f^2}\right) ds - \int_{S_0} ds \tag{7.78}$$

and substituting $n = n_{gr}$ gives

$$I_{1,gr} = \int_S \left(1 + 40.3 \frac{N_e}{f^2}\right) ds - \int_{S_0} ds. \tag{7.79}$$

The subscript 1 refers to the fact that this is the first order ionospheric error term. This concept will be explained further shortly, and higher order terms will be considered as well. As explained before, the factor $40.3 N_e/f^2$ is small for GPS frequencies, so that, as noted in Section 7.3.2, the curved path S does not diverge significantly from the straight (vacuum) path S_0. As a result, Equations 7.78 and 7.79 simplify as follows:

$$I_{1,ph} = -\int_{S_0} 40.3 \frac{N_e}{f^2} ds, \tag{7.80}$$

$$I_{1,gr} = \int_{S_0} 40.3 \frac{N_e}{f^2} ds, \tag{7.81}$$

where each expression has units of meters. Recalling that TEC $= \int_S N_e \, ds \approx \int_{S_0} N_e \, ds$, we finally obtain

$$I_{1,ph} = -40.3 \frac{\text{TEC}}{f^2}, \tag{7.82}$$

$$I_{1,gr} = 40.3 \frac{\text{TEC}}{f^2}. \tag{7.83}$$

Note that the phase advance term is equal in magnitude to the group delay term, but opposite in sign. This means that the carrier phase and pseudorange observables will differ from their vacuum values by the same amount in opposite directions. This phenomenon is known as *carrier-code divergence*.

In practice, the group delay of a GPS signal is expressed in terms of a time delay rather than a path length difference. Since GPS signals are assumed to travel at speed c, the group delay can be expressed as

$$t = \frac{I_{1,gr}}{c} = 1.34 \times 10^{-7} \frac{\text{TEC}}{f^2}, \tag{7.84}$$

measured in seconds. Figure 7.8 shows a family of curves relating Δt and f for various values of TEC. As an example, for a TEC value of 10^{17} electrons/m^2, a GPS signal on the L1 carrier frequency ($f = 1575.42$ MHz) will experience a delay of approximately 5.4 ns.

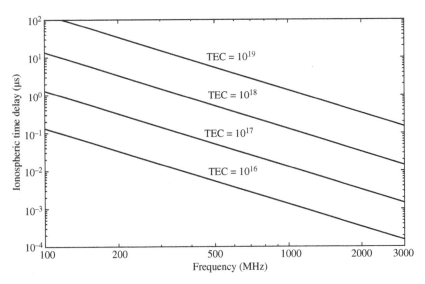

Figure 7.8 Ionospheric group delay vs. frequency for various values of TEC, where TEC is measured in units of electrons/m². Source: Modified from ITU Recommendation P.531-13 [173].

Similarly, the magnitude of the phase advance can be expressed in radians via the following formula:

$$\phi = \frac{2\pi}{\lambda}|I_{1,ph}| = \frac{2\pi f}{c}40.3\frac{\text{TEC}}{f^2} = 8.44 \times 10^{-7}\frac{\text{TEC}}{f}. \tag{7.85}$$

A family of curves relating $\Delta\phi$ to f for various values of TEC is given in Figure 7.9. For a TEC of 10^{17} electrons/m², a signal on L1 will experience approximately 53.6 radians of phase advance.

Ionospheric Doppler Shift
Recall that phase variations of a signal can be related to a change on the *instantaneous* frequency of the signal through

$$f_D = \frac{1}{2\pi}\frac{d\phi}{dt}. \tag{7.86}$$

Using Equation 7.85 in the aforementioned expression, we obtain:

$$f_D = \frac{1.34 \times 10^{-7}}{f}\frac{d\text{TEC}}{dt} \tag{7.87}$$

for a one-way traverse of the ionosphere so that the transmitted signal will be frequency shifted by this amount as it travels through the ionosphere. Note that f_D is equivalent to a Doppler shift that is dependent upon the rate of change of the TEC.

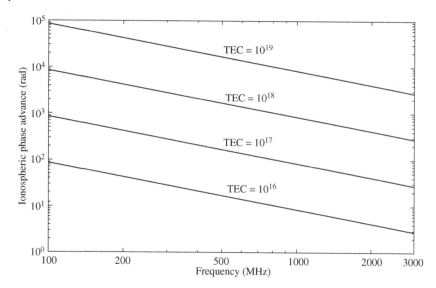

Figure 7.9 Ionospheric phase advance vs. frequency for various values of TEC, where TEC is measured in units of electrons/m². Source: Modified from ITU Recommendation P.531-13 [173].

As discussed in Section 7.3.1, the rate of TEC change is due partly to the movement of the satellite, which changes the propagation path, and partly to the change in the ionosphere itself.

It should be noted that for frequencies above 1 GHz, the Doppler shift induced by traversal of the ionosphere is negligible compared to the shift induced by the motion of the satellite. For example, for a typical TEC rate of change, of 0.1×10^{16} electrons/m²/s, $f_D \approx 0.085$ Hz on the L1 frequency ($f = 1575.42$ MHz). As a result, this effect can be safely ignored for GPS applications.

Higher Order Ionospheric Error Terms

The phase advance and group delay terms, $I_{1,gr}$ and $I_{1,ph}$ are known as first order ionospheric error terms and are inversely proportional to f^2. As implied by Equation 7.70, there are also higher order terms present, inversely proportional to f^3, f^4, etc. The first order term accounts for the majority of the error in a given GPS observable, while the higher order terms are several orders of magnitude smaller. For example, the zenith group delay (first order, $\propto 1/f^2$) induces an observation error of $\sim$1–30 m, while the second order ($\propto 1/f^3$) and third order ($\propto 1/f^3$) terms induce errors of $\sim$0–2 cm and $\sim$0–2 mm at zenith, respectively [176]. Thus, in most cases it is sufficient to consider only the first order term when accounting for ionospheric positioning error. However, for applications which require highly accurate positioning (sub-centimeter to sub-millimeter precision), it is necessary to consider the higher order ionospheric error terms.

At a given frequency, the first order error term is dependent solely on the TEC along the propagation path of a signal. Conversely, the higher order terms also depend on the effect of the anisotropy induced in the ionosphere by the geomagnetic field. As we did in Equation 7.72, we can use a Taylor series expansion of the Appleton–Hartree equation to determine this relationship mathematically. Previously, we assumed a collisionless, isotropic ionosphere to analyze the first order term. Now, we wish to consider the effects of anisotropy, so we will remove the simplification of an isotropic ionosphere. We will still assume the ionosphere is collisionless to simplify the discussion.

The Appleton–Hartree equation for a collisionless, anisotropic ionosphere is given by:

$$n_\pm^2 = 1 - \frac{2X(1-X)}{2(1-X) - Y_\perp^2 \pm \sqrt{Y_\perp^4 + 4(1-X)^2 Y_\parallel^2}}. \tag{7.88}$$

Here, $X = \left(\omega_p/\omega\right)^2$, $Y = \omega_c/\omega$, $Y_\perp = Y \sin\theta_B$, $Y_\parallel = Y \cos\theta_B$. ω_p is the plasma frequency, defined as before. In addition, $\omega_c = -eB_0/m_e$ is the cyclotron frequency (also known as the Larmor frequency or gyrofrequency), e is the elementary charge, B_0 is the magnitude of the geomagnetic field and m_e is the mass of an electron. θ_B is the angle between the signal's propagation vector $\vec{k}$ and the geomagnetic field vector $\vec{B}$. In the denominator, the plus and minus signs refer to the ordinary (o-) and extraordinary (x-) waves, respectively (see Section B.6 for details).

Using a Taylor series expansion of Equation 7.88 up to the fourth inverse power of frequency gives [177]:

$$n_\pm = 1 - \frac{1}{2}X \pm \frac{1}{2}XY\left|\cos\theta_B\right| - \frac{1}{4}X\left[\frac{1}{2}X + Y^2(1 + \cos^2\theta_B)\right]. \tag{7.89}$$

Terms beyond this have magnitude less than 10^{-9} and may be neglected [178]. Again, the plus and minus signs refer to the o- and x-waves, respectively. Recall from Section 7.2 that a RHCP wave propagates entirely as an x-wave for all angles $\left|\theta_B\right| < 90°$ and as an o-wave for all angles $\left|\theta_B\right| > 90°$. If we consider a RHCP wave, we can rewrite Equation 7.89 as

$$n_\pm^{RH} = 1 - \frac{1}{2}X - \frac{1}{2}XY\cos\theta_B - \frac{1}{4}X\left[\frac{1}{2}X + Y^2(1 + \cos^2\theta_B)\right]. \tag{7.90}$$

We see that the third term on the right hand side no longer has an absolute value nor a variable sign. By plugging in values of θ_B, we can confirm that the third term is negative (indicating an x-wave) for angles $\left|\theta_B\right| < 90°$ and is positive (indicating an o-wave) for angles $\left|\theta_B\right| > 90°$. This exactly matches the propagation of an RHCP wave.

The opposite case is true of LHCP waves. An LHCP wave propagates entirely as an o-wave for all angles $|\theta_B| < 90°$ and as an x-wave for all angles $|\theta_B| > 90°$. Thus we can write

$$n_{\pm}^{LH} = 1 - \frac{1}{2}X + \frac{1}{2}XY\cos\theta_B - \frac{1}{4}X\left[\frac{1}{2}X + Y^2(1 + \cos^2\theta_B)\right]. \tag{7.91}$$

Once again, by plugging in the numerical values of θ_B into this equation, we can verify that the sign of the third term on the right hand side always indicates the proper mode (ordinary or extraordinary) for a LHCP wave.

GPS signals are RHCP, so we will henceforth work from Equation 7.90. Substituting the expressions for X and Y, we get

$$n_{\pm}^{RH} = 1 - \frac{1}{2}\left(\frac{\frac{N_e e^2}{\varepsilon_0 m_e}}{(2\pi f)^2}\right) \pm \frac{1}{2}\left(\frac{\frac{N_e e^2}{\varepsilon_0 m_e}}{(2\pi f)^2}\right)\left(\frac{\frac{eB_0}{m_e}}{2\pi f}\right)\cos\theta_B$$
$$- \frac{1}{4}\left(\frac{\frac{N_e e^2}{\varepsilon_0 m_e}}{(2\pi f)^2}\right)\left[\frac{1}{2}\left(\frac{\frac{N_e e^2}{\varepsilon_0 m_e}}{(2\pi f)^2}\right) + \left(\frac{\frac{eB_0}{m_e}}{2\pi f}\right)^2 (1 + \cos^2\theta_B)\right]. \tag{7.92}$$

Plugging in the numerical values of e, ε_0, and m_e gives

$$n_{\pm}^{RH} = 1 - 40.3\frac{N_e}{f^2} \pm \frac{1}{2}\left(\frac{7527cN_e B_0 \cos\theta_B}{f^3}\right)$$
$$- \frac{1}{3}\left(\frac{2437N_e^2 + 4.74 \times 10^{22}N_e B_0^2(1 + \cos^2\theta_B)}{f^4}\right). \tag{7.93}$$

Recalling that $n_{\pm}$ is the phase index of refraction n_{ph}, we can use Equation 7.69 to obtain the group index:

$$n_{gr,\pm} = 1 + \frac{1}{2}X \mp XY|\cos\theta_B| + \frac{3}{4}X\left[\frac{1}{2}X + Y^2(1 + \cos^2\theta_B)\right]. \tag{7.94}$$

Here, the top and bottom signs refer to the o- and x-waves respectively. Using the same argument as for Equations 7.90 and 7.91, we can write equations for the special cases of RHCP and LHCP waves. Respectively, these expressions are:

$$n_{gr,\pm}^{RH} = 1 + \frac{1}{2}X + XY\cos\theta_B + \frac{3}{4}X\left[\frac{1}{2}X + Y^2(1 + \cos^2\theta_B)\right], \tag{7.95}$$

$$n_{gr,\pm}^{LH} = 1 + \frac{1}{2}X - XY\cos\theta_B + \frac{3}{4}X\left[\frac{1}{2}X + Y^2(1 + \cos^2\theta_B)\right]. \tag{7.96}$$

Once again, we will only consider the expression for RHCP waves. Substituting the expressions for X and Y into Equation 7.95, we see that

$$n_{gr,\pm}^{RH} = 1 + 40.3\frac{N_e}{f^2} \mp \frac{7527cN_e B_0 \cos\theta_B}{f^3}$$
$$+ \frac{2437N_e^2 + 4.74 \times 10^{22}N_e B_0^2(1 + \cos^2\theta_B)}{f^4}, \tag{7.97}$$

where c is the speed of light. We can use Equation 7.77 to determine the total path length difference, which is equivalent to the total ionospheric error. We can use Equations 7.93 and 7.97 for the phase advance and group delay terms, respectively. Following the same procedure used before for finding the first order ionospheric error, we obtain the following equations for a one way traversal of the ionosphere:

$$I_{ph} = -\frac{q}{f^2} - \frac{1}{2}\frac{s}{f^3} - \frac{1}{3}\frac{r}{f^4}, \tag{7.98}$$

$$I_{gr} = \frac{q}{f^2} + \frac{s}{f^3} + \frac{r}{f^4}, \tag{7.99}$$

where

$$q = 40.3 \times \text{TEC}, \tag{7.100}$$

$$s = 7527c \int_{S_0} N_e B_0 \cos\theta_B \, ds, \tag{7.101}$$

$$r = 2437 \int_{S_0} N_e^2 \, ds + 4.74 \times 10^{22} \int_{S_0} N_e B_0^2 \left(1 + \cos^2\theta_B\right) ds. \tag{7.102}$$

Recall from Figure 7.3 that northbound RHCP signals travel as x-waves while southbound RHCP signals travel as o-waves. For a northbound (x-wave) signal, $\theta_B < 90°$, so s is positive. Therefore, the second order group error (the second term of Equation 7.99) is positive, so the first order group delay is an *underestimation* of the true group error. Similarly, the second order phase error (the second term of Equation 7.98) is negative, meaning the first order phase advance is an *overestimation* of the true phase error.

For a southbound (o-wave) signal, the situation is reversed. $\theta_B > 90°$, so s is negative. The second order group error is now negative, so the first order group delay is an overestimation of the true group error. The second order phase error is now positive, meaning the first order phase advance is an underestimation of the true phase error.

The equation for r is dependent on $\cos^2\theta_B$, so regardless of the direction of propagation, r is positive. Thus, the third order group error is always positive, implying that the second order group delay is an underestimation. The third order phase error is always negative, implying that the second order phase advance is an overestimation.

The first order term accounts for 99.9% of total ionospheric error. Thus all higher order terms account for at most a few centimeters worth of positioning error. As noted before, for most applications, this minor contribution can be ignored; however, for applications requiring a very high degree of accuracy, higher order effects may need to be considered. For further information on higher order ionospheric delay terms, including modeling and correction techniques, the reader is encouraged to consult [176, 177].[3]

7.3.4 Dispersion

Since the ionosphere is dispersive, it follows that waves of different frequencies will propagate at different velocities through the ionosphere. As discussed in Section A.2, any information-carrying signal is necessarily comprised of multiple frequencies. As the wave travels, each constituent frequency experiences a slightly different time delay. Consider a signal with some bandwidth Δf and center frequency f_c. The upper frequency $f_c + \Delta f/2$ and the lower frequency $f_c - \Delta f/2$ will experience different group delays, say, t_U and t_L, respectively. The difference in group delay between the upper and lower frequencies is thus given by $\Delta t = t_U - t_L$. Dividing this group delay differential by the bandwidth, i.e. $\Delta t/\Delta f$ gives us a measure of the dispersion of the medium. In the limit, this becomes the derivative dt/df. Substituting in Equation 7.84 for t, we see that for a one way traverse of the ionosphere, group delay dispersion is given by

$$\frac{dt}{df} = -2.68 \times 10^{-7} \frac{\text{TEC}}{f^3}. \tag{7.103}$$

A GPS receiver cannot directly measure dispersion; instead, it must measure the group delay differential between the frequency components at the extreme ends of the bandwidth. The magnitude of the differential, measured in seconds, can be found by rearranging the aforementioned equation:

$$|\Delta t| = 2.68 \times 10^{-7} \Delta f \frac{\text{TEC}}{f^3}, \tag{7.104}$$

where Δf is the bandwidth and f is the carrier frequency of the wave.

Similarly, the phase advance of a signal was seen to have frequency dependence and therefore it is also subject to dispersion. This effect is separate from that of group delay dispersion and is instead called phase dispersion. Following the same procedure as earlier, we may take the derivative of Equation 7.85 with respect to frequency to obtain phase dispersion:

$$\frac{d\phi}{df} = -8.44 \times 10^{-7} \frac{\text{TEC}}{f^2}. \tag{7.105}$$

Again, the receiver can only measure the differential phase advance of the frequency components at extreme ends of the bandwidth. The magnitude of this differential, measured in radians, is given by

$$|\Delta\phi| = 8.44 \times 10^{-7} \Delta f \frac{\text{TEC}}{f^2}. \tag{7.106}$$

---

3 There are some errors in this paper. These errors have been directly addressed and corrected in [177–180].

Once again, Δf is the bandwidth and f is the carrier frequency. Note that at a given bandwidth, both the group delay and phase differentials will be smaller at higher frequencies. Therefore, the larger the bandwidth of the modulated signal, the larger the carrier frequency required to mitigate the effects of dispersion.

GPS signals are designed with this in mind. Consider, for example, the P(Y) code modulated on the L1 carrier frequency. The P(Y) code is modulated at a chipping rate of 10.23 Mcps, which corresponds to a bandwidth of 20.46 MHz. L1 is centered on 1575.42 MHz, so assuming an average TEC value of 10^{17} electrons/m^2, the group delay differential over the P(Y) bandwidth is approximately 0.14 ns. Similarly, for the same TEC, the phase differential over the P(Y) bandwidth is approximately 0.696 rad.

7.3.5 Faraday Rotation

The presence of the Earth's magnetic field creates anisotropy in the ionosphere, causing it to exhibit birefringence. In other words, a propagating wave will split in two, resulting in an ordinary wave and an extraordinary wave (see Section B.6 for details). Each of these waves experiences a different index of refraction. For a collisionless, anisotropic ionosphere, the Appleton–Hartree equation gives us the index of refraction for each case as

$$n_{\pm}^2 = 1 - \frac{2X(1-X)}{2(1-X) - Y_{\perp}^2 \pm \sqrt{Y_{\perp}^4 + 4(1-X)^2 Y_{\parallel}^2}}, \tag{7.107}$$

where the plus sign refers to the o-wave and the minus sign refers to the x-wave. In an anisotropic medium, there are only two characteristic polarizations or *eigenpolarizations* which remain unchanged as they propagate – all other polarizations will be modified as they travel. For a collisionless ionosphere ($Z = 0$), we see from Equation 7.37 that the polarization ratio is given by

$$R = \frac{-j}{Y_{\parallel}} \left[\frac{Y_{\perp}^2}{2(1-X)} \mp \sqrt{\frac{Y_{\perp}^4}{4(1-X)^2} + Y_{\parallel}^2} \right]. \tag{7.108}$$

Note that here, the top sign corresponds to the top sign of Equation 7.107, and the same is true of the bottom sign. In other words, the minus sign corresponds to the ordinary wave and the plus sign corresponds to the extraordinary wave. When $Y_{\perp}^2 \ll |Y_{\parallel}| |1 - X|$, we can use the quasi-longitudinal (Q_L) approximation, in which $Y_{\perp} = 0$. As discussed previously, at GPS frequencies this approximation is valid for virtually all angles of propagation. Thus we see that under this approximation, the polarization ratio reduces to $R = j$ for the ordinary wave, which corresponds to a RHCP. Similarly, the ratio reduces to $R = -j$ for the extraordinary wave, indicating a LHCP. This tells us that the two eigenpolarizations of the ionosphere are $R = \pm j$ – in other words, only circular polarizations are able to travel through the ionosphere unaffected.

RHCP and LHCP are orthogonal and can be chosen as basis polarizations. In other words, any arbitrary polarization, can be written as a linear combination of one RHCP and one LHCP wave (see Section A.3). However, via Equation 7.107, we see that these two polarizations will have different indices of refraction, so they will propagate at different velocities, inducing a relative phase difference between the two polarizations. For a linearly polarized wave, the net effect of this phase difference is to rotate the plane of polarization. For example, if a wave polarized in the y-direction enters the ionosphere, it may end up polarized in the x-direction by the time it exits.

We can quantify the effect of Faraday rotation by determining the path length difference between the ordinary (RHCP) wave and the extraordinary (LHCP) wave, which is given by

$$S_+ - S_- = \int_S (n_+ - n_-)ds. \tag{7.109}$$

To simplify the analysis, we assume quasi-longitudinal propagation wherein $Y_\perp = 0$, which reduces the Appleton–Hartree equation to

$$n_\pm^2 = 1 - \frac{X}{1 \pm |Y_\parallel|}. \tag{7.110}$$

Further, we can use the Taylor series approximation $\sqrt{1-x} = 1 - \frac{1}{2}x$ for very small x to obtain

$$n_\pm = 1 - \frac{1}{2}\frac{X}{1 \pm |Y_\parallel|}. \tag{7.111}$$

Using this expression, we see that

$$n_+ - n_- = 1 - \frac{1}{2}\frac{X}{1 + |Y_\parallel|} - \left(1 - \frac{1}{2}\frac{X}{1 - |Y_\parallel|}\right). \tag{7.112}$$

After some algebra, this equation simplifies to

$$n_+ - n_- = \frac{XY_\parallel}{1 - Y_\parallel^2}. \tag{7.113}$$

Recall that $Y_\parallel = \omega_c \cos\theta_B/\omega$, where ω_c is the cyclotron frequency, and θ_B is the angle between the propagating signal's wave vector and Earth's magnetic field vector. In the ionosphere, ω_c is typically $\mathcal{O}\left(10^6\right)$ Hz, so for signals whose frequencies are in the VHF band or higher, $Y_\parallel^2$ is negligibly small, and the expression reduces to

$$n_+ - n_- = XY_\parallel = \frac{\omega_p^2 \omega_c \cos\theta_B}{\omega^3}, \tag{7.114}$$

where ω_p is the plasma frequency and ω is the frequency of the propagating signal. Finally, substituting this into Equation 7.109, we get

$$\int_S (n_+ - n_-) ds = \frac{\omega_p^2}{\omega^3} \int_S N_e \omega_c \cos \theta_B \ ds. \tag{7.115}$$

The rotation of the plane of polarization for a one way traverse of the ionosphere is given by

$$\Omega = \frac{1}{2}(\Omega_+ - \Omega_-) = \frac{1}{2} k \int_S (n_+ - n_-) ds, \tag{7.116}$$

where $k = 2\pi/\lambda$ is the wavenumber of the propagating wave, or

$$\Omega = \frac{\pi}{\lambda} \frac{\omega_p^2}{\omega^3} \int_S N_e \omega_c \cos \theta_B \ ds. \tag{7.117}$$

Substituting $\lambda = c/f$, $\omega_p^2 = N_e e^2/(\varepsilon_0 m_e)$ and $\omega_c = -eB_0/m_e$, we see that

$$\Omega = -\frac{e^3}{8\pi^3 \varepsilon_0 m_e^2 f^2 c} \int_S N_e B_0 \cos \theta_B \ ds. \tag{7.118}$$

Commonly, $B_0 \cos \theta_B$ is replaced with B_{av}, the average value of the magnetic field. This is constant over the integration path and can be factored out. Substituting in numerical values where appropriate, and recalling that TEC $= \int_S N_e \ ds$, the magnitude of the Faraday rotation can be expressed as

$$|\Omega| = 2.36 \times 10^4 B_{av} \frac{\text{TEC}}{f^2}. \tag{7.119}$$

From this equation it can be seen that the Faraday rotation increases at higher values of TEC, and decreases at higher frequencies. Figure 7.10 shows Faraday rotation vs. frequency for selected TEC values. In this figure, B_{av} has been taken as 50,000 nT.

At GPS frequencies, the effect of Faraday rotation can be significant, especially for large TEC values. As an example, for a TEC of 10^{17} electrons/m^2 and B_{av} of 50,000 nT, a linearly polarized wave at the L1 carrier frequency ($f = 1575.42$ MHz) will experience about 0.0475 rad or 2.7° of rotation. As a more extreme example, for the same value of B_{av} and a TEC of 10^{18} electrons/m^2, a linearly polarized wave at the L5 frequency ($f = 1227.6$ MHz) will experience about 0.783 rad or 44.9° of rotation.

The reason Faraday rotation is so problematic is that most antennas are designed to receive only a specific polarization. Any wave component which has the same polarization as the antenna (which is said to be *co-polarized*) will be received, while any component with an orthogonal polarization (also called *cross-polarized*) will be rejected. For example, a vertically polarized antenna will receive vertically polarized waves and reject horizontally polarized waves. Similarly, a right circularly polarized antenna will receive RHCP waves and reject LHCP waves. Thus we

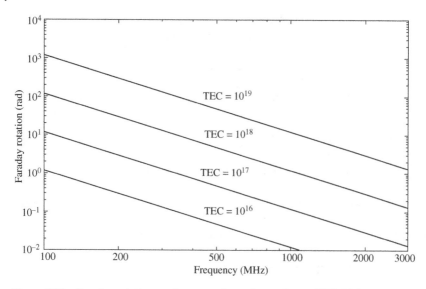

Figure 7.10 Faraday rotation vs. frequency for various values of TEC. TEC is measured in units of electrons/m^2 and B_{av} is taken to be 50,000 nT. Source: Modified from ITU Recommendation P.531-13 [173].

see that Faraday rotation poses a problem for linearly polarized signals traveling through the ionosphere. Consider a GPS receiver designed with a vertically polarized antenna. If a satellite transmits a vertically polarized signal, it will rotate as it traverses the ionosphere, and in general will arrive at the receiver with a different polarization – i.e. it may now have a horizontally polarized component. This horizontal component will be rejected by the receiver, reducing the received power and potentially preventing adequate reception of the signal. The power of a received signal in dB can be quantified in terms of the rotation angle Ω induced by Faraday rotation. This quantity, known as the cross-polarization discrimination or XPD, is given by the expression:

$$XPD = -20 \log_{10} (\tan \Omega). \tag{7.120}$$

For good system performance, the XPD should be at least 27–30 dB [181]. For a rotation of about 1°, the XPD is about 35 dB, which is sufficiently above the system requirements. However, performance drops off rapidly. Consider our previous example on L1, which experienced 2.7° of rotation. The XPD for this signal is 26.5 dB, below the system performance threshold. Similarly, our example L5 signal, which experienced 44.9° of rotation, would have an XPD of only 0.03 dB.

Clearly, Faraday rotation poses a serious problem for linearly polarized GPS signals. Fortunately, the solution to this problem is quite simple. We previously saw that the characteristic polarizations of the ionosphere at GPS frequencies are right and left circular polarizations. In other words, if a circularly polarized wave is sent through the ionosphere, the polarization will remain unchanged as the wave propagates. Therefore, GPS systems are designed to transmit and receive RHCP waves, which eschews the problem of Faraday rotation entirely.

7.3.6 Absorption

An electron in the presence of an electric field is accelerated via the Coulomb force. Some of the electric field's energy is converted into the electron's kinetic energy to accomplish this. As an electromagnetic wave travels through the ionosphere, its time-varying electric field will cause electrons to oscillate. Often, they will collide with molecules and ions and lose some of their energy. Macroscopically, this effect can be interpreted as absorption of the electromagnetic wave. In Section B.2.2, we show that an electron movement can be modeled as a damped harmonic oscillator, with a damping coefficient Γ which encapsulates the effects of absorption. In the ionosphere, Γ is equivalent to the electron collision frequency ν.

We can determine the absorption of the ionosphere by using the Appleton–Hartree equation. For simplicity, we will assume the quasi-longitudinal case, so $Y_T = 0$. For an anisotropic ionosphere with collisions, we see that the Appleton–Hartree equation becomes

$$n^2 = 1 - \frac{X}{1 \pm Y_\parallel - jZ}, \tag{7.121}$$

We can rationalize the denominator of the second term to obtain

$$n^2 = 1 - \frac{X(1 \pm Y_\parallel)}{(1 \pm Y_\parallel)^2 + Z^2} - j\frac{XZ}{(1 \pm Y_\parallel)^2 + Z^2}. \tag{7.122}$$

By expressing the index of refraction in terms of its real and imaginary parts, $n = n_R + j\kappa$, we see that $n^2 = n_R^2 - \kappa^2 - j2n_R\kappa$. We can equate the real and imaginary parts of this expression to those of Equation 7.122, i.e.

$$n_R^2 - \kappa^2 = 1 - \frac{X(1 \pm Y_\parallel)}{(1 \pm Y_\parallel)^2 + Z^2}, \tag{7.123}$$

$$2n_R\kappa = \frac{XZ}{(1 \pm Y_\parallel)^2 + Z^2}. \tag{7.124}$$

Rearranging Equation 7.124 in terms of κ gives:

$$\kappa = \frac{1}{2n_R} \frac{XZ}{(1 \pm Y_\parallel)^2 + Z^2}. \tag{7.125}$$

Recalling that the wavenumber k is related to n by the expression $k = n\omega/c$, we see that

$$k = \frac{\omega}{c}(n_R + j\kappa) = k_R + j\alpha, \tag{7.126}$$

where α is called the attenuation coefficient. Therefore, we see that

$$\alpha = \frac{\omega}{c} \frac{1}{2n_R} \frac{XZ}{(1 \pm Y_\parallel)^2 + Z^2}. \tag{7.127}$$

Substituting $Y_\parallel = \omega_\parallel/\omega$, $Z = v/\omega$, and $X = \omega_p^2/\omega^2$, where $\omega_p^2 = Ne^2/(\epsilon_0 m_e)$ and simplifying, we finally get

$$\alpha = \frac{e^2}{2\epsilon_0 m_e c} \frac{1}{n} \frac{Nv}{(\omega \pm \omega_\parallel)^2 + v^2}. \tag{7.128}$$

At VHF frequencies and above, $n \approx 1$, $\omega \gg \omega_\parallel$ and $\omega \pm \omega_\parallel \gg v$, so we can simplify this expression to

$$\alpha \approx \frac{e^2}{2\epsilon_0 m_e c} \frac{Nv}{\omega^2} = 5.3 \times 10^{-6} \frac{Nv}{\omega^2}. \tag{7.129}$$

We see that the absorption decreases at higher frequencies. At frequencies above 1 GHz, absorption is negligibly small – typically less than 0.01 dB. Therefore, ionospheric absorption can be safely ignored for GPS applications.

7.4 Scintillations

Perturbations in the ionosphere regularly give rise to irregularities in electron distribution. These irregularities cause random fluctuations in the refractive index along the propagation path of a signal, which in turn cause fluctuations in the amplitude, phase and angle of arrival of the signal. The recognizable twinkling of stars is evidence of scintillation of visible light as it passes through Earth's atmosphere. The ionospheric inhomogeneities that cause scintillations arise due to several mechanisms, such as $\vec{E} \times \vec{B}$ gradient drift and Rayleigh–Taylor instabilities. The former of these effects is discussed in Section 5.5.3, while the latter is briefly discussed in Section 6.4.4. Scintillations occur most strongly at equatorial latitudes – about 20° on either side of the magnetic equator – and at auroral and polar latitudes, i.e. above 55°–60°. Scintillations can occur at mid latitudes as well, but are typically much less intense, only becoming significant during magnetic storms.

Equatorial scintillations typically occur in the F layer, at altitudes of 200–1000 km, though occasionally a sporadic E layer can create scintillation effects in the 90–100 km range as well. F layer instabilities form shortly after sunset, causing scintillations which last from approximately 8 p.m. to 2 a.m. local time. On the other hand, sporadic E instabilities can form at any time of day. Scintillation effects are much stronger near the equator than anywhere else on Earth, causing deep fades in the VHF and UHF bands there.

At high latitudes, significant scintillations occur frequently as well. Like equatorial scintillations, those at high latitudes occur primarily in the F layer, and are strongest after sunset. Due to the complex coupling between the ionosphere, the magnetosphere, and the steady stream of charged particles from the solar wind, irregularities at high latitudes are correlated with magnetic disturbances. Thus high latitude scintillations are much more severe during periods of high magnetic activity, such as a magnetic storm.

A number of other factors affect the latitudinal variation of scintillations. For example, the presence of the Earth's magnetic field can cause the shape of the irregularities that form to differ with latitude. A thorough analysis of the global morphology of ionospheric scintillations can be found in [182]. Scintillations exhibit temporal variation as well. In addition to the diurnal variation mentioned previously, the severity of scintillations at all latitudes is directly correlated to the level of solar activity, and scintillations tend to be strongest just after the equinoxes. In [183] it is shown that the number of GPS cycle slips tracks directly with the temporal variation of scintillations.

7.4.1 Scale Size of Ionospheric Irregularities

Shortly after sunset, recombination in the bottomside ionosphere creates a layer of low electron density. Perturbations, such as atmospheric gravity waves, can create small seed disturbances which grow exponentially via Rayleigh–Taylor instability into bubbles of low electron density, which ascend through the upper layers of the ionosphere. The size of these bubbles, often called the scale size, can reach in excess of 100 km [44]. As the bubbles ascend, turbulent mixing at their perimeter produces many smaller bubbles and vortices with sizes as little as a few centimeters. It is these smaller irregularities, via diffraction and forward scattering, which cause scintillation of radio waves as they pass through the ionosphere.

Similar to large bubbles, small-scale irregularities will have lower electron density than the surrounding region of the ionosphere. Recall that the index of refraction of the ionosphere depends directly on the local electron density, so these irregularities will also have a different refractive index than their surroundings. Thus we see that under these conditions, the ionosphere behaves as an inhomogeneous medium.

Irregularities with a scale size below the first Fresnel zone[4] will cause wavefront diffraction and, as a result, produce phase and amplitude scintillations in the received signal. Irregularities with scale sizes progressively larger than the

---

4 See Section 3.5 for a discussion of Fresnel zones.

first Fresnel zone will begin to partially interfere destructively, until they reach a minimum at the second Fresnel zone. A successive pattern of constructive and destructive interferences will be formed by larger irregularities that span successively higher Fresnel zones but with the relative influence of higher order Fresnel zones becoming progressively weaker with the zone order. As a result, the first Fresnel zone provides a spatial filtering mechanism that determines the size of ionospheric irregularities that contribute to scintillation effects. When the size of the irregularities is much larger than the first Fresnel zone, diffractive effects become negligible and the wave merely refracts, causing phase fluctuations on the received signal but no significant scintillations. Note that the size of the Fresnel zone depends on the wavelength of the signal.

7.4.2 Statistical Description of Scintillations

The small-scale irregularities responsible for ionospheric scintillations are highly variable in both space and time and form over a spectrum of scale sizes. The highly complex behavior of the ionosphere makes it impossible to create a detailed deterministic model of scintillations; instead, scintillation effects are modeled probabilistically. A scintillated signal will arrive at the receiver with an amplitude that varies essentially randomly with time. The variation of the signal amplitude due to ionospheric scintillations is found to be well-described by the so-called Nakagami-m probability density function (pdf) [45]. Since the intensity of a signal is proportional to the square of the amplitude, we can also obtain the pdf of the random variable I that captures the fluctuations in the instantaneous *intensity* of a signal affected by scintillations. The pdf of I is given by

$$f(I) = \frac{m^m}{\Gamma(m)} I^{m-1} e^{-mI}, \tag{7.130}$$

where $\Gamma(m)$ is the standard gamma function. From the pdf, we can also obtain the cumulative distribution function (cdf), which is defined as the integral of the pdf. The cdf of I has a closed-form solution given by

$$F(I) = \int_0^I f(x)dx = \frac{\Gamma(m, mI)}{\Gamma(m)}. \tag{7.131}$$

The cdf of the intensity can be directly used to determine the fraction of time that I is above or below a certain threshold. For example, the probability that the signal's intensity is more than X dB below the mean is given by $F(10^{-X/10})$, and the fraction of time that the intensity is more than Y dB above the mean is given by $1 - F(10^{Y/10})$ [173].

In Equations 7.130 and 7.131, the mean value of I has been normalized to 1, $\Gamma(m)$ represents the gamma function, $\Gamma(m, mI)$ is the incomplete gamma function, and the shape parameter m is given by $1/(S_4)^2$. S_4 is a commonly used parameter known as the scintillation index and is defined as follows:

$$S_4 = \sqrt{\frac{\langle I^2 \rangle - \langle I \rangle^2}{\langle I \rangle^2}}. \tag{7.132}$$

The S_4 parameter has been found to be a very accurate predictor of the occurrence and severity of scintillations. Lower values of S_4 indicate weaker scintillations, while larger values indicate stronger ones. For convenience, the scintillation strength is classified as weak ($S_4 < 0.3$), moderate ($0.3 \leq S_4 \leq 0.6$), or strong ($S_4 > 0.6$). It is possible to express S_4 in terms of parameters related to the signal and the ionosphere [81] as

$$S_4 = B\lambda^3 hLK \sec^2 \chi \langle \Delta N^2 \rangle. \tag{7.133}$$

Here, B is a proportionality factor dependent on the geometry of the particular scenario under consideration and some fundamental constants. In addition, K is the outer scale size of the irregularity, λ is the wavelength of the signal, h is the altitude of the irregularity, L is the thickness of the irregularity, χ is the zenith angle, and $\langle \Delta N^2 \rangle$ is the mean square deviation of the electron density. Figures 7.11–7.13 show global maps of S_4 under various conditions, as modeled by GISM.

In an engineering context, it is often more useful to have a measure of the peak-to-peak power fluctuations of the signal I_4 in dB. An approximate relationship between S_4 and I_4 is

$$I_4 = 27.5 \, S_4^{1.26}. \tag{7.134}$$

Table 7.3 provides values of I_4 based on various S_4 values.

In addition to amplitude scintillations, which are accurately characterized by the random variable I and the S_4 index, phase scintillations ϕ are characterized as Gaussian distributed random variables with zero mean. The standard deviation of ϕ, σ_ϕ, is an accurate predictor of the behavior of phase scintillations. In the weak and moderate regimes, S_4 and σ_ϕ are highly correlated, and take on similar values (when σ_ϕ is measured in radians). The variance of ϕ, denoted σ_ϕ^2 can also be represented solely in terms of parameters related to the signal and the ionosphere [81]:

$$\sigma_\phi^2 = A\lambda^2 LK \sec \chi \langle \Delta N^2 \rangle. \tag{7.135}$$

Here, A is a proportionality constant dependent on geometrical factors and fundamental constants, and all other parameters are the same as in the expression for S_4.

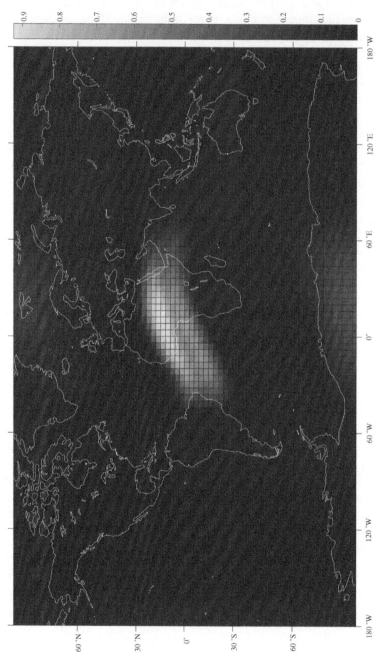

Figure 7.11 Global map of S_4 on the L1 carrier frequency, modeled by GISM for January 1st, 2000 at 24:00 UTC, assuming $\Phi = 150$ SFU.

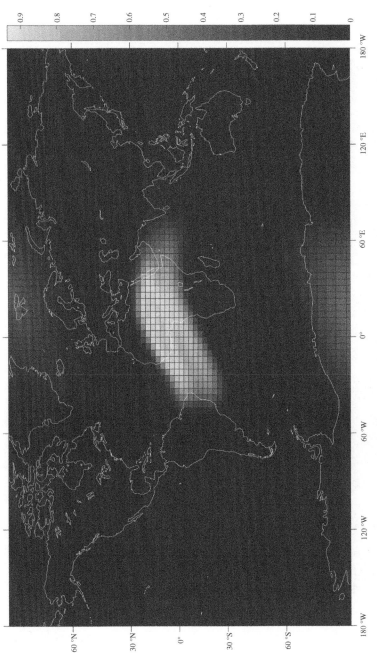

Figure 7.12   Global map of S_4 on the L1 carrier frequency, modeled by GISM for January 1st, 2000 at 24:00 UTC, assuming $\Phi = 150$ SFU.

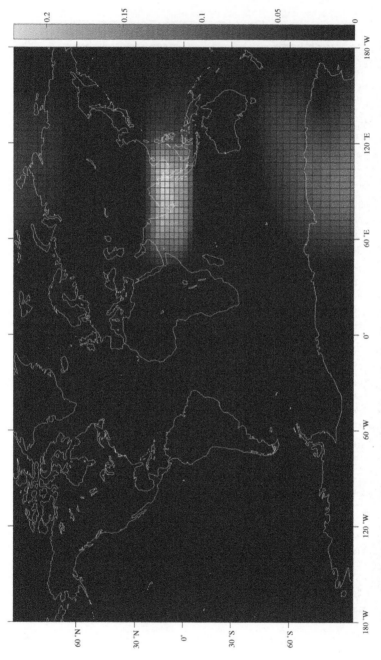

Figure 7.13 Global map of S_4 on the L1 carrier frequency, modeled by GISM for January 1st, 2000 at 24:00 UTC, assuming $\Phi = 150$ SFU.

Table 7.3 Empirical conversion between S_4 and I_4 values.

S_4	I_4 (dB)
0.1	1.5
0.2	3.5
0.3	6
0.4	8.5
0.5	11
0.6	14
0.7	17
0.8	20
0.9	24
1.0	27.5

7.4.3 Power Spectra of Scintillations

Since the ionosphere is highly dynamic, irregularities tend to fluctuate over time, causing ionospheric scintillations to fluctuate at a fading rate of about 0.1–1 Hz. Measurement data shows that the relative power of signal fluctuations due to scintillation rolls off with frequency. For signals with a carrier frequency of a few gigahertz such as GPS, the power spectrum is found to have a peak frequency at about 0.1 Hz and a roll-off at higher frequencies with slopes ranging from f^{-1} to f^{-6}. If no direct measurement data is available, the International Telecommunication Union (ITU) recommends assuming a slope of f^{-3}, which corresponds to the weak-to-moderate scintillation regime. Figure 7.14 shows a typical shape of a scintillation power spectrum. This plot includes the power spectra of a 4 GHz signal measured at six distinct time intervals over the course of one amplitude scintillation event in Taipei.

The six curves correspond to the following times:

A: 30 minutes before event onset.
B: At the beginning of the event.
C: One hour after onset.
D: Two hours after onset.
E: Three hours after onset.
D: Four hours after onset.

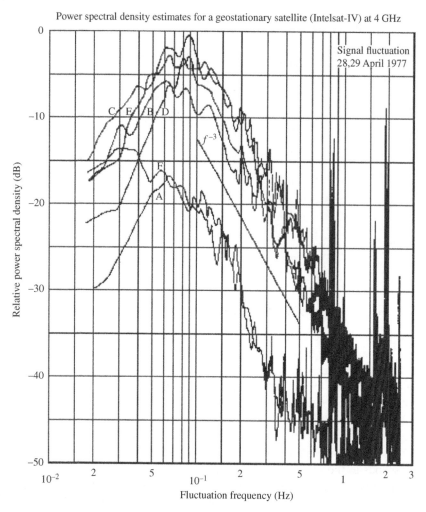

Power spectral density estimates for a geostationary satellite (Intelsat-IV) at 4 GHz

Figure 7.14 Power spectrum of a 4 GHz signal with respect to fading rate, as measured in Taipei on 28, 29 April 1977. Source: ITU Recommendation P.531-13 [173]. Reproduced with permission.

Both amplitude and phase scintillations decrease at higher fading rates, and in general follow similar slopes past the corner frequency of 0.1 Hz. However, at lower fading rates the power spectra of amplitude and phase scintillations diverge, with phase scintillations exhibiting much stronger fluctuation at lower frequencies. Figure 9 in [173] illustrates this divergence.

8

Predictive Models of the Ionosphere

8.1 Introduction

In this final chapter, we discuss several major predictive models relevant to the ionosphere. These models vary in scope and aims; some are concerned with efficient approximation of a single parameter, while others seek to provide a more holistic and comprehensive picture of the state of the ionosphere.

Unlike the troposphere, the ionosphere exhibits non-negligible frequency dispersion at global navigation satellite systems (GNSS) frequencies, so multi-frequency receivers can effectively mitigate group delay error by using the ionosphere-free combination method, which was discussed in Section 2.5.1. However, multi-frequency receivers are relatively expensive, so several predictive models have been developed to estimate and subtract ionospheric group delay for single-frequency receivers. Section 8.2 compares two major models for computing ionospheric group delay: the Klobuchar model and the NeQuick model. Just as was the case for tropospheric group delay models, these models must often be employed in real-time. For this reason, computational efficiency is a primary concern, which must be balanced against the accuracy of the model.

In a similar vein to group delay models, the Global Ionospheric Scintillation Model (GISM) focuses mainly on a specific propagation effect, in this case iono-spheric scintillations. Mathematically, scintillations may be modeled as though a wave were passing through a screen that changes the wave's phase. These so-called "phase screens" can be successively applied along a wave's propagation path to determine the amplitude of the scintillated wave, which in turn can be used to predict whether the scintillation levels are likely to cause signal outage. Section 8.3 provides a brief overview of the model and introduces a technique known as the multiple phase screen (MPS) method that allows the successive application of phase screens to achieve the aforementioned scintillation computation.

Tropospheric and Ionospheric Effects on Global Navigation Satellite Systems, First Edition.
Timothy H. Kindervatter and Fernando L. Teixeira.

Finally, in Section 8.4 the focus is turned to the International Reference Ionosphere (IRI), which is an empirical model that seeks to provide a highly comprehensive and accurate model of the ionosphere. The IRI is perhaps more accurately thought of as a collection of submodels, each of which computes a particular parameter of interest. The IRI's design is modular so that any given submodel can be updated in the event that a superior methodology is developed. Many submodels have been updated in this way since the inception of the IRI in the 1960s.

This book's coverage of the IRI attempts to be reasonably detailed, but it is necessarily limited, since the model's development has been the subject of vast number of papers and updates over the span of more than half a century. Significant attention is given to a few important parameters, such as electron density, electron and ion temperatures, and ion composition. In addition, numerous references have been provided to papers discussing various submodels for readers interested in reading further on the topic.

8.2 Group Delay Models for Single-Frequency GNSS Receivers

As seen in Section 7.3, the total electron content (TEC) of the ionosphere is directly responsible for a number of deleterious effects on radiowave propagation. At GPS frequencies, the only significant effects due to TEC are Faraday rotation and group delay. The issue of Faraday rotation can be easily solved by using right hand circularly polarized signals, which leaves group delay as the only major source of ionospheric error due to TEC that has yet to be accounted for.

There are two main ways to combat the issue of group delay. The first and most effective method is to use dual-frequency receivers. Due to the dispersive nature of the ionosphere, the two frequencies will experience different group delays. For a given GPS observable, this allows the receiver to construct a pair of linearly independent equations, which can then be used to obtain an expression for that observable, which is completely independent of group delay error. This technique is known as the ionosphere-free combination, and is discussed in detail in Section 2.5.1.

Unfortunately, dual-frequency receivers are relatively expensive, so although they provide the best possible mitigation of ionospheric error, they are cost-prohibitive for many applications. For this reason, many GPS receivers are single-frequency, which means an alternate method of group delay error mitigation must be employed. By modeling the behavior of the ionosphere, it is possible to predict the group delay error to a certain extent. A single-frequency GPS receiver can have an ionospheric model included in its software, allowing it to perform real-time corrections to its positioning measurement based on the ionospheric error predicted by the model.

Several widely-used models exist for the purpose of single-frequency ionospheric error correction. The ionosphere is highly dynamic, and as seen earlier in this chapter, effects such as scintillations can only be modeled statistically. Thus it is impossible to create a deterministic model to perfectly predict and remove ionospheric error. Highly sophisticated models are capable of reducing error to a greater extent, but this comes at the cost of higher computational complexity and larger amounts of data. The choice of model is therefore an important engineering consideration for GPS positioning applications. In this section, two widely used models will be discussed: the Klobuchar model and NeQuick.

8.2.1 Klobuchar Model

The Klobuchar model was originally published in a 1987 paper by John Klobuchar [184], and consists of an algorithm that allows a single-frequency receiver to estimate ionospheric group delay error and apply a correction to its positioning calculation. In the design of this algorithm, four key design considerations were heeded: (i) user computational complexity, (ii) the then-current knowledge of temporal and geographic variations in TEC, (iii) the number of coefficients to be transmitted in the GPS navigation message, and (iv) geographic areas where single-frequency receivers are likely to be used.

For GPS applications, positioning is often performed in real time, so it is desirable to have low computational complexity. Additionally, if a large number of coefficients were included in the navigation message, the message's total length would be increased. Since the navigation message is modulated onto the carrier at a data rate of only 50 bps, it would take a long time to broadcast this larger message. For these reasons, the Klobuchar model was designed to be a simple algorithm that runs quickly and only requires eight coefficients to be broadcast in the navigation message. However, this simplicity comes at the cost of reduced accuracy – the Klobuchar model is only able to correct approximately 50% rms of the ionospheric delay error. By including more coefficients, a larger percentage of the error could be corrected, but even a sophisticated model with many coefficients could only account for about 70–80% rms of the error, since some portion of the error is due to the random behavior of the ionosphere. The choice of eight coefficients and the resultant 50% rms error mitigation was considered an acceptable compromise.

As seen earlier in this chapter, ionospheric group delay is directly proportional to TEC. The behavior and temporal variation of TEC has been extensively studied, especially at mid latitudes where a large number of single frequency GPS users are located. At mid latitudes, TEC tends to vary smoothly and predictably with latitude, in contrast to the large TEC gradients seen at both high latitudes and equatorial latitudes. For these reasons, the Klobuchar model is designed with mid latitude users in mind, and works best in those regions.

TEC typically has a standard deviation of about 20–25% of the monthly mean value, and is found to reach a diurnal maximum at about 2 p.m. (1400 hours) local time at mid latitudes. The Klobuchar model is designed to fit daytime values of the monthly TEC average. Any variation from this monthly average cannot be accounted for by the model and must be accepted as observation error by a single frequency receiver. The diurnal variation of the average TEC is well approximated by a positive half cosine curve, as seen in Figure 8.1, which shows empirical data for the monthly average ionospheric delay as measured by a station in Jamaica.

For a given location, the cosine curve used to model the TEC will have a slightly different shape. The four parameters that are used to specify the shape of the cosine curve are its amplitude, period, phase, and DC offset.

The DC offset represents the nighttime value of the TEC, and is found to be essentially constant over a wide range of latitudes. The Klobuchar model uses the group delay calculated from this TEC, which is taken to be a constant 5 ns for the L1 carrier frequency. The phase of the curve's peak indicates the time of day when the

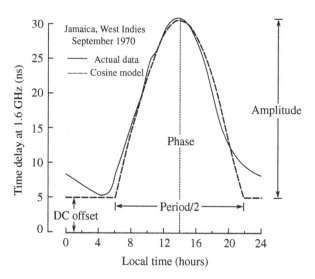

Figure 8.1 Diurnal variation of the monthly average ionospheric group delay for a measurement station in Jamaica, closely modeled by a positive half cosine curve. Source: Adapted from Figure 2 of [184]. Reproduced with permission of the Institute of Electrical and Electronics Engineers (IEEE).

maximum TEC occurs which, as mentioned previously, is consistently at about 2 p.m. local time. There is little variation in this time over a large range of latitudes, so the Klobuchar model takes the phase of the cosine curve as a constant 1400 hours local time.

The amplitude and period of the cosine curve are straightforward to interpret, representing the maximum TEC value and the amount of time between successive peaks, respectively. The daytime value of the delay will always be greater than the dc offset, so the positive half cosine must last for the entire sunlit portion of the day. The Klobuchar model sets a lower bound of 20 hours on the period, so that the positive half lasts for a minimum of 10 hours. In general, the period required for the cosine curve to provide a best fit to the observed data is significantly longer than 24 hours.

These two parameters are not constant, and instead must be treated as functions of latitude. It is important to note here that "latitude" refers to geomagnetic, not geographic, latitude. Both amplitude and period are represented by third degree polynomials, whose coefficients (α_n and β_n, for $n = 0, 1, 2, 3$) are precisely the eight coefficients broadcast by the navigation message. These coefficients are obtained from an empirical model of global ionospheric behavior derived by Bent [185]. Note that if polynomials of degree higher than three were used, a more accurate fit of both amplitude and period could be achieved. However, this would require more coefficients to be broadcast in the navigation message which, as discussed previously, is undesirable.

Since TEC varies with season and solar activity, the coefficients α_n and β_n are not constant in time, and must be periodically updated. The GPS Master Control Station (MCS), discussed in Chapter 1, updates the navigation message with new coefficients based on both the date and the solar flux value. One year is divided into 37 intervals, each with their own set of coefficients. Additionally, the MCS keeps a five day running average of measured solar flux and designates 10 grades, each of which is given its own set of coefficients. The MCS will update the navigation message with new coefficients at least once every ten days, and sometimes more frequently if the running mean solar flux changes dramatically within one ten day period. Some work has been done in developing a model specifically for the prediction of Klobuchar coefficients. Such a model would allow for accurate simulation of GPS positioning without the need for large amounts of empirical data. One such model is presented in [186].

Approximations Used in the Klobuchar Algorithm
Shortly we will present the Klobuchar algorithm, but in order to understand it fully, it is useful to first explicate several approximations that will be used. To simplify calculations, the Klobuchar algorithm uses the single layer ionosphere model as its basis. This model was discussed in Section 7.3.1, and it is suggested that the

reader familiarize themselves with the concepts presented therein, in particular Figure 7.4, as they will be referenced throughout this section.

Slant Factor   Nearly all GPS signals propagate along an oblique path to a receiver, so the slant TEC, rather than the vertical TEC must be used. Ionospheric models typically report the vertical TEC at a given location, so the slant TEC must be calculated. This can be accomplished using the unitless quantity F, called the slant factor. Using Equations 7.47 and 7.48, we see that the slant factor can be written as

$$F = \sec\left[\arcsin\left(\frac{a}{a + h_I}\sin\chi\right)\right]. \tag{8.1}$$

In the Klobuchar model, the mean ionospheric height h_I is taken to be 350 km, and F is expressed in terms of the elevation angle $E = 90 - \chi$. The radius of the Earth is $a = 6378$ km, so F can be rewritten as

$$F = \sec\left[\arcsin(0.9479\cos E)\right]. \tag{8.2}$$

Evaluating trigonometric operations is more computationally expensive than evaluating arithmetic operations, so to improve the speed of the Klobuchar algorithm, F is approximated by

$$F \approx 1 + 2\left(\frac{96 - E}{90}\right)^3. \tag{8.3}$$

This approximation produces a slant factor within 2% of the exact value for all elevation angles greater than 5°. Figure 8.2 compares the exact and approximate values of the slant factor, as given in the preceding equations.

Earth Centered Angle   Another quantity used in the Klobuchar algorithm is the Earth centered angle or geocentric angle between the receiver and the ionospheric piercing point (IPP), denoted ψ. ψ can be expressed as a function of the elevation angle. Consider the triangle in Figure 7.4 formed by the geocenter, the receiver location (R_x), and the IPP. The angles in this triangle sum to 180°:

$$90 + E + \chi' + \psi = 180. \tag{8.4}$$

Rearranging and substituting in Equation 7.48 for χ' gives

$$\psi = 90 - E - \arcsin\left(\frac{a}{a + h_I}\sin\chi\right). \tag{8.5}$$

Similar to the slant factor, the Klobuchar algorithm uses a simplified expression to compute ψ, measured in degrees, using only arithmetic operations:

$$\psi \approx \frac{445}{E + 20} - 4. \tag{8.6}$$

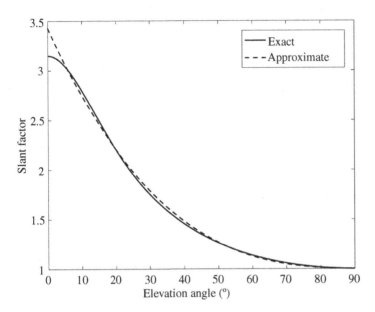

Figure 8.2 Comparison of exact and approximated values of slant factor F.

This approximation produces an error in the Earth centered angle of less than $0.2°$ for all elevation angles greater than $10°$, and less than $0.5°$ error for all elevation angles less than $10°$. Figure 8.3 shows both the exact and approximate values of Earth centered angle for various elevation angles.

Coordinates of the IPP A GPS receiver will typically have some estimate of its current geodetic coordinates: latitude ϕ_u and longitude λ_u. However, in the single layer ionosphere model, group delay is measured at the IPP. For oblique paths, the IPP will be at a different latitude and longitude than the receiver. The point on the surface of the Earth with the same latitude and longitude as the IPP is called the sub-ionospheric point, labeled SIP in Figure 7.4. The latitude and longitude of the sub-ionospheric point can be respectively expressed as follows:

$$\phi_I = \arcsin(\sin \phi_u \cos \psi + \cos \phi_u \sin \psi \cos A), \tag{8.7}$$

$$\lambda_I = \lambda_u + \arcsin\left(\frac{\sin \psi \sin A}{\cos \phi_I}\right), \tag{8.8}$$

where ψ is the Earth centered angle and A is the azimuth angle, measured with respect to the geographic north pole. However, since $h_I \ll a$, the difference in latitude and longitude will be small. Over such short distances, the curvature of the Earth is negligible, so it is appropriate to use the flat Earth approximation, which

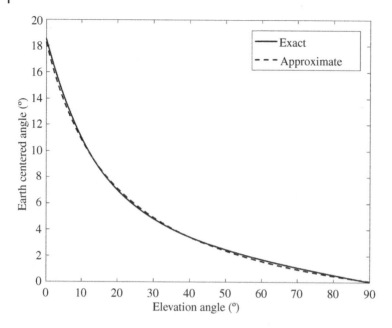

Figure 8.3 Comparison of exact and approximated values of Earth centered angle ψ.

simplifies the aforementioned expressions to

$$\phi_I \approx \phi_u + \psi \cos A, \tag{8.9}$$

$$\lambda_I \approx \lambda_u + \frac{\psi \sin A}{\cos \phi_I}. \tag{8.10}$$

These approximations hold for all latitudes smaller than 75°. In the Klobuchar algorithm, latitudes greater than 75° are simply defined as 75° to avoid large errors. This was deemed an acceptable compromise, since few single frequency GPS users will be located at latitudes higher than 75°. Nevertheless, some work has been done in refining the Klobuchar model for high latitudes, for example, in [187].

Conversion from Geodetic to Geomagnetic Coordinates TEC, and therefore group delay, vary with geomagnetic latitude rather than geodetic latitude. Therefore, it is necessary for the Klobuchar algorithm to include a conversion from geodetic to geomagnetic latitude. This conversion is given by

$$\sin \phi_m = \sin \phi \sin \phi_p + \cos \phi \cos \phi_p \cos \left(\lambda - \lambda_p \right), \tag{8.11}$$

where ϕ_m is the geomagnetic latitude, ϕ and λ are the geodetic coordinates to be converted, and ϕ_p and λ_p are the geodetic coordinates of the Earth's geomagnetic north pole. In the Klobuchar model, $\phi_p = 78.3°\text{N}$ and $\lambda_p = 291.0°\text{E}$, and the following approximation, in degrees, is used

$$\phi_m \approx \phi + 11.6 \cos(\lambda - 291) \tag{8.12}$$

is used. This approximation produces an error within 1° for all latitudes less than 40° and an error of less than 2° for all latitudes less than 65°. It should be noted that the geomagnetic north pole migrates over time, so its coordinates are no longer equal to those used in the Klobuchar model. According to the International Geomagnetic Reference Field-12 (IGRF-12) model [88], in 2017 the coordinates of the geomagnetic north pole are $\phi_m = 80.4°$ N and $\lambda_m = 287.2°$ E. The effect this has on the calculation of geomagnetic latitude is small, producing a value that is no more than $\sim 2.1°$ different than that estimated by Equation 8.12.

Local Time   In its calculations, the Klobuchar algorithm uses the local time at the IPP, which can be determined, in units of hours, in terms of its geodetic longitude and universal time as follows:

$$t = \frac{\lambda_I}{15} + t_{UTC}, \tag{8.13}$$

where λ_I is the geodetic longitude in degrees and t_{UTC} is the universal time in hours. t is always taken to be less than 24 hours, so if Equation 8.13 produces a value greater than 24, then 24 is subtracted from t.

Klobuchar Algorithm

The goal of the Klobuchar algorithm is to estimate ionospheric group delay as closely as possible. Recall that zenith ionospheric group delay is closely fitted by a positive half cosine curve with a dc offset. Mathematically, this allows us to write zenith group delay as

$$I_z = DC + A \cos\left(\frac{2\pi(t - t_0)}{P}\right), \tag{8.14}$$

where $DC = 5$ ns is the dc offset term for the L1 carrier frequency, A is the amplitude, $t_0 = 14$ hours is the time of peak delay, and P is the period. A and P can be determined in terms of the Klobuchar coefficients broadcast in the navigation message, as we will see shortly. In general, GPS signals travel on oblique paths, so the group delay must be multiplied by the slant factor F:

$$I = FI_z. \tag{8.15}$$

The following parameters are known to a GPS receiver: estimated geodetic latitude and longitude ϕ_u and λ_u, elevation angle E, azimuth angle A, and the eight Klobuchar coefficients α_n and β_n ($n = 0, 1, 2, 3$), which are broadcast in the navigation message. Using just these twelve parameters, it is possible to produce an estimate of ionospheric group delay via the Klobuchar algorithm. The steps of the algorithm are as follows (*note*: all angles are measured in semicircles and all times measured in seconds):

1. Calculate the Earth centered angle between the receiver and the IPP.
   Using Equation 8.6:

$$\psi = \frac{0.0137}{E + 0.11} - 0.022. \tag{8.16}$$

2. Calculate the latitude of the IPP.
   Using Equation 8.9 and capping latitude at $\pm 75°$:

$$\phi_I = \begin{cases} 0.416, & \phi_I > 0.416, \\ \phi_u + \psi \cos A, & -0.416 \leq \phi_I \leq 0.416, \\ -0.416, & \phi_I < 0.416. \end{cases}$$

3. Calculate the longitude of the IPP.
   Using Equation 8.10:

$$\lambda_I = \lambda_u + \frac{\psi \sin A}{\cos \phi_I}. \tag{8.17}$$

4. Convert the latitude of the IPP to geomagnetic coordinates.
   Using Equation 8.12

$$\phi_m = \phi_I + 0.064 \cos \left(\lambda_I - 1.617 \right). \tag{8.18}$$

5. Find the local time at the IPP.
   Using Equation 8.13, restricting t to between 0 and 86,400 seconds:

$$t = \begin{cases} 43200\lambda_I + t_{GPS}, & 0 \leq t < 86400, \\ 43200\lambda_I + t_{GPS} + 86400, & t < 0, \\ 43200\lambda_I + t_{GPS} - 86400, & t \geq 86400. \end{cases}$$

6. Calculate the amplitude of the ionospheric delay:
   The ionospheric delay is lower bounded by the dc offset, which represents the nighttime value. Thus, when $A < 0$, A is set to zero.

$$A = \begin{cases} \sum_{n=0}^{3} \alpha_n \phi_m^n, & A \geq 0, \\ 0, & A < 0. \end{cases}$$

7. Calculate the period of the ionospheric delay. The period is lower bounded by 72,000 seconds:

$$P = \begin{cases} \sum_{n=0}^{3} \beta_n \phi_m^n, & A \geq 72000, \\ 72000, & A < 72000. \end{cases}$$

8. Calculate the phase of the ionospheric delay.

   Using the argument of the cosine function in Equation 8.14:

$$X = \frac{2\pi(t - 50400)}{P}.$$

(8.19)

9. Calculate the slant factor.

$$F = 1.0 + 16.0(0.53 - E)^3.$$

(8.20)

10. Calculate the ionospheric delay.

    Using Equations 8.15 and 8.14, as well as all of the information found in the preceding steps of the algorithm, and expanding the cosine function as a Taylor series, truncated to three terms, we finally find that the estimated ionospheric delay is given by:

$$I = \begin{cases} F \times \left[5 \times 10^{-9} + \sum_{n=0}^{3} \alpha_n \phi_m^n \left(1 - \frac{X^2}{2} + \frac{X^4}{24} \right) \right], & |X| \leq 1.57, \\ F \times 5 \times 10^{-9}, & |X| > 1.57. \end{cases}$$

Note that the estimated ionospheric delay found in this section corresponds only to the L1 carrier frequency. However, from this value it is simple to obtain the estimated delays experienced by other frequencies. For any frequency f, the ionospheric delay on f is given by

$$I_f = \left(\frac{f_{L1}}{f} \right)^2 I_{L1}.$$

(8.21)

For example, the delay experienced by the L2 carrier frequency is

$$I_f = \left(\frac{1575.42 \times 10^6}{1227.60 \times 10^6} \right)^2 I_{L1} = 1.647 \, I_{L1}.$$

(8.22)

This correction can be used to apply the Klobuchar algorithm to any GNSS carrier frequency, including Galileo and GLONASS frequencies.

8.2.2 NeQuick

The NeQuick model is a three-dimensional electron density profiler that has been adopted by a number of organizations as a modeling standard and has been implemented in a variety of ionospheric applications. For example, the International Telecommunication Union, Radiocommunication Sector (ITU-R) uses NeQuick as a recommended method for simulating and mapping both vertical and slant TEC, and the Rutherford-Appleton Laboratory in the UK uses it for TEC forecasting. It is a quick-run model, which also makes it particularly suited to real-time error correction in GNSS, and has been adopted by the European Space Agency (ESA) for mitigation of ionospheric delay error in the

Galileo satellite constellation. Additionally, NeQuick has been implemented as a subroutine in other models, such as the GISM, which will be discussed in detail later in this text. NeQuick has undergone numerous revisions since its inception. At the time of this writing, the current version of the model is NeQuick 2, whose FORTRAN source code is provided by the International Centre for Theoretical Physics (ICTP).

An overview of the analytical formulation of NeQuick 2 will be provided shortly. However, this model is the culmination of decades of revisions, which makes it difficult to concisely encapsulate. Thus, the reader may wish to refer to the relevant literature to obtain the full context of this model's development. To that end, an effort has been made to outline a short history of the relevant literature with citations to some of the most influential papers, to which the reader may refer for a deeper exploration of the subject.

The NeQuick model has its roots in a model developed by Di Giovanni and Radicella in 1990 [188]. This model, commonly referred to as the DGR model, was formulated based on three basic criteria for electron density models developed several years prior by Dudeney and Kressman [189]. These three criteria were: (i) the electron density as a function of height $N(h)$ should be continuous in its first and second derivatives, (ii) all input parameters should be available from data routinely scaled from ionograms, and (iii) the computational complexity should be as low as possible. The DGR model satisfies these criteria through the use of a function known as the Epstein layer function, as described by Rawer in his 1982 paper [190].

The E, F1, and F2 regions of the ionosphere are each described by their own Epstein layer function. The peak of each region is used as an "anchor point" for the algorithm, and the Epstein layer for a given region of the ionosphere can be described fully by knowing only the height h_m and the electron density N_m at each of these three anchor points, as well as a thickness parameter B. Each of these three unknowns can in turn be determined in terms of the ionogram parameters f_oE, f_oF1, f_oF2, and $M(3000)F2$, which will be discussed in greater detail shortly. The entire ionosphere up to the F2 peak (i.e. the bottomside ionosphere) can be modeled as the sum of the three Epstein layers:

$$N(h) = N_E(h) + N_{F1}(h) + N_{F2}(h). \tag{8.23}$$

In their 1995 paper, Radicella and Zhang introduced a refinement of the DGR model [191]. It was found that greater accuracy could be achieved by modeling the asymmetry in thickness between the top and bottom portions of each ionospheric layer. The basic approach of the DGR model remains unchanged, but the bottomside ionosphere is now modeled with five semi-Epstein layers corresponding to the lower and upper E regions, the lower and upper F1 regions, and the lower F2 region. In addition to this more nuanced description of the bottomside ionosphere,

Radicella and Zhang's paper introduced a brand new formulation for the topside F2 region, which had previously proved challenging to model. The topside F2 region can also be modeled by a semi-Epstein layer, but its thickness parameter must be scaled by an empirically determined shape parameter k, which has two alternative and seasonally dependent formulations.

Building upon the modified DGR model, Hochegger et al. developed a family of new ionospheric models in 2000, as described in [192, 193]. The three models – dubbed NeQuick, COSTprof, and NeUoG-plas – each describe the ionosphere in similar, yet slightly different ways, and are intended for different applications. The NeQuick model is designed to be simple and run quickly, while the other two models are more sophisticated, accounting for the effects of diffusion in the topside ionosphere as well as a magnetic field aligned formulation of the plasmasphere. Only NeQuick will be discussed further; for more information on COSTprof and NeUoG-plas, the reader is encouraged to refer to the previously referenced papers.

The NeQuick model is very similar to the modified DGR model, using five semi-Epstein layers for the bottomside ionosphere and a sixth semi-Epstein layer with a modified thickness parameter for the topside F2 layer. However, NeQuick also introduces a few novel modifications of its own. It was found that the E and F1 Epstein layers sometimes produced secondary maxima at the F2 layer peak, making it difficult to ensure that the electron density at that height corresponded exactly to $f_o F2$. To remedy this, NeQuick introduced a new feature: a "fadeout" parameter ξ is included in the E and F1 Epstein layers to ensure that their values decrease to zero by the time they reach the F2 peak. In addition, the thickness parameter of the topside F2 layer was modified to include a height dependence.

Over the next few years, it became clear that NeQuick would occasionally produce plots with abnormally large TEC values, sharp gradients in the E and F1 regions, and strange electron density structures. Two papers followed with revisions that would ultimately solve these problems, the first of which was by Leitinger et al. in 2005 [194]. In this paper, three major changes were made to the original NeQuick model. First, the previous formulation of the F1 peak height $h_m F1$ was determined to be too complicated, and was simplified. Secondly, a new description of $f_o F1$ was introduced to ensure that its value does not get too close to $f_o F2$. Finally the expressions for the thickness parameters $B_{F1,top}$, $B_{F1,bottom}$, and $B_{E,top}$ were simplified.

The second paper, by Coïsson et al. in 2006 [195], provided a refined description of the topside F2 layer shape parameter k. Utilizing a large amount of TEC and ionosonde data, including topside sounding data from the ISIS 2 satellite, the authors were able to develop an empirical relation for k that is valid year-round, as opposed to the previous formulation, which gave two separate relations for k with seasonal dependence. With the implementation of these revisions, the updated model was named NeQuick 2. An analytical formulation of NeQuick 2 and a comparison of its performance as compared with NeQuick are presented by Nava et al. [196].

Analytical Formulation of the NeQuick 2 Model

At the time of this writing, the NeQuick 2 model is the most recent revision of NeQuick. The algorithm takes inputs of position (geographic latitude and longitude), time (month and time of day in UT) and solar flux (either F10.7 flux Φ in SFU or sunspot number R). The output is the electron density at the given location and time. In this section, an analytical description of the model is given, based on the work in [196].

The ionosphere can be described as a sum of Epstein layers corresponding to the E, F1 and F2 regions of the ionosphere. In general, an Epstein layer function has the form

$$Eps_1(x) = \frac{A}{(1 + \exp(x))^2} \exp(x). \tag{8.24}$$

Figure 8.4 shows the shape of this equation for $A = 1$.

For a given region i of the ionosphere, the electron density as a function of height can be expressed using an Epstein layer. We let $A = 4N_{mi}$ so that the peak of the function gives N_{mi}, the peak electron density of region i. The function is also shifted so that it is centered on h_{mi}, the height of peak electron density in region i. Finally, the function is scaled by a parameter B_i, known as the thickness

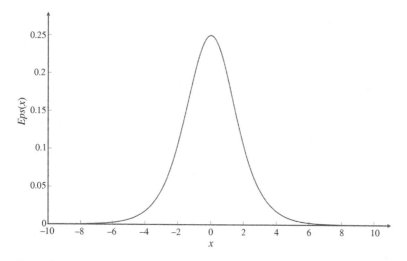

Figure 8.4 Plot of a general Epstein layer function.

parameter, which determines the width of the function's peak. Using all of this information, we can write the Epstein layer for region i as

$$N(h) = \frac{4N_{mi}}{\left(1 + \exp\left(\dfrac{h - h_{mi}}{B_i}\right)\right)^2} \exp\left(\frac{h - h_{mi}}{B_i}\right). \tag{8.25}$$

Every term in the Epstein layer equation be determined from just four ionogram parameters: the critical frequencies of each layer (f_oE, f_oF1, and f_oF2) and a parameter known as the ionospheric transition parameter or ionospheric propagation factor, denoted $M(3000)F2$. $M(3000)F2$ is defined as the ratio of the maximum usable frequency at a distance of 3000 km, $MUF(3000)$, to the F2 layer critical frequency f_oF2 [197], i.e.

$$M(3000)F2 = \frac{MUF(3000)}{f_oF2}. \tag{8.26}$$

The concepts of critical frequency and maximum usable frequency are discussed in Section 6.2

A simple model developed by Titheridge [198] is used to compute f_oE. Mathematically, it is given by

$$(f_oE)^2 = (a_e\sqrt{\Phi})^2 \cos(\chi_{eff})^{0.6}, \tag{8.27}$$

where a_e is a seasonal coefficient given by the values in Table 8.1, Φ is the solar flux of 10.7 cm radiation measured in solar flux units (SFU), and χ_{eff} is the solar zenith angle that is in turn given by

$$\chi_{eff} = \begin{cases} \chi, & \chi \leq 86.23°, \\ 90° - 0.24° \exp\left(20° - 0.2\chi\right), & \chi > 86.23°. \end{cases}$$

The case $\chi \leq 86.23°$ corresponds to daytime, while $\chi > 86.23°$ corresponds to nighttime. The use of an exponential function ensures that f_oE is continuous and differentiable.

Table 8.1 Values of the seasonal coefficient used in the Titheridge model.

a_e	Month, Northern Hemisphere	Month, Southern Hemisphere
1.131	Jan, Feb, Nov, Dec	May, Jun, Jul, Aug
1.112	Mar, Apr, Sep, Oct	Mar, Apr, Sep, Oct
1.093	May, Jun, Jul, Aug	Jan, Feb, Nov, Dec

A model for f_oF1 was included in the revisions made by Leitinger et al. [194]. In this model, f_oF1 is given by

$$f_oF1 = \begin{cases} 1.4f_oE, & f_oE \geq 2, \\ 0, & f_oE < 2, \\ 0.85(1.4f_oE), & 1.4f_oE > 0.85f_oF2. \end{cases}$$

NeQuick 2 calculates both f_oF2 or $M(3000)F2$ based on ITU-R coefficients (formerly called Comité Consultatif International des Radiocommunications [CCIR] coefficients), which are empirically determined by a worldwide network of sounding stations [199]. In order to perform this calculation for a given location, the modified dip latitude (modip) must also be known. Included with the source code for NeQuick 2 are several files containing the necessary parameters: twelve files (one for each month of the year) contain the ITU-R coefficients for calculating both f_oF2 and $M(3000)F2$ and one additional file contains modip values. It should be noted that modip files must be updated every five years to account for the natural variation of the Earth's magnetic field [200]. Functions included in the NeQuick 2 source code automatically perform all necessary calculations based on the user's inputs.

Using Equation 6.30, measured in 10^{11} electrons/m^3, the peak electron densities of the E, F1 and F2 regions are respectively given by the expressions:

$$N_mE = 0.124(f_oE)^2, \tag{8.28}$$

$$N_mF1 = 0.124(f_oF1)^2, \tag{8.29}$$

$$N_mF2 = 0.124(f_oF2)^2, \tag{8.30}$$

where f_oE, f_oF1, and f_oF2 are measured in MHz. The heights of the E, F1, and F2 region peaks, measured in km, are respectively defined in the NeQuick 2 model to be

$$h_mE = 120, \tag{8.31}$$

$$h_mF1 = \frac{h_mE + h_mF2}{2}, \tag{8.32}$$

$$h_mF2 = \frac{1490MF}{M + \Delta M} - 176, \tag{8.33}$$

where

$$M = M(3000)F2, \tag{8.34}$$

$$MF = M\sqrt{\frac{0.0196M^2 + 1}{1.2967M^2 - 1}}, \tag{8.35}$$

and

$$\Delta M = \begin{cases} \dfrac{0.253}{\dfrac{f_oF2}{f_oE} - 1.215} - 0.012, & f_oE \neq 0, \\[2em] -0.012, & f_oE = 0. \end{cases}$$

The expression for h_mF1 is one of the revisions introduced by Leitinger et al. [194]. The expression for h_mF2 is based on work by Dudeney et al. (see Equation 56 in [201], for further information, see also [202–204]).

Bottomside Ionosphere A profile of ionospheric electron density in the bottomside ionosphere (i.e. below the F2 peak) as a function of height h can be expressed via:

$$N(h) = N_E(h) + N_{F1}(h) + N_{F2}(h), \tag{8.36}$$

where

$$N_E(h) = \frac{4N_m^*E}{\left(1 + \exp\left(\dfrac{h - h_mE}{B_E}\xi(h)\right)\right)^2} \exp\left(\frac{h - h_mE}{B_E}\xi(h)\right), \tag{8.37}$$

$$N_{F1}(h) = \frac{4N_m^*F1}{\left(1 + \exp\left(\dfrac{h - h_mF1}{B_{F1}}\xi(h)\right)\right)^2} \exp\left(\frac{h - h_mF1}{B_{F1}}\xi(h)\right), \tag{8.38}$$

$$N_{F2}(h) = \frac{4N_mF2}{\left(1 + \exp\left(\dfrac{h - h_mF2}{B_{F2}}\right)\right)^2} \exp\left(\frac{h - h_mF2}{B_{F2}}\right). \tag{8.39}$$

The Epstein layer functions have tails that extend past the region they are meant to describe. Thus, the electron density in the E layer will be primarily given by $N_E(h)$, but there will also be a smaller contribution from the tails of $N_{F1}(h)$ and $N_{F2}(h)$. Therefore, for the model to be accurate, at the peak height for each iono-spheric region the sum of the Epstein layers must equal the values predicted by Equations 8.28–8.30. In other words,

$$N_mE = N_m^*E - N_{F1}(h_mE) - N_{F2}(h_mE), \tag{8.40}$$

$$N_mF1 = N_m^*F1 - N_E(h_mF1) - N_{F2}(h_mF1), \tag{8.41}$$

$$N_mF2 = N_m^*F2 - N_E(h_mF2) - N_{F1}(h_mF2). \tag{8.42}$$

A fadeout parameter $\xi(h)$, introduced by Hochegger et al. [192], is included in the E and F1 Epstein layers, causing them to reach a value of zero at h_mF2. Therefore, the last two terms in Equation 8.42 vanish, leaving $N_mF2 = N_m^*F2$. This is why

Equation 8.39 has N_mF2 rather than N_m^*F2 in its numerator. The fadeout parameter used to accomplish this is given by the expression

$$\xi(h) = \exp\left(\frac{10}{1 + |h - h_mF2|}\right).$$ (8.43)

As per the improved DGR model [191], the bottomside ionosphere is split into five semi-Epstein layers: the bottomside E layer, the topside E layer, the bottomside F1 layer, the topside F1 layer, and the bottomside F2 layer[1] The topside F2 layer is treated differently and will be discussed shortly.

The difference between the bottomside and topside of a given ionospheric region is entirely encapsulated by the thickness parameter. Thus, Equations 8.40–8.42 remain unchanged and the thickness parameters are given in km by

$$B_{E,bot} = 5,$$ (8.44)

$$B_{E,top} = \max\left(0.5(h_mF1 - h_mE), 7\right),$$ (8.45)

$$B_{F1,bot} = 0.5(h_mF1 - h_mE),$$ (8.46)

$$B_{F1,top} = 0.3(h_mF2 - h_mF1),$$ (8.47)

$$B_{F2,bot} = \frac{0.385N_mF2}{\left(\dfrac{dN}{dh}\right)_{max}}.$$ (8.48)

The expressions for $B_{E,top}$, $B_{F1,bot}$, and $B_{F1,top}$ were revisions made by Leitinger et al. [194]. The derivative of electron density with respect to height is given by the empirical formula [205]

$$\ln\left(\left(\frac{dN}{dh}\right)_{max}\right) = -3.567 + 1.714\ln(f_oF2) + 2.02\ln(M(3000)F2).$$ (8.49)

Topside Ionosphere The topside ionosphere is also modeled by a semi-Epstein layer:

$$N_{F2,top}(h) = \frac{4N_mF2}{\left(1 + \exp\left(\dfrac{h - h_mF2}{H}\right)\right)^2} \exp\left(\frac{h - h_mF2}{H}\right).$$ (8.50)

This equation only differs from Equation 8.39 by the thickness parameter. However, this thickness parameter is different from those of the bottomside ionosphere. Rather than being constant, H is a function of height and given in km by

$$H = kB_{2,bot}\left[1 + \frac{rg(h - h_mF2)}{rkB_{2,bot} + g(h - h_mF2)}\right],$$ (8.51)

---

1 The dividing line between the bottomside and the topside of a region is the peak of that region.

where $r = 100$ and $g = 0.125$. k is a shape parameter for the topside ionosphere, and is given by an empirical relation developed by Coïsson et al. [195]:

$$k = 3.22 - 0.0538 f_o F2 - 0.006\ 64 h_m F2 + 0.113 \frac{h_m F2}{B_{2,bot}} + 0.002\ 57 R_{12},$$

$$k \geq 1, \tag{8.52}$$

where R_{12} is the smoothed 12-month running average of the sunspot number. The restriction that $k \geq 1$ is informed by experimental data.

NeQuick G Model

The NeQuick model was initially developed to use the monthly mean solar flux as an input. Later, NeQuick was adopted for the purposes of ionospheric delay correction in the Galileo system, and daily values of TEC can vary significantly from the mean. Therefore, it became necessary to modify the NeQuick model slightly in order to implement it in Galileo. This modified version is known as NeQuick G.

The original NeQuick model uses the monthly mean of the F10.7 index Φ, while NeQuick G uses a different parameter Az, also known as the effective ionization level. Az, measured in SFU, is given by the expression

$$Az = a_0 + a_1 \mu + a_2 \mu^2, \tag{8.53}$$

where a_0, a_1, and a_2 are coefficients determined from TEC data measured by a global network of reference stations. These coefficients are updated at least once a day and are broadcast in the Galileo navigation message. μ is the modified dip latitude, or modip, given by the expression

$$\mu = \arctan\left(\frac{I}{\sqrt{\cos\phi}}\right), \tag{8.54}$$

where ϕ is the geodetic latitude and I is here the magnetic dip angle at the receiver's position at an altitude of 300 km [200].

Other than this change, NeQuick G is identical to the NeQuick 2 model described previously. A Galileo receiver can use NeQuick G to compute slant TEC via the following procedure [206]:

1. The receiver calculates Az from the local modip μ and the coefficients a_0, a_1, and a_2. Modip is calculated based on an a priori estimate of the receiver's position.
2. The electron density $N(h)$ is calculated for a single point h along the satellite-receiver path using the NeQuick model, with Az used in place of Φ for the solar flux.
3. Steps 1 and 2 are repeated for many points along the propagation path. The number of points used depends on the total length of the path and provide a trade-off between integration error and computation time.

4. All electron density values along the path are integrated to obtain the slant TEC.
5. The range error induced by group delay, in meters, can be calculated based on the equation

$$I_{gr} = 40.3 \frac{10^{16} \text{TEC}}{f^2}, \tag{8.55}$$

where TEC is the slant TEC measured in TECU and f is the carrier frequency of the signal in Hz.

Note that, like the Klobuchar algorithm, NeQuick G can be implemented in any GNSS signal (e.g. GPS or GLONASS) by simply inputting the signal's carrier frequency into Equation 8.55.

8.3 Global Ionospheric Scintillation Model

The GISM provides the statistical characteristics of signals that have traversed the ionosphere. As discussed in Section 7.4, scintillations are fluctuations in the phase, amplitude, and angle of arrival of an electromagnetic wave. In the ionosphere, scintillations are caused by inhomogeneities in the refractive index, which arise due to mechanisms such as $\vec{E} \times \vec{B}$ gradient drift and Rayleigh-Taylor instabilities.

The GISM takes inputs of solar flux, year, day of year, local time, latitude, longitude, and altitude. From this information, the model is able to calculate the scintillation index, fade durations, and the probability distribution of the signal's received intensity [207]. In this section, a brief overview of the GISM is presented.

The GISM simulates propagation of an electromagnetic wave in the ionosphere. This simulation requires several components. First the medium must be modeled, which in turn requires models of electron density and geomagnetic parameters. Second, the line of sight of the wave is determined via ray tracing. Finally, the line of sight is divided into successive layers, each separated by a so-called *phase screen*. Scintillations in the wave can be modeled by considering the repeated scattering of the wave from one phase screen to the next. This technique is known as the multiple phase screen (MPS) method.

We will not discuss the medium modelling portion of the GISM in detail. However, in brief: the electron density is provided by the NeQuick model, which is included as a subroutine in the GISM source code. The NeQuick model is discussed in Section 8.2.2. Geomagnetic parameters – in particular the magnetic field components, vertical intensity, declination, and inclination – are computed using a Schmidt quasi-normalized spherical harmonic model of the Earth's magnetic field [208]. For a discussion on Schmidt quasi-normalization, see [209].

The remainder of this section will focus on the two main steps of the GISM propagation model: determination of the line of sight via ray-tracing and calculation of scintillations via the MPS method.

8.3.1 Ray Tracing in the Ionosphere

In order to simulate the propagation of a wave through the ionosphere, first the wave's propagation path must be determined. This calculation must take into account the bending of the wave due to refraction. This can be accomplished via the use of a ray tracing algorithm. A simple ray tracing algorithm is the repeated application of Snell's law to a stratified ionosphere. In such an algorithm, the ionosphere is divided into distinct layers, each with their own constant refractive index. At the boundaries between layers, an electromagnetic wave refracts according to the familiar Snell's law:

$$n_1 \sin \theta_1 = n_2 \sin \theta_2, \tag{8.56}$$

where n_1 and n_2 are the refractive indices of the two adjacent layers, and θ_1 and θ_2 are the angles of the wave's propagation direction with respect to the interface normal. As the number of layers used to approximate the ionosphere is increase, the accuracy of the ray path increases.

Despite its simplicity and intuitiveness, the Snell's law ray tracing algorithm is not well-suited to the case of the ionosphere. The assumption of stratified layers makes it difficult to account for anisotropy and inhomogeneities in the refractive index, both of which are key concerns in the ionosphere. Instead of Snell's law, a different ray tracing algorithm known as the Jones–Stephenson algorithm is employed in the GISM. This algorithm uses a set of equations known as the Haselgrove equations to efficiently and accurately trace the path of a ray through the ionosphere.

The Haselgrove equations are a set of six differential equations, which are given in spherical coordinates as follows:

$$\frac{r'}{c} = \frac{1}{n^2} \left(v_r - n \frac{dn}{dv_r} \right), \tag{8.57}$$

$$\frac{\theta'}{c} = \frac{1}{n^2 r} \left(v_\theta - n \frac{dn}{dv_\theta} \right), \tag{8.58}$$

$$\frac{\phi'}{c} = \frac{1}{n^2 r \sin \theta} \left(v_\phi - n \frac{dn}{dv_\phi} \right), \tag{8.59}$$

$$\frac{v_r'}{c} = \frac{1}{n} \frac{dn}{dr} + v_\theta \frac{\theta'}{c} + \sin \theta v_\phi \frac{\phi'}{c}, \tag{8.60}$$

$$\frac{v_\theta'}{c} = \frac{1}{r} \left(\frac{1}{n} \frac{dn}{d\theta} - v_\theta \frac{r'}{c} + r \cos \theta v_\theta \frac{\phi'}{c} \right), \tag{8.61}$$

$$\frac{v_\phi'}{c} = \frac{1}{r \sin \theta} \left(\frac{1}{n} \frac{dn}{d\phi} - \sin \theta v_\phi \frac{r'}{c} - r \cos \theta v_\phi \frac{\theta'}{c} \right). \tag{8.62}$$

The derivation of these equations is beyond the scope of this text. The interested reader may consult sources such as [210], [211, pp. 407–409], [212, 213] for a further exploration of the subject.

In order to compute the Haselgrove equations, the index of refraction of the ionosphere is needed. This is given by the Appleton–Hartree equation, which was discussed in Section 7.2. For convenience, a simplified version of the equation (for a collisionless ionosphere) is reproduced here:

$$n_{\pm}^2 = 1 - \frac{2X(1-X)}{2(1-X) - Y_{\perp}^2 \pm \sqrt{Y_{\perp}^4 + 4(1-X)^2 Y_{\parallel}^2}}, \tag{8.63}$$

where the top sign corresponds to propagation of an ordinary wave and the bottom sign corresponds to propagation of an extraordinary wave. The quantities X, $Y_{\parallel}$, and $Y_{\perp}$ are defined as follows:

$$X = \frac{\omega_p^2}{\omega^2}, \quad Y_{\parallel} = \frac{\omega_{\parallel}}{\omega}, \quad Y_{\perp} = \frac{\omega_{\perp}}{\omega}. \tag{8.64}$$

Here, ω is the frequency of the wave, ω_p is the plasma frequency of the ionosphere, and $\omega_{\parallel}$ and $\omega_{\perp}$ are respectively the components of the ionosphere's cyclotron frequency that are parallel and perpendicular to the geomagnetic field. These quantities are discussed in greater detail in Section 7.2.

The plasma frequency depends on the electron density of the ionosphere, which is provided in the GISM by the NeQuick model. The cyclotron frequency depends on the magnitude of the geomagnetic field, which is provided in the GISM by a spherical harmonic model. With this information, it is possible to determine the index of refraction, and thus the bending of the ray path at any desired point in the ionosphere.

8.3.2 Multiple Phase Screen Method

Once the line of sight has been determined for a wave traversing the ionosphere, it is possible to employ a technique known as the MPS method to determine how the wave is scintillated as it travels. The following derivation is adapted from [214].

The wave evolves according the wave equation, which in general must be solved in vector form. However, consider the case where fluctuations in the ionosphere have a characteristic scale size much larger than the wave's wavelength and temporal variation much longer than the wave's period. Under these conditions, the ionosphere may be approximated as locally isotropic and homogeneous, which means that the wave's evolution may instead be modeled by the scalar wave equation:

$$\nabla^2 E + k^2 \left[1 + \varepsilon_1(\vec{\mathbf{r}}, t)\right] E = 0. \tag{8.65}$$

Here, E is the electric field, $\varepsilon_1(\vec{r}, t)$ is the fluctuating part of the relative dielectric permittivity due to electron density fluctuations, and $k^2 = k_0^2 \langle \varepsilon \rangle$, where k_0 is the wave's vacuum wavenumber and $\langle \varepsilon \rangle$ is the average dielectric permittivity. If we label the wave's direction of propagation as the z-axis in a Cartesian coordinate system, solutions to the scalar wave equation are of the form:

$$E(\vec{r}, t) = u(\vec{r}, t) e^{-jkz}. \tag{8.66}$$

If we substitute this into Equation 8.65, we obtain:

$$\nabla^2 \left(u(\vec{r}, t) e^{-jkz} \right) + k^2 \left[1 + \varepsilon_1(\vec{r}, t) \right] u(\vec{r}, t) e^{-jkz} = 0. \tag{8.67}$$

Recalling that in Cartesian coordinates, the Laplacian is given by:

$$\nabla^2 = \frac{\partial^2}{\partial x^2} + \frac{\partial^2}{\partial y^2} + \frac{\partial^2}{\partial z^2}, \tag{8.68}$$

the first term of Equation 8.67 becomes:

$$\left(\frac{\partial^2}{\partial x^2} + \frac{\partial^2}{\partial y^2} + \frac{\partial^2}{\partial z^2} \right) \left(u(\vec{r}, t) e^{-jkz} \right), \tag{8.69}$$

which evaluates to:

$$\frac{\partial^2 u(\vec{r}, t)}{\partial x^2} e^{-jkz} + \frac{\partial^2 u(\vec{r}, t)}{\partial y^2} e^{-jkz} + \frac{\partial}{\partial z} \left(\frac{\partial u(\vec{r}, t)}{\partial z} e^{-jkz} - jku(\vec{r}, t) e^{-jkz} \right). \tag{8.70}$$

This expression can be further simplified to yield:

$$\left(\frac{\partial^2 u(\vec{r}, t)}{\partial x^2} + \frac{\partial^2 u(\vec{r}, t)}{\partial y^2} \right) e^{-jkz} + \left(\frac{\partial^2 u(\vec{r}, t)}{\partial z^2} - 2jk \frac{\partial u(\vec{r}, t)}{\partial z} - k^2 u(\vec{r}, t) \right) e^{-jkz}. \tag{8.71}$$

Substituting this back into Equation 8.67 and rearranging slightly, we obtain:

$$\left(\frac{\partial^2 u(\vec{r}, t)}{\partial x^2} + \frac{\partial^2 u(\vec{r}, t)}{\partial y^2} \right) e^{-jkz} + \left(\frac{\partial^2 u(\vec{r}, t)}{\partial z^2} - 2jk \frac{\partial u(\vec{r}, t)}{\partial z} - k^2 u(\vec{r}, t) \right) e^{-jkz}$$
$$= -k^2 \left[1 + \varepsilon_1(\vec{r}, t) \right] u(\vec{r}, t) e^{-jkz}, \tag{8.72}$$

which can be simplified to

$$\frac{\partial^2 u(\vec{r}, t)}{\partial x^2} + \frac{\partial^2 u(\vec{r}, t)}{\partial y^2} + \frac{\partial^2 u(\vec{r}, t)}{\partial z^2} - 2jk \frac{\partial u(\vec{r}, t)}{\partial z} = -k^2 \varepsilon_1(\vec{r}, t) u(\vec{r}, t). \tag{8.73}$$

If we assume that the variation of the field is primarily in the plane perpendicular to the wave's propagation direction, we may apply the slowly varying envelope approximation (SVEA), which states that:

$$2k \left| \frac{\partial u}{\partial z} \right| \gg \left| \frac{\partial^2 u}{\partial z^2} \right|. \tag{8.74}$$

This holds for a small range of angles around the direction of propagation, so this is also known as the *paraxial approximation*. Under this assumption, the third term of Equation 8.73 is negligible with respect to the fourth term, so the equation simplifies to:

$$\frac{\partial^2 u(\vec{r}, t)}{\partial x^2} + \frac{\partial^2 u(\vec{r}, t)}{\partial y^2} - 2jk\frac{\partial u(\vec{r}, t)}{\partial z} = -k^2\varepsilon_1(\vec{r}, t)u(\vec{r}, t). \tag{8.75}$$

Commonly, the so-called transverse Laplacian, given by

$$\nabla_\perp^2 = \frac{\partial^2}{\partial x^2} + \frac{\partial^2}{\partial y^2} \tag{8.76}$$

is used. Therefore, Equation 8.75 can be rewritten as:

$$\nabla_\perp^2 u(\vec{r}, t) - 2jk\frac{\partial u(\vec{r}, t)}{\partial z} = -k^2\varepsilon_1(\vec{r}, t)u(\vec{r}, t). \tag{8.77}$$

Although this equation accurately describes the evolution of the wave through the ionosphere, it is difficult to solve with arbitrary $\varepsilon_1(\vec{r}, t)$. However, it has been shown, for example, in [215, 216], that the phase changes caused by the fluctuating permittivity $\varepsilon_1(\vec{r}, t)$ are well-modeled by assuming that the effect is concentrated to a thin irregular slab. This slab, often called a *phase changing screen* (or often simply a *phase screen*), diffracts the wave and modifies its amplitude and phase in a predictable way. For an in-depth discussion of this topic, the reader may refer to [216].

If the thickness of the phase screen L is small and the outer scale size of the ionospheric irregularities r_0 is large, then the transverse Laplacian term in Equation 8.77 is negligible [214]. In particular, when $k/L \gg 1/r_0^2$, this approximation is valid.[2] In this case, Equation 8.77 reduces to:

$$-2jk\frac{\partial u(\vec{r}, t)}{\partial z} = -k^2\varepsilon_1(\vec{r}, t)u(\vec{r}, t). \tag{8.78}$$

Treating t as a fixed parameter, we can rearrange this equation to obtain:

$$\frac{du(\vec{r}, t)}{u(\vec{r}, t)} = -j\frac{k}{2}\varepsilon_1(\vec{r}, t)dz. \tag{8.79}$$

Let $\vec{\rho}$ be a vector in the plane perpendicular to the propagation direction z. We can rewrite $u(\vec{r}, t)$ as $u(\vec{\rho}, z, t)$ and $\varepsilon_1(\vec{r}, t)$ as $\varepsilon_1(\vec{\rho}, z, t)$. If we suppose that the phase-changing screen extends along the z-axis from 0 to z, then we can integrate

---

2 This condition is equivalent to the size of the first Fresnel zone $\sqrt{\lambda L}$ of the diffracting screen being much smaller than the outer scale size r_0 [214].

both sides of this equation over the width of the screen as follows:

$$\int_0^z \frac{du(\vec{\rho}, z', t)}{u(\vec{\rho}, z', t)} = \int_0^z -j\frac{k}{2}\varepsilon_1(\vec{\rho}, z', t)dz'. \tag{8.80}$$

This integral evaluates to:

$$u\left(\vec{\rho}, z, t\right) = u\left(\vec{\rho}, 0, t\right) \exp\left(-j\frac{k}{2}\int_0^z \varepsilon_1(\vec{\rho}, z', t)dz'\right), \tag{8.81}$$

where z' is a dummy variable. Note that the amplitude of the wave leaving the phase screen (i.e. at location z) is the same as it was when it entered the phase screen (at location 0). Thus, the amplitude remains constant and only the phase of the wave changes as it traverses the phase screen. This can be seen more clearly if we rewrite Equation 8.81 in the form:

$$u(\vec{\rho}, z, t) = A_0\, e^{-j\phi(\vec{\rho}, z, t)}, \tag{8.82}$$

where $A_0 = u\left(\vec{\rho}, 0, t\right)$ is the wave amplitude and $\phi(\vec{\rho}, z, t) = \frac{k}{2}\int_0^z \varepsilon_1(\vec{\rho}, z', t)dz'$ is the phase change induced by the phase screen. It can be shown that the phase screen induces a phase change of [217]:

$$\phi(\vec{\rho}, z, t) = -\lambda r_e \Delta N_T(\vec{\rho}, z, t), \tag{8.83}$$

where λ is the wavelength of the wave, r_e is the classical electron radius, and ΔN_T is the fluctuation of the electron density over the width of the slab.

Beyond the screen, the wave is assumed to propagate in free space. Here, there are no permittivity fluctuations, so $\varepsilon_1(\vec{r}, t) = 0$, and Equation 8.77 reduces to:

$$\nabla_\perp^2 u(\vec{r}, t) - 2jk\frac{\partial u(\vec{r}, t)}{\partial z} = 0. \tag{8.84}$$

In the far field, the amplitude of a diffracted wave can be found using the Fresnel diffraction formula. Using Equation 8.82 as an initial condition, Equation 8.84 has solution [217]:

$$u(\vec{\rho}, z, t) = \frac{jkA_0}{2\pi z}\int_{-\infty}^{\infty}\int_{-\infty}^{\infty} \exp\left(-j\left[\phi(\vec{\rho}', z, t) + \frac{k}{2z}|\vec{\rho} - \vec{\rho}'|^2\right]\right) d^2\vec{\rho}'. \tag{8.85}$$

In practice, a series of phase screens are used sequentially as they wave travels through the ionosphere. The steps of this algorithm are as follows:

1. The phase change due to the first phase screen is calculated. The complex amplitude of the wave as it exits the phase screen is given by Equation 8.81.
2. The Fourier transform of this complex amplitude is taken.

3. This result is then used as the input to the Fresnel diffraction formula (Equation 8.85) to determine the wave's evolution in free space.
4. The wave is propagated in free space until it reaches the subsequent phase screen.
5. The aforementioned steps are repeated for each phase screen.

The aforementioned algorithm, known as the MPS method, can make use of the fast Fourier transform (FFT) for efficient computation [207].

8.4 International Reference Ionosphere

The IRI is an empirical model that aims to provide the most thorough and accurate representation of the ionosphere possible. The IRI was originally launched as a joint initiative between the Committee on Space Research (COSPAR) and the International Union of Radio Science (URSI) in 1968. In the decades since the IRI's inception, these two organizations have continued to work together to maintain and continually improve the IRI. A team of more than 60 experts from around the world, known as the IRI working group, meet annually to discuss strategies for improving the model.

There are several key objectives that the IRI working group aims to achieve through its continued development of the IRI. Summarily, these objectives are [218]:

1. To develop a standard set of ionospheric parameters based primarily on experimental data. Theoretical descriptions may be used to check for internal consistency as well as bridge any data gaps.
2. To identify discrepancies between different data sources and consider the reliability of available data.
3. To continually update the IRI as additional data becomes available.

It was decided that the IRI should be based first and foremost on experimental data. This approach is advantageous because the accuracy of the model does not depend on the evolving theoretical understanding of the ionosphere. At the time of the IRI's conception, the ionosphere and the complex physical processes that dictate its behavior were still poorly understood. A model based on this limited understanding would have been inaccurate and would likely have quickly become obsolete. Conversely, even after half a century of advances in ionospheric theory, the IRI continues to reproduce phenomena that theoretical models are unable to account for. For example, in 2006 a four maxima structure was observed in the longitudinal variation of the F-peak electron density – a phenomenon that contemporary theoretical models had not been designed to account for. By contrast, it was discovered that IRI had already reproduced the phenomenon by virtue of its basis in empirical data.

A number of different data sources are used to form the basis of the IRI. These sources include a global network of ground-based ionosondes, incoherent scatter radars (ISRs), topside sounders, and in-situ measurements from instruments aboard rockets. Together, these sources report measurements from below, within, and above the ionosphere, providing a nuanced picture of its structure and behavior in both space and time. However, reliance on experimentally available data comes with one major downside: in locations where data is sparse or missing entirely, the accuracy of the model diminishes significantly. For example, the highest concentration of ground-based ionosonde stations is in mid-latitude regions. The relatively sparse data at low and high latitudes, in conjunction with the complex physical phenomena that occur there, have historically made the IRI less reliable for modeling the ionosphere in these locations. Similarly, it is challenging to obtain electron density profiles over the ocean, since ground-based ionosonde data is entirely unavailable there.

Despite these challenges, the IRI remains one of the most comprehensive and reliable models available for simulation of the ionosphere. The IRI working group endeavors to further improve not only its efficacy, but also its availability. At the time of this writing, the FORTRAN source code for the most recent version of the model, IRI-2016, is freely available for download from the official IRI web pages: irimodel.org. Also available are data services and files containing important input parameters, older versions of the source code, and versions written for other languages such as MATLAB. Finally, the site conveniently aggregates the numerous papers and IRI workshop reports regarding the decades' worth of improvements made to the model.

The IRI is, in reality, a collection of distinct data-based models with unique analytical formulations. Each of these smaller models is more narrowly focused, typically describing the behavior of a single parameter. Further, since the ionosphere is so dynamic, a given parameter (e.g. electron density) may be modeled differently in each layer. In many cases, multiple alternative models are included for the same parameter, and the user may choose which one they wish to use. Together, these smaller models provide a comprehensive picture of the ionosphere. This modularity is one of the IRI's greatest assets, because it allows individual pieces of the model to be updated and replaced without obsoleting the entire model.

In this section, an overview of the most recent version of the IRI will be presented. This overview is not meant to be comprehensive; rather it is intended to highlight some of the basic features and meaningful revisions of the model, to give the reader some familiarity with the subject. From there, the reader may consult the many references provided on irimodel.org should they desire a deeper treatment of the material. In addition, some code comments from the FORTRAN implementation of the IRI model have been reproduced (with permission[3]) in this section. Various configuration options for the IRI are referenced throughout this chapter, and the reader may find it useful to periodically consult these comments for context.

8.4.1 Data Sources, Inputs, and Outputs

The IRI is an empirical model, which means its efficacy is dependent on assimilating a large amount of data from diverse sources. Good coverage in latitude, longitude, time of day, season, and level of solar activity is desirable so that the model can accurately represent the ionosphere under a wide range of conditions. Table 8.2 (adapted from Section 3.1.1 of [219] and Table 1 of [220]) enumerates the major data sources upon which the current iteration of the IRI is based:

By default, the IRI only requires a handful of inputs; all of the parameters required to run the various subroutines are either calculated internally (e.g. magnetic parameters via the IGRF subroutine) or read from regularly updated text files provided on irimodel.org (e.g. the file ig_rz.dat, which contains the ionospheric and solar indices IG_{12} and R_{12} dating back to Jan 1958). Required inputs are those related to the date, time of day, and location on Earth.

In addition to the default inputs, a number of alternate options are available, allowing the user to customize their simulation in a variety of ways. For example, users may choose to input their own values for certain parameters such as

---

Table 8.2 Major data sources used in the realization of IRI 2016.

Data source	Measured ionospheric parameter
Global network of ionosondes	Peak plasma frequencies ⇒ Peak electron densities
Incoherent scatter radars (ISR)	E valley, F2 peak height, F2 layer and topside density, electron and ion temperatures
Topside sounders aboard Alouette 1 and 2 and ISIS 1 and 2 satellites	Topside electron density
In-situ measurements from various satellite missions, including AE-C, -D, -E, DE-2, AEROS-A, -B, IK-24, ISIS 1 and 2, and Intercosmos 19, 24, and 25	Electron and ion temperatures, ion composition
In-situ rocket measurements	D- and E-layer electron densities and ion compositions
GNSS satellites	Total electron content (TEC)
Radio occultation from GPS satellites to LEO satellites, e.g. COSMIC and CHAMP	Electron density
Ground-based absorption measurements	D- and E-layer variability

peak layer height or sunspot number, rather than relying on internal models or reference data files. Certain conditions may be toggled on and off, such as geomagnetic storm models, and calculation of certain outputs may be suppressed to expedite the simulation. In a number of cases, multiple options exist for modeling a given parameter and the user is given the option to choose which of these formulations they wish to use. All of these choices are governed by one of the required inputs, JF, which is a boolean array whose elements each correspond to an option that may be toggled.

In what follows, commented lines from the IRI 2016 FORTRAN source code are reproduced verbatim. These lines indicate the default inputs of the main IRI subroutine, the format of JF, and additional inputs needed when certain options are chosen.

```
C*********************************************************************
C********* INTERNATIONAL REFERENCE IONOSPHERE (IRI). *************
C*********************************************************************
C***************** ALL-IN-ONE SUBROUTINE  *************************
C*********************************************************************
C
C
SUBROUTINE IRI_SUB(JF,JMAG,ALATI,ALONG,IYYYY,MMDD,DHOUR,
&    HEIBEG,HEIEND,HEISTP,OUTF,OARR)
C-------------------------------------------------------------------
C
C INPUT:  JF(1:50)      true/false switches for several options
C         JMAG          =0 geographic  = 1 geomagnetic coordinates
C         ALATI,ALONG   LATITUDE NORTH AND LONGITUDE EAST IN DEGREES
C         IYYYY         Year as YYYY, e.g. 1985
C         MMDD (-DDD)   DATE (OR DAY OF YEAR AS A NEGATIVE NUMBER)
C         DHOUR         LOCAL TIME (OR UNIVERSAL TIME + 25) IN DECIMAL
C                            HOURS
C         HEIBEG,       HEIGHT RANGE IN KM; maximal 100 heights, i.e.
C          HEIEND,HEISTP    int((heiend-heibeg)/heistp)+1.le.100
C
C    JF switches to turn off/on (.true./.false.) several options
C
C  i   .true.              .false.         standard version
C
C  1   Ne computed         Ne not computed                      t
C  2   Te, Ti computed     Te, Ti not computed                  t
C  3   Ne & Ni computed    Ni not computed                      t
C  4   B0,B1 - Bil-2000    B0,B1 - other models jf(31)      false
C  5   foF2 - CCIR         foF2 - URSI                      false
C  6   Ni - DS-1995 & DY-1985  Ni - RBV-2010 & TBT-2015     false
C  7   Ne - Tops: f10.7<188    f10.7 unlimited                  t
C  8   foF2 from model     foF2 or NmF2 - user input            t
C  9   hmF2 from model     hmF2 or M3000F2 - user input         t
C  10  Te - Standard       Te - Using Te/Ne correlation         t
C  11  Ne - Standard Profile   Ne - Lay-function formalism      t
C  12  Messages to unit 6  to messages.txt on unit 11           t
C  13  foF1 from model     foF1 or NmF1 - user input            t
C  14  hmF1 from model     hmF1 - user input (only Lay version)t
C  15  foE  from model     foE or NmE - user input              t
C  16  hmE  from model     hmE - user input                     t
C  17  Rz12 from file      Rz12 - user input                    t
C  18  IGRF dip, magbr, modip  old FIELDG using POGO68/10 for 1973 t
C  19  F1 probability model    only if foF1>0 and not NIGHT     t
C  20  standard F1         standard F1 plus L condition         t
C (19,20) = (t,t) f1-prob, (t,f) f1-prob-L, (f,t) old F1, (f,f) no F1
C  21  ion drift computed  ion drift not computed           false
C  22  ion densities in %  ion densities in m-3                 t
C  23  Te_tops (Bil-1985)  Te_topside (TBT-2012)            false
C  24  D-region: IRI-1990  FT-2001 and DRS-1995                 t
C  25  F107D from APF107.DAT   F107D user input (oarr(41))      t
C  26  foF2 storm model    no storm updating                    t
C  27  IG12 from file      IG12 - user                          t
C  28  spread-F probability    not computed                 false
C  29  IRI01-topside       new options as def. by JF(30)    false
```

```
C   30      IRI01-topside corr.      NeQuick topside model             false
C (29,30) = (t,t) IRIold, (f,t) IRIcor, (f,f) NeQuick
C   31      B0,B1 ABT-2009           B0 Gulyaeva-1987 h0.5                t
C (4,31)  = (t,t) Bil-00, (f,t) ABT-09, (f,f) Gul-87, (t,f) not used
C   32      F10.7_81 from file       F10.7_81 - user input (oarr(46))     t
C   33      Auroral boundary model on/off    true/false                false
C   34      Messages on              Messages off                         t
C   35      foE storm model          no foE storm updating             false
C   36      hmF2 w/out foF2_storm    with foF2-storm                      t
C   37      topside w/out foF2-storm with foF2-storm                      t
C   38    . turn WRITEs off in IRIFLIP  turn WRITEs on                    t
C   39      hmF2 (M3000F2)           new models                        false
C   40      hmF2 AMTB-model          Shubin-COSMIC model                  t
C   41      Use COV=F10.7_365        COV=f(IG12) (IRI before Oct 2015)    t
C   42      Te with PF10.7 dep.      w/o PF10.7 dependance                t
C   43      B0 from model            B0 user input                        t
C   44      B1 from model            B1 user input                        t
C   45
C
C    ....
C   50
C
C   -----------------------------------------------------------------------
C
C   Depending on the jf() settings additional INPUT parameters may
c   be required:
C
C         Setting              INPUT parameter
C   -----------------------------------------------------------------------
C   jf(8)  =.false.     OARR(1)=user input for foF2/MHz or NmF2/m-3
C   jf(9)  =.false.     OARR(2)=user input for hmF2/km or M(3000)F2
C   jf(10) =.false.     OARR(15),OARR(16)=user input for Ne(300km),
C      Ne(400km)/m-3. Use OARR()=-1 if one of these values is not
C      available. If jf(23)=.false. then Ne(300km), Ne(550km)/m-3.
C   jf(13) =.false.     OARR(3)=user input for foF1/MHz or NmF1/m-3
C   jf(14) =.false.     OARR(4)=user input for hmF1/km
C   jf(15) =.false.     OARR(5)=user input for foE/MHz or NmE/m-3
C   jf(16) =.false.     OARR(6)=user input for hmE/km
C   jf(17) =.flase.     OARR(33)=user input for Rz12
C   jf(25) =.false.     OARR(41)=user input for daily F10.7 index
C   jf(27) =.false.     OARR(39)=user input for IG12
C   jf(43) =.false.     OARR(10)=user input for B0
C   jf(44) =.false.     OARR(87)=user input for B1
```

The elements JF(4), JF(6), JF(23), JF(24), JF(29), JF(30), JF(31), and JF(40) allow the user to select different modeling options. The shorthand used to denote each option therein is based on the papers in which the models were originally proposed. For the reader's convenience, citations of each of these papers is provided in Table 8.3.

The outputs of the IRI are also conveniently detailed as comments in the FOR-TRAN source code. This portion of the code has been reproduced on the following pages. It can be seen that by default, important ionospheric parameters such as electron density, electron and ion temperatures, and positive ion compositions are

Table 8.3 IRI model option abbreviations and citations for their corresponding papers.

IRI shorthand	Original paper citation
B0, B1 – Bil-2000	[218]
Ni – DS-1995 and DY-1985	[221, 222]
Ni – RBV-2010 and TBT-2015	[223, 224]
Te_tops (Bil-1985)	[225, 226]
Te_topside (TBT-2012)	[227]
D-region: IRI-1990	[219]
FT-2001 and DRS-1995	[228, 229]
IRI01-topside	[219, 230]
IRI01-topside corr.	[231]
NeQuick topside model	[196]
B0, B1 ABT-2009	[232]
B0 Gulyaeva-1987 h0.5	[233]
hmF2 AMTB-model	[234]
Shubin-COSMIC model	[235, 236]

output. By modifying JF, various additional outputs such as vertical ion drift, storm time conditions, auroral boundaries, and spread-F probability can be selected.

```
C   OUTPUT:  OUTF(1:20,1:1000)
C                OUTF(1,*)   ELECTRON DENSITY/M-3
C                OUTF(2,*)   NEUTRAL TEMPERATURE/K
C                OUTF(3,*)   ION TEMPERATURE/K
C                OUTF(4,*)   ELECTRON TEMPERATURE/K
C                OUTF(5,*)   O+ ION DENSITY/% or /M-3 if jf(22)=f
C                OUTF(6,*)   H+ ION DENSITY/% or /M-3 if jf(22)=f
C                OUTF(7,*)   HE+ ION DENSITY/% or /M-3 if jf(22)=f
C                OUTF(8,*)   O2+ ION DENSITY/% or /M-3 if jf(22)=f
C                OUTF(9,*)   NO+ ION DENSITY/% or /M-3 if jf(22)=f
C                AND, IF JF(6)=.FALSE.:
C                OUTF(10,*)  CLUSTER IONS DEN/% or /M-3 if jf(22)=f
C                OUTF(11,*)  N+ ION DENSITY/% or /M-3 if jf(22)=f
C                OUTF(12,*)
C                OUTF(13,*)
C   if(jf(24)    OUTF(14,1:11) standard IRI-Ne for 60,65,..,110km
C      =.false.)       12:22) Friedrich (FIRI) model at these heights
C                      23:33) standard Danilov (SW=0, WA=0)
C                      34:44) for minor Stratospheric Warming (SW=0.5)
C                      45:55) for major Stratospheric Warming (SW=1)
```

```
C                    56:66) weak Winter Anomaly (WA=0.5) conditions
C                    67:77) strong Winter Anomaly (WA=1) conditions
C            OUTF(15-20,*)  free
C
C     OARR(1:100)    ADDITIONAL OUTPUT PARAMETERS
C
C    #OARR(1)  = NMF2/M-3           #OARR(2)  = HMF2/KM
C    #OARR(3)  = NMF1/M-3           #OARR(4)  = HMF1/KM
C    #OARR(5)  = NME/M-3            #OARR(6)  = HME/KM
C     OARR(7)  = NMD/M-3             OARR(8)  = HMD/KM
C     OARR(9)  = HHALF/KM           #OARR(10) = B0/KM
C     OARR(11) =VALLEY-BASE/M-3      OARR(12) = VALLEY-TOP/KM
C     OARR(13) = TE-PEAK/K           OARR(14) = TE-PEAK HEIGHT/KM
C    #OARR(15) = TE-MOD(300KM)      #OARR(16) = TE-MOD(400KM)/K
C     OARR(17) = TE-MOD(600KM)       OARR(18) = TE-MOD(1400KM)/K
C     OARR(19) = TE-MOD(3000KM)      OARR(20) = TE(120KM)=TN=TI/K
C     OARR(21) = TI-MOD(430KM)       OARR(22) = X/KM, WHERE TE=TI
C     OARR(23) = SOL ZENITH ANG/DEG  OARR(24) = SUN DECLINATION/DEG
C     OARR(25) = DIP/deg             OARR(26) = DIP LATITUDE/deg
C     OARR(27) = MODIFIED DIP LAT.   OARR(28) = Geographic latitude
C     OARR(29) = sunrise/dec. hours  OARR(30) = sunset/dec. hours
C     OARR(31) = ISEASON (1=spring)  OARR(32) = Geographic longitude
C    #OARR(33) = Rz12                OARR(34) = Covington Index
C    #OARR(35) = B1                  OARR(36) = M(3000)F2
C    $OARR(37) = TEC/m-2            $OARR(38) = TEC_top/TEC*100.
C    #OARR(39) = gind (IG12)         OARR(40) = F1 probability
C    #OARR(41) = F10.7 daily         OARR(42) = c1 (F1 shape)
C     OARR(43) = daynr               OARR(44) = equatorial vertical
C     OARR(45) = foF2_storm/foF2_quiet        ion drift in m/s
C    #OARR(46) = F10.7_81            OARR(47) = foE_storm/foE_quiet
C     OARR(48) = spread-F probability
C     OARR(49) = Geomag. latitude    OARR(50) = Geomag. longitude
C     OARR(51) = ap at current time  OARR(52) = daily ap
C     OARR(53) = invdip/degree       OARR(54) = MLT-Te
C     OARR(55) = CGM-latitude        OARR(56) = CGM-longitude
C     OARR(57) = CGM-MLT             OARR(58) = CGM lat eq. aurl bodry
C     OARR(59) = CGM-lati(MLT=0)     OARR(60) = CGM-lati for MLT=1
C     OARR(61) = CGM-lati(MLT=2)     OARR(62) = CGM-lati for MLT=3
C     OARR(63) = CGM-lati(MLT=4)     OARR(64) = CGM-lati for MLT=5
C     OARR(65) = CGM-lati(MLT=6)     OARR(66) = CGM-lati for MLT=7
C     OARR(67) = CGM-lati(MLT=8)     OARR(68) = CGM-lati for MLT=9
C     OARR(69) = CGM-lati(MLT=10)    OARR(70) = CGM-lati for MLT=11
C     OARR(71) = CGM-lati(MLT=12)    OARR(72) = CGM-lati for MLT=13
C     OARR(73) = CGM-lati(MLT=14)    OARR(74) = CGM-lati for MLT=15
C     OARR(75) = CGM-lati(MLT=16)    OARR(76) = CGM-lati for MLT=17
C     OARR(77) = CGM-lati(MLT=18)    OARR(78) = CGM-lati for MLT=19
C     OARR(79) = CGM-lati(MLT=20)    OARR(80) = CGM-lati for MLT=21
C     OARR(81) = CGM-lati(MLT=22)    OARR(82) = CGM-lati for MLT=23
C     OARR(83) = Kp at current time  OARR(84) = magnetic declination
C     OARR(85) = L-value             OARR(86) = dipole moment
C               # INPUT as well as OUTPUT parameter
C               $ special for IRIWeb (only place-holders)
C-----------------------------------------------------------------
```

```
C*******************************************************************
C*** THE ALTITUDE LIMITS ARE:  LOWER (DAY/NIGHT)  UPPER        ***
C***     ELECTRON DENSITY           60/80 KM      1500 KM      ***
C***     TEMPERATURES                  60 KM      2500/3000 KM ***
C***     ION DENSITIES                100 KM      1500 KM      ***
C*******************************************************************
C*******************************************************************
C*********              INTERNALLY              ****************
C*********         ALL ANGLES ARE IN DEGREE     ****************
C*********         ALL DENSITIES ARE IN M-3     ****************
C*********         ALL ALTITUDES ARE IN KM      ****************
C*********         ALL TEMPERATURES ARE IN KELVIN ****************
C*********         ALL TIMES ARE IN DECIMAL HOURS ****************
C*******************************************************************
C*******************************************************************
```

8.4.2 Important Functions

A number of mathematical functions and concepts are particularly suited to the IRI's approach to modeling ionospheric phenomena. The IRI is fundamentally an inverse problem, relying on experimental datasets as a basis for more generalized predictions. Such problems often prove challenging to solve, but reliable tools such as spherical harmonic expansions have been used to great effect in fields such as geophysics.

Additionally, as a computer model, the IRI is limited to dealing with discrete mathematics. Many of the convenient analytical solutions utilized in theoretical models, which rely on concepts such as continuity and differentiability, simply cannot be implemented perfectly by a computer. However, these solutions can be closely approximated using only discrete quantities; for example, the continuous variation of electron density with height can be approximated via a "skeleton profile" of constant gradients, as we will see shortly.

These concepts, among a few others, continually appear in the IRI, so it is worth spending some time to understand them first.

Epstein Functions

In Section 8.2.2 a function, denoted $\text{Eps}_1(x)$, was introduced. In that section, the aforementioned function was referred to as the Epstein layer function. In reality, this is just one of a family related functions introduced by Epstein. Of special interest are the Epstein transition function, $\text{Eps}_{-1}(x)$, the Epstein step function, $\text{Eps}_0(x)$, and the previously discussed Epstein peak function, $\text{Eps}_1(x)$. These three functions are given by:

$$\text{Eps}_{-1}(x) = \ln(1 + e^x), \tag{8.86}$$

$$\text{Eps}_0(x) = \frac{1}{(1 + e^{-x})},$$ (8.87)

$$\text{Eps}_1(x) = \frac{e^x}{(1 + e^x)^2}.$$ (8.88)

In the context of atmospheric applications, the argument of an Epstein function is given by

$$x = \frac{h - h_m}{B}.$$ (8.89)

This centers the Epstein function at h_m, and scales its width by the thickness parameter B. The shape of these functions can be seen in Figure 8.5.

Note that, in general, the remaining members of the Epstein family of functions may be obtained by repeated differentiation.

$$\text{Eps}_{i+1}(x) = \frac{d}{dx} \text{Eps}_i(x).$$ (8.90)

The fact that Epstein functions are everywhere differentiable is one of the key reasons that they are so useful for constructing analytical descriptions of ionospheric profiles. When solving radiowave propagation problems, a discontinuity in

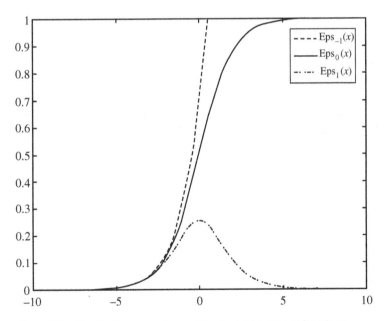

Figure 8.5 The first three members of the Epstein family of functions.

the electron density profile's derivative introduces spurious reflections. Further, if more than one discontinuity exists in the profile's derivative, multiple spurious reflections result and may interfere with one another. Not only does this make the propagation problem more difficult to compute, but it also means that the solution may not accurately model the true physical situation.

Booker Function

One approach to constructing an ionospheric profile was proposed by Henry Booker in 1977 [237]. In this approach, the profile is first approximated by a number of segments with constant gradient (in other words, a series of straight lines). Booker called this the "skeleton profile" and an example can be seen in Figure 8.6.

The skeleton profile is a good starting point, but its derivative is simply a series of constant values that discontinuously jump wherever two line segments meet. To remove these discontinuities, the Epstein step function can be used to smoothly transition from one constant value to another. For example, a transition

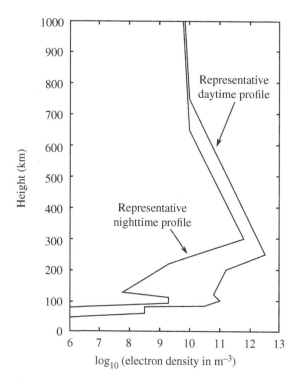

Figure 8.6 Two example ionospheric skeleton profiles. Source: Booker [237]. Reproduced with Permission of Elsevier.

from height h_1 to height h_2 is described by the following function (denoted EPSTEP in the IRI source code):

$$\text{EPSTEP}(h) = h_1 + (h_2 - h_1)\text{Eps}_0(h, h_m, B) = h_1 + \frac{h_2 - h_1}{1 + \exp\left(-\dfrac{h - h_m}{B}\right)}. \tag{8.91}$$

Thus, the derivative of the profile can be represented as a sum of EPSTEP functions. Finally, in order to obtain the function for the original profile, its derivative must be integrated. Integration over the range h_0 to h produces

$$B(h) = (h - h_0)DN_1 + \sum_{j=1}^{M}(DN_{j+1} - DN_j)B_j[\text{Eps}_{-1}(h, h_j, B_j)$$

$$- \text{Eps}_{-1}(h_0, h_j, B_j)], \tag{8.92}$$

where DN_j is the constant gradient of the jth subsection of the skeleton profile, h_j is the boundary of the jth subsection, B_j is the thickness parameter of the jth subsection, and M is the total number of subsections. This function is commonly known as the Booker function, after its creator. Using Equation 8.86 and applying logarithm rules, we can write the Booker function more explicitly as

$$B(h) = (h - h_0)DN_1 + \sum_{j=1}^{M}(DN_{j+1} - DN_j)B_j \ln\left(\frac{1 + \exp\left(\dfrac{h - h_j}{B_j}\right)}{1 + \exp\left(\dfrac{h_0 - h_j}{B_j}\right)}\right). \tag{8.93}$$

When this approach is applied to the skeleton profiles in Figure 8.6, it can be seen that the resulting is profiles very closely resemble their underlying skeletons, but no longer contain any sharp transitions. The smoothed profiles are shown in Figure 8.7.

Spherical Harmonics

At its most basic, the IRI is an inverse problem. An inverse problem is one in which the outputs of a system, such as the ionosphere, are known (e.g. via empirical measurements), and an underlying model must be ascertained.[4] The task of forming a model that is able to generalize the data is a challenging task. However, powerful tools can make such problems tractable in a number of situations. Among these tools are spherical harmonics.

---

4 This type of problem is termed inverse because it is the opposite of the "forward problem," in which an event or process is followed to its causal conclusion to predict a set of outputs.

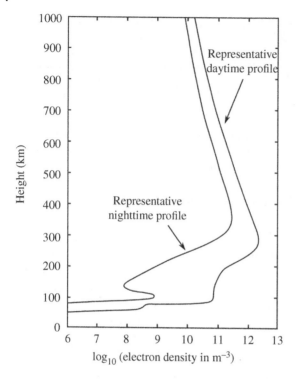

Figure 8.7 Smoothed ionospheric profiles constructed using the Booker function. Source: Booker [237]. Reproduced with Permission of Elsevier.

Spherical harmonics are solutions to the Laplace equation in spherical coordinates.[5] They form an orthonormal basis for the Hilbert space of square-integrable functions on a sphere (such as the Earth), and thus any linear combination of spherical harmonics may be used to construct such a function. Note that this is the spherical analogue to how any periodic function on the real numbers can be expressed via a Fourier series.

Spherical harmonic analysis has been a powerful tool for geophysical applications in general, and has commonly been used to study potential fields such as the Earth's gravitational and magnetic fields. Analytically, a time-dependent function on the surface of a sphere may be represented by the following equation:

$$f(\theta, \lambda, t) = \sum_{n=0}^{N} \sum_{m=0}^{n} P_n^m(\cos \theta) \left[g_n^m(t) \cos\left(m\lambda\right) + h_n^m(t) \sin(m\lambda) \right]. \tag{8.94}$$

---

5 *Note*: Any function whose second partial derivatives are continuous and which also satisfies Laplace's equation, i.e. $\nabla^2 f = 0$, is known as a harmonic function. Hence, any functions that satisfy these properties in spherical coordinates are known as spherical harmonics.

Here, θ is colatitude (i.e. the complement of latitude), λ is longitude, t is time, and P_n^m are Legendre polynomials of degree n and order m. g_n^m and h_n^m are the spherical harmonic coefficients, also known as the Gauss coefficients, and are the parameters that define the model (in the parlance of linear algebra, these are the coordinates with respect to our basis or, equivalently, the weights of our linear combination). The derivation of Equation 8.94 is beyond the scope of this book, but may be found in a number of textbooks on mathematical physics.

Applied to the case of the ionosphere, Equation 8.94 gives the value of a function (e.g. electron temperature, F2 layer peak height, etc.) above some point on the surface of the Earth. Thus it follows that the inverse problem of constructing an analytical representation of a given parameter over the whole globe could be achieved if a sufficiently large grid of points with good coverage in latitude and longitude were used. Such a grid would produce an overdetermined system of equations, which could be solved via least squares. Obtaining such a grid for iono-spheric parameters is challenging, because large swaths of the Earth's surface are oceans, meaning ionosonde measurements are not available there. Many authors have taken different approaches to solving this problem, as we will see in the upcoming sections.

8.4.3 Characteristic Heights and Electron Densities

Perhaps the most important parameter modeled by the IRI is the electron density of the ionosphere. As discussed in Section 6.3, the electron density can be roughly organized into layers, each of which behaves uniquely. Accordingly, the IRI uses different models for different altitude ranges of the ionosphere. Figure 8.8 shows the organization of the IRI's electron density profile.

The models for each of these regions are defined in terms of certain character-istic parameters of the ionosphere, and each of these parameters is modeled in the IRI. This section will summarize the models that are used in the most recent version of the IRI. Note that all heights throughout this section are given in units of kilometers.

Peak Electron Densities N_m

Within a given layer of the ionosphere, there will be some maximum electron density. For some layer of the ionosphere, this maximum electron density is denoted N_m. Recall that Equation 6.30 relates the critical frequency f_0 of an ionospheric layer to its peak electron density. Critical frequencies for the E, F1, and F2 layers are easily obtained from ionograms, and the peak densities for these layers are given by Equations 8.28–8.30, which are reproduced in the following

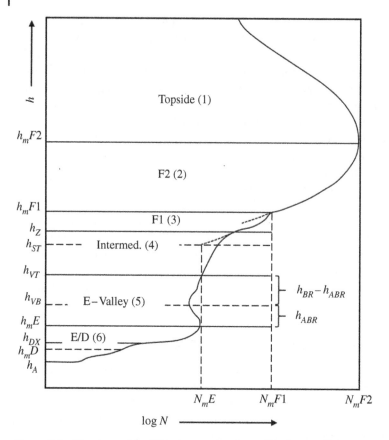

Figure 8.8 Structure of the IRI's electron density profile. Source: Bilitza et al. [219], Figure 3 (p. 51)/National Aeronautics and Space Administration/Public Domain.

text for convenience (assuming each critical frequency has units of MHz):

$$N_mE = 0.124(f_oE)^2, \tag{8.95}$$

$$N_mF1 = 0.124(f_oF1)^2, \tag{8.96}$$

$$N_mF2 = 0.124(f_oF2)^2. \tag{8.97}$$

Here, each peak density N_m has units of 10^{11} electrons/m^3. In order to achieve good coverage in time and location, each critical frequency must in turn be modeled. Global ionogram data is readily available from a large network of ionosondes, providing a good database for the development of empirical models. Such models are also included in the IRI, and will be discussed shortly, but first it is important to address the absence of the D layer peak density N_mD from the preceding equations.

At D layer altitudes, neutral molecules are abundant, causing significant attenuation of the lower frequencies that would otherwise be reflected in this region. This means that ionograms cannot be used to reliably determine the critical frequency, and thus the peak electron density, of the D layer. Instead, the IRI uses in situ rocket measurements to characterize the electron density. Using data from D layer rocket measurements, Mechtley and Bilitza developed an empirical model for $N_m D$ dependent on solar activity and time of day [238]. This relationship is given by the equation

$$N_m D = (6.05 + 0.088 R_{12}) \exp\left(\frac{-0.1}{\cos^{2.7}(\chi)}\right), \tag{8.98}$$

where $N_m D$ is measured in 10^8 electrons/m^3, R_{12} is the 12-month running mean of the solar sunspot number, and χ is the solar zenith angle. $N_m D$ is given a minimum (nighttime) value of 4×10^8 electrons/m^3, and is reset to this value if Equation 8.98 provides a smaller result.

E Layer Critical Frequency $f_o E$

An empirical model for the E layer critical frequency $f_o E$ was developed by Kouris and Muggleton using 45 globally distributed ionosonde stations over a period of 11 years [239, 240]. The mathematical formulation of their model is as follows:

$$(f_o E)^4 = ABCD, \tag{8.99}$$

where

$$A = 1 + 0.0094(F10.7_{12} - 66), \tag{8.100}$$

$$B = \cos^m \chi_{noon}, \tag{8.101}$$

$$m = \begin{cases} -1.93 + 1.92 \cos\phi, & |\phi| < 32°, \\ 0.11 - 0.49 \cos\phi, & |\phi| \geq 32°, \end{cases}$$

$$C = \begin{cases} 23 + 116 \cos\phi, & |\phi| < 32°, \\ 92 + 35 \cos\phi, & |\phi| \geq 32°, \end{cases}$$

$$D = \cos^n \chi_a, \tag{8.102}$$

$$n = \begin{cases} 1.2, & |\phi| > 12°, \\ 1.31, & |\phi| \leq 12°. \end{cases}$$

Here, $F10.7_{12}$ is the 12-month running mean of the solar 10.7 cm radio flux, ϕ is the geodetic latitude, χ_{noon} is the solar zenith angle at noon, and χ_a is a modified zenith angle given by the expression

$$\chi_a = \chi - 3\mathrm{Eps}_{-1}(\chi, 89.98, 3) = \chi - 3\ln\left(1 + \exp\left(\frac{h - 89.98}{3}\right)\right). \tag{8.103}$$

f_oE is always kept at a value greater than or equal to a minimum value given by

$$f_oE_{min} = 0.121 + 0.0015(F10.7_{12} - 66). \qquad (8.104)$$

If Equation 8.99 produces a value less than f_oE_{min}, it is instead replaced by f_oE_{min}.

F1 Layer Critical Frequency f_oF1

DuCharme et al. presented an empirical model of the F1 layer critical frequency using ionosonde data from 39 globally distributed stations over a time period of 15 years. Their model is given by the following expressions:

$$f_oF1 = f_s\cos^n\chi, \qquad (8.105)$$

where

$$f_s = f_0 + \frac{(f_{100} - f_0)R_{12}}{100}, \qquad (8.106)$$

$$f_0 = 4.35 + 0.0058\,|\Psi| - 0.00012\Psi^2, \qquad (8.107)$$

$$f_{100} = 5.348 + 0.011\,|\Psi| - 0.00023\Psi^2, \qquad (8.108)$$

$$n = 0.093 + 0.0046\,|\Psi| - 0.000054\Psi^2 + 0.0003R_{12}, \qquad (8.109)$$

where Ψ is the magnetic dip latitude in degrees and χ and R_{12} are defined the same as for f_oE. Note that dip latitude is a different quantity from modified dip latitude, which was introduced in previous sections. Dip latitude is given by the equation:

$$\Psi = \arctan\left(\frac{\tan I}{2}\right), \qquad (8.110)$$

where I is the magnetic dip angle.

F2 Layer Critical Frequency f_oF2

As discussed in Chapter 5, unlike the lower layers of the ionosphere, the F2 layer cannot be described as a Chapman layer. It is more difficult to model, which is reflected in the complexity of the two modeling options offered in the IRI for determination of f_oF2. Both models utilize a method originally proposed by Jones and Gallet [199] to globally extrapolate measurements from only a handful of ionospheric stations. In essence, this method entails the use of Fourier series to express the diurnal variation of f_oF2 representative of a given station within a given month. Repeating this process for each station creates a grid of f_oF2 values on the surface of the Earth. With this grid as a basis, spherical harmonics are used to create an analytical function which models the value of f_oF2 at every point on Earth's surface. The details of this method are beyond the scope of this text, but the reader may consult Section 8.4.2 for a brief primer on spherical harmonics, and Jones and Gallet [199] for a more thorough treatment of the full process.

Jones and Gallet's model is sufficient for many purposes, but was found to be lacking in places where ionosonde data was not available (e.g. over the ocean). Rush et al. [241, 242] addressed this issue by using theoretical considerations to create additional points to include in the spherical harmonic analysis. The density of positive ions in the F2 layer can be determined via the continuity equation (see Section 5.5.5), and since the electron density and positive ion density are equal in the ionosphere, this information can then be used to determine f_oF2 via Equation 8.97. At F2 layer altitudes, the ionosphere is very complex, depending on various plasma transport phenomena such as diffusion, neutral winds and electromagnetic drift, as discussed in Section 5.5. Thus Rush et al. needed to model of each of these phenomena in order to obtain reasonable predictions of ion density. The details of their model will not be presented here, but may be found in [241, 242]. This model was found to have better performance than Jones and Gallet's model, and has been adopted as the default option in the current iteration of IRI.

The user may choose which of these two models to use by changing the value of JF(5) (see Section 8.4.1). By default, this parameter is set to false, which uses the model developed by Rush et al. It should be noted that this model is referred to as "fof2-URSI," since its authors are members of the International Union of Radio Science (URSI) Working Group G.5. Conversely, Jones and Gallet's model may be selected by setting JF(5) to true. This model, which was adopted by the CCIR, is denoted "fof2-CCIR."

Since both of these models depend on spherical harmonic and Fourier series expansions, a large number of coefficients is required for their computation. Both models use the same number of coefficients: two sets of 988 coefficients for each month, resulting in a total of 23,712 coefficients. Each set of 988 coefficients describes the spatial and diurnal variation of f_oF2 in a given month for a given solar activity. Files containing these coefficients, organized by month, are available from irimodel.org. The files are named ccirXX.dat and ursiXX.dat, where XX is the month number plus 10 (e.g. March = 13). Within each file are two sets of 988 coefficients; one set corresponds to low solar activity ($R_{12} = 0$) and the other to high solar activity ($R_{12} = 100$). To obtain the coefficients for intermediate solar activities, the IRI source code linearly interpolates between these values [219, p. 52].

D Layer Characteristic Heights h_A, h_mD, h_{DX}

The default model used by the IRI for determining D region electron density makes use of three characteristic heights: h_A, h_mD, and h_{DX}. The values of these heights were determined via an analysis of in situ rocket data presented by Mechtley and Bilitza in 1974 [238]. Each of the three heights was given a constant daytime value and a different, yet still constant, nighttime value.

The lowermost of these three points, h_A, designates the bottom of the D layer and thus the bottom of the entire IRI electron density profile. This height was determined to be 65 km during the day and 80 km at night. Above h_A, the electron density gradually increases, eventually reaching a value of $N_m D$. The height at which $N_m D$ occurs is denoted $h_m D$ and is modeled using a daytime value of 80 km and a nighttime value of 88 km.

Just above $h_m D$, the electron density exhibits a sudden steep gradient, causing electron density values to increase greatly over only a few kilometers. This dramatic increase appears as a cliff-like feature in the electron density profile, so it has been termed the D ledge [243]. The height at which the ledge occurs is denoted h_{DX}, and in the IRI is given a value of 85.6 km during the day and 92.5 km at night.

E Layer Characteristic Heights $h_m E$, h_{VB}, h_{VT}

Above the D region, the electron density continues to increase up to a value of $N_m E$; this height is the E layer peak height, denoted $h_m E$. In IRI-1995, based on a large amount of ionosonde and ISR measurements, $h_m E$ was assigned the constant value of 110 km. This value has not been changed since, and is still used in the current iteration of the IRI.

Above $h_m E$, many ionograms indicate the existence of a "valley" in the electron density profile. That is, as height increases above $h_m E$ the electron density decreases to a value below $N_m E$, reaches a local minimum, and then increases back up to $N_m E$ again. This feature, commonly referred to as the E valley, can clearly be seen in Figure 8.8. The height of the valley base, h_{VB}, is defined as the point at which the local minimum of the E valley occurs. Similarly, the height of the valley's top, h_{VT} is the altitude at which the electron density reaches $N_m E$ once again.

In the IRI, h_{VB} and h_{VT} are quantified with respect to $h_m E$ rather than the ground. Put another way, the IRI instead models h_{ABR}, the distance between $h_m E$ and h_{VB}, and h_{BR}, the width of the valley (or equivalently the distance between $h_m E$ and h_{VT}). These two values are important for the calculation of the electron density in the E valley region, as will be seen in Section 8.4.4.

Intermediate Region and F1 Layer Characteristic Heights h_{ST}, h_Z, $h_m F1$

The intermediate region of the ionosphere is the transition region between the E and F1 layers. This region extends from the top of the E valley up to a point h_Z, which is given by the expression

$$h_Z = \frac{(h_{ST} + h_m F1)}{2}. \tag{8.111}$$

In other words, h_Z is the midpoint between the other two characteristic heights in this region: h_{ST} and h_mF1. Both of these heights are defined as the points at which the F1 electron density function reaches a certain value.

Note that the F1 electron density function – henceforth N_{F1} – is only one portion of the full electron density profile. Recall that in Section 8.2.2, Epstein functions were used to model each layer's electron density individually and their sum represented the entire electron density profile. A similar approach is used in the IRI, although the analytical representation of a given layer is not an Epstein function. The exact formulation of these functions will be discussed shortly in Section 8.4.4, but for now it suffices to know that each layer exhibits a peak shape that closely resembles an Epstein layer function (see $\text{Eps}_1(x)$ in Figure 8.5).

N_{F1} is maximized at the F1 layer peak density N_mF1. The height at which this occurs is defined as h_mF1. Below h_mF1, electron density decreases, but since the full profile is the sum of multiple layers' densities, this dropoff is slower than that of N_{F1} alone. This can be seen in Figure 8.8, where a dotted line is used to show the theoretical extension of N_{F1}, assuming the electron densities of all other layers are neglected (note that the same is done for the F2 layer). The point at which N_{F1} equals N_mE is defined as h_{ST}.

F2 Layer Characteristic Height h_mF2

IRI-2016, the most recent version of the IRI, includes two models for calculation of h_mF2. These models, much like the ones for f_oF2, must use Fourier series and spherical harmonics in order to adequately capture the complex global behavior of h_mF2. Much like in the f_oF2 models discussed previously, Fourier series expansions are used to match the diurnal variation for a given station, and spherical harmonics are used to extend this definition to the whole surface of the Earth.

The first of these models, developed by Altadill et al. in 2013 (henceforth referred to as the AMTB model [234]), utilizes data from globally distributed ionosondes. Thus, the same problem arose as in Jones and Gallet's f_oF2 model: ionosonde data is sparse or nonexistent in certain parts of the world, such as over oceans. In these regions, the model would perform poorly, so it is essential to extrapolate the data to obtain additional points. To accomplish this, Altadill et al. used a zeroth order approximation, simply assuming that longitudinal variation and diurnal variation were equivalent. Thus, the diurnal variation of h_mF2 at a given station is directly translated to longitudinal variation. The diurnal profile is sampled hourly and mapped to points along the same magnetic latitude as the source station, resulting in 24 points spaced 15° apart in longitude.

The AMTB model is only intended to predict h_mF2 under magnetically quiet conditions. Thus, it was necessary to remove any ionosonde data from periods of intense magnetic activity, such as magnetic storms. To accomplish this, a monthly average representative profile (MARP) is adopted for each station. Each station's MARP is calculated by excluding the 25% of electron density profiles with the greatest deviation from the monthly mean.

A spherical harmonic expansion is applied with terms up to order 4 in latitude and up to 8 in longitude, as well as a Fourier series expansion with terms up to order 2 to reflect the annual variation. These coefficients represent one full set; two full sets of coefficients are provided, resulting in a total of 610 coefficients. Each set corresponds to a certain level of solar activity: one for low solar activity ($R_{12} = 15$) and another for high solar activity ($R_{12} = 120$). Linear interpolation is used to obtain coefficients for intermediate values.

For more details on this modeling approach, the reader is encouraged to read the original paper [234]. It should be noted that Altadill et al. used this exact same approach in the development of a model for the F2 electron density. This model, which is also used in the IRI, is discussed in some detail in Section 8.4.4.

The second h_mF2 model offered by IRI-2016 was presented by Shubin et al. (henceforth, the SMF2 model), and uses radio occultation (RO) measurements as empirical basis. RO data from the CHAMP, COSMIC, and GRACE satellites was used for analysis. The overwhelming majority of this data – more than 3,500,000 profiles – were from the COSMIC constellation, while CHAMP and GRACE account for about 300,000 and 100,000 profiles respectively. Combined, these datasets span the years 2001–2012, giving coverage of approximately one full solar cycle. However, the COSMIC dataset is limited to the years 2006–2012, which is approximately half of a full solar cycle. During this period of time, the minimum of solar cycle 24 occurred, so low solar activity data is significantly overrepresented in the SMF2 data base. To supplement the RO data's lack of high solar activity data, both bottomside and topside ionosonde data was also included in the analysis (SDMF2 model). In particular, profiles from 62 globally distributed ionosonde stations, as well as from the IK-19 satellite, were considered.

Profiles obtained from RO data have the potential to be very noisy, or at times may indicate that the absolute maximum electron density occurs well outside the range of plausible values. Shubin et al. developed an algorithm to detect and discard profiles of dubious quality, which resulted in the removal of about 10% of the original data. For additional validation, the modeled h_mF2 values were compared against those predicted by ionosonde data, as well as by the h_mF2 model used in IRI-2012 (the most recent iteration of the IRI available at the time). The SMF2 model found good agreement with ionosonde data, and outperformed IRI-2012 [244].

To model global variation, a spherical harmonic expansion using terms up to order 12 in modified dip latitude and up to order 8 in longitude, was employed.

Similarly, a Fourier series with terms up to order 3 was used to express diurnal variation. To express these expansions, a number of coefficients were determined for different times of day, times of year, and solar activities.

The aforementioned expansions require a set of 149 coefficients to fully describe the global variation of h_mF2. A set of coefficients was determined for each hour of the day in UTC, and this process was repeated for each month of the year. This whole process was repeated twice: once for low solar activity ($F10.7_{81} < 80$) and once for high solar activity ($F10.7_{81} > 120$).[6] This results in a total of $149 \times 24 \times 12 \times 2 = 85,824$ coefficients. For intermediate solar activity values, a logarithmic relationship is used which reflects a saturation effect that was observed at high solar activities [245]. For further information on the SMF2 and SDMF2 models, one may refer to [235, 236, 244].

To express each of these models, the IRI source code must have access to the coefficients of the Fourier and spherical harmonic expansions. The coefficients of the AMTB model are initialized as an array within one of the IRI subroutines. On the other hand, the SDMF2 model uses many more coefficients, so this data is instead included alongside the IRI source code as separate text files. These files have the naming convention mcsatXX.dat, where XX is the month number plus 10.

Of these two models, the AMTB model was found to produce more accurate results [245], and so has been adopted as the default choice in the IRI. If JF(40) (see Section 8.4.1) is instead set to false, the SDMF2 model is used instead. In addition, if JF(39) is changed to true, the F2 model used in older versions of the IRI may be chosen instead. This legacy model is less accurate than either of the two newer models, so it is not recommended for h_mF2 modeling. This model will not be discussed further here, but the interested reader may consult [246].

8.4.4 Electron Density

The IRI electron density model is subdivided into six regions: the topside, the F2 layer, the F1 layer, the intermediate region, the E valley, and the E bottomside/D region. Each has its own separate mathematical formulation, many of which have been improved and updated over the years. Figure 8.8 shows a sketch of this structure, and it can be seen that several characteristic points such as the layer peak heights (e.g. h_mF2) are used as the dividing lines between region. In the following subsections, each layer will be discussed separately. The header for each subsection includes the upper and lower boundaries of the layer it concerns.[7] Throughout

---

6 $F10.7_{81}$ is the 81-day running average of the 10.7 cm solar radio flux.
7 The profile sections are addressed from the top of the ionosphere down, so the boundaries of a given section are listed in the same fashion – that is, upper bound first and lower bound second.

this section, the electron density function for a given region will be denoted with a numeric subscript. This subscript corresponds to the numbering system used in Figure 8.8. That is, the electron density function of the topside region will be denoted $N_1(h)$, that of the $F2$ layer will be denoted $N_2(h)$, and so on.

Topside (1000 km to $h_m F2$)
At the time of this writing, the IRI source code contains two options for modeling the topside electron density, both of which were introduced as part of the 2007 version of the IRI. The first of these options is an evolution of the original topside formulation which dates back several decades. This formulation, originally proposed by the Rawer in [230], was based primarily on data from the Alouette 1 satellite's topside sounder, with some in situ measurements from the AE-C and DE-2 satellites and profiles from the Jicamarca ISR [247].

The topside density profile is normalized to the F2 peak density $N_m F2$, and a Booker function is used to construct profile from a skeleton of constant gradients, as discussed in the Section 8.4.2. Thus, the topside density can be written:

$$\frac{N_1(h)}{N_m F2} = \exp\left(-\frac{1}{\alpha}\left[(x - x_0)TG_1 + \sum_{j=1}^{2}(TG_{j+1} - TG_j)TC_j \ln\left(\frac{1 + \exp\left(\frac{x - TX_j}{TC_j}\right)}{1 + \exp\left(\frac{x_0 - TX_j}{TC_j}\right)}\right)\right]\right). \tag{8.112}$$

Here, the variable x is a transformation of h such that the height range $h = h_m F2$ to $h = 1000$ km corresponds to $x = x_0$ to $x = x_0 + 700$. This can be accomplished using the following scaling:

$$x - x_0 = \alpha(h - h_m F2), \tag{8.113}$$

$$\alpha = \frac{700}{1000 - h_m F2}, \tag{8.114}$$

$$x_0 = 300 - \delta. \tag{8.115}$$

The parameter δ will be addressed momentarily. Note that, excluding the factor of $-1/\alpha$, the argument of the outermost exponential function is Equation 8.93 (i.e. a Booker function) with $M = 2$, $DN_j = TG_j$, $B_j = TC_j$, and $h_j = TX_j$. TG, TX, and TC are known as Epstein parameters, and their values are given in Table 8.4.

Table 8.4 Epstein parameter values for the IRI topside profile.

j	TG	TX	TC
1	$-\xi$	300	100
2	0	394.5	β
3	η	—	—

The coefficients ξ, β, and η are functions of the geomagnetic latitude ϕ_m, the monthly average solar radio flux $F10.7$, and the F2 critical frequency f_oF2. They can be written as follows [219, p. 54]:

$$\xi = \beta = t_0 + t_1 T(\phi_m) + t_2 R(F10.7) + t_3 T(\phi_m) R(F10.7) + t_4 f_o F2 \\ + t_5 f_o F2 T(\phi_m) + t_6 f_o F2^2, \tag{8.116}$$

$$\eta = t_0 + t_1 T_m(\phi_m) + t_2 R(F10.7) + t_3 T(\phi_m) R(F10.7) + t_4 f_o F2 \\ + t_5 f_o F2 T(\phi_m) + t_6 f_o F2^2, \tag{8.117}$$

where

$$T(\phi_m) = \cos^2(\phi_m), \tag{8.118}$$

$$T_m(\phi_m) = \frac{\exp\left(-\dfrac{\phi_m}{15}\right)}{\left(1 + \exp\left(-\dfrac{\phi_m}{15}\right)\right)^2}, \tag{8.119}$$

$$R(F10.7) = \begin{cases} \dfrac{F10.7 - 40}{30}, & F10.7 < 193, \\ 5.1, & F10.7 \geq 193, \end{cases}$$

and the parameters t_i were fitted to the topside densities predicted by the Bent model (a prominent ionospheric model at the time of the IRI's original development) [248]. This was accomplished via a nonlinear least squares fitting procedure. Tabulated values for these parameters may be found in Table 2 of Section 3.2.1.3 of the IRI 1990 report [219, p. 55].

Recall the correction term δ from Equation 8.115. This term can be expressed in terms of ξ, β, and η:

$$\delta = \frac{\dfrac{\eta}{1+Z} - \dfrac{\xi}{2}}{\dfrac{\eta}{\beta}\dfrac{Z}{(1+Z)^2} + \dfrac{\xi}{400}}, \tag{8.120}$$

where

$$Z = \exp\left(\frac{94.45}{\beta}\right).\tag{8.121}$$

The preceding equations constitute the original IRI topside formulation in its entirety. This model served as a solid basis for ionospheric inquiry, but in the years following its release it became apparent that it had significant shortcomings. These shortcomings were attributed to the limited data upon which the original model was based. Bilitza et al. set out to improve the topside formulation by using a large amount of additional data [231]. In this study, over 150,000 topside profiles from the Alouette 1 and 2 and ISIS 1 and 2 satellites were considered. When this data was compared with the topside profile predicted by the original IRI topside model, it was found that the predicted electron density values deviated significantly from the measured values. Average ratios of measured values to predicted values at various modified dip latitudes (modip) is shown for heights above the F2 peak in Figure 8.9.

It can be seen in Figure 8.9 that at altitudes greater than about 1000 km above the F2 peak, the IRI overestimates electron densities (ratio < 1) in all modip ranges. At altitudes less than 1000 km above the F2 peak, there is a strong dependence on modip, with low latitudes exhibiting smaller ratios. At low latitudes in this altitude

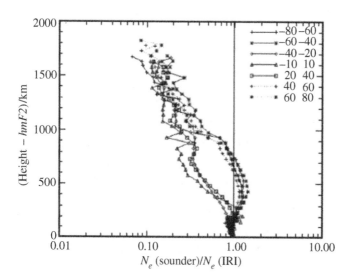

Figure 8.9 Average ratios of measured Alouette/ISIS electron densities to densities predicted by the original IRI topside formulation. Each curve represents a separate modip range, and each range is depicted in the top right. Source: Bilitza [231]. Reproduced with Permission of Elsevier.

range, the IRI overestimates electron densities (ratio < 1) while at high latitudes the IRI underestimates densities (ratio > 1).

Bilitza et al. proposed that more accurate results could be obtained if the IRI values at a given altitude, modip and time of day were simply multiplied by a scaling factor equal to the measured-to-predicted ratio under those same conditions. It would have been infeasible to calculate this ratio as a continuous function of altitude, modip and time of day, so instead Bilitza et al. chose a few values for each of these parameters and interpolated between them to obtain a good estimate of the ratio's dependence on them.

The chosen altitudes of 450, 800, and 1500 km are all measured with respect to the F2 peak (e.g. 450 km represents 450 km above the F2 peak). These altitudes are used as anchor points of a Booker function, which is used to approximate the electron density ratio as a smooth function of height. Seven modip values $-70°$, $-50°$, $-30°$, $0°$, $30°$, $50°$, and $70°$ – are used, and intermediate values are obtained via linear interpolation, as seen in Figure 8.10. Two times of day – noon and midnight – were chosen to illustrate the variation between daytime and nighttime ratios.

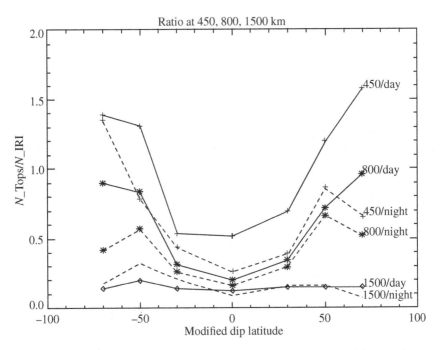

Figure 8.10 Values of the measured-to-predicted ratio of electron densities at various altitudes, modips, and times of day. Source: Bilitza [231]. Reproduced with Permission of Elsevier.

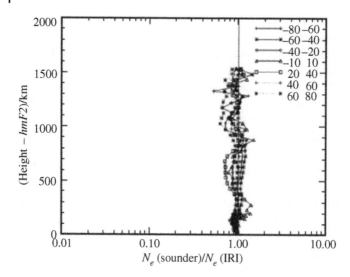

Figure 8.11 Average ratios of measured electron densities to those predicted by the improved IRI topside model for various modip ranges. Source: Bilitza [231]. Reproduced with Permission of Elsevier.

It was found that simply multiplying the predicted IRI topside profiles by the appropriate electron density ratio produced remarkably more accurate results. It can be seen in Figure 8.11 that this corrected IRI topside model produces values that conform much more closely to measured data.

At the beginning of this section, it was mentioned that the current version of the IRI offers two distinct modeling options. We have just finished discussing the first of these two. The second option is the same formulation used in the NeQuick 2 model. This formulation is even more accurate than the corrected topside model discussed in this section and as such is taken as the default option in the current iteration of the IRI. Since this formulation was already discussed in detail in Section 8.2.2, and may also be found in [196], it will not be repeated here. If desired, the user may also choose to use the model described in this section, either in its original form or its newer, corrected form. The two versions are included as entirely separate options which may be selected by changing values of the JF boolean input array (see Section 8.4.1), namely, JF(29) and JF(30).

F2 Region (h_mF2 to h_mF1)

The electron density of the F2 layer is modeled differently than that of the topside. In this region, the electron density is normalized to the F2 peak density N_mF2.

This normalized density is modeled by the equation:

$$\frac{N_2(h)}{N_mF2} = \frac{\exp\left(-X^{B_1}\right)}{\cosh(X)},$$ (8.122)

where

$$X = \frac{h_mF2 - h}{B_0}.$$ (8.123)

B_0 is known as the thickness parameter, and B_1 is called the shape parameter. These parameters, which vary with time and location, must also be determined. To accomplish this, several models are included in the IRI source code, which the user may choose among using JF(4) and JF(31). By default, the most recent version of the IRI uses a formulation developed by Altadill et al. in a 2009 paper [232], though models from older versions of the IRI are retained as alternative options. These legacy models, namely, those described by Bilitza et al. [218] and by Gulyaeva in [233], are less accurate and it is recommended that users utilize the default option instead. For this reason, they will not be discussed in detail here.

The current formulation by Altadill et al. will, however, be treated in some detail in what follows. This model is based on empirical data obtained from 27 globally distributed ionosondes. These stations offer coverage in latitude from 78°N to 52°S and coverage in longitude from 121°W to 167°E. Hourly sampled profiles from 1998 to 2006 – i.e. nearly a full solar cycle – were considered to ensure a wide range of solar activity was represented.

Altadill et al. used a MARP to encapsulate the conditions for a given station during a given month. This model is only concerned with quiet ionospheric conditions. Thus, "extreme" profiles, defined as the 25% of profiles in a given month which have the greatest deviation from the monthly average, are excluded when calculating the MARP. In this way, profiles that are most likely related to disturbed ionospheric conditions are removed.

In order to obtain a globally applicable model, it is necessary to extrapolate the available data beyond the measurement stations' locations. Altadill et al. chose to accomplish this through the use of spherical harmonics. It is recommended that the reader consult the spherical harmonics portion of Section 8.4.2 for some background. Additionally, another model developed by Altadill et al. for predicting h_mF2 is also included in the current iteration of the IRI, and was briefly discussed in Section 8.4.3. There are a number of similarities between these two models' approaches, so the reader may find it helpful to consult that section before continuing.

In essence, spherical harmonics may be used to produce a global model of a parameter's behavior using empirical data. For this technique to be effective, the empirical data must have good coverage in both latitude and longitude. Unfortunately, one of the most reliable and widely available sources of electron density measurements – ionosondes – cannot be used to obtain electron density data over the ocean. This unfortunately leaves large gaps in longitudinal coverage.

To solve this problem, Altadill et al. reasoned that under quiet ionospheric conditions, the ionosphere's behavior would simply vary diurnally, and thus local time differences could be mapped to longitude differences. To implement this approximation, the authors selected a number of stations which provided good latitudinal coverage. They then drew a parallel through each of their chosen stations and created fictitious points spaced 15° apart in longitude, for a total of 24 points per parallel. Since the stations they chose were distributed nicely in latitude, this approach provided the grid of points needed for the application of spherical harmonics.

It should be noted that when drawing parallels through stations, the choice of geographic vs. magnetic coordinates is an important consideration. Certain physical processes in the ionosphere, such as photoionization and heating, are mainly determined by geographic location. Conversely, transport effects such as electromagnetic drift and neutral winds are heavily influenced by Earth's magnetic field and are instead better described in magnetic coordinates. Altadill et al. tested various coordinate systems when developing their models for B_0 and B_1, and found that B_0 is best modeled using parallels in magnetic dip coordinates, while geographic coordinates were better suited to draw the parallels for B_1.

The diurnal variations of B_0 and B_1 were analyzed in various latitude ranges, seasons and levels of solar activity. Altadill et al. observed various trends in this data, which informed the number of terms they chose to use in their expansion. The maximum degree of expansion – N in Equation 8.94 – was chosen to be 6 for both B_0 and B_1, while the maximum order of the expansion – M in Equation 8.94 – in longitude (or equivalently, in local time) was set to 4 for B_0 but only to 2 for B_1. The time dependence of the spherical harmonic coefficients was captured by a two-degree Fourier expansion, which can be expressed as follows:

$$\begin{bmatrix} g_n^m(t) \\ h_n^m(t) \end{bmatrix} = \sum_{q=0}^{2} \begin{bmatrix} ga_{q,n}^m \\ ha_{q,n}^m \end{bmatrix} \cos\left(\frac{2\pi qt}{12}\right) + \begin{bmatrix} gb_{q,n}^m \\ hb_{q,n}^m \end{bmatrix} \sin\left(\frac{2\pi qt}{12}\right). \tag{8.124}$$

The details of these choices and the rationale behind them are elaborated upon in Sections 3.1 and 3.2 of their paper [232].

A separate set of expansion coefficients was obtained for each year, resulting in nine sets total. It was found that the expansion coefficients could be well modeled as a linear function of solar activity, so analytical functions were developed to allow the coefficients to be calculated for arbitrary solar activity levels. The yearly average Sunspot number R_{12} was used as a proxy for solar activity. In total, the models for B_0 and B_1 are given by 430 and 230 coefficients, respectively. Figures 8.12 and 8.13 show contour plots of B_0 and B_1 as functions of latitude and longitude for various seasons and levels of solar activity.

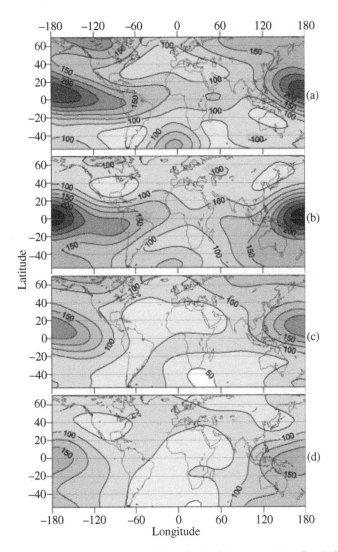

Figure 8.12 Contour plots of B_0 at 0 UT. High solar activity: R = 119.6. Low solar activity: R = 15.2. (a) July, high solar activity. (b) January, high solar activity. (c) July, low solar activity. (d) January, low solar activity. Source: Altadill et al. [232]. Reproduced with Permission of Elsevier.

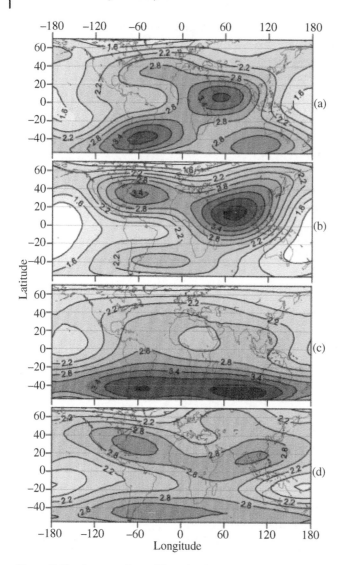

Figure 8.13 Contour plots of B_1 at 0 UT. High solar activity: R = 119.6. Low solar activity: R = 15.2. (a) July, high solar activity. (b) January, high solar activity. (c) July, low solar activity. (d) January, low solar activity. Source: Altadill et al. [232]. Reproduced with Permission of Elsevier.

F1 Region (h_mF1 to h_z)

The electron density model for the F1 region was introduced in IRI-2001 and remains the default option in the current iteration of the IRI. In this model, developed by Reinisch and Huang [249], the electron density function that describes the F1 layer is dependent on the F2 formulation, which was discussed previously. In particular, Equation 8.122 is modified to obtain the description of the F1 density. This is accomplished using the expression:

$$N_3(h) = N_2(h^*). \tag{8.125}$$

Here, N_3 is the electron density of the F1 layer, N_2 is the electron density of the F2 layer (as given by Equation 8.122), and h^* is given by the equation:

$$h^* = h_mF1 \left[1 - \left(\frac{h_mF1 - h}{h_mF1} \right)^{1+D_1} \right], \tag{8.126}$$

where D_1 is a thickness parameter which depends on the time of day, and is determined as follows:

$$D_1 = \begin{cases} 2.5C_1, & t_{SR} = t_{SS}, \\ 2.5C_1 \cos\left(\frac{(t-12)\pi}{t_{SS} - t_{SR}} \right), & t_{SR} \neq t_{SS}, \\ 0, & D_1 < 0. \end{cases}$$

In these expressions, t is the local time, while t_{SR} and t_{SS} are the local time of sunrise and sunset, respectively, as seen from the altitude of the F1 layer. All of these quantities are measured in hours. The parameter C_1 is defined in terms of an Epstein step function that depends on modified dip latitude (modip).

$$C_1 = 0.09 + \frac{0.11}{1 + \exp\left(-\frac{|\mu| - 30}{10} \right)}. \tag{8.127}$$

The absolute value of modip, denoted $|\mu|$, is never given a value smaller than 18. For calculation at lower latitudes, a value of 18 is still used.

The thickness of the F1 layer is determined by D_1, as seen in Figure 8.14. The time dependence of D_1 on a positive half-cosine curve reflects the diurnal variation of the electron density at F1 altitudes. The curve is centered at local noon, when the electron density peaks, and is nonzero for a time equal to the number of hours of sunlight ($t_{SS} - t_{SR}$). In other words, during the day $D_1 \neq 0$, with a peak at local noon, and at night $D_1 = 0$. Note that when $D_1 = 0$,

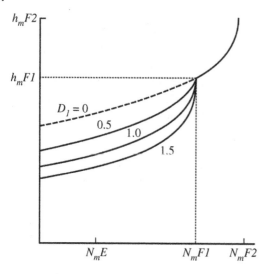

Figure 8.14 F1 layer electron density profiles representing different values of D_1. Source: Adapted from Figure 3a of [249]. Reproduced with Permission of Elsevier.

Equation 8.126 reduces to h and thus $N_3 = N_2$. In other words, at night, the F1 layer disappears and the F2 layer formulation is used to describe the electron density down to h_Z. This reflects our observations of the ionosphere wherein the F1 and F2 layers, distinct during the day, merge into a single F layer at night, becoming indistinguishable from one another.

Unlike other layers of the ionosphere, the F1 layer will sometimes fail to appear, even during the day. The IRI includes a model for a parameter measuring the probability that the F1 layer will occur. This model was developed by Scotto et al. and has been included as an option in all versions of the IRI since IRI-2001 [250]. Measured in the percentage of days per month that the F1 layer typically appears, the probability of F1 occurrence is given by the equation

$$P_{F1} = \left[0.5 + 0.5\cos(\chi)\right]^\gamma, \tag{8.128}$$

where

$$\gamma = a + b\phi_m + c\phi_m^2, \tag{8.129}$$

ϕ_m is geomagnetic latitude and

$$a = 2.9798 + 0.0853993R_{12}, \tag{8.130}$$

$$b = 0.01069 - 0.0021967R_{12}, \tag{8.131}$$

$$c = -0.000256409 + 0.0000146678R_{12}. \tag{8.132}$$

Recall that R_{12} is the 12-month running mean of the sunspot number.

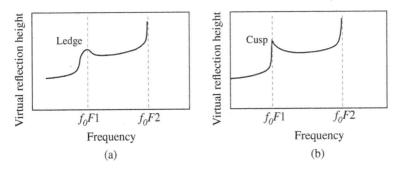

Figure 8.15 Sketch of ionograms displaying the L condition (a) and a normal F1 layer (b).

In addition to the probability of a normal F1 layer, this model includes a method for predicting the probability of a phenomenon known as the L condition. The L condition refers to cases in which the F1 layer occurs, but its ionogram displays a rounded "ledge" rather than a sharper, more pointed "cusp." Figure 8.15 displays a sketch the difference between these shapes.

Scotto et al. found that the probability of the L condition's occurrence is nearly independent of geomagnetic latitude and sunspot number. They found that this probability was well-modeled by assigning γ a constant value of 2.36. In other words, the expression describing the L condition probability is:

$$P_L = \left[0.5 + 0.5\cos(\chi)\right]^{2.36}. \tag{8.133}$$

Intermediate Region (h_z to h_{VT})

In the same paper by Reinisch and Huang that described the F1 layer electron density, a model for the intermediate region was also proposed. This model is quite similar to that of the F1 region, and is described as follows:

$$N_4(h) = N_2(h^{**}), \tag{8.134}$$

where N_4 is the electron density in the intermediate region, N_2 is the electron density in the $F2$ region, and h^{**} is given by:

$$h^{**} = h_m F1 \left[1 - \left(\frac{h_m F1 - \eta}{h_m F1}\right)^{1+D_1}\right]. \tag{8.135}$$

The parameter η assumes different values based on the relative values of h_{ST} and h_{VT}:

$$\eta = \begin{cases} h, & h_{ST} = h_{VT}, \\ h_z + \frac{T}{2} - \left[T\left(\frac{T}{4} + h_z - h\right)\right]^{\frac{1}{2}}, & h_{ST} > h_{VT}, \\ h_z + \frac{T}{2} + \left[T\left(\frac{T}{4} + h_z - h\right)\right]^{\frac{1}{2}}, & h_{ST} < h_{VT}. \end{cases}$$

T is defined as:

$$T = \frac{(h_Z - h_{ST})^2}{h_{ST} - h_{VT}}, \quad h_{ST} \neq h_{VT}. \tag{8.136}$$

All other parameters are defined the same way as in Reinisch and Huang's F1 layer model, which was discussed previously.

E Valley (h_{VT} to $h_m E$)

The E valley has a shape very unlike the layers above and therefore requires a substantially different mathematical formulation. In the IRI, the electron density in this region is given by a fifth order power series:

$$\frac{N_5(h)}{N_m E} = 1 + E_1 x^2 + E_2 x^3 + E_3 x^4 + E_4 x^5, \tag{8.137}$$

where

$$x = h - h_m E. \tag{8.138}$$

As of IRI-95, $h_m E$ is set at a constant 110 km. The coefficients E_i of this power series are given in terms of a number of characteristic parameters of the E valley region. The formulas[8] for these coefficients are as follows:

$$E_1 = Z_1 - h_{ABR}(E_2 + h_{ABR}(E_3 + h_{ABR}E_4)), \tag{8.139}$$

$$E_2 = -\frac{2Z_1}{h_{ABR}} - 2h_{ABR}E_3 - 3h_{ABR}^2 E_4, \tag{8.140}$$

$$E_3 = \frac{Z_1(2h_{BR} - 3h_{ABR})}{(h_{ABR}(h_{ABR} - h_{BR})^2)} - (2h_{ABR} + h_{BR})E_4, \tag{8.141}$$

$$E_4 = \frac{2(Z_1(h_{BR} - 2h_{ABR})h_{BR} + \left(\frac{DLN}{2h_{BR}}\right)(h_{ABR} - h_{BR})h_{ABR})}{(h_{ABR}h_{BR}(h_{ABR} - h_{BR})^3)}. \tag{8.142}$$

Recall that h_{BR} denotes the total width of the E valley ($h_{VT} - h_m E$) and h_{ABR} denotes the distance between the E layer peak height and the height of the E valley base ($h_{VB} - h_m E$). Figure 8.8 depicts these quantities. The parameter Z_1 is dependent on the depth of the valley through:

$$Z_1 = \begin{cases} -\dfrac{DP}{100 h_{ABR}^2}, & \text{Daytime,} \\[3mm] \dfrac{\ln(1 - DP)}{h_{ABR}^2}, & \text{Nighttime,} \end{cases}$$

---

8 Note that these equations are similar, but not identical, to those given in [233]. The equations presented here are those given in the TAL subroutine of IRI-2016, which may be found in the file "irifun.for" available on irimodel.org.

Table 8.5 Values of E valley characteristic parameters.

Local time	Modip (°)	H_{ABR} (km)	H_{BR} (km)	DP (%)	DLN (km^{-1})
Midnight	18	28	45	81	0.06
	45	28	67	81	0.06
Noon	18	0	0	0	0
	45	10.5	17.8	Winter: 10	Summer: 0.01
				Else: 5	Else: 0.016

where DP denotes the depth of the E valley as a percentage, given by the equation:

$$DP = \frac{100(N_m E - N_{VB})}{N_m E}. \tag{8.143}$$

In other words, if we assume an E layer peak density of $N_m E = 1 \times 10^{11}$ electrons/m^3, a DP of 10% would correspond to a valley base density of $N_{VB} = 9 \times 10^{10}$. The parameter DLN in Equation 8.142 is the logarithmic derivative of the electron density, evaluated at the top of the E valley, and can be found using the formula:

$$DLN = \left. \frac{d \ln N}{dh} \right|_{h=h_{VT}}. \tag{8.144}$$

The E valley characteristic parameters h_{ABR}, h_{BR}, DP, and DLN, were determined via ISR measurements [251]. Table 8.5 shows their values at selected times of day and modips.

These values may be interpolated in both time and modip to obtain intermediate values. At low latitudes, the E valley is occasionally observed during the day, but not often enough to be included in a MARP. Thus, all these parameters are given values of zero and it is assumed that the E valley does not occur during the day at low latitudes. Note that when the E valley is given a depth of zero, the F1 layer function is extended down to $h_m E$.

E Bottomside and D Region ($h_m E$ to h_A)

At heights below $h_m E$, it becomes infeasible to measure electron density using remote sensing techniques such as ionosondes or ISRs. This is due to the relatively low electron density at these heights, in conjunction with the high density of neutral particles which attenuate the incident signal. Together, these effects weaken the received signal and cause the electron density profile to be difficult, if not impossible, to read at these heights.

Table 8.6 Values of D layer characteristic parameters.

Time of day	Geographic latitude	H_A (km)	h_mD (km)	h_{DX} (km)	F_1 (km)	F_2 (km^2)	F_3^a (km^3)	F_3^b (km^3)
Day	Low	65	80	85.6	0.02	-2×10^{-4}	9.37×10^{-3}	4.89×10^{-4}
	High	65	80	85.6	0.05	-1.25×10^{-3}	8.18×10^{-3}	1.707×10^{-4}
Night	All	80	88	92.5	0.05	-1.25×10^{-3}	8.79×10^{-3}	1.22×10^{-2}

Instead of remote sensing data, in situ measurements from rocket flights in the lower ionosphere is used as the empirical data set for modeling. The default model in the IRI is one developed by Mechtley and Bilitza, and is based on such rocket measurements [238, 252]. In this model, which begins at the height h_A, the electron density in the D layer is given by the equation:

$$\frac{N_{6,bot}(h)}{N_mD} = \exp\left(F_1 x + F_2 x^2 + F_3 x^3\right),$$ (8.145)

where

$$x = h - h_mD.$$

The coefficients F_i of the third order polynomial are assigned different values depending on the time of day and geographic latitude. F_3 is given two sets of values F_3^a and F_3^b, which correspond to the value of F_3 above and below h_mD, respectively. This is necessary because there is a dramatic change in scale height (see Equation 5.16) at h_mD. The values of F_i may be seen in Table 8.6.

Just above h_mD, the electron density gradient increases sharply, causing the electron density profile to climb much more quickly. This steep transition region, which connects the D and E layers, extends from the height h_{DX} up to the E peak height h_mE. At heights within the transition region, electron density is modeled differently than in the lower D layer. This equation describing its behavior in this region is as follows:

$$\frac{N_{6,top}(h)}{N_mE} = \exp\left(-D_1(h_mE - h)^K\right).$$ (8.146)

The parameters D_1 and K are defined such that $N(h)$ and its first derivative are continuous at h_{DX}. That is to say, $N_{6,bot}(h_{DX}) = N_{6,top}(h_{DX})$ and $N'_{6,bot}(h_{DX}) = N'_{6,top}(h_{DX})$. These constraints give:

$$D_1 = \frac{N'(h_{DX})}{K(h_m E - h_{DX})^{K-1} N(h_{DX})}, \tag{8.147}$$

$$K = \frac{-(h_m E - h_{DX}) N'(h_{DX})}{\ln\left(\dfrac{N(h_{DX})}{N_m E}\right) N(h_{DX})}. \tag{8.148}$$

Here, $N(h_{DX})$ denotes the electron density at height h_{DX} and $N'(h_{DX})$ denotes its derivative at the same height.

In addition to this default model, as of IRI-2001 an alternate modeling option may be selected by changing the value of JF(24), an element of the JF boolean input array (see Section 8.4.1). The alternate option utilizes two separate D layer models: one developed by Friedrich and Torkar and another by Danilov and Smirnova. These models will not be discussed in detail here, but may be read about in [221, 228], respectively.

8.4.5 Electron Temperature

IRI-2016 provides two modeling options for electron temperature, selectable via JF(23). These two models – namely, the model by Bilitza et al. proposed in 1985 [225, 226] and a newer one introduced by Truhlik et al. in 2012 [227] – have similar formulations. Both models accurately predict the variation of electron temperature in latitude, longitude, altitude, time of day, and season. However, only the model by Truhlik et al. sufficiently describes variation with solar activity, and because of this key advantage it has been chosen as the default option in the IRI. The older model by Bilitza et al. will not be discussed further, but its formulation is detailed in Section 3.3.1 of [219].

As their empirical data base, Truhlik and coworkers used a large number of in situ measurements from numerous satellites.[9] These measurements span nearly half a century (i.e. more than four full solar cycles), latitudes from +80° to −90°, and an altitude range from approximately 100 km to over 4000 km. In all cases, measurements were taken at least every 12 hours, though many of the satellites took measurements 24 hours per day. In total, over 9 million data points across 20 satellites were considered in the formulation of the model [227].

Several of the mathematical tools used in modeling electron temperature are similar to the ones described in the preceding sections of this chapter. For example, to obtain describe global variation, a spherical harmonic expansion is used, much

---

9 Table 1 of [253] shows detailed information on these data sources.

like in the models discussed for the various F2 layer parameters (i.e. f_oF2, h_mF2, B_0, B_1). Additionally, in order to construct an altitude profile, several key heights were chosen as anchor points for Epstein functions, in a fashion similar to the IRI electron density model.

At its most basic, electron temperature is modeled by summing two terms: $T_{e,0}$, the core electron density model, and $T_{e,PF_{10.7}}$, the solar activity term. In other words, the total electron temperature T_e is given by:

$$T_e = T_{e,0} + T_{e,PF_{10.7}}. \tag{8.149}$$

The apparent simplicity of this formulation belies the complexity involved in modeling $T_{e,0}$ and $T_{e,PF_{10.7}}$. Each will be discussed separately in what follows.

The Core Model ($T_{e,0}$)

The electron density core model is further divided into submodels describing various seasons and altitude ranges. Satellite measurements were grouped into one of three seasons – winter, summer, and equinox – each one a 90 day period centered on its corresponding solstice or equinox. For spring and autumn, all data from both hemispheres for all six equinoctial months was combined. For summer and winter, data from both hemispheres was also combined, grouping northern summer with southern summer and vice versa. For example, northern hemisphere data from the three months surrounding the June solstice would be paired with southern hemisphere data from the three months surrounding the December solstice.

Based on both the altitude distribution of the available satellite data, as well as the need for good coverage throughout the ionosphere, Truhlik et al. chose five heights as anchor points for their electron temperature profile. The following bins were established, centered on each of the anchor heights: $300 \pm 40\,\text{km}$, $550 \pm 50\,\text{km}$, $850 \pm 90\,\text{km}$, $1400 \pm 150\,\text{km}$, and $2000 \pm 300\,\text{km}$.

Global variation of electron temperature is modeled via a spherical harmonic expansion. The coordinates of the expansion are the magnetic local time (MLT) and latitude. If the latitude coordinate is chosen carefully to reflect the shape of the Earth's magnetic field, longitudinal variation can be reduced to a second order effect. Truhlik et al. accomplish this by using a parameter called invariant dip latitude as the latitude coordinate. Invariant dip latitude is a coordinate which is defined such that it smoothly transitions between two other commonly used latitude systems. In particular, invariant dip latitude (invdip) has values similar to the dip latitude Ψ at low latitudes and values similar to the invariant latitude Λ at high latitudes (these two quantities will be described shortly). The equation describing invdip is as follows:

$$\vartheta = \frac{\alpha\Lambda + \beta\Psi}{\alpha + \beta}, \tag{8.150}$$

where $\alpha = \sin^3|\Psi|$ and $\beta = \cos^3\Lambda$.

Invariant latitude is the latitude at which a given magnetic field line intersects the surface of the Earth. Recalling Equation 5.5, we see that a given L-shell reaches Earth's surface (i.e. $r = 1$ Earth radius) at a latitude given by:

$$1 = L\cos^2\phi_m. \tag{8.151}$$

This special latitude is commonly denoted Λ, to distinguish it from any arbitrary magnetic latitude ϕ_m. Therefore, after rearranging, we see that invariant latitude can be written as:

$$\Lambda = \arccos\left(\sqrt{\frac{1}{L}}\right). \tag{8.152}$$

Dip latitude is defined in terms of the magnetic dip angle I:

$$\Psi = \arctan\left(\frac{\tan I}{2}\right). \tag{8.153}$$

With invdip and MLT as coordinates, Truhlik et al. created a unique grid for each of the height and season groups mentioned previously. Each of these grids was defined by a minimum of 9 by 18 bins, and the electron temperature data within each bin was averaged to obtain a representative value. The base-10 logarithm of these values was then used as the basis for a spherical harmonic expansion of order 8, as can be seen in the following equation:

$$\log_{10}T_{e,0} = \sum_{l=1}^{8}\left(a_l^0 P_l^0(\cos\theta) + \sum_{m=1}^{l}\left[a_l^m\cos(m\lambda) + b_l^m\sin(m\lambda)\right]P_l^m(\cos\theta)\right). \tag{8.154}$$

Here, P_l^m is the set of associated Legendre polynomials, θ is the invariant dip colatitude (the complement of invdip), and λ is the MLT (expressed in radians as a proxy for longitude). θ ranges from 0 to π and λ ranges from 0 to 2π. The coefficients a_l^m and b_l^m are the spherical harmonic coefficients which uniquely define the expansion. Each height range and season are given their own set of expansion coefficients.

The Solar Activity Term (T_e,$PF_{10.7}$)

The true innovation of the model introduced by Truhlik et al. was its ability to accurately predict the dependence of electron temperature on solar activity. For decades prior, such a description had eluded researchers. As opposed to many other ionospheric parameters, which typically increase along with solar activity, electron temperature does not follow such a straightforward and predictable pattern, making it difficult to isolate the dependence on solar activity. The primary electron heat transfer processes in the ionosphere were well known, namely: heating via expulsion of electrons during photoionization, cooling via collisions with ions and neutral molecules, and heat conduction along magnetic field lines. However, each of these processes is dependent on myriad parameters, including not only solar irradiance, but also neutral particle temperature, and the density of electrons, ions, and neutrals. Further, these processes offset one another, meaning that their net effect can cause electron temperature to increase, decrease, or stay the same depending on the location on Earth, time of day, and season.

Truhlik et al. found that a parameter which they called $PF_{10.7}$ had good correlation with electron temperature. This parameter was defined as:

$$PF_{10.7} = \frac{1}{2}(F_{10.7,d} + F_{10.7,81}),\tag{8.155}$$

where $F_{10.7,d}$ is the daily average of the 10.7 cm solar radio flux, and $F_{10.7,81}$ is the average of the same parameter over an 81 day period. Three solar activity levels were defined (all in SFU): $PF_{10.7} < 100$ (low), $110 \le PF_{10.7} \le 180$ (medium), and $PF_{10.7} > 180$ (high). The difference ΔT_e of a given bin's value and the low activity bin's value was normalized with respect to the low activity bin. Thus, the normalized value of the low activity bin is always zero, while the middle and high activity bins will in general have nonzero normalized values ($\Delta T_{e,m}$ and $\Delta T_{e,h}$, respectively).

Values of $\Delta T_{e,m}$ and $\Delta T_{e,h}$ were then plotted against invariant dip latitude for each of the heights defined for the core model (five total). A similar set of plots was produced for daytime (defined as 13:00 MLT $\pm$ 3 hours) and nighttime (defined as 1:00 MLT $\pm$ 2.5 hours) during both equinox and solstice. This resulted in a total of 20 separate sets of axes, each of which contains both $\Delta T_{e,m}$ and $\Delta T_{e,h}$. These plots may be seen in Figures 8.16 and 8.17, wherein the zero level is defined as the median value of the low solar activity bin. The rows of numbers in these plots correspond to the median solar activity values in each bin, with the top row representing high activity, the middle row medium activity, and the bottom row low activity.

In order to obtain the dependence of electron temperature on solar activity for any value of $PF_{10.7}$, quadratic interpolation was performed for the interval between the low activity bin median $PF_{10.7,l}$ and the high activity bin median $PF_{10.7,h}$. Outside of this interval, linear extrapolation was used. Mathematically, these operations are described as follows:

$$T_{e,PF_{10.7}} = \begin{cases} \dfrac{\Delta T_{e,m}(PF_{10.7} - PF_{10.7,l})(PF_{10.7} - PF_{10.7,h})}{(PF_{10.7,m} - PF_{10.7,l})(PF_{10.7,m-PF_{10.7,h}})} \\ \quad + \dfrac{\Delta T_{e,h}(PF_{10.7} - PF_{10.7,l})(PF_{10.7} - PF_{10.7,m})}{(PF_{10.7,h} - PF_{10.7,l})(PF_{10.7,h} - PF_{10.7,m})}, & PF_{10.7,l} \le PF_{10.7} \le PF_{10.7,h}, \\[2ex] \dfrac{\Delta T_{e,m}(PF_{10.7} - PF_{10.7,l})}{(PF_{10.7,m} - PF_{10.7,l})}, & PF_{10.7} < PF_{10.7,l}, \\[2ex] \dfrac{(\Delta T_{e,h} - \Delta T_{e,m})(PF_{10.7} - PF_{10.7,h})}{(PF_{10.7,h} - PF_{10.7,m})} + \Delta T_{e,h}, & PF_{10.7} > PF_{10.7,h}. \end{cases}$$

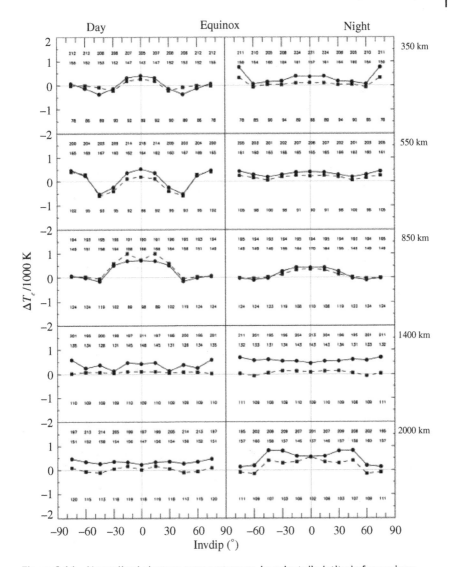

Figure 8.16 Normalized electron temperature vs. invariant dip latitude for equinox. Solid line = $\Delta T_{e,h}$, dashed line = $\Delta T_{e,m}$. Source: Truhlik et al. [227]. Reproduced with permission of Springer under the Creative Commons 4.0 license.

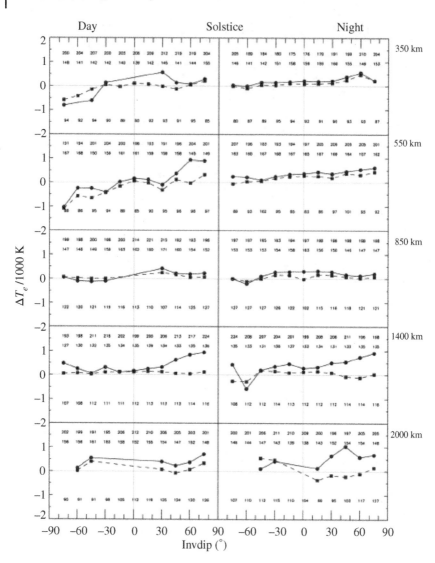

Figure 8.17 Normalized electron temperature vs. invariant dip latitude for solstice. Solid line = $\Delta T_{e,h}$, dashed line = $\Delta T_{e,m}$. Source: Truhlik et al. [227]. Reproduced with permission of Springer under the Creative Commons 4.0 license.

A continuous profile in height is obtained in the same manner as described in Section 8.4.2, wherein a skeleton profile is generated using regions of constant gradient, and Epstein step functions are used to generate smooth transitions between each region. Finally, a continuous diurnal variation is achieved via cosine function as follows:

$$
T_{e,PF10.7} = \frac{1}{2}(T_{e,PF10.7,day} - T_{e,PF10.7,night}) \left[1 - \cos\frac{2\pi}{24}(t_m - 1)\right] + T_{e,PF10.7,night},
$$
(8.156)

where t_m is local magnetic time in hours. Note that this function has a maximum value of $T_{e,PF10.7,day}$ at 13:00 MLT and a minimum value of $T_{e,PF10.7,night}$ at 1:00 MLT, just as defined earlier.

8.4.6 Ion Temperature

The IRI ion temperature model has remained unchanged for some time. The current formulation was introduced in IRI-1986 and described in [219]. This formulation uses a Booker function do describe variation in altitude, much like the IRI electron density model. The reader may consult Section 8.4.2 for a brief primer on this topic. In particular, the ion temperature function is given by:

$$
T_i = T_{i,0} + (h - h_0)DT_1 + \sum_{j=1}^{3}(DT_{j+1} - DT_j)B_j \ln\left(\frac{1 + \exp\left(\dfrac{h - h_j}{B_j}\right)}{1 + \exp\left(\dfrac{h_0 - h_j}{B_j}\right)}\right),
$$
(8.157)

which describes a skeleton profile with sections of constant gradient DT_j that are smoothly connected by Epstein step functions (encapsulated by the natural logarithm term). The boundaries and transition thicknesses of these sections are given by the heights h_j and the thickness parameters B_j, respectively. The gradients DT_j are given by the expression

$$
DT_j = \frac{T_i(h_j) - T_i(h_{j-1})}{h_j - h_{j-1}},
$$
(8.158)

where $T_i(h_j)$ is the ion temperature at height h_j. The parameters h_j, $T_i(h_j)$, and B_j can therefore be used to fully describe the ion temperature profile. These parameters are defined differently for each section of the profile, so we will now discuss each of these sections in turn. Note that while there are only four sections, we actually need to specify five values for j (0–4) because certain terms have index $j + 1$ or $j - 1$. These five values correspond to the boundaries of the four sections.

The ion temperature profile begins at the height h_0. Below this point, the ion temperature is assumed to be equal to the neutral temperature T_n. Between h_0 and h_1, T_i varies linearly from T_n up to the value $T_{i,m}$. Using AEROS satellite data, Bilitza found that $T_{i,m}$ varies with magnetic latitude in the following way [254]:

$$
T_{i,m} = \begin{cases} 1240 - 1400 \; \dfrac{\exp\left(\dfrac{\phi_m}{11.11}\right)}{\left(1 + \exp\left(\dfrac{\phi_m}{11.11}\right)\right)^2}, & \text{day,} \\[4mm] 1200 - 300 \cos^{1/2} Z, & \text{night,} \end{cases}
$$

where ϕ_m is magnetic latitude and Z is given by:

$$
Z = 0.47 |\phi_m| + 0.024 \phi_m^2. \tag{8.159}
$$

The height h_0 is defined as the height at which the neutral temperature profile intersects the point $h_1/T_{i,m}$ and h_1 is defined to be a constant 430 km. The skeleton profile contains discontinuities at the heights h_1, h_2, and h_3, where the sections of constant gradient meet one another. At each of these "joints" in the skeleton, smoothing is performed via an Epstein step function, characterized by some transition thickness. At h_1, this transition thickness is described using the thickness parameter B_1, which is given a value of 10.

Above h_1, T_i is modeled to be independent of latitude, simply taking on a constant value. This value is dependent on the time of day, and is given simply as:

$$
DT_2 = \begin{cases} 3, & \text{day,} \\ 0, & \text{night.} \end{cases}
$$

The ion temperature increases in this fashion up to the height h_2, which is defined as the point at which the ion and electron temperature profiles meet each other. At this height, the transition thickness is given by $B_2 = 10$.

It is assumed that the electron and ion temperatures are equal above h_2. This upper region of the ion temperature profile is split into two sections, which may take on different gradient values. The lower of these two sections begins at h_2 and extends up to $h_3 = 1400$ km, and from there the upper section extends up to $h_4 = 3000$ km. The thickness parameter at h_3 is given by $B_3 = 20$. In the event that $h_2 > 1400$ km, h_3 is instead defined to be 3000 km, and h_4 is omitted.

A summary of the relevant parameters from this section is presented in Table 8.7

Table 8.7 Parameters used to construct the ion temperature profile in the IRI.

j	h_j	$T_i(h_j)$	B_j
0	$h\left(T_n = \dfrac{h_1}{T_{i,m}}\right)$	$T_n(h_0)$	—
1	430 km	$T_{i,m}$	10
2	$h(T_i = T_e)$	$T_e(h_2)$	10
3	1400 km	$T_e(h_3)$	20
4	3000 km	$T_e(h_4)$	—

8.4.7 Ion Composition

We have seen in Chapter 5 that the number of electrons and the number of positive ions present in the ionosphere at any given time are roughly equal. Since any given electron is indistinguishable from another, a single parameter – namely, electron density – is sufficient to model phenomena which depend on them. Conversely, a number of different positive ion species exist in the ionosphere, and the differences in their physical and chemical properties are relevant to certain ionospheric phenomena. Thus it is useful to quantify not only ion density but also *ion composition* in the ionosphere, which describes the relative numbers of different ion species as percentages of the total ionospheric ion content.

Historically, ion composition was difficult to model because relatively little reliable data exists to use as a basis. Ion composition cannot be measured via ionosondes; instead, measurements are typically made with an ISR or in situ during a rocket flight, using an instrument such as a retarding potential analyzer (RPA) or ion mass spectrometer (IMS). Inherently, these methods have poor coverage in space, time or both. Only a handful of ISRs exist in the world, meaning global coverage must be supplemented by rocket missions, which are expensive and infrequent.

This dearth of available data is worsened by the fact that many of the measurements are unreliable, incomplete, or simply incorrect. RPAs and IMSes frequently fail to measure all constituent ion species during an experiment, making it impossible to determine ion composition from that data. Additionally, calibration issues can contaminate datasets, making them unsuitable for modeling purposes.

To solve this problem, Richards et al. proposed a new model, dubbed the ion density calculator (IDC), which used theoretical considerations to improve upon the accuracy afforded by empirical measurements alone [223]. In particular, they utilized chemistry principles which underlie the Field Line Interhemispheric Plasma (FLIP) model, a well validated and comprehensive model that solves the continuity, momentum, and thermal equations for ions and electrons in the ionosphere.[10] The details of these considerations are beyond the scope of this text, but the ultimate goal of these calculations is to determine the equilibrium values of the major ion species present at different heights in the ionosphere.

The IDC model solves for the equilibrium values of O^+, N_2^+, O_2^+, NO^+, and N^+. For most of these species, it is straightforward to calculate an equilibrium value via a series of chemical reactions. Unfortunately, above 180 km the ground state of O^+ cannot reach equilibrium since diffusion effects become important at this height. Without this information, the number of O^+ ions cannot be determined and the full specification of ion composition remains incomplete. However, Richards et al. realized that since the total ion number density is equal to the electron number density, it is possible to solve for the O^+ number density as follows:

$$N_{O^+} = N_e - N_{N_2^+} - N_{O_2^+} - N_{NO^+} - N_{N^+} , \tag{8.160}$$

where N_j is the number density of chemical species j. Richards et al. validated the results of the IDC model against those of the FLIP model, as well as data from the AE-C satellite mission (which had been thoroughly verified for correctness). The IDC results, as well as detailed discussion of their validation, may be found in [223].

While the IDC model is quite successful at calculating ion composition in the bottomside ionosphere, it is not useful for modeling the topside. As altitude increases, the heavier molecular ion species begin to dwindle in number, allowing atomic species to comprise a greater fraction of the total ion composition. In the topside ionosphere, the most prevalent ions are O^+, N^+, H^+, and He^+ – quite a different landscape than the one analyzed by the IDC model.

To calculate ion composition at topside heights, the IRI includes another model, which is separate and independent from the IDC. The topside model included in the 2016 revision of the IRI was developed by Truhlik et al. [224], the authors who also contributed the IRI's current description of electron temperature. The approach taken by both of these models is similar in many ways, so it is recommended that the reader familiarize themselves with the concepts in the "Core Model" portion of Section 8.4.5 before reading the remainder of this section.

---

10 Although FLIP could be used to calculate ion composition, it is a large and computationally costly model that is not suitable for many applications. Richards et al. sought to provide a model that was more limited in scope, but also more efficient.

Additionally, if the reader is not yet familiar with spherical harmonics, it may be helpful to read Section 8.4.2.

The data that forms the basis of the topside ion composition model was collected via various satellites, including the AE-C, AE-E, IK-24, and C/NOFS missions. To account for seasonal variation, Truhlik et al. grouped this data into three seasons: summer, winter, and equinox. Further, each season's data was divided into height bins. Based on the available data, it was determined that two sets of height bins were required: one for the AE data and one of the IK-24 data,[11] which were chosen as follows:

- IK-24: 550 ± 80 km, 900 ± 100 km, 1500 ± 150 km, and 2250 ± 250 km.
- AE: 400 ± 50 km, 550 ± 70 km, 750 ± 90 km, and 1000 ± 125 km.

Global variation was described in terms of invariant dip latitude (invdip) and MLT. Recall that invdip is described via the equation:

$$\vartheta = \frac{\alpha\Lambda + \beta\Psi}{\alpha + \beta}, \tag{8.161}$$

where Λ is the invariant latitude and Ψ is the dip latitude (see Section 8.4.5 for details). Here, α and β are defined differently than they are in the electron temperature model:

$$\alpha = 2 - \frac{1}{\exp\left(\frac{\Psi - 25}{2}\right) + 1} + \frac{1}{\exp\left(\frac{-\Psi - 25}{2}\right) + 1}, \tag{8.162}$$

$$\beta = -1 + \frac{1}{\exp\left(\frac{\Lambda - 25}{2}\right) + 1} + \frac{1}{\exp\left(\frac{-\Lambda - 25}{2}\right) + 1}. \tag{8.163}$$

By defining these parameters in this fashion, invdip takes on values close to Ψ for latitudes equatorward of $25°$. Similarly, poleward of $25°$, values of invdip tend towards Λ.

For each group of data (as defined by season and height bins), the common logarithm of a given ion species' number density was defined in terms of spherical harmonics. In particular, a spherical harmonic expansion of order 6 was employed, as given by the equation:

$$\log_{10}(n_j) = a_0^0$$
$$+ \sum_{l=1}^{6} \left(a_l^0 P_l^0(\cos\theta) + \sum_{m=1}^{l} [a_l^m \cos(m\lambda) + b_l^m \sin(m\lambda)]P_l^m \cos\theta \right), \tag{8.164}$$

---

11 The original paper by Truhlik et al. [224] does not explicitly specify the bins used for the C/NOFS data. However, the C/NOFS orbit has a perigee of 405 km and an apogee of 853 km, which aligns more closely with the AE bins.

where n_j is the number density of ion species j, and j is O^+, N^+, H^+, or He^+. The invariant dip colatitude (the complement of invdip), denoted θ, takes on values from 0 to π. The MLT, λ, is also defined in radians so that it can function as a proxy for longitude, and ranges from 0 to 2π. P_l^m is the set of associated Legendre polynomials of degree l and order m, and the coefficients a_l^m and b_l^m are the spherical harmonic coefficients, which uniquely define the expansion. Each data group is modeled by a separate set of coefficients, which were determined via least squares fitting of the empirical data.

Once each individual ion density n_j is determined, they can be expressed as a fraction of the total ion density. At topside heights, ions other than O^+, N^+, H^+, and He^+ occur in negligible quantities, so it may be assumed that:

$$N_i = N_{O^+} + N_{N^+} + N_{H^+} + N_{He^+}, \tag{8.165}$$

where N_i is the total topside ion content. With this information, the full specification of ionospheric ion composition is complete. Together, the IDC model by Richards et al. and the topside specification by Truhlik et al. comprise the default ion composition model in IRI-2016. Legacy models from previous IRI iterations are also included. Another pair of models – a bottomside specification by Danilov and Smirnova [221] and a topside specification by Danilov and Yaichnikov [222] – is provided as well, and may be selected by changing JF(6) (see Section 8.4.1).

8.4.8 Additional Parameters

In Sections 8.4.1–8.4.7, we discussed the formulation of each of the major ionospheric parameters available in the IRI. In addition to these primary outputs, the IRI also provides models for a number of lesser-used parameters. Most of these outputs are turned off by default, but the user may enable them by toggling certain elements in the JF boolean array (see Section 8.4.1). In this section, we will only give a very brief overview of these optional parameters, though in each case references will be provided for readers who wish to explore further.

Equatorial Vertical Ion Drift

Due to the interplay of several plasma transport processes, the behavior of the F2 layer is highly complex, causing large scale movement of electrons and ions. At equatorial latitudes, these combined processes give rise to a phenomenon known as the equatorial fountain (also known as the equatorial anomaly or Appleton anomaly). Due to this effect, plasma near the equator will rise vertically up to a certain height, at which point they are then guided by the geomagnetic field to higher latitudes (for further details, see Section 6.4.1).

In order to characterize the movement of ionospheric plasma near the equator, ions' vertical drift velocities may be modeled. One such empirical model, developed by Scherliess and Fejer, has been adopted in the IRI. The data used to develop the model originated from two main sources: the Jicamarca ISR, and an ion drift meter (IDM) aboard the AE-E satellite. The model successfully captures diurnal, seasonal, solar cycle, and longitudinal variation, and has been validated against a number of ground-based vertical drift measurements. For a full description of this model's formulation and validation, one may consult [255]. This model is turned off by default in the current iteration of the IRI, but the user may turn it on by changing JF(21) to true.

Spread F Probability

Occasionally, ionograms will exhibit a "fuzzy," indistinct shape at F region heights. This phenomenon, known as spread F, occurs if the electron density in that region is highly inhomogeneous, causing ionosonde signals to be reflected irregularly. The causes of spread F are highly complex (see Section 6.4.4 for details), so its occurrence is commonly modeled statistically. In the IRI, a model developed by Abdu and Souza is used, in which spread F occurrence (SFO) is modeled as the percentage of one month during which spread F can be expected to occur. The model, which was based on ionosonde data, predicts the variation of SFO with respect to time of day, season, solar activity and latitude. For further information on this model, the reader is encouraged to refer to [256]. By default, the IRI will not compute spread F probability. To turn on this functionality, the user must change JF(28) to true.

Auroral Boundaries

At Earth's magnetic poles, the magnetic field lines become nearly perpendicular to the surface of the Earth. Charged particles are guided along the field lines deeper into the atmosphere, where they commonly collide with neutral atoms and molecules. These collisions can ionize the neutrals – a process that is known as corpuscular ionization. This extra ionization occurs most strongly at the poles, meaning there is a region in which the standard IRI electron density model will underestimate the true value. The regions in which this additional ionization occurs are known as the auroral ovals (each magnetic pole exhibits its own oval). It is important to quantify the equatorward boundaries of the ovals so that a corrective model can be applied at latitudes poleward of the boundaries.

The determination of auroral boundaries is not straightforward. The auroral ovals are not symmetric in longitude, since typical diurnal variation in electron density causes an irregular shape. In addition, the extent of the ovals depends on magnetic activity, making the location of the boundaries more difficult to predict. Zhang et al. [257] proposed a model that solves these problems. They showed that by using far ultraviolet (FUV) emissions from N_2 molecules, it was possible to determine the mean electron energy E_0 and electron energy flux Q in the auroral region. These quantities were found to correlate strongly with several known features of the auroral ovals. Further, the authors determined that a threshold value of 0.2 ergs/s/cm² for the electron energy flux gave a good representation of the auroral boundaries. That is, an auroral boundary is equivalent to the set of magnetic latitude and magnetic longitude points defining the level curve $Q = 0.2$ ergs/s/cm². This choice is justified in the original paper [257], which also contains a more detailed description of how this model was implemented for use in the IRI. This model is turned off by default in IRI-2016, but may be turned on by changing the value of JF(33) to true.

Storm Time Conditions

The parameters discussed during the preceding sections of this chapter describe the state of the ionosphere during magnetically quiet conditions. During magnetic storms, these quantities can vary significantly from their typical values. Unless accounted for, these discrepancies could render the IRI useless for prediction of ionospheric conditions. This problem has been the subject of much research, with the goal of improving the robustness of the IRI during storm-time conditions.

Of particular interest are the electron densities of the uppermost layers of the ionosphere. During magnetic storms, highly energetic charged particles bombard the ionosphere, which can ionize neutral atoms and molecules through corpuscular ionization. This additional ionization is not accounted for by the base IRI model, so researchers have sought to introduce corrective terms that increase the predicted electron density, but only during magnetic storms.

Currently, the IRI includes models that account for storm-time deviation of both the F2 and E layer electron densities. A model for the former was devised by Araujo-Pradere et al., in which the ratio of observed f_oF2 to monthly mean f_oF2 was observed using data collected from 75 globally distributed ionosonde stations during 43 magnetic storm events. The data showed a dependence on the magnetic index a_p (see Section 6.4.3), a parameter commonly used to quantify the severity of magnetic disturbance. This model has proven quite successful in providing accurate electron density values during storm-time conditions. For further information on the specification and validation of this model, the reader may refer to [258].

A model for the E layer storm-time electron density correction was introduced by Mertens et al. This model uses data from the Sounding of the Atmosphere using Broadband Emission Radiometry (SABER) instrument, which is capable of measuring molecular emission spectra in the wavelength range of 1.27–17 µm. The SABER instrument, which is installed aboard the Thermosphere-Ionosphere-Mesosphere Energetics and Dynamics (TIMED) satellite, has been collecting data since its launch in 2001. Mertens et al. used data from the SABER mission to look for trends that occur during magnetically disturbed conditions. They found that radiation in the 4.3 µm band was significantly enhanced during magnetic storms – a phenomenon that they attributed to vibrationally excited NO^+ ions produced via particle precipitation. The authors demonstrated the viability of using the volume emission rate of this radiation as a proxy for E region electron density during magnetic storms. For further information on this approach, and a detailed description of the model included in the IRI, the reader may consult [259, 260].

These two models, although successful, are only necessary for calculations involving storm-time conditions. Therefore, to avoid unnecessary computation in the much more common case of quiet-time ionospheric conditions, the models are turned off by default in the current iteration of the IRI. If required, these models may be turned on by changing various elements of the JF boolean array (see Section 8.4.1). The F region model is controlled using JF(26), JF(36), and JF(37), and the E region model is controlled using JF(35).

Appendices

Appendix A

Review of Electromagnetics Concepts

A.1 Electromagnetic Waves

A.1.1 Maxwell's Equations and the Wave Equation

Global navigation satellite systems (GNSS) signal propagation is based on electromagnetic waves. Electromagnetic fields and waves are governed by Maxwell's equations [261, 262] which read as

$\nabla \cdot \vec{D} = \rho_v,$	(Gauss's law for electricity),
$\nabla \cdot \vec{B} = 0,$	(Gauss's law for magnetism),
$\nabla \times \vec{E} = -\dfrac{\partial \vec{B}}{\partial t},$	(Faraday's law),
$\nabla \times \vec{H} = \dfrac{\partial \vec{D}}{\partial t} + \vec{J},$	(Ampere's law),

where $\vec{E}$ is the electric field intensity in V/m, $\vec{D}$ is the electric flux density (or electric displacement field) in C/m^2, $\vec{H}$ is the magnetic field intensity in H/m, $\vec{B}$ is the magnetic flux density in wb/m^2, ρ_v is the volume charge density in C/m^3, and $\vec{J}$ is the current density in A/m^2.

We also have the constitutive relations:

$$\vec{D} = \varepsilon \vec{E}, \tag{A.1}$$

$$\vec{B} = \mu \vec{H}, \tag{A.2}$$

$$\vec{J} = \vec{J}_i + \sigma \vec{E}, \tag{A.3}$$

where the current density is split into impressed and induced current density terms, respectively, ε is the permittivity, μ is the permeability, and σ is the conductivity. These are medium dependent parameters. In a vacuum, $\varepsilon = \varepsilon_0 \approx 8.854 \times 10^{12}$ F/m, $\mu = \mu_0 = 4\pi \times 10^{-7}$ H/m, and $\sigma = 0$. In linear and isotropic media, we

Tropospheric and Ionospheric Effects on Global Navigation Satellite Systems, First Edition.
Timothy H. Kindervatter and Fernando L. Teixeira.
© 2022 The Institute of Electrical and Electronics Engineers, Inc. Published 2022 by John Wiley & Sons, Inc.

can write $\varepsilon = \varepsilon_r \varepsilon_0$ and $\mu = \mu_r \mu_0$, where ε_r is relative permittivity and μ_r is the relative permeability.

Wave Equation in a Source Free Region

We next derive the wave equation from Maxwell's equations. For simplicity, we will consider the case of a source-free (i.e. $\rho_v = 0, \vec{J}_l = 0$) and lossless (i.e. $\sigma = 0$) region of space. We may use Equations A.1 and A.2 to rewrite Maxwell's equations solely in terms of $\vec{E}$ and $\vec{B}$. This gives us Maxwell's equations in the following form:

$$\nabla \cdot \vec{E} = 0, \tag{A.4}$$

$$\nabla \cdot \vec{B} = 0, \tag{A.5}$$

$$\nabla \times \vec{E} = -\frac{\partial \vec{B}}{\partial t}, \tag{A.6}$$

$$\nabla \times \vec{B} = \mu\varepsilon\frac{\partial \vec{E}}{\partial t}. \tag{A.7}$$

If we take the curl of both sides of Equation A.6, we see that

$$\nabla \times \left(\nabla \times \vec{E}\right) = -\frac{\partial}{\partial t}\nabla \times \vec{B}.$$

On the left side, we may use the vector identity

$$\nabla \times \left(\nabla \times \vec{V}\right) = \nabla\left(\nabla \cdot \vec{V}\right) - \nabla^2\vec{V}$$

and on the right side, we may replace $\nabla \times \vec{B}$ using Equation A.7. This gives us

$$\nabla\left(\nabla \cdot \vec{E}\right) - \nabla^2\vec{E} = -\mu\varepsilon\frac{\partial^2\vec{E}}{\partial t^2}.$$

From Equation A.4, we see that the first term on the left hand side vanishes, leaving us with

$$\nabla^2\vec{E} = \mu\varepsilon\frac{\partial^2\vec{E}}{\partial t^2}. \tag{A.8}$$

This is the wave equation that describes the evolution of the electric field in time and space. By taking the curl of both sides of Equation A.7, we could get an analogous expression for the evolution of the magnetic field. Going forward, we will only consider the electric field, but know that the observations in the following sections also apply to the magnetic field.

Velocity of a Traveling Wave

The wave equation can be written as

$$\nabla^2\vec{u} = \frac{1}{v^2}\frac{\partial^2\vec{u}}{\partial t^2}.$$

It may be verified that $\vec{u}(x, t) = \vec{a}f(x - vt)$ is a possible solution of the wave equation, where $\vec{a}$ is an arbitrary constant vector and $f(\cdot)$ is a second-order differentiable but otherwise arbitrary function of the argument. By substituting different values of x and t, we conclude that this solution represents a wave traveling in the positive x direction with velocity v. Likewise, $\vec{u}(x, t) = \vec{a}f(x + vt)$, $\vec{u}(y, t) = \vec{a}f(y \pm vt)$, and $\vec{u}(z, t) = \vec{a}f(z \pm vt)$ are possible solutions of the wave equation was well. This type of solution is called a d'Alembert solution, and can be further generalized for waves $\vec{u}(x, y, z, t)$ traveling along an arbitrary direction. For our electric field wave, this tells us that:

$$\frac{1}{v^2} = \mu\epsilon \Rightarrow v = \frac{1}{\sqrt{\mu\epsilon}}. \tag{A.9}$$

In a vacuum, where $\epsilon = \epsilon_0$ and $\mu = \mu_0$, we see that $v = 1/\sqrt{\mu_0\epsilon_0} = c$, where c denotes the speed of light. This allows us to re-write Equation A.9:

$$v = \frac{1}{\sqrt{\mu\epsilon}} = \frac{c}{\sqrt{\mu_r\epsilon_r}}.$$

Both ϵ_r and μ_r are intrinsic properties of a medium. In a vacuum, ϵ_r and μ_r both equal 1, but in other media these parameters will assume different values, based on the medium's atomic and molecular structure. Most media, including the troposphere, are non-magnetic, meaning that $\mu_r = 1$. In this case, it is appropriate to make the simplification:

$$v = \frac{c}{\sqrt{\epsilon_r}}. \tag{A.10}$$

The index of refraction n of a medium may be defined as the ratio of a wave's velocity in a vacuum to its velocity in that medium. This is given by the familiar expression:

$$n = \frac{c}{v}. \tag{A.11}$$

From Equation A.10, we see that $n = \sqrt{\epsilon_r}$. In most materials, $\epsilon_r > 1$, which gives us an index of refraction greater than 1, telling us that the wave propagates slower than c. However, there are some special media in which sometimes $\epsilon_r < 1$, such as in the ionosphere at some frequencies. The wave propagates through such a medium at a velocity greater than c. Note that this does not violate relativity because it refers to the *phase velocity*, which is not the velocity of transmission of information, the latter being the *group velocity*. To understand this further, we must delve a bit deeper into the physics of wave propagation.

A.1.2 Plane Wave Solutions

In general, the field from a bounded source (for example, any antenna) produces a spherical wave at distances sufficient far from the transmitter. In the vicinity of a receiver that is far away from a transmitter, this spherical wave may be locally approximated as a plane wave, i.e. it will exhibit nearly planar wavefronts. Plane waves are also important because they can be used to synthetize more general types of waves, in a similar way as Fourier components can synthetize more general types of signals. Because of their importance, we turn our attention to plane waves next. We first examine solutions of the wave equation in the form

$$\vec{E}(x,y,z,t) = \underline{\vec{E}}(x,y,z)f(t).$$ (A.12)

If we rewrite Equation A.8 to reflect this, we can see that

$$\nabla^2 \underline{\vec{E}}(x,y,z)f(t) = \mu\varepsilon\frac{\partial^2}{\partial t^2}\underline{\vec{E}}(x,y,z)f(t).$$

Using separation of variables, i.e. by grouping all the time dependent terms on the right, we get

$$\nabla^2 \underline{\vec{E}}(x,y,z) = \mu\varepsilon\underline{\vec{E}}(x,y,z)\frac{1}{f(t)}\frac{\partial^2 f(t)}{\partial t^2}.$$

Now we see that the left hand side is independent of t and dependens only on x, y, and z. But both sides are equal, which tells us that the time-dependent factor on the right hand side should be equal to some constant value, which we will call $-\omega^2$, i.e.

$$\frac{1}{f(t)}\frac{\partial^2 f(t)}{\partial t^2} = -\omega^2.$$

The reason for this choice, as we will see in a moment, is that ω is the angular frequency of the wave in radians, and may be related to the frequency in Hz by the expression $\omega = 2\pi f$.

The time-dependent equation earlier is simpler, so we will consider that first. We see that we now have a homogeneous ordinary differential equation (ODE):

$$\frac{d^2 f(t)}{dt^2} = -\omega^2 f(t).$$ (A.13)

The simplest solution to this ODE is of the form $f(t) = e^{\pm j\omega t}$, which may be verified by substituting it into the differential equation.

Solving the Helmholtz Equation

Now we must turn our attention to the spatially dependent equation, more commonly referred to as the Helmholtz equation, which is more complicated.

$$\nabla^2 \underline{\vec{E}}(x,y,z) = -\omega^2\mu\varepsilon\underline{\vec{E}}(x,y,z).$$

The collection of coefficients on the right hand is a quantity known as the propagation constant (also sometimes called the wavenumber). We denote this quantity with the letter k.

$$k^2 \equiv \omega^2\mu\varepsilon \Rightarrow k = \omega\sqrt{\mu\varepsilon}.$$ (A.14)

The wavenumber k is measured in units of radians per meter, and is related to the wavelength λ by the expression $k = 2\pi/\lambda$.

Recall that $\underline{\vec{E}}(x, y, z)$ is a vector quantity and therefore we can expand it in Cartesian components as

$$\nabla^2 \left(\hat{x}\underline{E}_x + \hat{y}\underline{E}_y + \hat{z}\underline{E}_z \right) = -k^2 \left(\hat{x}\underline{E}_x + \hat{y}\underline{E}_y + \hat{z}\underline{E}_z \right). \tag{A.15}$$

The vector partial differential equation (PDE) A.15 may be separated into three scalar PDEs. Each scalar component, $\underline{E}_x, \underline{E}_y$, and $\underline{E}_z$ may vary with x, y, and z, resulting in the following set of equations:

$$\nabla^2 \underline{E}_x(x, y, z) = -k^2 \underline{E}_x(x, y, z), \tag{A.16}$$

$$\nabla^2 \underline{E}_y(x, y, z) = -k^2 \underline{E}_y(x, y, z),$$

$$\nabla^2 \underline{E}_z(x, y, z) = -k^2 \underline{E}_z(x, y, z).$$

These equations are formally identical so their general solutions are all of the same form. We will solve the equation for $\underline{E}_x$. Recall that the Laplacian operator, ∇^2, is given by:

$$\nabla^2 = \frac{\partial^2}{\partial x^2} + \frac{\partial^2}{\partial y^2} + \frac{\partial^2}{\partial z^2}.$$

Therefore, Equation A.16 may be expanded to

$$\frac{\partial^2 \underline{E}_x(x, y, z)}{\partial x^2} + \frac{\partial^2 \underline{E}_x(x, y, z)}{\partial y^2} + \frac{\partial^2 \underline{E}_x(x, y, z)}{\partial z^2} = -k^2 \underline{E}_x(x, y, z). \tag{A.17}$$

We may further separate our solution as follows:

$$\underline{E}_x(x, y, z) = f(x)g(y)h(z). \tag{A.18}$$

For each partial derivative in Equation A.17, two of the three single-variable functions in Equation A.18 will remain constant, allowing us to factor them out of the derivative.

$$gh\frac{\partial^2 f}{\partial x^2} + fh\frac{\partial^2 g}{\partial y^2} + fg\frac{\partial^2 h}{\partial z^2} = -k^2 fgh.$$

If we divide both sides by fgh, we get:

$$\frac{1}{f}\frac{\partial^2 f}{\partial x^2} + \frac{1}{g}\frac{\partial^2 g}{\partial y^2} + \frac{1}{h}\frac{\partial^2 h}{\partial z^2} = -k^2.$$

We have a constant value on the right side. This tells us that each term on the left hand side must also be constant, otherwise they would not add up to a constant value. Therefore we may write the three following equations:

$$\frac{1}{f(x)}\frac{\partial^2 f(x)}{\partial x^2} = -k_x^2 \Rightarrow \frac{\partial^2 f(x)}{\partial x^2} = -k_x^2 f(x),$$

$$\frac{1}{g(y)}\frac{\partial^2 g(y)}{\partial y^2} = -k_y^2 \Rightarrow \frac{\partial^2 g(y)}{\partial y^2} = -k_y^2 g(y),$$

$$\frac{1}{h(z)}\frac{\partial^2 h(z)}{\partial z^2} = -k_z^2 \Rightarrow \frac{\partial^2 h(z)}{\partial z^2} = -k_z^2 h(z).$$

These three equations are homogeneous ODEs of the exact same form as Equation A.13. Therefore, the general solutions to each of these three equations are respectively

$$f(x) = c_1 e^{-jk_x x} + c_2 e^{jk_x x},$$

$$g(y) = c_3 e^{-jk_y y} + c_4 e^{jk_y y},$$

$$h(z) = c_5 e^{-jk_z z} + c_6 e^{jk_z z},$$

where c_1–c_6 are complex-valued constants determined by the boundary conditions. First consider $f(x)$. The two complex exponentials in this solution represent a pair counter-propagating waves in the x-direction. The first term represents a wave traveling in the positive x-direction, while the second term represents a wave traveling in the negative x direction. As seen from Equation A.18, $\underline{E}_x(x, y, z)$ is simply the product of these three solutions. Since each solution has two terms, we have eight possible terms. Let us consider one of them, e.g.

$$\underline{E}_x(x, y, z) = c_1 e^{-jk_x x} c_3 e^{-jk_y y} c_5 e^{-jk_z z} = \underline{E}_{0x} e^{-j(k_x x + k_y y + k_z z)}. \tag{A.19}$$

This is a complex-valued solution (a phasor quantity). A real-valued solution for the electric field can be simply obtained by multiplying the aforementioned by $e^{j\omega t}$ and taking the real part of the resultant expression. Letting $\underline{E}_{0x} = E_{0x} e^{j\phi}$, where ϕ and E_{0x} are the (real-valued) phase and magnitude of $\underline{E}_{0x}$, the resultant solution is a traveling wave of the form:

$$E_x(x, y, z, t) = E_{0x} \cos(\omega t - k_x x - k_y y - k_z z + \phi), \tag{A.20}$$

whose propagation direction is defined by the vector

$$\vec{k} = \hat{x}k_x + \hat{y}k_y + \hat{z}k_z.$$

We see that the magnitude of this vector is

$$|\vec{k}| = k_x^2 + k_y^2 + k_z^2 = k^2. \tag{A.21}$$

By performing the dot product of $\vec{k}$ with the unit position vector, $\vec{r} = \hat{x}x + \hat{y}y + \hat{z}z$, we get

$$\vec{k} \cdot \vec{r} = k_x x + k_y y + k_z z,$$

which is the expression found in the argument of the exponential function in Equation A.19. Therefore we may simplify this equation to

$$\underline{E}_x(x, y, z) = \underline{E}_{0x} e^{-j\vec{k}\cdot\vec{r}}. \tag{A.22}$$

Recall that we previously neglected solutions to the y and z components of $\vec{\underline{E}}(x, y, z)$. Their solutions are analogous to Equation A.22. Therefore $\vec{\underline{E}}(x, y, z)$ may be fully expressed in a column vector form as

$$
\vec{\underline{E}}(x, y, z) = \begin{bmatrix} \underline{E}_x(x, y, z) \\ \underline{E}_y(x, y, z) \\ \underline{E}_z(x, y, z) \end{bmatrix} = \begin{bmatrix} \underline{E}_{0x} e^{-j\vec{k}\cdot\vec{r}} \\ \underline{E}_{0y} e^{-j\vec{k}\cdot\vec{r}} \\ \underline{E}_{0z} e^{-j\vec{k}\cdot\vec{r}} \end{bmatrix},
$$

where $\underline{E}_{0x}$, $\underline{E}_{0y}$, and $\underline{E}_{0z}$ are arbitrary complex-valued amplitudes. Finally, to get the real-valued final expression for the electric field as a function of time, we take the real part of Equation A.12, which leads to

$$
\vec{E}(x, y, z, t) = \begin{bmatrix} E_{0x} \cos\left(\omega t - \vec{k}\cdot\vec{r} + \phi_x\right) \\ E_{0y} \cos\left(\omega t - \vec{k}\cdot\vec{r} + \phi_y\right) \\ E_{0z} \cos\left(\omega t - \vec{k}\cdot\vec{r} + \phi_z\right) \end{bmatrix}. \tag{A.23}
$$

A.1.3 Constraints Via Maxwell's Equations

Not all solutions represented by Equation A.23 are physically realizable. In addition, Maxwell's equations require the relationship between $\vec{E}$ and $\vec{H}$ to satisfy a few key conditions in a vacuum:

- $\vec{E}, \vec{H}$, and $\vec{k}$ must be mutually orthogonal to one another.
- $\vec{E}$ and $\vec{H}$ must propagate in the same direction at the same frequency.
- $\vec{E}$ and $\vec{H}$ must be in-phase.
- The magnitudes of $\vec{E}$ and $\vec{H}$ must differ by a factor of $\sqrt{\mu}/\sqrt{\varepsilon}$.

We will see how all of these results follow naturally for solutions to the wave equation which satisfy Maxwell's equations.

1. *Mutual orthogonality of $\vec{E}, \vec{H}$, and $\vec{k}$*: The electric and magnetic fields that compose an electromagnetic wave are always orthogonal to each other, as well as to the direction of propagation. These results follow naturally for any solutions to the wave equation which also satisfy Maxwell's equations. We start with Gauss's law for electricity.

$$
\nabla \cdot \vec{E} = 0.
$$

Evaluating this expression using an expression for $\vec{E}$ in Cartesian coordinates such as found in Equation A.23,

$$
\nabla \cdot \vec{E} = \frac{\partial E_x}{\partial x} + \frac{\partial E_y}{\partial y} + \frac{\partial E_z}{\partial z} = 0. \tag{A.24}
$$

For complex-valued plane waves, we can make the substitution

$$\nabla = \begin{bmatrix} -jk_x \\ -jk_y \\ -jk_z \end{bmatrix} = -j\vec{k}, \tag{A.25}$$

so that Equation A.24 can be rewritten as

$$-j\vec{k} \cdot \vec{E} = 0.$$

From this, it is evident that $\vec{k} \cdot \vec{E} = 0$, which tells us that the propagation vector is indeed perpendicular to the electric field.

As mentioned previously, the solutions to the Helmholtz equation for the magnetic field are of the same form as for the electric field. An identical derivation may be performed using Gauss's law for magnetism ($\nabla \cdot \vec{B} = 0$) and the solutions to the magnetic field wave equation so that $\vec{k} \cdot \vec{B} = \vec{k} \cdot \mu \vec{H} = 0$.

From Maxwell's equations, we also have

$$-\frac{1}{j\omega\mu} \left(\nabla \times \vec{E} \right) = \vec{H},$$

by again expressing the Del operator for plane waves in terms of $\vec{k}$, we get

$$-\frac{1}{j\omega\mu} \left(-j\vec{k} \times \vec{E} \right) = \vec{H}.$$

Since $\vec{k} = \hat{k}k$, where $\hat{k}$ is the unit vector in the direction of $\vec{k}$ and k is the vector magnitude, we can factor out $-jk$ to give us

$$\frac{k}{\omega\mu} \left(\hat{k} \times \vec{E} \right) = \vec{H}. \tag{A.26}$$

We know that the cross product of two vectors produces a third vector that is perpendicular to both of them. At this point, it is evident that $\vec{H}$ must be perpendicular to not only $\vec{k}$ as we saw before, but also $\vec{E}$.

We may perform a similar derivation starting with Ampere's law to obtain

$$-\frac{k}{\omega\varepsilon} \left(\hat{k} \times \vec{H} \right) = \vec{E},$$

which confirms the mutual orthogonality.

2. *Equivalence of $\vec{k}$, ω, and ϕ*: Solutions to the wave equation can be written as complex exponentials. Therefore, any partial derivative of such a solution with respect to x, y, z, or t will always have the same form, but be multiplied by some constant. In other words, the argument of the complex exponential remains unchanged when differentiated. This means that the derivative of a given solution has the same wave vector $\vec{k}$, frequency ω and initial phase ϕ as the solution itself. Via Maxwell's equations, $\vec{E}$ and $\vec{H}$ can always be expressed in terms of one another's derivatives, which tells us that they must have the same values for $\vec{k}$, ω, and ϕ.

For plane waves in vacuum, it is a simple exercise to substitute the previously obtained solutions into Maxwell's equations and obtain $\omega^2 \mu \varepsilon = k^2$. This general dependency between the wavenumber $k(\omega)$ and the frequency $\omega(k)$ is termed *dispersion relation*. In more complex media such as the ionosphere, the dispersion relation has a more complicated expression.

3. *Relationship between field magnitudes*: Using Equations A.26 and A.14, we see that

$$\frac{k}{\omega\mu}\left(\hat{k}\times\vec{E}\right) = \frac{\omega\sqrt{\mu\varepsilon}}{\omega\mu}\left(\hat{k}\times\vec{E}\right) = \sqrt{\frac{\varepsilon}{\mu}}\left(\hat{k}\times\vec{E}\right) = \vec{H}$$

and

$$-\frac{\omega\sqrt{\mu\varepsilon}}{\omega\varepsilon}\left(\hat{k}\times\vec{H}\right) = -\sqrt{\frac{\mu}{\varepsilon}}\left(\hat{k}\times\vec{H}\right) = \vec{E}.$$

We may define

$$\eta \equiv \sqrt{\frac{\mu}{\varepsilon}} \tag{A.27}$$

as the wave impedance of the medium, measured in ohms. In free space,

$$\eta = \eta_0 = \sqrt{\frac{\mu_0}{\varepsilon_0}} \approx 377 \ \Omega.$$

Thus we see that:

$$\vec{H} = \frac{1}{\eta}\left(\hat{k}\times\vec{E}\right), \tag{A.28}$$

$$\vec{E} = -\eta\left(\hat{k}\times\vec{H}\right).$$

In terms of magnitudes:

$$\frac{|\vec{E}|}{|\vec{H}|} = \eta.$$

A.1.4 Poynting Vector

The power density flow of an electromagnetic field is denoted by a quantity known as the Poynting vector. We may derive this quantity from Maxwell's equations. By taking the dot product of $\vec{H}$ with Faraday's law and $\vec{E}$ with Ampere's law we obtain

$$\vec{H}\cdot\left(\nabla\times\vec{E}\right) = -\vec{H}\cdot\frac{\partial\vec{B}}{\partial t},$$

$$\vec{E}\cdot\left(\nabla\times\vec{H}\right) = \vec{E}\cdot\frac{\partial\vec{D}}{\partial t} + \vec{E}\cdot\vec{J}.$$

Taking the difference of these two equations results in

$$\vec{E} \cdot \left(\nabla \times \vec{H} \right) - \vec{H} \cdot \left(\nabla \times \vec{E} \right) = \vec{E} \cdot \vec{J} + \vec{E} \cdot \frac{\partial \vec{D}}{\partial t} + \vec{H} \cdot \frac{\partial \vec{B}}{\partial t}.$$

On the left hand side, we may use the vector identity

$$\vec{E} \cdot \left(\nabla \times \vec{H} \right) - \vec{H} \cdot \left(\nabla \times \vec{E} \right) = \nabla \cdot \left(\vec{H} \times \vec{E} \right)$$

and on the right hand side, we have

$$\vec{E} \cdot \frac{\partial \vec{D}}{\partial t} + \vec{H} \cdot \frac{\partial \vec{B}}{\partial t} = \frac{1}{2} \left(\vec{E} \cdot \frac{\partial \vec{D}}{\partial t} + \vec{D} \cdot \frac{\partial \vec{E}}{\partial t} + \vec{H} \cdot \frac{\partial \vec{B}}{\partial t} + \vec{B} \cdot \frac{\partial \vec{H}}{\partial t} \right)$$

$$= \frac{\partial}{\partial t} \left[\frac{1}{2} (\vec{E} \cdot \vec{D}) + \frac{1}{2} (\vec{H} \cdot \vec{B}) \right]. \tag{A.29}$$

Combining these expressions into one equation:

$$\nabla \cdot \left(\vec{H} \times \vec{E} \right) = \frac{\partial}{\partial t} \left[\frac{1}{2} (\vec{E} \cdot \vec{D}) + \frac{1}{2} (\vec{H} \cdot \vec{B}) \right] + \vec{E} \cdot \vec{J}.$$

Utilizing the constitutive relations and recalling that $(\vec{H} \times \vec{E}) = -(\vec{E} \times \vec{H})$, we may rewrite this equation as

$$-\nabla \cdot \left(\vec{E} \times \vec{H} \right) = \frac{\partial}{\partial t} \left[\frac{1}{2} \varepsilon \vec{E}^2 + \frac{1}{2} \mu \vec{H}^2 \right] + \vec{E} \cdot \vec{J}. \tag{A.30}$$

We define

$$\vec{S} = \vec{E} \times \vec{H}$$

as the Poynting vector, which is a quantity that measures the magnitude and direction of the wave's power flux density (rate of power transfer per unit area), and has units of W/m^2. Further, we may define

$$W_E = \frac{1}{2} \left(\vec{E} \cdot \vec{D} \right) = \frac{1}{2} \varepsilon \vec{E}^2 \tag{A.31}$$

and

$$W_M = \frac{1}{2} \left(\vec{H} \cdot \vec{B} \right) = \frac{1}{2} \mu \vec{H}^2$$

as the electric and magnetic energy densities in J/m^3, respectively. The current $\vec{J}$ may be split up into the impressed source current $\vec{J}_i$ and the conduction current $\sigma \vec{E}$ to give

$$\vec{E} \cdot \vec{J} = \vec{E} \cdot \vec{J}_i + \sigma \vec{E} \cdot \vec{E},$$

where the first term represents the power exchanged with the impressed source and the second term power dissipation due to ohmic losses. Using all of these definitions, we may rewrite Equation A.30 as follows:

$$-\nabla \cdot \vec{S} = \frac{\partial}{\partial t} \left[W_E + W_M \right] + \vec{E} \cdot \vec{J}_i + \sigma \vec{E} \cdot \vec{E}.$$

Each of these terms represents part of the total power in the system. Therefore this equation, which is sometimes called Poynting's theorem, represents the statement of conservation of energy in electromagnetism.

Time-Averaged Poynting Vector

It is often useful to work with the time-averaged Poynting vector, which is a time-independent quantity, rather than the instantaneous value of the time-varying vector $\vec{S}$. To derive this quantity, it is helpful to work with the time-harmonic forms of $\vec{E}$ and $\vec{H}$:

$$\vec{E}(t) = \text{Re}\{\underline{\vec{E}}e^{j\omega t}\},$$

$$\vec{H}(t) = \text{Re}\{\underline{\vec{H}}e^{j\omega t}\},$$

so that we can rewrite $\vec{S}(t)$ as

$$\vec{S}(t) = \vec{E}(t) \times \vec{H}(t) = \text{Re}\{\underline{\vec{E}}e^{j\omega t}\} \times \text{Re}\{\underline{\vec{H}}e^{j\omega t}\}. \tag{A.32}$$

Since $\vec{E}$ and $\vec{H}$ are complex numbers, their real parts can be obtained by adding them to their complex conjugates and dividing by two.

$$\text{Re}\{\underline{\vec{E}}e^{j\omega t}\} = \frac{1}{2}\left(\underline{\vec{E}}e^{j\omega t} + \underline{\vec{E}}^*e^{-j\omega t}\right),$$

$$\text{Re}\{\underline{\vec{H}}e^{j\omega t}\} = \frac{1}{2}\left(\underline{\vec{H}}e^{j\omega t} + \underline{\vec{H}}^*e^{-j\omega t}\right).$$

Equation A.32 then becomes

$$\vec{S}(t) = \frac{1}{4}\left(\underline{\vec{E}}\times\underline{\vec{H}}^* + \underline{\vec{E}}^*\times\underline{\vec{H}} + \underline{\vec{E}}\times\underline{\vec{H}}e^{j2\omega t} + \underline{\vec{E}}^*\times\underline{\vec{H}}^*e^{-j2\omega t}\right),$$

which can be simplified to

$$\vec{S}(t) = \frac{1}{2}\text{Re}\{\underline{\vec{E}}\times\underline{\vec{H}}^*\} + \frac{1}{2}\text{Re}\{\underline{\vec{E}}\times\underline{\vec{H}}e^{j2\omega t}\}.$$

To find the average value over one period $T = 2\pi/\omega$, we evaluate the integral

$$\langle\vec{S}\rangle = \frac{1}{T}\int_0^T\left[\frac{1}{2}\text{Re}\left\{\underline{\vec{E}}\times\underline{\vec{H}}^*\right\} + \frac{1}{2}\text{Re}\left\{\underline{\vec{E}}\times\underline{\vec{H}}e^{j2\omega t}\right\}\right]dt.$$

The average value of a complex exponential over one period is zero, so the second term vanishes, leaving

$$\langle\vec{S}\rangle = \frac{1}{2}\text{Re}\{\underline{\vec{E}}\times\underline{\vec{H}}^*\}.$$

Using Equation A.28, we can write $\langle\vec{S}\rangle$ in an alternative form:

$$\langle\vec{S}\rangle = \frac{|\underline{\vec{E}}|^2}{2\text{Re}\{\eta\}}\hat{k}.$$

A.2 Phase and Group Velocity

A.2.1 Phase Velocity

Until now, we have considered very general solutions to the wave equation. While these general solutions are broadly applicable, they can also be difficult to visualize. So for the sake of simplicity, we will analyze a concrete example. Consider an electric field traveling in the $+z$ direction, with only an x-component, of the form:

$$\vec{E} = \hat{x}\underline{E}_0 e^{-j(\vec{k}\cdot\vec{r})}e^{j\omega t} = \hat{x}\underline{E}_0 e^{-jkz}e^{j\omega t} = \hat{x}\underline{E}_0 e^{-j(kz-\omega t)},$$

with real-valued amplitude E_0. The real part of this equation represents the instantaneous field vector:

$$\vec{E}(x,y,z,t) = \text{Re}\left\{\vec{E}(x,y,z)e^{j\omega t}\right\} = \hat{x}E_0 \cos{(kz-\omega t)}, \tag{A.33}$$

which is a monochromatic (a.k.a. pure tone or unmodulated), (co)sinusoidal function. The phase of this signal is:

$$\phi(x,t) = kx - \omega t.$$

A wave is commonly depicted as a series of wavefronts propagating through space. A wavefront may be thought of as a surface of constant phase. In other words, at each wavefront the derivative of the phase with respect to time should equal zero, i.e.

$$\frac{\mathrm{d}\phi}{\mathrm{d}t} = k\frac{\mathrm{d}x}{\mathrm{d}t} - \omega = 0 \Rightarrow \frac{\mathrm{d}x}{\mathrm{d}t} = \frac{\omega}{k}.$$

Therefore, the quantity $\frac{\mathrm{d}x}{\mathrm{d}t}$ is called the phase velocity v_p, and is a measure of how quickly a wavefront travels through space,

$$v_p = \frac{\omega}{k}.$$

A.2.2 Modulated Signals and Group Velocity

For the sake of the following discussion, we need not concern ourselves with the vector nature of the waves as long as the waves all propagate in the same direction and share a plane of oscillation. We consider two plane waves of the form

$$E_1 = E_0 \cos(k_1 x - \omega_1 t),$$
$$E_2 = E_0 \cos(k_2 x - \omega_2 t).$$

The superposition of these waves may be written as:

$$E = E_0[\cos(k_1 x - \omega_1 t) + \cos(k_2 x - \omega_2 t)]. \tag{A.34}$$

Since Maxwell's equations are linear, the superposition of two waves is also a valid wave solution. Using the trigonometric identity

$$\cos \alpha + \cos \beta = 2 \cos \left(\frac{\alpha + \beta}{2} \right) \cos \left(\frac{\alpha - \beta}{2} \right),$$

we can write Equation A.34 as:

$$E = 2E_0 \cos \left(\frac{(k_1 + k_2)x - (\omega_1 + \omega_2)t}{2} \right) \cos \left(\frac{(k_1 - k_2)x - (\omega_1 - \omega_2)t}{2} \right).$$

Let us define the mean frequency and modulation frequency respectively as:

$$\overline{\omega} = \frac{1}{2}(\omega_1 + \omega_2), \quad \omega_m = \frac{1}{2}(\omega_1 - \omega_2).$$

Similarly, we can define the mean wavenumber and modulation wavenumber respectively as:

$$\overline{k} = \frac{1}{2}(k_1 + k_2), \quad k_m = \frac{1}{2}(k_1 - k_2).$$

This allows us to rewrite Equation A.2.2 in the more concise form:

$$E = 2E_0 \cos(k_m x - \omega_m t) \cos(\overline{k}x - \overline{\omega}t). \tag{A.35}$$

This is an amplitude modulated wave with envelope

$$E_0(x, t) = 2E_0 \cos(k_m x - \omega_m t). \tag{A.36}$$

In other words,

$$E = E_0(x, t) \cos(\overline{k}x - \overline{\omega}t). \tag{A.37}$$

The carrier signal is the cosine term, with the mean frequency and wavenumber of the two constituent signals.

It is important to note that, for the sake of simplicity, this discussion has assumed the superposition of only two waves. In general, a modulated signal may be represented by the superposition of an arbitrary number of pure tone signals, potentially with different amplitudes and phases. In general, the resultant wave may be represented as a Fourier series. The derivation of this is beyond the scope of this text, but a thorough examination may be found in [263, pp. 302–320]. The case presented earlier may also be represented as a simple Fourier series with only two terms.

In a dispersive medium, the individual frequencies which comprise a signal each may travel at a different speed. The envelope created by the superposition of these waves will therefore travel at a different speed than each constituent wave. It is useful to develop the concept of the group velocity, which is the speed at which the envelope of the signal propagates. If we can distinguish some feature of the envelope which has constant phase (e.g. the leading edge of a pulse), we can take that feature's velocity as the velocity of the group of waves as a whole.

In this case, the phase velocity of the signal still equals ω/k as we found before, since is dependent only upon the carrier signal, which is represented by the cosine term in Equation A.37. Therefore, $v_p = \overline{\omega}/\overline{k}$. The envelope of the wave is solely expressed by Equation A.36. The phase of the envelope is

$$\phi = k_m x - \omega_m t,$$

so for a feature with constant phase, such as the leading edge of a pulse, the time derivative of the phase will equal zero. This allows us to solve for the velocity as before:

$$\frac{d\phi}{dt} = k_m \frac{dx}{dt} - \omega_m = 0 \Rightarrow \frac{dx}{dt} = \frac{\omega_m}{k_m}.$$

But recall our definitions of k_m and ω_m:

$$\frac{\omega_m}{k_m} = \frac{\frac{1}{2}\left(\omega_1 - \omega_2\right)}{\frac{1}{2}\left(k_1 - k_2\right)} = \frac{\Delta\omega}{\Delta k}.$$

So we can define the group velocity, the propagation velocity of the signal's envelope, as:

$$v_g = \frac{\Delta\omega}{\Delta k} \Rightarrow \lim_{\Delta k \to 0} \frac{\Delta\omega}{\Delta k} = \frac{d\omega}{dk},$$

which becomes the derivative in the limit when the two frequencies approach each other. As seen before, ω is a related to k through the dispersion relation $\omega(k)$, which is dependent on the medium of propagation. In general, v_g may be greater than, less than, or equal to v_p, depending on the dispersion relation. However, it can be shown from relativistic considerations that $v_g < c$.

A.2.3 Group Index of Refraction

We have now introduced two different velocities associated with a modulated wave, so which velocity is referred to in the equation for the index of refraction? Going back through the derivation, it can be easily concluded that this velocity is in fact the phase velocity, v_p. Consequently, we may rewrite the index of refraction more explicitly as

$$n = \frac{c}{v_p} = c\frac{k}{\omega}. \tag{A.38}$$

As noted, the group velocity of the wave in general may be different from the phase velocity. As such, it makes sense to define a group index of refraction, n_g, given by

$$n_g = \frac{c}{v_g} = c\frac{dk}{d\omega}. \tag{A.39}$$

It should be noted that while Equation A.39 is sometimes called the group index, Equation A.38 is *not* called the phase index. We may continue to refer to this quantity simply as the index of refraction, as we have in the past.

A.2.4 Relationship Between Phase and Group Velocities

Group velocity and phase velocity are intrinsically related. Using Equation A.38, and recalling that in general n is a function of k, we can see that

$$v_g = \frac{d\omega}{dk} = \frac{d}{dk}\left(\frac{c}{n}k\right) = \frac{c}{n} - \left(\frac{c\,k}{n^2}\right)\frac{dn}{dk}. \tag{A.40}$$

Factoring out c/n and recalling that this quantity is simply v_p, we get

$$v_g = v_p\left[1 - \frac{k}{n}\frac{dn}{dk}\right]. \tag{A.41}$$

In the case of a vacuum, where the refractive index is not a function of frequency, and therefore not a function of k, $\frac{dn}{dk} = 0$, and the whole expression reduces to $v_g = v_p$. Since the phase velocity in a vacuum is c, this tells us that

$$v_p v_g = \frac{\omega}{k}\frac{d\omega}{dk} = c^2. \tag{A.42}$$

This may be easily verified as follows. We may rewrite Equation A.38 in terms of ω, giving us $\omega = kc/n$. In a vacuum, $n = 1$, leaving us with $\omega = kc$. Bearing in mind that $\omega/k = v_p = c$ in vacuum, it follows that

$$\left(\frac{\omega}{k}\right)\frac{d\omega}{dk} = \left(\frac{\omega}{k}\right)\frac{d}{dk}(kc) = \frac{\omega}{k}c = c^2.$$

The index of refraction of the troposphere is very close to 1, so electromagnetic waves will travel with speed close to c. In media with higher refractive indices, the speed of propagation will be slower. In general for a wave propagating in a medium with given permittivity ε and permeability μ, $v_p v_g = v^2$, where $v = 1/\sqrt{\mu\varepsilon}$ is the velocity predicted by the wave equation.

A.3 Polarization

We have seen that in general an electromagnetic wave is described vectorially. The polarization state of the wave is determined by the relative amplitude and phase of the electric field components. The several different types of polarization will be elaborated upon in this section.

Throughout this section we will consider electromagnetic waves traveling in the +z-direction without loss of generality. The electric field must oscillate perpendicularly to the direction of propagation, so $\vec{E}$ only has x- and y-components. The instantaneous field vector is given by

$$\vec{E}(t) = \hat{x}E_x + \hat{y}E_y = \text{Re}\left[\hat{x}\underline{E}_{0x}e^{j(\omega t - kz)} + \hat{y}\underline{E}_{0y}e^{j(\omega t - kz)}\right]$$

$$= \hat{x}E_{0x}\cos\left(\omega t - kz + \phi_x\right) + \hat{y}E_{0y}\cos\left(\omega t - kz + \phi_y\right),$$

where again $\underline{E}_{0x}, \underline{E}_{0y}$ are complex-valued and E_{0x}, E_{0y} real-valued. We may choose any plane perpendicular to the z-axis to analyze the polarization of this wave, but for simplicity and without loss of generality we will choose the $z = 0$ plane.

For convenience, we may also define a few parameters. The polarization ratio P is given by

$$P = \frac{E_{0y}e^{j\phi_y}}{E_{0x}e^{j\phi_x}} = |P|\,e^{j\delta},$$

where δ is the relative phase difference between the two components:

$$\delta = \phi_y - \phi_x.$$

If we choose the phase of E_x as our zero reference, we see that

$$E_x = E_{0x}\cos\left(\omega t - kz\right),$$

$$E_y = E_{0y}\cos\left(\omega t - kz + \delta\right).$$

A.3.1 Linear Polarization

First consider the case where $E_{0y} = 0$. The instantaneous field vector then becomes

$$\vec{E}(t) = \hat{x}E_{0x}\cos\left(\omega t + \phi_x\right).$$

We see that this wave oscillates exclusively in the x-direction. Figure A.1a illustrates this visually. This wave is said to be linearly polarized in the x-direction or *horizontally polarized*. Similarly, we may consider the case where $E_{0x} = 0$, which gives the instantaneous field:

$$\vec{E}(t) = \hat{y}E_{0y}\cos\left(\omega t + \phi_y\right).$$

This case may be seen in Figure A.1b. This wave is said to be linearly polarized in the y-direction or *vertically polarized*. In general, both E_{0x} and E_{0y} may be nonzero, giving the instantaneous field vector

$$\vec{E}(t) = \hat{x}E_{0x}\cos\left(\omega t + \phi_x\right) + \hat{y}E_{0y}\cos\left(\omega t + \phi_y\right).$$

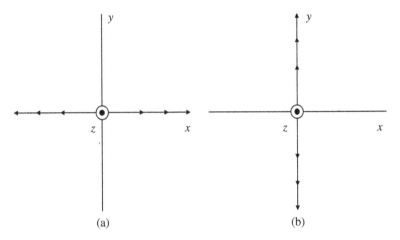

Figure A.1 The arrows indicate the tip of the electric field vector at different time instants. (a) Horizontally Polarized Wave. (b) Vertically Polarized Wave.

In the case where the x and y-components are in phase, i.e. $\phi_x = \phi_y$ ($\delta = 0$), then

$$E_x = E_{0x} \cos(\omega t - kz),$$

$$E_y = E_{0y} \cos(\omega t - kz),$$

so we may write $\vec{E}$ as

$$\vec{E}(t) = (\hat{x} E_{0x} + \hat{y} E_{0y}) \cos(\omega t - kz).$$

This wave is linearly polarized. It oscillates along neither the x- nor the y-axis, but instead at an angle θ. This angle is typically defined with respect to the x-axis and is given by

$$\theta = \arctan\left(\frac{E_{0y}}{E_{0x}}\right) = \arctan|P|.$$

Thus we can see that the angle of polarization is dependent solely on the relative amplitudes of the x and y-components of the wave (Figure A.2).

It is easy to see that this case reduces to polarization strictly in the x and y-directions when θ equals 0 and $\frac{\pi}{2}$, respectively.

In the case where the x and y-components are 180° out of phase, i.e. $\delta = \phi_y - \phi_x = \pm\pi$, we see that

$$E_x = E_{0x} \cos(\omega t - kz),$$

$$E_y = E_{0y} \cos(\omega t - kz \pm \pi),$$

and since $\cos(\phi \pm \pi) = -\cos\phi$, $\vec{E}$ becomes

$$\vec{E}(t) = (\hat{x} E_{0x} - \hat{y} E_{0y}) \cos(\omega t - kz).$$

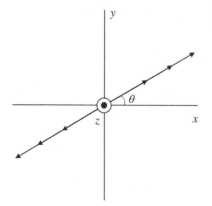

Figure A.2 Linearly polarized wave at an angle θ with respect to the x axis.

This wave is also linearly polarized, and oscillates at an angle θ defined by

$$\theta = \arctan\left(-\frac{E_{0y}}{E_{0x}}\right) = \arctan\left(-|P|\right).$$

A.3.2 Circular Polarization

We will continue to observe waves propagating in the $+z$-direction, focusing on the $z = 0$ plane. Consider the case where the magnitudes of the x and y-components are equal, but they are 90° out of phase. In other words

$$|P| = 1 \Rightarrow E_{0x} = E_{0y} = E_0,$$

$$\delta = \phi_y - \phi_x = \pm\frac{\pi}{2}.$$

Right-Hand Circular Polarization

First let's consider the case where the x-component leads the y component:

$$\phi_x = 0,$$

$$\phi_y = -\frac{\pi}{2}.$$

Thus the electric field is written as

$$\vec{E}(t) = \hat{x}\left|E_x\right|\cos(\omega t) + \hat{y}\left|E_y\right|\cos\left(\omega t - \pi/2\right)$$

$$= \hat{x}E_0\cos(\omega t) + \hat{y}E_0\sin(\omega t).$$

We see that this vector has a magnitude of

$$|E| = \sqrt{E_x^2 + E_y^2} = \sqrt{E_0^2\left(\cos^2(\omega t)\right) + \sin^2(\omega t)} = E_0$$

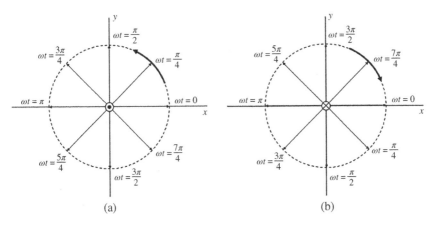

Figure A.3 Right-hand circularly polarized wave propagating (a) toward the observer and (b) away from the observer.

and is directed at an angle of

$$\theta = \arctan\left(\frac{E_y}{E_x}\right) = \arctan\left(\frac{E_0 \sin(\omega t)}{E_0 \cos(\omega t)}\right) = \arctan(\tan(\omega t)) = \omega t.$$

The angle that the electric field vector makes with respect to the x-axis is a function of time. The vector rotates in the xy-plane with a rate of ω, the angular frequency. This type of polarization is called *right-handed* because if one uses their right thumb to point in the direction of propagation, the direction that their fingers will curl is the same as the direction of rotation of the electric field vector. Therefore, if the wave is propagating toward an observer, the $\vec{E}$ vector will appear to rotate counter-clockwise, as seen in Figure A.3a. Conversely, if the wave is propagating away from the observer, the $\vec{E}$ vector will appear to rotate clockwise, as seen in Figure A.3b.

Left-Hand Circular Polarization

We will now consider the opposite case, where the x-component lags the y component.

$$\phi_x = 0,$$

$$\phi_y = +\frac{\pi}{2}.$$

In this case, the electric field becomes

$$\vec{E}(t) = \hat{x}\,|E_x|\cos(\omega t) + \hat{y}\,|E_y|\cos\left(\omega t + \frac{\pi}{2}\right)$$

$$= \hat{x}E_0\cos(\omega t) - \hat{y}E_0\sin(\omega t).$$

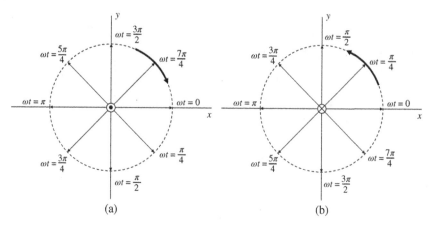

Figure A.4 Left-hand circularly polarized wave propagating (a) toward the observer and (b) away from the observer.

The vector's magnitude is the same as before:

$$|\vec{E}(t)| = \sqrt{E_x^2 + E_y^2} = \sqrt{E_0^2\cos^2(\omega t) + E_0^2\sin^2(\omega t)} = E_0.$$

In this case, the angle becomes

$$\theta = \arctan\left(\frac{E_y}{E_x}\right) = \arctan\left(\frac{-E_0\sin(\omega t)}{E_0\cos(\omega t)}\right) = \arctan(-\tan(\omega t)) = -\omega t,$$

which is similar the right-handed case. The only difference is the negative sign, which indicates that this wave rotates in the opposite direction. Appropriately, this polarization is termed *left-handed*. If one points their left thumb in the direction of propagation, their fingers will curl in the direction of rotation. A wave propagating toward an observer will see the $\vec{E}$ vector rotating clockwise, while a wave propagating away from the observer will appear to rotate counter-clockwise. These two cases are shown in Figure A.4a,b respectively.

A.3.3 Elliptical Polarization

So far we have only observed special cases of polarization in which certain amplitude and phase relationships held true. It is possible to develop a general treatment of polarization regardless of the relationships between these quantities. We will see that this corresponds to an elliptically polarized wave.

We will start with the equations for the x and y-components of the wave, as before:

$$E_x = E_{0x}\cos(\omega t - kz), \tag{A.43}$$

$$E_y = E_{0y}\cos(\omega t - kz + \delta). \tag{A.44}$$

If we square both sides of Equation A.43 and rearrange a bit, we see that

$$\left(\frac{E_x}{E_{0x}}\right)^2 = \cos^2(\omega t - kz) = 1 - \sin^2(\omega t - kz)$$

$$\Rightarrow \sin^2(\omega t - kz) = 1 - \left(\frac{E_x}{E_{0x}}\right)^2. \tag{A.45}$$

Turning our attention to Equation A.44, we see that we may divide by E_{0y} and then use the trigonometric identity

$$\cos(\alpha + \beta) = \cos\alpha\cos\beta - \sin\alpha\sin\beta,$$

which allows us to write

$$\frac{E_y}{E_{0y}} = \cos(\omega t - kz)\cos\delta - \sin(\omega t - kz)\sin\delta.$$

If we solve Equation A.43 for $\cos(\omega t - kz)$, we may make a substitution to obtain

$$\frac{E_y}{E_{0y}} - \frac{E_x}{E_{0x}}\cos\delta = -\sin(\omega t - kz)\sin\delta.$$

We can square both sides of this equation, and then make use of Equation A.45, giving us

$$\left(\frac{E_y}{E_{0y}} - \frac{E_x}{E_{0x}}\cos\delta\right)^2 = \sin^2(\omega t - kz)\sin^2\delta = \left[1 - \left[\frac{E_x}{E_{0x}}\right]^2\right]\sin^2\delta.$$

If we expand this equation, we see that

$$\left(\frac{E_y}{E_{0y}}\right)^2 - 2\left(\frac{E_x}{E_{0x}}\right)\left(\frac{E_y}{E_{0y}}\right)\cos\delta + \left(\frac{E_x}{E_{0x}}\right)^2(\cos^2\delta + \sin^2\delta) = \sin^2\delta,$$

which we can finally simplify to

$$\left(\frac{E_y}{E_{0y}}\right)^2 + \left(\frac{E_x}{E_{0x}}\right)^2 - 2\left(\frac{E_x}{E_{0x}}\right)\left(\frac{E_y}{E_{0y}}\right)\cos\delta = \sin^2\delta. \tag{A.46}$$

This is the equation for an ellipse in terms of the variables E_x and E_y, whose semi-major axis is tilted at an angle ϕ with respect to the x-axis (shown in Figure A.5). This angle is given by

$$\tan(2\phi) = \frac{2E_{0x}E_{0y}}{E_{0x}^2 - E_{0y}^2}\cos\delta. \tag{A.47}$$

The lengths of the principal axes of this tilted ellipse may be written in terms of ϕ.

$$a^2 = E_{0x}^2\cos^2\phi + E_{0y}^2\sin^2\phi + 2E_{0x}E_{0y}\cos\delta\cos\phi\sin\phi, \tag{A.48}$$

$$b^2 = E_{0x}^2\sin^2\phi + E_{0y}^2\cos^2\phi - 2E_{0x}E_{0y}\cos\delta\cos\phi\sin\phi. \tag{A.49}$$

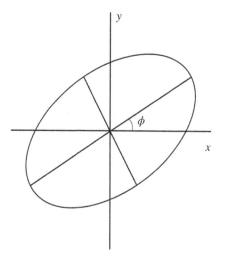

Figure A.5 Ellipse oriented at an angle ϕ with respect to the x-axis.

In these expressions, a denotes the semi-major axis of the ellipse and b denotes the semi-minor axis. From Equations A.47–1.49, we see that the shape and orientation of the ellipse are dependent on the amplitudes of the components, E_{0x} and E_{0y} as well as their relative phase δ.

When $\delta = \pm n\pi$, for $n = 0, \pm 1, \pm 2, \ldots$, Equation A.46 reduces to

$$\left(\frac{E_x}{E_{0x}} \pm \frac{E_y}{E_{0y}}\right)^2 = 0 \Rightarrow \frac{E_x}{E_{0x}} = \pm\frac{E_y}{E_{0y}} \Rightarrow E_y = \pm\frac{E_{0y}}{E_{0x}}E_x,$$

which is the equation for a line with slope $|E_{0y}/E_{0x}| = |P|$. This is exactly the case we studied in Section A.3.1, which simply tells us that linear polarization is a special case of an elliptical polarization. Similarly, when $E_{0x} = E_{0y} = E_0$ and $\delta = \pm m\pi/2$, for $m = \pm 1, \pm 2, \ldots$, Equation A.46 reduces to

$$\left(\frac{E_x}{E_0}\right)^2 + \left(\frac{E_y}{E_0}\right)^2 = 1,$$

which is the equation for a circle. A circularly polarized wave can also be seen as a special case of an elliptical polarization.

Figures A.6 and A.7 demonstrate two different polarization ratios, and how the shape of the polarization changes for different phase relationships. Note that as a wave passes through a linear polarization, the ellipse shifts from one handedness to another.

A.3.4 Jones Vectors and Decomposing Polarizations

Jones vector notation is a concise notation for polarization, named after its creator, R.C. Jones. Since the electric field of a plane wave always oscillates in the plane

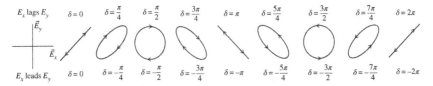

Figure A.6 Polarization ellipses for $E_{0x} = E_{0y}$ with various phase relationships δ.

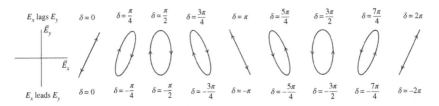

Figure A.7 Polarization ellipses for $E_{0x} = \frac{1}{2}E_{0y}$ with various phase relationships δ.

perpendicular to propagation, the polarization state may be uniquely expressed as a two-vector:

$$\vec{E} = \begin{bmatrix} E_x \\ E_y \end{bmatrix} = \begin{bmatrix} E_{0x}e^{j\phi_x} \\ E_{0y}e^{j\phi_y} \end{bmatrix}. \tag{A.50}$$

In the case where $E_{0y} = 0$, we get a horizontally polarized wave

$$\vec{E} = \begin{bmatrix} E_{0x}e^{j\phi_x} \\ 0 \end{bmatrix}$$

and when $E_{0x} = 0$, we get a vertically polarized wave

$$\vec{E} = \begin{bmatrix} 0 \\ E_{0y}e^{j\phi_y} \end{bmatrix}.$$

By normalizing these vectors, we obtain

$$\hat{x} = \begin{bmatrix} 1 \\ 0 \end{bmatrix}, \quad \hat{y} = \begin{bmatrix} 0 \\ 1 \end{bmatrix},$$

where $\hat{x}$ and $\hat{y}$ are the unit vectors for horizontal and vertical polarizations, respectively. These polarizations are orthogonal to each other, i.e. $\hat{x} \cdot \hat{y} = 0$. Returning to Equation A.50, if we impose the constraints $E_{0x} = E_{0y}$ and $\delta = -\pi/2$, we see that

$$\vec{E} = \begin{bmatrix} E_{0x}e^{j\phi_x} \\ E_{0x}e^{j(\phi_x - \frac{\pi}{2})} \end{bmatrix} = \begin{bmatrix} E_{0x}e^{j\phi_x} \\ -jE_{0x}e^{j\phi_x} \end{bmatrix}.$$

Normalizing results in the unit vector

$$\vec{R} = \frac{1}{\sqrt{2}} \begin{bmatrix} 1 \\ -j \end{bmatrix},$$

which represents a right-hand circularly polarized wave. Using a similar approach with $\delta = \pi/2$ results in the unit vector

$$\vec{L} = \frac{1}{\sqrt{2}} \begin{bmatrix} 1 \\ j \end{bmatrix},$$

which represents a left-hand circularly polarized wave. These two vectors are also orthogonal. For complex vectors, orthogonality is shown via the more general expression $\vec{A} \cdot \vec{B}^* = 0$, in which case

$$\hat{L} \cdot \hat{R}^* = \frac{1}{2} \left[1(1)^* + j(-j)^* \right] = 0.$$

Choosing a Basis

So far, we have found four different unit vectors which may be used to represent a polarization. Since a Jones vector is a column matrix of rank 2, we only need two linearly independent vectors to form a basis. It makes the most sense to choose a basis where both elements are linear or both elements are circular (as opposed to one of each). This is useful in media such as the ionosphere for example, where the characteristic modes may correspond to circularly polarized waves. If we do this, we can show that the remaining vectors may be expressed in terms of the basis vectors. In other words, if we choose $\hat{x}$ and $\hat{y}$ as our basis, then

$$\hat{R} = \frac{1}{\sqrt{2}} \begin{bmatrix} 1 \\ -j \end{bmatrix} = \frac{1}{\sqrt{2}} \begin{bmatrix} 1 \\ 0 \end{bmatrix} - j\frac{1}{\sqrt{2}} \begin{bmatrix} 0 \\ 1 \end{bmatrix} = \frac{1}{\sqrt{2}} \left(\hat{x} - j\hat{y} \right), \tag{A.51}$$

$$\hat{L} = \frac{1}{\sqrt{2}} \begin{bmatrix} 1 \\ j \end{bmatrix} = \frac{1}{\sqrt{2}} \begin{bmatrix} 1 \\ 0 \end{bmatrix} + j\frac{1}{\sqrt{2}} \begin{bmatrix} 0 \\ 1 \end{bmatrix} = \frac{1}{\sqrt{2}} \left(\hat{x} + j\hat{y} \right). \tag{A.52}$$

Similarly, if we choose our basis to be $\hat{R}$ and $\hat{L}$, we see that

$$\hat{x} = \begin{bmatrix} 1 \\ 0 \end{bmatrix} = \frac{1}{2} \begin{bmatrix} 1 \\ -j \end{bmatrix} + \frac{1}{2} \begin{bmatrix} 1 \\ j \end{bmatrix} = \frac{1}{\sqrt{2}} \left(\hat{R} + \hat{L} \right), \tag{A.53}$$

$$\hat{y} = \begin{bmatrix} 0 \\ 1 \end{bmatrix} = \frac{j}{2} \begin{bmatrix} 1 \\ -j \end{bmatrix} - \frac{j}{2} \begin{bmatrix} 1 \\ j \end{bmatrix} = \frac{j}{\sqrt{2}} \left(\hat{R} - \hat{L} \right). \tag{A.54}$$

These four unit vectors are most commonly chosen as basis vectors because they are the simplest to work with. However, any two orthogonal Jones vectors may be chosen as a basis for polarization states. For example, the unit vector for a linearly polarized wave at an angle of 45° is

$$\hat{D} = \frac{1}{\sqrt{2}} \begin{bmatrix} 1 \\ 1 \end{bmatrix}.$$

The vector orthogonal to this is a linearly polarized wave at an angle of −45°, and is given by

$$\hat{A} = \frac{1}{\sqrt{2}} \begin{bmatrix} 1 \\ -1 \end{bmatrix}.$$

Another example is a pair of orthogonal vectors representing elliptical polarizations:

$$\hat{e}_1 = \frac{1}{\sqrt{5}} \begin{bmatrix} 1 \\ 2j \end{bmatrix}, \quad \hat{e}_2 = \frac{1}{\sqrt{5}} \begin{bmatrix} 2 \\ -j \end{bmatrix}.$$

These two pairs of vectors also form bases for the polarization space. The basis vectors need not be unit vectors (though unit vectors are easiest to work with), and in general any pair of vectors which satisfy $\vec{A} \cdot \vec{B}^* = 0$ may be chosen as a basis.

Decomposing Polarizations

Earlier, we showed that unit vectors representing polarization states could be written in terms of a chosen basis. In general any arbitrary polarization can be written as a linear combination of two orthogonal linear polarizations. Using $\hat{x}$ and $\hat{y}$ as our basis:

$$\vec{E} = \begin{bmatrix} E_{0x}e^{j\phi_x} \\ E_{0y}e^{j\phi_y} \end{bmatrix} = \begin{bmatrix} E_{0x}e^{j\phi_x} \\ 0 \end{bmatrix} + \begin{bmatrix} 0 \\ E_{0y}e^{j\phi_y} \end{bmatrix}$$

$$= E_{0x}e^{j\phi_x} \begin{bmatrix} 1 \\ 0 \end{bmatrix} + E_{0y}e^{j\phi_y} \begin{bmatrix} 0 \\ 1 \end{bmatrix} = E_{0x}e^{j\phi_x}\hat{x} + E_{0y}e^{j\phi_y}\hat{y}. \tag{A.55}$$

We may also show that any polarization can be written as the superposition of one left-hand and one right-hand circular polarization. Using Equation A.55 and substituting in Equations A.53 and A.54 results in:

$$\vec{E} = E_{0x}e^{j\phi_x} \left(\frac{1}{2} \begin{bmatrix} 1 \\ -j \end{bmatrix} + \frac{1}{2} \begin{bmatrix} 1 \\ j \end{bmatrix} \right) - jE_{0y}e^{j\phi_y} \left(-\frac{1}{2} \begin{bmatrix} 1 \\ -j \end{bmatrix} + \frac{1}{2} \begin{bmatrix} 1 \\ j \end{bmatrix} \right).$$

We see that we may now group like terms as follows:

$$\vec{E} = \frac{\left(E_{0x}e^{j\phi_x} + jE_{0y}e^{j\phi_y}\right)}{2} \begin{bmatrix} 1 \\ -j \end{bmatrix} + \frac{\left(E_{0x}e^{j\phi_x} - jE_{0y}e^{j\phi_y}\right)}{2} \begin{bmatrix} 1 \\ j \end{bmatrix}. \tag{A.56}$$

Both of the vectors in Equation A.56 have magnitude $\sqrt{2}$, so we can turn them into unit vectors by multiplying through by $1/\sqrt{2}$.

$$\vec{E} = \frac{\left(E_{0x}e^{j\phi_x} + jE_{0y}e^{j\phi_y}\right)}{\sqrt{2}} \begin{bmatrix} \frac{1}{\sqrt{2}} \\ \frac{-j}{\sqrt{2}} \end{bmatrix} + \frac{\left(E_{0x}e^{j\phi_x} - jE_{0y}e^{j\phi_y}\right)}{\sqrt{2}} \begin{bmatrix} \frac{1}{\sqrt{2}} \\ \frac{j}{\sqrt{2}} \end{bmatrix}.$$

These vectors are simply $\hat{R}$ and $\hat{L}$, so we may finally write $\vec{E}$ as

$$\vec{E} = \left(E_{0R}\hat{R} + E_{0L}\hat{L}\right),$$

where E_{0R} and E_{0L} are the representation of $\vec{E}$ (complex-valued coefficients) in the $\hat{R}$ and $\hat{L}$ basis.

A.4 Derivation of Rayleigh Scattering

Rayleigh scattering is a phenomenon that occurs when an electromagnetic wave is scattered by a particle whose size is much smaller than the wavelength of the wave. We will derive the equation for the intensity of the light scattered by such a particle. In this derivation, we will use a water droplet as an illustrative example of a spherical scatterer.[1] The derivation will proceed in three major steps. First, we will derive the electric potential due to a single electric dipole. Second, we will obtain an expression for the total effective dipole moment of the water droplet, which depends directly on the magnitude of the incident electric field. Third, we will show that if the external electric field is time-varying, it causes the effective dipole to re-radiate an electromagnetic wave.

Even with the simplifying assumption of a uniform electric field, the full solution to this problem is quite involved. To make the derivation more manageable, we will first introduce some further simplifications. As we go along, the complexity will be reintroduced until we arrive at the full solution. The simplifications we will make are as follows:

1. We begin by considering only a single molecule, as opposed to the numerous molecules which comprise the water droplet.
2. Once multiple molecules are reintroduced, we will need to compute the electric field inside the water droplet. In reality, the internal field is the superposition of the external field with an electric field created by all of the molecule's dipole moments. We will initially ignore the external field, and then add it back in later.
3. For the majority of the derivation, we will assume a static (i.e. non-time-varying) external electric field. This will be reintroduced toward the end of the derivation when we discuss re-radiation.

A.4.1 Electric Potential of an Ideal Dipole

In the presence of a static electric field, the electron cloud of an atom or molecule will displace slightly from the nuclei, creating a charge separation over a very small distance.[2] In other words, the medium will exhibit a net *polarization*[3] that will be aligned with the external electric field. The induced charge separation can be

---

1 Note that there is nothing special about the choice of water as the medium for the particle. Other spherical particles also adhere to the principles discussed in this section. The exact behavior will differ slightly depending on the material's electric susceptibility (which is related to its index of refraction). Susceptibility is discussed in detail in Section B.2.
2 In the case of polar molecules such as water, a preferential *orientation* will be induced as well.
3 Note that this refers to material polarization, which is an entirely distinct phenomenon from wave polarization.

approximated as an ideal electric dipole – i.e. two point charges, one with charge $+q$ and the other with charge $-q$, separated by a distance d.

To obtain the electric field seen by a distant observer, we can start with the electric potential of an ideal dipole:

$$V = \frac{kq}{r_1} + \frac{k(-q)}{r_2}, \tag{A.57}$$

where V is the dipole potential, $k = 1/(4\pi\varepsilon_0)$, ε_0 is the permittivity of free space, r_1 is the distance of the charge $+q$ to the observer, and r_2 is the distance of the charge $-q$ to the observer. It will also be useful to define another distance, r, from the center of the dipole to the observer. The problem setup is illustrated in Figure A.8.

The electric field around the dipole depends on the angle θ, which is measured with respect to the dipole's axis. In general, this relationship is quite complex, especially near the dipole. However, for a distant observer (i.e. $r \gg d$), we can derive a simple approximate expression.

The distances r_1 and r_2 depend on θ, and in general will not be identical. It is more useful to rewrite them in terms of r, which is fixed at the center of the dipole and does not depend θ. From Figure A.8, it can be seen that:

$$r_1 = r - \frac{d}{2}\cos\theta \tag{A.58}$$

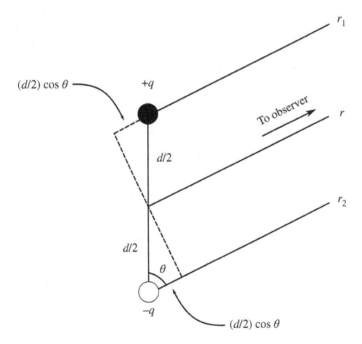

Figure A.8 An ideal dipole whose center is at a distance r from an observer.

and

$$r_2 = r + \frac{d}{2}\cos\theta. \tag{A.59}$$

With these expressions, Equation A.57 becomes:

$$V = \frac{kq}{r - \frac{d}{2}\cos\theta} - \frac{kq}{r + \frac{d}{2}\cos\theta} = \frac{kq}{r}\left(\frac{1}{1 - \frac{d}{2r}\cos\theta} - \frac{1}{1 + \frac{d}{2r}\cos\theta}\right). \tag{A.60}$$

Using Taylor's series, it is easy to show that $(1 + \epsilon)^n \approx 1 + n\epsilon$ for small ϵ.

$$\frac{1}{(1 + \epsilon)} \approx (1 - \epsilon) \tag{A.61}$$

and

$$\frac{1}{(1 - \epsilon)} \approx (1 + \epsilon). \tag{A.62}$$

Since $r \gg d$, $d/2r$ is very small and we can therefore apply the preceding approximations in Equation A.60 to obtain:

$$V \approx \frac{kq}{r}\left(1 + \frac{d}{2r}\cos\theta - 1 + \frac{d}{2r}\cos\theta\right) = \frac{kqd}{r^2}\cos\theta. \tag{A.63}$$

The quantity qd is called the *dipole moment*, and is commonly denoted p. Using this, and expanding k, we may obtain the final expression for the dipole's electric potential:

$$\boxed{V = \frac{p}{4\pi\varepsilon_0 r^2}\cos\theta.} \tag{A.64}$$

Note that the dipole potential has a faster decay with distance, $1/r^2$, that the $1/r$ potential produced by an isolated point charge.

A.4.2 Effective Dipole Moment of a Spherical Scattering Particle

At this point, we will now reintroduce the other molecules which comprise the water droplet. All of the molecules will polarize in the presence of the external electric field,[4] and each will develop a dipole moment p. Dipole moments sum

---

4 Note that we are still ignoring the magnitude of the external field at this point. It is important to know that the external field is responsible for the droplet's polarization, but it would complicate calculations to include its contribution to the internal field at this point. We will first calculate the internal field solely due to the dipoles, and then consider the external field later by invoking superposition.

vectorially, but since the external electric field is uniform, all the vectors point in the same direction. Therefore, the equivalent dipole moment of the water droplet, p_{tot}, is simply the sum of each individual dipole moment p_i:

$$p_{tot} = \sum_i p_i. \tag{A.65}$$

The total sphere of dipoles can therefore be replaced by a single dipole with equivalent dipole moment p_{tot}. Note that this equivalent dipole is also aligned along the direction of the external electric field. We are assume this direction to be the z axis without loss of generality. We can next determine the electric potential on the surface of the water droplet. We will denote this quantity V_s to distinguish it from the more general expression given in Equation A.64:

$$V_s = \frac{p_{tot}}{4\pi\varepsilon_0 a^2} \cos\theta. \tag{A.66}$$

Note that the origin of our coordinate system is at the center of the water droplet, so the distance to the droplet's surface is its radius. We have denoted the droplet radius as a. We can simplify this expression a bit by introducing a quantity known as the polarization density, denoted P, which is simply the total dipole moment per unit volume. Since we have assumed a spherical water droplet, it follows that:

$$p_{tot} = \frac{4\pi a^3}{3} P. \tag{A.67}$$

Substituting this into Equation A.66 and simplifying, we obtain:

$$V_s = \frac{Pa}{3\varepsilon_0} \cos\theta. \tag{A.68}$$

It is convenient to convert to cylindrical coordinates with a z-axis aligned with the dipole axis. In these coordinates, $z = r\cos\theta$, so at the droplet's surface where $r = a$ we have:

$$V_s = \frac{Pz}{3\varepsilon_0}. \tag{A.69}$$

Equation A.69 specifies the electric potential on every point of the water droplet's surface. We are interested in determining the field inside the water droplet as well. Any solution for the electric potential inside the water droplet must satisfy two conditions:

1. The electric potential V must be a solution to Laplace's equation, i.e. $\nabla^2 V = 0$.
2. On the surface of the water droplet, the electric potential must equal the value in Equation A.69. That is, at the surface, $V = V_s$.

One solution is to simply let $V = V_s$ *everywhere* within the water droplet. This clearly fulfills condition 2. The Laplacian, represented in cylindrical coordinates, is:

$$\nabla^2 f = \frac{1}{\rho} \frac{\partial}{\partial \rho} \left(\rho \frac{\partial f}{\partial \rho} \right) + \frac{1}{\rho^2} \frac{\partial^2 f}{\partial \theta^2} + \frac{\partial^2 f}{\partial z^2}. \tag{A.70}$$

V_s only depends on z, so the first two terms of $\nabla^2 V_s$ are zero. The third term also evaluates to zero, since the first derivative of V_s with respect to z is a constant value $P/3\varepsilon_0$. Therefore, V_s satisfies condition 1 as well, and is a valid solution for the potential inside the water droplet.

We may now apply a result known as the uniqueness theorem (which is beyond the scope of this text, but may be found in numerous books on electromagnetics, such as [1, pp. 119, 120] and [264, pp. 133, 134]). The uniqueness theorem states that if a potential V is specified on some closed surface S, then any solution to Laplace's equation within the volume enclosed by S is the *only* solution. This means that, since we just showed that V_s is a possible value for the potential inside the water droplet, it is in fact the *only* possible value for the potential inside the water droplet. Thus, as we proceed, we can be confident that our use of V_s within the sphere is justified.

The electric field is related to the electric potential by the equation $\vec{E} = -\nabla V$, so it follows that the internal electric field produced by the droplet's material polarization is given by:

$$\vec{E} = \hat{z} E_{pol} = -\nabla V_s = -\hat{z} \frac{\partial}{\partial z} \left(\frac{Pz}{3\varepsilon_0} \right) = -\hat{z} \frac{P}{3\varepsilon_0}, \tag{A.71}$$

where, similar to the Laplacian, the only nonzero component of the gradient is the z-component. Equation A.71 tells us that the electric field within the water droplet is uniform, and it points in the $-z$ direction.

Recall that we have been ignoring the magnitude of the external electric field up to this point. We are now ready to reintroduce its effect on the total internal field. Due to the superposition principle, the total electric field within the droplet is simply the sum of the electric field due to the dipoles (i.e. Equation A.71) and the external field. Since all these fields are aligned in the z direction, we can work with scalar quantities only and write:

$$E_{in} = E_{pol} + E_{ext}. \tag{A.72}$$

The polarization density is related to the total electric field through the equation:

$$P = \chi \varepsilon_0 E_{in} = (\varepsilon_r - 1) \varepsilon_0 E_{in}. \tag{A.73}$$

For further details, see Appendix B.2. We may now combine Equations A.71 and A.73 to obtain:

$$E_{pol} = -\frac{(\varepsilon_r - 1)\varepsilon_0}{3\varepsilon_0}E_{in} = -\frac{(\varepsilon_r - 1)}{3}E_{in}. \tag{A.74}$$

Substituting this into Equation A.72 and simplifying, we obtain an expression for the internal field in terms of the external field.

$$E_{in} = \left(\frac{3}{2 + \varepsilon_r}\right)E_{ext}. \tag{A.75}$$

This, in conjunction with Equation A.73, allows us to write the dipole moment per unit volume in the water droplet:

$$P = 3\varepsilon_0\left(\frac{\varepsilon_r - 1}{2 + \varepsilon_r}\right)E_{ext}. \tag{A.76}$$

Using Equation A.67, we obtain the effective dipole moment of a spherical scatterer:

$$\boxed{p_{tot} = 4\pi a^3 \varepsilon_0 \left(\frac{\varepsilon_r - 1}{2 + \varepsilon_r}\right)E_{ext}.} \tag{A.77}$$

Thus we see that the effective electric dipole for the water droplet depends directly on the magnitude of the external field.

A.4.3 Re-radiation by a Scattering Particle

We will now reintroduce a time-varying external field. A time-harmonic electric field can be expressed as:

$$E_{ext} = E_0\cos(\omega t), \tag{A.78}$$

where E_0 is the amplitude of the wave, ω is its frequency, and t is time. In the presence of a varying external field, each of the dipoles within the water droplet will be driven at the incident field's frequency. Therefore, via Equation A.77, the effective dipole of the water droplet will also oscillate over time:

$$p_{tot} = 4\pi a^3 \varepsilon_0\left(\frac{\varepsilon_r - 1}{2 + \varepsilon_r}\right)E_0\cos(\omega t). \tag{A.79}$$

A time-varying dipole can be modeled in one of two ways. The most natural way to think of it is two charges of fixed magnitude q separated by a distance d that oscillates over time. This is similar to how one might model two spheres connected by a spring. Mathematically, this model would be expressed as follows:

$$p_{tot}(t) = qd(t) = q(d_0 \cos(\omega t)). \tag{A.80}$$

However, dealing with moving charges is difficult, so we will instead consider the equivalent model of a dipole with fixed length, but whose charges vary sinusoidally:

$$p_{tot}(t) = q(t)d = (q_0 \cos(\omega t))d. \tag{A.81}$$

This model is very similar to the static dipole case we derived previously. Note that the magnitude of the oscillating dipole is:

$$p_0 = q_0 d = 4\pi a^3 \varepsilon_0 \left(\frac{\varepsilon_r - 1}{2 + \varepsilon_r} \right) E_0. \tag{A.82}$$

The time-varying dipole will re-radiate an electromagnetic wave. We wish to determine the intensity of this wave as seen in the so-called *far-field* regime – that is, at some distance r from the dipole such that $r \gg \lambda$. In our example, $\lambda \approx 19$ cm is the wavelength of the incident L1 GPS signal. The intensity of an electromagnetic wave can be expressed in terms of the magnitudes of its component electric and magnetic fields:

$$I \propto \vec{E} \times \vec{B}, \tag{A.83}$$

where I is the intensity of the wave, $\vec{E}$ is the electric field vector, and $\vec{B}$ is the magnetic field vector. $\vec{E}$ and $\vec{B}$ can, in turn be derived from the electric scalar potential V and the magnetic vector potential $\vec{A}$. In particular, the following equations relate these quantities[5]:

$$\vec{E} = -\nabla V - \frac{\partial \vec{A}}{\partial t}, \tag{A.84}$$

$$\vec{B} = \nabla \times \vec{A}. \tag{A.85}$$

We will now derive V and $\vec{A}$ induced by p_0.

We can determine the electric scalar potential (henceforth, simply "potential") at some distance r from the center of the dipole, as before. However, now that the dipole moment varies sinusoidally, we need to consider the time it takes for the variation to propagate out to a distance r. Any change in the electric potential at a distance r at time t must have originated at some earlier time $t' = t - r/c$. Therefore, the potential looks like a modified version of Equation A.57:

$$V = \frac{kq_0 \cos\left(\omega\left(t - \frac{r_1}{c}\right)\right)}{r_1} - \frac{kq_0 \cos\left(\omega\left(t - \frac{r_2}{c}\right)\right)}{r_2}. \tag{A.86}$$

---

5 For details, see [1, Chapters 2,5].

As before, we will assume that the dipole is at the origin and the potential is being observed at some distance much larger than the separation of the dipole charges, i.e. $r \gg d$. Therefore, Equations A.58 and A.59 hold. Substituting these relationships in for r_1 and r_2, we obtain:

$$V = \frac{kq_0 \cos\left(\omega\left(t - \frac{1}{c}\left(r - \frac{d}{2}\cos\theta\right)\right)\right)}{r - \frac{d}{2}\cos\theta} - \frac{kq_0 \cos\left(\omega\left(t - \frac{1}{c}\left(r + \frac{d}{2}\cos\theta\right)\right)\right)}{r + \frac{d}{2}\cos\theta}.$$

$$(A.87)$$

Letting $t' = t - r/c$ and simplifying the previous expression a bit, we obtain:

$$V = \frac{kq_0}{r}\left[\frac{\cos\left(\omega t' + \frac{\omega d}{2c}\cos\theta\right)}{1 - \frac{d}{2r}\cos\theta} - \frac{\cos\left(\omega t' - \frac{\omega d}{2c}\cos\theta\right)}{1 + \frac{d}{2r}\cos\theta}\right].$$

$$(A.88)$$

Using the trigonometric identity $\cos(a \pm b) = \cos a \cos b \mp \sin a \sin b$, we can write:

$$\cos\left(\omega t' \pm \frac{\omega d}{2c}\cos\theta\right) = \cos\left(\omega t'\right)\cos\left(\frac{\omega d}{2c}\cos\theta\right)$$

$$\mp \sin\left(\omega t'\right)\sin\left(\frac{\omega d}{2c}\cos\theta\right).$$

$$(A.89)$$

At this point, we will apply the approximation that d is much smaller than the wavelength of the dipole oscillation, λ. The dipole oscillation is driven by the incident electromagnetic wave. In our example, this is the L1 GPS signal with $\lambda \approx 19\,cm$. This is much larger than d, which is approximately the size of an atom or molecule (on the order of tens to hundreds of picometers). Since $\lambda = 2\pi c/\omega$, it follows that if $\lambda \gg d$, then $c/\omega \gg d$. This means that:

$$\frac{\omega d}{2c} \approx 0.$$

$$(A.90)$$

For small α, $\cos\alpha \approx 1$ and $\sin\alpha \approx \alpha$ to first order. Therefore, we can simplify Equation A.89 to:

$$\cos\left(\omega t' \pm \frac{\omega d}{2c}\cos\theta\right) = \cos\left(\omega t'\right) \mp \sin\left(\omega t'\right)\frac{\omega d}{2c}\cos\theta.$$

$$(A.91)$$

We can substitute this into Equation A.88, which gives us:

$$V = \frac{kq_0}{r}\left(\frac{\cos\left(\omega t'\right) - \sin\left(\omega t'\right)\frac{\omega d}{2c}\cos\theta}{1 - \frac{d}{2r}\cos\theta} - \frac{\cos\left(\omega t'\right) + \sin\left(\omega t'\right)\frac{\omega d}{2c}\cos\theta}{1 + \frac{d}{2r}\cos\theta}\right).$$

$$(A.92)$$

Now, recalling that since $r \gg d$, Equations A.61 and A.62 hold, and we may use these approximations to obtain:

$$V \approx \frac{kq_0}{r} \left(\left[\cos(\omega t') - \sin(\omega t') \frac{\omega d}{2c} \cos\theta \right] \left[1 + \frac{d}{2r} \cos\theta \right] \right.$$
$$\left. - \left[\cos(\omega t') + \sin(\omega t') \frac{\omega d}{2c} \cos\theta \right] \left[1 - \frac{d}{2r} \cos\theta \right] \right). \tag{A.93}$$

Expanding and simplifying this expression, we end up with:

$$V \approx \frac{kq_0 d \cos\theta}{r} \left(\frac{1}{r} \cos(\omega t') - \frac{\omega}{c} \sin(\omega t') \right). \tag{A.94}$$

We are interested in determining the potential seen in the far-field regime, where $r \gg \lambda$. In this regime, it is also true that $r \gg c/\omega$, so Equation A.94 is dominated by its second term. With this simplification (and also using $p_0 = q_0 d$ and $k = 1/4\pi\varepsilon_0$), we obtain:

$$V = -\frac{p_0 \cos\theta\omega}{4\pi\varepsilon_0 cr} \sin(\omega t'). \tag{A.95}$$

Now that we have derived V, all that remains is to derive $\vec{A}$, and then put everything together. $\vec{A}$ can be expressed in terms of the current induced by the time-varying dipole. We will assume a coordinate system in which the dipole axis lies in the z-direction. Thus, the current $\vec{I}$ is given by:

$$\vec{I} = \frac{dq}{dt} = -q_0 \omega \sin(\omega t)\hat{z}. \tag{A.96}$$

The general equation for the magnetic vector potential (henceforth, simply magnetic potential) $\vec{A}$ that arises due to a one-dimensional current source is [1, p. 245]:

$$\vec{A} = \frac{\mu_0}{4\pi} \int \frac{\vec{I}}{r} dl, \tag{A.97}$$

where μ_0 is the permeability of free space, $\vec{I}$ is the current, r is the distance from the current source, and dl is a differential length element.

The magnetic potential experienced at r must have propagated there at speed c. With this in mind, if we integrate Equation A.96 along the length of the dipole, with the origin at its center (i.e. from $-d/2$ to $d/2$), we get:

$$\vec{A} = \frac{\mu_0}{4\pi} \int_{-d/2}^{d/2} \frac{-q_0 \omega \sin(\omega t') \hat{z}}{r} dz. \tag{A.98}$$

This evaluates to:

$$\vec{A} = -\frac{\mu_0 (q_0 d) \omega}{4\pi r} \sin(\omega t') \hat{z}. \tag{A.99}$$

Finally, with $p_0 = q_0 d$, we have:

$$\boxed{\vec{A} = -\frac{\mu_0 p_0 \omega}{4\pi r} \sin\left(\omega t'\right) \hat{z}.}$$ (A.100)

Now we are in a position to use Equations A.84 and A.85. We will start by building up $\vec{E}$ from ∇V and $d\vec{A}/dt$. We will use spherical coordinates; recall that in spherical coordinates the gradient is given by:

$$\nabla V = \frac{dV}{dr}\hat{r} + \frac{1}{r}\frac{dV}{d\theta}\hat{\theta} + \frac{1}{r\sin\theta}\frac{dV}{d\phi}\hat{\phi}.$$ (A.101)

Applying this to Equation A.95, we get:

$$\nabla V = -\frac{p_0\omega}{4\pi\varepsilon_0 c}\left[\cos\theta\left(-\frac{1}{r^2}\sin\left(\omega t'\right) - \frac{\omega}{rc}\cos\left(\omega t'\right)\right)\hat{r}\right.$$
$$\left. - \frac{\sin\theta}{r^2}\sin\left(\omega t'\right)\hat{\theta}\right].$$ (A.102)

Recalling that we are concerned with the far-field regime, where $r \gg c/\omega$, the first and last terms vanish, leaving:

$$\nabla V \approx \frac{p_0\omega^2}{4\pi\varepsilon_0 c^2}\left(\frac{\cos\theta}{r}\right)\cos\left(\omega t'\right)\hat{r}.$$ (A.103)

The time derivative of $\vec{A}$ is:

$$\frac{\partial\vec{A}}{\partial t} = -\frac{\mu_0 p_0 \omega^2}{4\pi r}\cos\left(\omega t'\right)\hat{z}.$$ (A.104)

In spherical coordinates, $\hat{z} = \cos\theta\hat{r} - \sin\theta\hat{\theta}$, so Equation A.104 becomes:

$$\frac{\partial\vec{A}}{\partial t} = -\frac{\mu_0 p_0 \omega^2}{4\pi r}\cos\left(\omega t'\right)(\cos\theta\,\hat{r} - \sin\theta\,\hat{\theta}).$$ (A.105)

Using Equations A.103 and A.105, and noting that $\varepsilon_0 c^2 = 1/\mu_0$, we can compute $\vec{E}$:

$$\vec{E} = -\nabla V - \frac{\partial\vec{A}}{\partial t} = -\frac{\mu_0 p_0 \omega^2}{4\pi}\left(\frac{\sin\theta}{r}\right)\cos\left(\omega t'\right)\hat{\theta}.$$ (A.106)

Next, we will compute $\vec{B}$, which is equal to $\nabla\times\vec{A}$. In spherical coordinates, the curl of $\vec{A}$ reduces to:

$$\nabla\times\vec{A} = \frac{1}{r}\left(\frac{\partial(rA_\theta)}{\partial r} - \frac{\partial A_r}{\partial\theta}\right)\hat{\phi},$$ (A.107)

where A_θ is the component of $\vec{A}$ in the θ-direction, and A_r is the component of $\vec{A}$ in the r-direction. The $\hat{r}$ and $\hat{\theta}$ components vanish because they depend on A_ϕ and $d\vec{A}/d\phi$, both of which are zero. Applying the curl to Equation A.100, we get:

$$\nabla\times\vec{A} = -\frac{\mu_0 p_0 \omega}{4\pi r}\left(\frac{\omega}{c}\sin\theta\cos\left(\omega t'\right) + \frac{\sin\theta}{r}\sin\left(\omega t'\right)\right)\hat{\phi}.$$ (A.108)

Once again applying the far-field approximation, the second term vanishes, leaving:

$$\vec{B} = \nabla \times \vec{A} = -\frac{\mu_0 p_0 \omega^2}{4\pi c} \left(\frac{\sin\theta}{r}\right) \cos\left(\omega t'\right) \hat{\boldsymbol{\phi}}. \tag{A.109}$$

From $\vec{E}$ and $\vec{B}$, we can compute the Poynting vector of the corresponding electromagnetic wave:

$$\vec{S} = \frac{1}{\mu_0}(\vec{E} \times \vec{B}) = \frac{\mu_0}{c} \left[\frac{p_0 \omega^2}{4\pi} \left(\frac{\sin\theta}{r}\right) \cos\left(\omega t'\right)\right]^2 \hat{r}. \tag{A.110}$$

The time average of the Poynting vector over one complete cycle is:

$$\langle \vec{S} \rangle = \frac{1}{2} S_0 \hat{r}, \tag{A.111}$$

where S_0 is the magnitude of the vector. The time-averaged power density of an electromagnetic wave is also referred to as its intensity. Therefore, the intensity the re-radiated wave is:

$$\langle \vec{S} \rangle = \frac{1}{2} \left(\frac{\mu_0 p_0^2 \omega^4}{16\pi^2 c}\right) \left(\frac{\sin^2\theta}{r^2}\right) \hat{r}. \tag{A.112}$$

Using the relationship $\omega = 2\pi c/\lambda$, we can write:

$$\langle \vec{S} \rangle = \left(\frac{\mu_0 p_0^2}{32\pi^2 c}\right) \left(\frac{2\pi c}{\lambda}\right)^4 \left(\frac{\sin^2\theta}{r^2}\right) \hat{r}. \tag{A.113}$$

Recalling Equation A.82, we can substitute in for p_0:

$$\langle \vec{S} \rangle = \left(\frac{\mu_0 \pi^2 c^3}{2\lambda^4}\right) \left(4\pi a^3 \varepsilon_0 \left(\frac{\varepsilon_r - 1}{2 + \varepsilon_r}\right) E_0\right)^2 \left(\frac{\sin^2\theta}{r^2}\right) \hat{r}. \tag{A.114}$$

Simplifying, we obtain:

$$\langle \vec{S} \rangle = \frac{1}{2} \mu_0 \varepsilon_0^2 c^3 E_0^2 \left(\frac{16\pi^4 a^6}{\lambda^4}\right) \left(\frac{\varepsilon_r - 1}{2 + \varepsilon_r}\right)^2 \left(\frac{\sin^2\theta}{r^2}\right) \hat{r}. \tag{A.115}$$

We may now use the fact that $\mu_0 \varepsilon_0 = 1/c^2$ to finally obtain an expression for the intensity of the re-radiated (i.e. scattered) wave:

$$\langle \vec{S} \rangle = \frac{1}{2} \varepsilon_0 c E_0^2 \left(\frac{16\pi^4 a^6}{\lambda^4}\right) \left(\frac{\varepsilon_r - 1}{2 + \varepsilon_r}\right)^2 \left(\frac{\sin^2\theta}{r^2}\right) \hat{r}. \tag{A.116}$$

The intensity of the external electromagnetic wave incident on the water droplet is $I_{ext} = \frac{1}{2}\varepsilon_0 c E_0^2$. This means that the scattered intensity is proportional to the external intensity. The second term in Equation A.116 is a geometric factor, which depends on the relationship between the radius of the scatterer (a) and the wavelength (λ). The third term is a material factor, which takes into account the ratio of the scatterer's permittivity to that of the surrounding medium (ε_r). The third term takes into account the inverse-square law ($1/r^2$) as well as the asymmetry of a dipole's radiation pattern ($\sin^2\theta$).

Appendix B

Electromagnetic Properties of Media

B.1 Introduction

We have seen that the constitutive parameters ε, μ, and σ are fundamentally important to the propagation of electromagnetic waves. In the following sections, it will become apparent that there is a great deal more complexity involved with the definition of these parameters than we have explored thus far. These complexities may give rise to effects that dramatically change the behavior of electromagnetic waves.

The nature of the constitutive parameters is dependent upon the atomic structure of a medium. For Earth-space links such as global navigation satellite systems (GNSS), the medium of propagation is the atmosphere. The material properties of the troposphere and ionosphere are highly dependent on their respective permittivities and conductivities, which give rise to a number of propagation effects. On the other hand, they are both non-magnetic, meaning their permeabilities are equivalent to that of a vacuum, so no magnetic propagation effects arise. Magnetization and its connection to permeability are completely analogous to the relationship between dielectric polarization and permittivity. But since magnetic phenomena are not relevant to atmospheric propagation, we have chosen to forego a detailed analysis of permeability for the purposes of this book.

B.2 Dielectric Polarization

Before adding layers of complexity upon the concept of permittivity, it is valuable to examine its most simple incarnation in more detail. To do this, we must introduce the concepts of induced dielectric polarization and susceptibility.

Tropospheric and Ionospheric Effects on Global Navigation Satellite Systems, First Edition.
Timothy H. Kindervatter and Fernando L. Teixeira.

B.2.1 Induced Dielectric Polarization

In Section A.1.1, we supplied the equation $\vec{D} = \varepsilon\vec{E}$, the constitutive relation between the electric flux density $\vec{D}$ and the electric field intensity $\vec{E}$. But where does this relationship come from? And what exactly is the electric flux density? An analysis of propagation from the perspective of atomic theory provides insight into these questions.

From atomic theory, we know that matter is comprised of atoms which have positively charged nuclei and negatively charged orbiting electrons. Under ordinary circumstances, each atom is electrically neutral, but in the presence of an electric field, the center of the electron cloud will displace slightly from the nucleus.[1] Both the center of the electron cloud and the nucleus can be pictorially represented as point charges with same magnitude and opposite signs, $-q$ and $+q$, respectively. Under the influence of an electric field, these two point charges become separated by a distance and as a result induce a dipole moment given by the equation

$$\vec{p} = q\vec{r},$$

where $\vec{p}$ is the dipole moment, q is the charge, and $\vec{r}$ is the displacement vector between the two point charges, pointing from the negative to the positive charge.

Since a large number of atoms will be affected simultaneously by the electric field, it is more useful to consider the vector sum of all of these (microscopic) dipoles within a given volume $\mathcal{V}$, given by

$$\vec{P} = \frac{1}{\mathcal{V}} \sum_{i}^{n} q_i\vec{r},$$

where n is the total number of atoms within $\mathcal{V}$. $\vec{P}$ is called the dipole moment per unit volume, or the polarization vector, and is directly proportional to the applied electric field:

$$\vec{P} = \chi\varepsilon_0\vec{E}. \tag{B.1}$$

The proportionality factor χ is called the electric susceptibility. This quantity will be explored in more depth shortly. The electric flux density $\vec{D}$ is given by

$$\vec{D} = \varepsilon_0\vec{E} + \vec{P},$$

so using Equation B.1, we see that

$$\vec{D} = \varepsilon_0\vec{E} + \chi\varepsilon_0\vec{E} = \varepsilon_0(1 + \chi)\vec{E}.$$

---

1 A similar picture can be drawn at the molecular level as well. In the presence of an external electric field, the electron cloud of molecules in general may become distorted (distortion polarization). In addition, polar molecules may become preferentially oriented (orientation polarization).

Using the definition

$$\varepsilon_r \equiv (1 + \chi), \tag{B.2}$$

we can finally write

$$\vec{D} = \varepsilon_0 \varepsilon_r \vec{E} = \varepsilon \vec{E}.$$

B.2.2 Electric Susceptibility

In Section B.2.1, we looked closely at the constitutive relation between $\vec{D}$ and $\vec{E}$. Along the way, we discovered that electric permittivity is dependent on a proportionality factor which we called χ, the electric susceptibility of the material. Since the permittivity of a material is so fundamental in determining its propagation characteristics, it is important to have a deep understanding of its origins. In this section, we provide a derivation of the susceptibility as a function of frequency based on Ref. [265].

Consider an atom or molecule in the presence of a time-harmonic electromagnetic field. The electrons in the presence of such a field may be modeled as damped harmonic oscillators (analogous to tiny masses on springs). Denoting $\vec{r}$ as the displacement vector of electron from its equilibrium position, the electron movement will obey the following differential equation

$$m_e \left(\frac{d^2\vec{r}}{dt^2} + \Gamma \frac{d\vec{r}}{dt} + \omega_0^2 \vec{r} \right) = \vec{F}_L,$$

where m_e is the electron mass, Γ is the damping coefficient, ω_0 is the natural frequency (also known as the resonance frequency), and $\vec{F}_L$ is the force exerted on the electron by the external electric field. From the Lorentz force law, we have

$$\vec{F}_L = q\vec{E} + q\vec{v} \times \vec{B},$$

where q is the charge. In the case of an electron, $q = -e$, where $e = 1.602 \times 10^{-19}$ C is the elementary charge. We first assume that the magnetic force exerted on the electron is negligible compared with the electric force. The more general case will be considered later on. In this case, the Lorentz force may be simplified to

$$\vec{F}_L = -e\vec{E}.$$

Assuming a time-harmonic electric field in the x-direction with frequency ω, we may write it in phasor form as

$$\vec{E} = \hat{x}\underline{E}_0 e^{-j\omega t}.$$

In this Appendix we adopt the opposite sign convention in the time-harmonic exponent from that found elsewhere in this text. This is the convention more commonly used in the literature related to the topics of this Appendix. Letting $-j \rightarrow j$

will make the two conventions equivalent. Solving the differential equation using this function will result in an electron displacement that is also time-harmonic:

$$\vec{r} = \hat{x}\underline{x}_0 e^{-j\omega t}.$$

Plugging these expressions into our differential equation results in

$$\underline{x}_0 m_e \left(-\omega^2 - j\omega\Gamma + \omega_0^2 \right) = -e\underline{E}_0,$$

which may be simply rearranged to solve for x_0:

$$\underline{x}_0 = \frac{-e\underline{E}_0}{m_e \left(\omega_0^2 - \omega^2 - j\omega\Gamma \right)}.$$

As seen before, the polarization vector may be defined as $\vec{P} = -Ne\vec{r}$, where N is the number of atoms per unit volume. Here, $\vec{r} = \vec{x}$, so we may write $\vec{P}$ as

$$\vec{P} = \hat{x}\frac{Ne^2\underline{E}_0}{m_e \left(\omega_0^2 - \omega^2 - j\omega\Gamma \right)} e^{-j\omega t}.$$

Recalling that $\vec{E} = \hat{x}\underline{E}_0 e^{-j\omega t}$, we see that the proportionality factor between $\vec{P}$ and $\vec{E}$ is

$$\frac{Ne^2}{m_e(\omega_0^2 - \omega^2 - j\omega\Gamma)}.$$

Now we recall that the relationship between $\vec{P}$ and $\vec{E}$ is given by $\vec{P} = \varepsilon_0 \chi \vec{E}$. Therefore,

$$\chi(\omega) = \frac{Ne^2}{\varepsilon_0 m_e(\omega_0^2 - \omega^2 - j\omega\Gamma)}. \tag{B.3}$$

B.3 Lossy and Dispersive Media

The result obtained in Equation B.3 leads very naturally into a discussion about dispersion and absorption (also referred to as attenuation or loss). All media other than a vacuum exhibit both dispersion and absorption to some degree.

B.3.1 Absorption

We notice that χ is a complex number, which indicates that n is also complex in general. Therefore we can express the complex index of refraction as

$$n = n_R + j\kappa, \tag{B.4}$$

where n_R is the real part, which influences the phase velocity, and κ is called the extinction coefficient, which induces attenuation of the electromagnetic wave in a lossy medium (a lossy medium is one for which $\kappa \neq 0$). We can explain this mathematically by noticing that the propagation constant of a wave in a medium is

$k = 2\pi n/\lambda_0$ where λ_0 is the wavelength in a vacuum. Substituting in Equation B.4, we get

$$k = \frac{2\pi}{\lambda_0}(n_R + j\kappa) = k_R + j\alpha,$$

where α is known as the attenuation coefficient. When we substitute this into the expression for a plane wave traveling through this medium along the positive z direction, we get

$$\vec{E} = \hat{x}\underline{E}_0 e^{j(kz-\omega t)} = \hat{x}\underline{E}_0 e^{-\alpha z} e^{j(k_R z - \omega t)}.$$

This expression is very similar to the plane wave solution in a lossless medium (i.e. $\kappa = 0$), but we now have a factor of $e^{-\alpha z}$ out front, indicating exponential attenuation in space along the direction of propagation.

B.3.2 Dispersion

We can use susceptibility to obtain a more explicit expression for the index of refraction. It is useful to separate the real and imaginary parts. We can do this by rationalizing the denominator of Equation B.3, which gives us

$$\chi(\omega) = \frac{Ne^2}{\varepsilon_0 m_e}\left[\frac{(\omega_0^2 - \omega^2)}{(\omega_0^2 - \omega^2)^2 + (\omega\Gamma)^2} + j\frac{\omega\Gamma}{(\omega_0^2 - \omega^2)^2 + (\omega\Gamma)^2}\right].$$

Bearing in mind that $n = \sqrt{1 + \chi}$, we can see that we can express the index of refraction in terms of χ as:

$$n^2 = 1 + \frac{Ne^2}{\varepsilon_0 m_e}\left[\frac{(\omega_0^2 - \omega^2)}{(\omega_0^2 - \omega^2)^2 + (\omega\Gamma)^2} + j\frac{\omega\Gamma}{(\omega_0^2 - \omega^2)^2 + (\omega\Gamma)^2}\right]. \tag{B.5}$$

Because the index of refraction is a function of frequency, every frequency of electromagnetic radiation will travel at a different speed. We call this phenomenon dispersion. A familiar example of dispersion at optical frequencies is a prism. White light is made up of all visible frequencies, and since each of these frequencies (colors) travels at a different speed through the prism, they will also refract at different angles, causing them to split apart into a rainbow. This effect is not exclusive to visible light. Dispersion can cause errors in GNSS observables that must be accounted for.

B.3.3 Graphical Analysis

Plotting the real and imaginary parts of the index of refraction vs. frequency lends us some additional insight. To do this, we must take the square root of both sides of

Equation B.5. However, this makes the right side of the equation extremely complicated. Therefore, we may use a Taylor approximation to greatly simplify our expression for n.

$$\sqrt{1+\chi} \approx 1 + \frac{\chi}{2}.$$

Higher order terms have been neglected in the series expansion. This is justified because in gases, molecules are sparse and χ is small. We may now write n as:

$$n \approx 1 + \frac{Ne^2}{2\varepsilon_0 m_e} \left[\frac{(\omega_0^2 - \omega^2)}{(\omega_0^2 - \omega^2)^2 + (\omega\Gamma)^2} + j \frac{\omega\Gamma}{(\omega_0^2 - \omega^2)^2 + (\omega\Gamma)^2} \right]$$

so that

$$n_R \approx 1 + \frac{Ne^2}{2\varepsilon_0 m_e} \frac{(\omega_0^2 - \omega^2)}{(\omega_0^2 - \omega^2)^2 + (\omega\Gamma)^2}, \tag{B.6}$$

$$\kappa \approx \frac{Ne^2}{2\varepsilon_0 m_e} \frac{\omega\Gamma}{(\omega_0^2 - \omega^2)^2 + (\omega\Gamma)^2}. \tag{B.7}$$

These two functions may be seen in Figure B.1. A lot of information is contained in this single graph, so we will break things down case by case.

Case 1: $\omega \ll \omega_0$

If the frequency of the incident wave is negligible compared with the resonance frequency of the medium, we may assume $\omega \approx 0$, which reduces n_R to the static case:

$$n_R = 1 + \frac{Ne^2}{2\varepsilon_0 m_e \omega_0^2}.$$

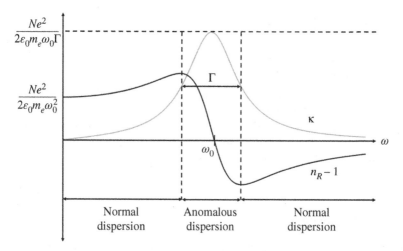

Figure B.1 Real and imaginary parts of index of refraction vs. frequency.

If ω_0 is very large, then the second term is negligible and n_R reduces to 1. In addition, κ reduces to zero when $\omega = 0$. Even for small positive values of ω, the denominator includes a term $\mathcal{O}\left(\omega_0^4\right)$, which will make κ negligibly small so that it may still be approximated as zero.

Case 2: $\omega \approx \omega_0$

In the vicinity of ω_0, n_R begins to change very rapidly. By inserting $\omega = \omega_0$ into the numerator of Equation B.6, we can easily see that $n_R = 1$ at the resonance frequency. As we keep increasing the frequency slightly past ω_0, the second term becomes negative, which means that n_R becomes less than 1. An index of refraction less than 1 corresponds to a phase velocity greater than c. Recall from our discussion in Appendix A that the phase of a wave carries no information, so its phase velocity is not bounded by c. Instead, the envelope of the wave carries the information, and can be characterized using the group velocity, which has its own group index of refraction n_g and *is* bounded by c.

The graph of n_R also makes it clear that there is a region in which $\frac{dn_R}{d\omega}$ is negative. This phenomenon is called anomalous dispersion, as opposed to normal dispersion which occurs when $\frac{dn_R}{d\omega}$ is positive. We can determine the interval in which anomalous dispersion occurs by finding the locations of the local maximum and minimum. This is done by setting the derivative equal to zero.

$$\frac{dn_R}{d\omega} = \frac{Ne^2}{\varepsilon_0 m_e} \frac{\omega\left(\omega_0^4 - \omega_0^2\left[2\omega^2 + \Gamma^2\right] + \omega^4\right)}{\left((\omega_0^2 - \omega^2)^2 + (\omega\Gamma)^2\right)^2}. \tag{B.8}$$

The zeroes of this function are found to be

$$\omega = 0,$$

$$\omega = \pm\sqrt{\omega_0^2 - \omega_0\Gamma},$$

$$\omega = \pm\sqrt{\omega_0^2 + \omega_0\Gamma}.$$

The solution $\omega = 0$ is trivial, and in Case 1 we have already discussed the behavior of the function in the limit of low frequency. Negative values of ω are not physically meaningful solutions, so we discard them. This leaves us with $\omega_1 = \sqrt{\omega_0^2 - \omega\Gamma}$ and $\omega_2 = \sqrt{\omega_0^2 + \omega\Gamma}$, where ω_1 corresponds to the lower bound of the interval shown in Figure B.1, and ω_2 corresponds to the upper bound. We can rewrite these expressions as $\frac{\omega_1}{\omega_0} = \sqrt{1 - \frac{\Gamma}{\omega_0}}$ and $\frac{\omega_1}{\omega_0} = \sqrt{1 + \frac{\Gamma}{\omega_0}}$, respectively. Using a Taylor series, these expressions can be simplified to $\frac{\omega_1}{\omega_0} = 1 - \frac{\Gamma}{2\omega_0}$ and $\frac{\omega_1}{\omega_0} = 1 + \frac{\Gamma}{2\omega_0}$. Multiplying back through by ω_0 gives us two frequencies which represent the upper and lower bounds of the region of anomalous dispersion:

$$\omega_1 = \omega_0 - \frac{\Gamma}{2}, \quad \omega_2 = \omega_0 + \frac{\Gamma}{2}.$$

We see that the difference of these two frequencies is $\omega_2 - \omega_1 = \Gamma$, which is the result seen in Figure B.1.

The extinction coefficient κ also changes significantly in the vicinity of the resonance frequency. We see in Equation B.7 that the numerator is $\mathcal{O}(\omega)$ while the denominator is $\mathcal{O}(\omega^4)$. So for all frequencies except those very close to the resonance frequency, κ will be small and will cause less pronounced absorption. When $\omega_0 \approx \omega$, $\omega_0^2 - \omega^2 \approx 0$, and κ increases dramatically, causing a large amount of absorption. The value of κ peaks at the resonance frequency, where $\omega_0^2 - \omega^2 = 0$ and the whole expression reduces to

$$\kappa = \frac{Ne^2}{2\varepsilon_0 m_e \omega_0 \Gamma}. \tag{B.9}$$

Case 3: $\omega \gg \omega_0$

At frequencies much higher than the resonance frequency, ω_0 is negligible compared with ω and may be ignored. The denominator of the second term of n_R is $\mathcal{O}(\omega^4)$ while the numerator is only $\mathcal{O}(\omega^2)$. Therefore as $\omega \to \infty$, $n_R \to 1$. n_R is still less than one at frequencies higher then ω_0, but the higher the frequency, the closer n_R is to 1, theoretically reaching 1 again in the limit of very high frequencies.

Similarly, the denominator of the expression for κ increases much more rapidly than the numerator for large ω, reducing the expression to 0 in the limit of very high frequencies (Table B.1).

Tropospheric dispersion and absorption: At GNSS frequencies, the troposphere is an example of Case 1. The resonance frequencies of the gases and water vapor in the troposphere are on the order of tens to hundreds of GHz, which are at least one order of magnitude larger than the frequency of GNSS signals. Therefore, the dispersive and absorptive effects in the troposphere at GNSS frequencies are relatively weak.

Ionospheric dispersion and absorption: In the ionosphere, electrons are dissociated by ionizing solar radiation. These free electrons are not subject to a restoring force when influenced by an electric field, so their resonance frequencies are $\omega_0 = 0$. If Figure B.1 were centered at $\omega_0 = 0$, only the portion of the graph to the right of ω_0 would be meaningful. Therefore, any frequency sent through the ionosphere is greater than the resonance frequency, resulting in a negative index of refraction and a phase velocity greater than c. Thus, the ionosphere is an example of Case 3.

The ionosphere is strongly dispersive, meaning different GNSS frequencies (e.g. L1, L2, and L5) all experience different group delays and phase advances as

Table B.1 Summary of index of refraction behavior.

	n_R	κ	$\dfrac{dn}{d\omega}$
$\omega \approx 0$	≈ 1	0	0
$0 < \omega < \sqrt{\omega_0^2 - \omega\Gamma}$	> 1	Increasing	+
$\omega = \sqrt{\omega_0^2 - \omega\Gamma}$	> 1	Increasing	0
$\sqrt{\omega_0^2 - \omega\Gamma} < \omega < \omega_0$	> 1	Increasing	−
$\omega = \omega_0$	1	Peak	−
$\omega_0 < \omega < \sqrt{\omega_0^2 + \omega\Gamma}$	< 1	Decreasing	−
$\sqrt{\omega_0^2 + \omega\Gamma}$	< 1	Decreasing	0
$\sqrt{\omega_0^2 + \omega\Gamma} < \omega < \infty$	< 1	Decreasing	+
$\omega \to \infty$	1	0	0

the wave travels through it. Knowing the precise frequency dependence of this effect allows us to use a linear combination of GNSS observables known as an ionosphere-free combination. This allows for mitigation of the ionospheric error term in the pseudorange and carrier phase measurements in dual-frequency receivers. We also note that peak absorption occurs at $\omega = 0$, so the lower the frequency, the more ionospheric attenuation the signal will experience.

B.3.4 Multiple Resonances

Our analysis thus far has assumed only a single resonance frequency and damping ratio for the entire medium. In reality, a single atom or molecule may be associated with more than one resonance frequency and damping ratio, and a given medium may be made up of many molecules of different types. We may model each resonance as a separate oscillator, each of which adds independently to the final result. To reflect this, we can rewrite our susceptibility equation as a summation of multiple oscillators:

$$\chi(\omega) = \frac{e^2}{\varepsilon_0 m_e} \sum_i \frac{N_i}{(\omega_i^2 - \omega^2 - j\omega\Gamma_i)},$$

where N_i is the number of atoms per unit volume with associated resonance frequency ω_i and damping ratio Γ_i. Figure B.2 depicts the frequency response of a hypothetical material with multiple resonance frequencies ω_1, ω_2, and ω_3. We see that the region around each resonance behaves as discussed previously.

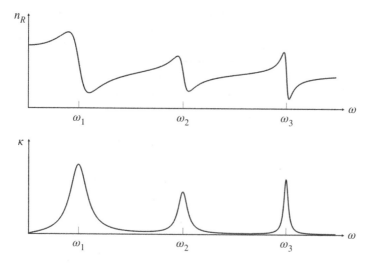

Figure B.2 Real and imaginary parts of the index of refraction with multiple resonances.

B.4 Conducting Media

B.4.1 Time-Varying Conduction Current

In Section B.2, we considered only the effects of bound electrons in the presence of an electric field. To do this, we assumed the number of unbound charges (related to the conductivity) in the medium was zero ($\sigma = 0$). In this section, we will do the opposite – that is, we will ignore the effects of bound electrons and solely consider the effects of unbound charges. The derivation that follows has been adapted from [266, pp. 160–163].

Consider the motion of free electrons in a conducting medium in the presence of an electric field. The equation of motion for these electrons is similar to that of bound charges but with the difference that there is no restoring force. Therefore, the equation of motion is written as

$$m_e \frac{d^2\vec{r}}{dt^2} + m_e \tau^{-1} \frac{d\vec{r}}{dt} = -e\vec{E}, \tag{B.10}$$

where we once again assumed the magnetic force term to be negligible and denote the damping coefficient as τ^{-1}.

The current density $\vec{J}$ due to the movement of the electrons can be written as

$$\vec{J} = -N_e \, e \frac{d\vec{r}}{dt},$$

where N_e is the number density of electrons (number of electrons per unit volume). If we substitute this expression into Equation B.10, we see that

$$\frac{d\vec{J}}{dt} + \tau^{-1}\vec{J} = \frac{N_e e^2}{m_e}\vec{E}. \tag{B.11}$$

The general solution to this differential equation is related to the associated homogeneous equation

$$\frac{d\vec{J}}{dt} + \tau^{-1}\vec{J} = 0.$$

Solving this equation gives the expression for a transient current with time constant τ known as the *relaxation time*:

$$\vec{J} = \vec{J}_0\, e^{-t/\tau}.$$

On the other hand, if we assume a static regime with static field $\vec{E}$, the derivative of $\vec{J}$ vanishes, leaving

$$\vec{J} = \frac{N_e e^2}{m_e}\tau\vec{E}.$$

By defining

$$\sigma \equiv \frac{N_e e^2}{m_e}\tau \tag{B.12}$$

we obtain the familiar expression for the Ohm's law at a point

$$\vec{J} = \sigma\vec{E}.$$

Note that this expression only applies for the static case. To generalize this expression to include time-varying fields, assume that $\vec{E}$ is a time harmonic field $\vec{E} = \underline{\vec{E}}\, e^{-j\omega t}$. $\vec{J}$ will have a similar form. When these expressions are substituted into Equation B.11, we get

$$\left(-j\omega + \tau^{-1}\right)\vec{J} = \frac{N_e e^2}{m_e}\vec{E}.$$

Using Equation B.12, we see that

$$\left(-j\omega + \tau^{-1}\right)\vec{J} = \tau^{-1}\sigma\vec{E}.$$

Multiplying through by τ and rearranging gives us

$$\vec{J} = \frac{\sigma}{1 - j\omega\tau}\vec{E}, \tag{B.13}$$

which reduces to $\vec{J} = \sigma\vec{E}$ for $\omega = 0$.

B.4.2 Propagation in Conducting Media

To analyze propagation in conducting media, we return to the wave equation. In our original derivation of the wave equation in Section A.1.1, we assumed a source-free, nonconducting medium, so that $\vec{J} = 0$. If we instead retain $\vec{J}$ in our derivation of the wave equation, we obtain the expression

$$\nabla^2 \vec{E} = \mu\varepsilon \frac{\partial^2 \vec{E}}{\partial t^2} + \mu \frac{\partial \vec{J}}{\partial t}.$$

Assuming time-harmonic solutions and using Equation B.13, we can write the wave equation entirely in terms of $\vec{E}$:

$$\nabla^2 \vec{E} = \mu\varepsilon \frac{\partial^2 \vec{E}}{\partial t^2} + \frac{\mu\sigma}{1 - j\omega\tau} \frac{\partial \vec{E}}{\partial t}.$$

A solution to this equation is $\vec{E}_0\, e^{j(kz - \omega t)}$. When this solution is plugged into the wave equation, we see that

$$k^2 = \omega^2 \mu\varepsilon + \frac{j\omega\mu\sigma}{1 - j\omega\tau}.$$

Substituting in Equation B.12 for σ and simplifying the expression gives

$$n^2 = 1 - \frac{\omega_N^2}{\omega^2 + j\omega\tau^{-1}}, \tag{B.14}$$

where we have defined a quantity ω_N called the *plasma frequency*:

$$\omega_N^2 = \frac{N_e e^2}{m_e \varepsilon}.$$

By splitting the aforementioned expression for n^2 into its real and imaginary parts and rationalizing the denominator:

$$n^2 = 1 - \omega_N^2 \left[\frac{\omega^2}{\omega^4 + \omega^2\tau^{-2}} - \frac{j\omega\tau^{-1}}{\omega^4 + \omega^2\tau^{-2}} \right] = 1 - \omega_N^2 \left[\frac{1}{\omega^2 + \tau^{-2}} - j \frac{\frac{1}{\omega\tau}}{\omega^2 + \tau^{-2}} \right].$$

Using a Taylor series approximation $\sqrt{1 - x} \approx 1 - x/2$, we can write

$$n \approx 1 - \frac{\omega_N^2}{2} \left[\frac{1}{\omega^2 + \tau^{-2}} - j \frac{\frac{1}{\omega\tau}}{\omega^2 + \tau^{-2}} \right].$$

Since $n = n_R + j\kappa$, it follows

$$n_R \approx 1 - \frac{1}{2} \left[\frac{\omega_N^2}{\omega^2 + \tau^{-2}} \right],$$

$$\kappa \approx \frac{1}{2} \left[\frac{\omega_N^2}{\omega^3\tau + \omega\tau^{-1}} \right].$$

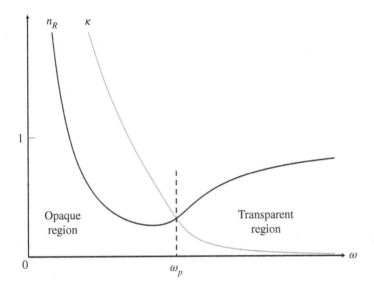

Figure B.3 Relationship between n and κ for a conducting medium.

The relationship between n_R and κ is shown in Figure B.3, where it may be seen that for frequencies below the plasma frequency, the real part of the index of refraction is smaller than the imaginary part. In this region the medium is opaque, absorbing, and reflecting incident waves. For frequencies above the plasma frequency, this relationship reverses, and the medium becomes more transparent. This effect can be shown by considering the dispersion relation of the medium. Consider Equation B.14. In good conductors, the relaxation times of electrons are fairly large, so τ^{-1} will be very small. This reduces the expression to

$$n^2 = 1 - \frac{\omega_N^2}{\omega^2}. \tag{B.15}$$

Since $k = n\omega/c$, we have that

$$k^2 \frac{c^2}{\omega^2} = 1 - \frac{\omega_N^2}{\omega^2},$$

which is the dispersion relation of a conducting medium. This equation is commonly written in the form

$$\omega^2 = c^2 k^2 + \omega_N^2.$$

To observe the propagation of a wave in this medium, let's rearrange the equation in terms of k:

$$k = \frac{\omega}{c} \sqrt{1 - \frac{\omega_N^2}{\omega^2}}.$$

For the case where $\omega > \omega_N$, the quantity under the square root is positive, leading to a real-valued k. Thus, waves at frequencies above the plasma frequency propagate normally. On the other hand, when $\omega < \omega_N$, the quantity under the square root is negative, so k becomes imaginary.

$$k = j\frac{\sqrt{\omega_N^2 - \omega^2}}{c} \equiv j|k|. \tag{B.16}$$

This tells us that in this case the electric field can be written as

$$\vec{E} = \underline{E}_0 e^{j(kz-\omega t)} = \underline{E}_0 e^{j(j|k|z-\omega t)} = \underline{E}_0 e^{-|k|z} e^{-j\omega t}.$$

This as an evanescent wave whose intensity decays with distance. This means that when the wave arrives at the boundary of such medium, it will penetrate only a relatively short distance past the boundary on a length scale given by $1/|k|$ (the so-called skin depth). The wave will be totally reflected because no energy is transmitted into the medium by this process, which can be verified by calculating the average Poynting vector of the wave inside said medium:

$$\langle \vec{S} \rangle = \frac{|\vec{E}|^2}{2\,\mathrm{Re}\{\eta\}}\hat{k}.$$

Recalling that $\eta = \sqrt{\frac{\mu}{\varepsilon}}$ and $k = \omega\sqrt{\mu\varepsilon}$, we see that

$$\langle \vec{S} \rangle = \frac{|\vec{E}|^2}{2}\mathrm{Re}\{k\}\hat{k}.$$

Since Equation B.16 tells us that k is purely imaginary, we see that $\mathrm{Re}\{k\}$ and thus $\langle \vec{S} \rangle$ equal zero. Therefore no energy propagates into the medium in the limit of large values of τ (or conductivity).[2]

The inability of waves to propagate through such a medium is interpreted as opacity. Many metals have plasma frequencies in the ultraviolet region, which is why metals are typically opaque to visible light. The ionosphere has a plasma frequency on the order of 10^6–10^7 Hz. GNSS signals, which propagate around 10^9 Hz, are well above the plasma frequency of the ionosphere, so they are allowed to propagate, rather than being reflected. On the other hand, electromagnetic signals launched from ground transmitters with frequencies lower than the plasma frequency of the ionosphere can be used to establish ground-to-ground communication links by exploiting ionospheric reflections [45].

---

2 For smaller values of τ, an incident wave onto the boundary of such a medium will be partially reflected and partially transmitted. The transmitted portion of the wave will then be subject to absorption as it travels within the medium.

B.4.3 Combined Effects of Dispersion and Conduction

In Sections B.2, B.3, B.4.1, and B.4.2, we ignored either the bound charges or the free charges in order to simplify calculations. In reality, both of these effects are always present to some extent (except in a vacuum). In good conductors, the contribution of bound charges is negligible and may be ignored. Similarly, in good dielectrics, the contribution of free charges is small enough to be neglected. However, in materials between these two extremes (e.g. semiconductors), both types of charges must be accounted for. In this case, the general expression for n^2 can be written as

$$n^2 = 1 - \frac{\omega_N^2}{\omega^2 + j\omega\tau^{-1}} + \frac{e^2}{\varepsilon_0 m_e} \sum_i \frac{N_i}{(\omega_i^2 - \omega^2 - j\omega\Gamma_i)}.$$

B.5 Kramers–Kronig Relations

As seen before, the electric susceptibility of a medium can be written as

$$\vec{P}(\omega) = \varepsilon_0 \chi(\omega)\vec{E}(\omega),$$

where $\chi(\omega)$ is represented by the expression in Equation B.3. From Fourier transform theory, we know that a product in the frequency domain is equivalent to a convolution in the time domain:

$$X(\omega)Y(\omega) \overset{F}{\Leftrightarrow} x(t) * y(t), \tag{B.17}$$

so that

$$\vec{P}(t) = \varepsilon_0 \chi(t) * \vec{E} = \varepsilon_0 \int_{-\infty}^{\infty} \chi(t - t')\vec{E}(t')dt'.$$

The electric susceptibility may be thought of as the impulse response relating the electric field (the input) to the polarization vector in the medium (the output). Given the impulse response of the system $\chi(t)$, we may express the frequency response via a Fourier transform:

$$\chi(\omega) = \int_{-\infty}^{\infty} \chi(t)e^{j\omega t}dt.$$

Since the response of this system must be causal, this tells us that

$$\chi(t) = 0 \ \forall t < 0,$$

which simplifies the bounds of the Fourier integral to

$$\chi(\omega) = \int_{0}^{\infty} \chi(t)e^{j\omega t}dt.$$

This is equivalent to keeping the lower bound at $-\infty$ and multiplying $\chi(t)$ by the Heaviside step function $u(t)$:

$$\chi(\omega) = \int_{-\infty}^{\infty} \chi(t)u(t)e^{j\omega t}\,dt.$$

This is clearly the Fourier transform of $\chi(t)u(t)$, or $\mathcal{F}\{\chi(t)u(t)\}$. Once again using the convolution theorem, we can write

$$\mathcal{F}\{\chi(t)u(t)\} = \frac{1}{2\pi}\chi(\omega) * \mathcal{F}\{u(t)\},$$

where the Fourier transform of the Heaviside function is given by

$$\mathcal{F}\{u(t)\} = \pi\delta(\omega) + \frac{1}{j\omega}$$

and $\delta(\omega)$ is the Dirac delta function. Performing the convolution results in

$$\chi(\omega) = \frac{1}{2\pi}\left(\int_{-\infty}^{\infty} \pi\chi(\omega')\delta(\omega-\omega')d\omega' + \int_{-\infty}^{\infty} \frac{\chi(\omega')}{j(\omega-\omega')}d\omega'\right).$$

Utilizing the sifting property of the Dirac delta function, we see that

$$\chi(\omega) = \frac{1}{2\pi}\left(\pi\chi(\omega) + \int_{-\infty}^{\infty} \frac{\chi(\omega')}{j(\omega-\omega')}d\omega'\right).$$

Grouping all $\chi(\omega)$ terms on the left hand side results in

$$\chi(\omega) = \frac{1}{\pi}\int_{-\infty}^{\infty} \frac{\chi(\omega')}{j(\omega-\omega')}d\omega',$$

which describes the invariance of a causal system under a Hilbert transformation. We may split this integral into

$$\chi(\omega) = \frac{1}{\pi}\int_{-\infty}^{0} \frac{\chi(\omega')}{j(\omega-\omega')}d\omega' + \frac{1}{\pi}\int_{0}^{\infty} \frac{\chi(\omega')}{j(\omega-\omega')}d\omega'.$$

By frequency reversing the first integral, we may change the bounds to be positive:

$$\chi(\omega) = \frac{1}{\pi}\int_{0}^{\infty} \frac{\chi(-\omega')}{j(\omega+\omega')}d\omega' + \frac{1}{\pi}\int_{0}^{\infty} \frac{\chi(\omega')}{j(\omega-\omega')}d\omega'.$$

The susceptibility is real-valued in the time domain. Real valued functions satisfy $\chi(t) = \chi^*(t)$. Taking the Fourier transform: $\mathcal{F}\{\chi^*(t)\} \Leftrightarrow \chi^*(-\omega)$. Thus we see that $\mathcal{F}\{\chi(t)\} = \mathcal{F}\{\chi^*(t)\} \Leftrightarrow \chi(\omega) = \chi^*(-\omega)$. Finally, conjugating both sides gives $\chi^*(\omega) = \chi(-\omega)$. By substituting this result into our integral equation, we may now merge the two integrals and find a common denominator to yield

$$\chi(\omega) = \frac{1}{\pi}\int_{0}^{\infty} \frac{\chi^*(\omega')(\omega-\omega') + \chi(\omega')(\omega+\omega')}{j(\omega^2-\omega'^2)}d\omega'.$$

In terms of the real and imaginary components of $\chi(\omega)$ we have

$$\chi(\omega) = \chi_{Re}(\omega) + j\chi_{Im}(\omega),$$

$$\chi^*(\omega) = \chi_{Re}(\omega) - j\chi_{Im}(\omega).$$

Substituting in these expressions and performing some algebra, we can split the last integral into its real and imaginary parts as follows:

$$\chi_{Re}(\omega) + j\chi_{Im}(\omega) = \frac{2}{\pi}\int_0^\infty \frac{\omega'\chi_{Im}(\omega')}{(\omega^2 - \omega'^2)}d\omega' + j\frac{2}{\pi}\int_0^\infty \frac{-\omega\chi_{Re}(\omega')}{(\omega^2 - \omega'^2)}d\omega'.$$

Finally, matching the real and imaginary parts on both sides of the equation, we obtain the standard form of the pair of formulas known as the Kramers–Kronig relations.

$$\chi_{Re}(\omega) = \frac{2}{\pi}\int_0^\infty \frac{\omega'\chi_{Im}(\omega')}{(\omega^2 - \omega'^2)}d\omega', \qquad (B.18)$$

$$\chi_{Im}(\omega) = -\frac{2}{\pi}\int_0^\infty \frac{\omega\chi_{Re}(\omega')}{(\omega^2 - \omega'^2)}d\omega'. \qquad (B.19)$$

The most important observation to make from these equations is that the real and imaginary parts of the susceptibility are dependent upon each other. This implies the slightly subtler point that if the real part of the susceptibility is nonzero, then the imaginary part must also be nonzero, and vice versa. Since the susceptibility is nonzero in any medium other than a vacuum, this tells us that all media necessarily have complex susceptibility in the frequency domain. As we have seen before in Equation B.2, the relative permittivity of a material is dependent upon the susceptibility. And further, Equations A.9 and A.11 tell us that the index of refraction is dependent on the permittivity. Therefore, any medium with a complex susceptibility has a complex permittivity and a complex index of refraction. We have also seen in Section B.3 that the real part of the index of refraction corresponds to the phenomenon of dispersion, and the imaginary part of the index of refraction is responsible for absorption (also known as attenuation or loss). Therefore, the Kramers–Kronig relations tell us that dispersion and loss are intrinsically linked – if a material absorbs electromagnetic radiation at all, it *must* be dispersive, and vice versa. This result has been obtained simply by imposing the condition of causality upon our physical system. Put another way, it is impossible to formulate a causal theory of electromagnetism unless absorption and dispersion go hand-in-hand.

The Kramers–Kronig relations can be alternatively derived by invoking techniques from complex analysis and performing appropriate contour integrals on the complex ω plane [267]. Indeed, Kramers–Kronig relations were originally derived in this manner. Here, we presented instead an alternative derivation adapted from [268, pp. 20–27] from a linear systems theory standpoint, which is likely more accessible to the majority of engineers.

B.6 Anisotropic Media

In Section B.2.2, we compared the oscillation of a bound electron to a mass on a spring. In doing so, we made the implicit assumption that this electron would oscillate exactly the same way in all directions, i.e. *isotropically*. However, many media are anisotropic, meaning that the electron will oscillate differently in one direction than it will in another [261, 262, 269]. In the mass-spring analogy, this would be the same as having springs of different stiffness in different directions. This means that we can no longer express the susceptibility as a scalar. Instead, we must now express it as a 3×3 tensor:

$$\begin{bmatrix} P_x \\ P_y \\ P_z \end{bmatrix} = \epsilon_0 \begin{bmatrix} \chi_{xx} & \chi_{xy} & \chi_{xz} \\ \chi_{yx} & \chi_{yy} & \chi_{yz} \\ \chi_{zx} & \chi_{zy} & \chi_{zz} \end{bmatrix} \begin{bmatrix} E_x \\ E_y \\ E_z \end{bmatrix}.$$

This may be written more compactly as

$$\vec{P} = \epsilon_0 \overline{\overline{\chi}} \vec{E}.$$

In this case the permittivity must also be written as a tensor, known as the dielectric tensor:

$$\overline{\overline{\epsilon}} = \epsilon_0 (\overline{\overline{I}} + \overline{\overline{\chi}}),$$

where $\overline{\overline{I}}$ is the 3×3 identity matrix, and ϵ relates $\vec{D}$ and $\vec{E}$ so that

$$\begin{bmatrix} D_x \\ D_y \\ D_z \end{bmatrix} = \begin{bmatrix} \epsilon_{xx} & \epsilon_{xy} & \epsilon_{xz} \\ \epsilon_{yx} & \epsilon_{yy} & \epsilon_{yz} \\ \epsilon_{zx} & \epsilon_{zy} & \epsilon_{zz} \end{bmatrix} \begin{bmatrix} E_x \\ E_y \\ E_z \end{bmatrix}.$$

An index notation is often preferred to make the equations look more compact. In this case,

$$D_i = \sum_{j=x,y,z} \epsilon_{ij} E_j.$$

We may simplify even further using Einstein's summation notation, in which the summation symbol is omitted, and summation over repeated indices is implied. In this case,

$$D_i = \epsilon_{ij} E_j.$$

B.6.1 Dielectric Tensor Properties

Using Equation A.31 in conjunction with Einstein's summation notation and assuming a non-dispersive medium for simplicity, the energy density stored in the electric field may be written as

$$W_e = \frac{1}{2}\vec{E} \cdot \vec{D} = \frac{1}{2}E_i \varepsilon_{ij} E_j.$$

The energy density is always positive. Therefore, the previous expression should provide a positive quantity for any $\vec{E}$. In other words, the matrix representing the dielectric tensor $\overline{\overline{\varepsilon}}$ should be positive-definite by definition. In addition, if the matrix representing the dielectric tensor is symmetric, i.e. $\varepsilon_{ij} = \varepsilon_{ji}$ then the corresponding medium is called *reciprocal*. Otherwise we call the medium non-reciprocal. Note that isotropic media are a special case of reciprocal media.

We have assumed non-dispersive media to simplify the discussion. Therefore, all the ε_{ij} are real-valued. In the more general case of dispersive media ε_{ij} are complex-valued when represented in the Fourier (phasor) domain. In this case, $\varepsilon_{ij} = \varepsilon_{ji}$ still implies a reciprocal medium [262] but an anisotropic medium is loss-less if and only if $\varepsilon_{ij} = \varepsilon_{ji}^*$, where the star superscript denotes complex conjugation [261, 262, 270]. In other words, a lossless anisotropic medium implies an Hermitian permittivity tensor.

In the context of GNSS propagation, the troposphere is an example of a lossy and reciprocal medium because in general $\varepsilon_{ij} = \varepsilon_{ji}$ but $\varepsilon_{ij} \neq \varepsilon_{ji}^*$ there. It should be clear that we retrieve these two conditions for any isotropic medium with complex-valued permittivity. On the other hand, the magnetized ionosphere is in general a non-reciprocal medium. We will discuss the non-reciprocal properties of the ionosphere later in this Appendix when considering the gyrotropic susceptibility tensor.

The dielectric tensor is always diagonalizable, meaning the associated eigenvectors may always be chosen such that they form a basis for the underlying vector space. If we choose our coordinate system to coincide with these eigenvectors, the dielectric tensor reduces to a diagonal matrix of the form

$$\overline{\overline{\varepsilon}} = \begin{bmatrix} \varepsilon_{xx} & 0 & 0 \\ 0 & \varepsilon_{yy} & 0 \\ 0 & 0 & \varepsilon_{zz} \end{bmatrix}.$$

The diagonal entries of this matrix represent the eigenvalues of the dielectric tensor. These eigenvalues are called the *principal permittivities* of the medium, and the eigenvectors to which they correspond lie along the *principal axes* of our coordinate system.

It is possible to further simplify the dielectric tensor by considering symmetries in the structure of the medium. We may classify media into one of the three following categories:

Isotropic	$\overline{\overline{\varepsilon}} = \begin{bmatrix} \varepsilon & 0 & 0 \\ 0 & \varepsilon & 0 \\ 0 & 0 & \varepsilon \end{bmatrix}$	$\varepsilon = \varepsilon_{xx} = \varepsilon_{yy} = \varepsilon_{zz}$
Uniaxial	$\overline{\overline{\varepsilon}} = \begin{bmatrix} \varepsilon_x & 0 & 0 \\ 0 & \varepsilon_x & 0 \\ 0 & 0 & \varepsilon_z \end{bmatrix}$	$\varepsilon_x = \varepsilon_{xx} = \varepsilon_{yy}$ $\varepsilon_z = \varepsilon_{zz}$
Biaxial	$\overline{\overline{\varepsilon}} = \begin{bmatrix} \varepsilon_x & 0 & 0 \\ 0 & \varepsilon_y & 0 \\ 0 & 0 & \varepsilon_z \end{bmatrix}$	$\varepsilon_x = \varepsilon_{xx}$ $\varepsilon_y = \varepsilon_{yy}$ $\varepsilon_z = \varepsilon_{zz}$

We are already familiar with isotropic media, in which the permittivity is the same in all directions. In uniaxial media, two axes have the same permittivity, while the third axis differs (typically, the differing axis is chosen as the z-axis). The most general case is a biaxial medium, in which all three axes exhibit different permittivities.

B.6.2 Wave Equation in Anisotropic Media

How exactly does the tensor nature of an anisotropic medium's permittivity affect propagation? As always, we begin with the wave equation. For simplicity, we will choose the coordinate system to coincide with the principal axes of the medium. If we assume a source-free medium, the wave equation becomes

$$\nabla \times \left(\nabla \times \vec{E} \right) = -\mu \overline{\overline{\varepsilon}} \frac{\partial^2 \vec{E}}{\partial t^2}.$$

Assuming a time-harmonic plane wave, this expression can be written as

$$\vec{k} \times \left(\vec{k} \times \vec{E} \right) + \omega^2 \mu \overline{\overline{\varepsilon}} \vec{E} = 0,$$

which may be split into three scalar equations as follows:

$$\left(\omega^2 \mu \varepsilon_x - k_y^2 - k_z^2 \right) E_x + k_x k_y E_y + k_x k_z E_z = 0,$$
$$k_y k_x E_x + \left(\omega^2 \mu \varepsilon_y - k_x^2 - k_z^2 \right) E_y + k_y k_z E_z = 0,$$
$$k_z k_x E_x + k_z k_y E_y + \left(\omega^2 \mu \varepsilon_z - k_x^2 - k_y^2 \right) E_z = 0.$$

We may represent this system of equations in matrix form as

$$\begin{bmatrix} \left[\omega^2 \mu \varepsilon_x - k_y^2 - k_z^2 \right] & k_x k_y & k_x k_z \\ k_y k_x & \left[\omega^2 \mu \varepsilon_y - k_x^2 - k_z^2 \right] & k_y k_z \\ k_z k_x & k_z k_y & \left[\omega^2 \mu \varepsilon_z - k_x^2 - k_y^2 \right] \end{bmatrix} \begin{bmatrix} E_x \\ E_y \\ E_z \end{bmatrix} = \begin{bmatrix} 0 \\ 0 \\ 0 \end{bmatrix}.$$

We shall call the above 3×3 matrix $\overline{\overline{A}}$. Nontrivial solutions, i.e. solutions other than $\vec{E} = 0$, exists only when the determinant of $\overline{\overline{A}}$ is equal to zero:

$$\det\left(\overline{\overline{A}}\right) = \begin{vmatrix} \left(\omega^2\mu\varepsilon_x - k_y^2 - k_z^2\right) & k_x k_y & k_x k_z \\ k_y k_x & \left(\omega^2\mu\varepsilon_y - k_x^2 - k_z^2\right) & k_y k_z \\ k_z k_x & k_z k_y & \left(\omega^2\mu\varepsilon_z - k_x^2 - k_y^2\right) \end{vmatrix} = 0.$$

This is in fact the dispersion relation of the medium, and represents a two-dimensional surface in the three-dimensional domain (k_x, k_y, k_z) called a $\vec{k}$ surface or a wave vector surface. It may be difficult to visualize this surface in 3D, so we will consider the projection of this surface onto the xy-plane of the three-dimensional domain (or the (k_x, k_y) plane). In this case, $k_z = 0$, and the dispersion relation reduces to

$$\left(\omega^2\mu\varepsilon_z - k_x^2 - k_y^2\right)\left[\left(\omega^2\mu\varepsilon_x - k_y^2\right)\left(\omega^2\mu\varepsilon_y - k_x^2\right) - k_x^2 k_y^2\right] = 0. \tag{B.20}$$

We see that this is a product of two expressions. If either of these quantities equals zero, the expression is valid. Therefore

$$\left(\omega^2\mu\varepsilon_z - k_x^2 - k_y^2\right) = 0, \tag{B.21}$$

$$\left(\omega^2\mu\varepsilon_x - k_y^2\right)\left(\omega^2\mu\varepsilon_y - k_x^2\right) - k_x^2 k_y^2 = 0. \tag{B.22}$$

Equation B.21 gives the expression for a circle of radius $\omega^2\mu\varepsilon_z$. This can be seen more clearly if it is written as

$$k_x^2 + k_y^2 = \omega^2\mu\varepsilon_z.$$

Similarly, Equation B.22 gives the expression for an ellipse. When expanded and rearranged, we get the expression

$$\frac{k_x^2}{\omega^2\mu\varepsilon_y} + \frac{k_y^2}{\omega^2\mu\varepsilon_x} = 1.$$

Analogous solutions may be found for the xz and yz planes, telling us that there are actually two $\vec{k}$ surfaces – one sphere and one ellipsoid. What this means physically is that for any given propagation direction $\vec{k}$, there are two pairs[3] of valid wavenumbers associated with the wave. Since the phase velocity is dependent on the wavenumber, this also tells us that the wave propagates at two pairs of phase velocities. Each of these phase velocity pairs is associated with a specific polarization of the $\vec{E}$ vector. These polarizations are the eigenvectors of $\overline{\overline{A}}$, and are called the *characteristic polarizations*. It turns out that these polarizations

---

3 Since all of the components k_i are squared, we actually get four distinct wavenumbers. One pair $\pm|\vec{k}_1|$ on the circle and another pair $\pm|\vec{k}_2|$ on the ellipse. The negative values simply refer to propagation in the opposite direction. Similarly, there are two pairs of phase velocities, with the negative values representing propagation in the opposite direction [271].

are always orthogonal to each other as well as to the direction of propagation. The derivation of this result is not presented in this text, but may be found in [272, pp. 799–802]. Since the phase velocity can assume two different magnitudes, that means there are two different indices of refraction as well. This means that in general anisotropic media will cause double refraction – the two characteristic polarizations will refract at slightly different angles. This double refraction phenomenon is often called *birefringence.*[4]

B.6.3 Optical Axes

While it is true in general that a wave propagating in an anisotropic medium will experience birefringence, there are special propagation directions for which this will not occur. There are a few points at which the $\vec{k}$ surfaces intersect, and the two wavenumbers equal each other. It is simplest to use the circular and elliptical projections we developed earlier to observe this in two dimensions.

In Figure B.4a, we see that for most propagation directions, the propagation vector will intersect the $\vec{k}$ surface at four points, which represent the two pairs of wavenumbers predicted by the dispersion relation. However, there are two special directions where the pairs have the same magnitudes – in other words where the circle and the ellipse intersect. These are shown in Figure B.4b,c. In these directions, the phase velocities of the two characteristic polarizations are equal, so birefringence does not occur. We call these two lines the optical axes of the medium. This is where the term *biaxial* comes from. The case of uniaxial media is shown in Figure B.5. We see that the circle and the ellipse only intersect at two locations instead of four in this case. This tells us that there is only one optical axis, so any wave not traveling parallel to the optical axis will experience birefringence.

B.6.4 Index Ellipsoid

Since $\varepsilon_r = n^2$, we can write the dielectric tensor as

$$\overline{\overline{\varepsilon}} = \varepsilon_0 \begin{bmatrix} n_x^2 & 0 & 0 \\ 0 & n_y^2 & 0 \\ 0 & 0 & n_z^2 \end{bmatrix}.$$

We have seen in Section B.6.2 that for any given propagation direction, the medium supports two orthogonal polarizations (eigenpolarizations). In general, these polarizations travel at different phase velocities and thus have different

---

4 There are a few anisotropic media which are an exception to this rule and do not exhibit birefringence, see e.g. [273].

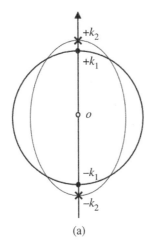

(a)

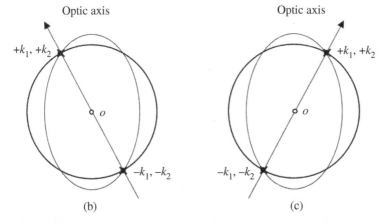

(b)   (c)

Figure B.4 Propagation in a biaxial medium (a) along neither optical axis, (b) along one optical axis, (c) along the other optical axis.

indices of refraction. There is a useful tool we may use to find the two eigenpolarizations and their corresponding indices of refraction, called the index ellipsoid. To derive this, we write

$$W_e = \frac{1}{2}\vec{E} \cdot \vec{D} = \frac{1}{2}\left(\frac{D_x^2}{\varepsilon_0 n_x^2} + \frac{D_y^2}{\varepsilon_0 n_y^2} + \frac{D_z^2}{\varepsilon_0 n_z^2} \right),$$

and rearranging a bit, we get

$$\frac{x^2}{n_x^2} + \frac{y^2}{n_y^2} + \frac{z^2}{n_z^2} = 1. \tag{B.23}$$

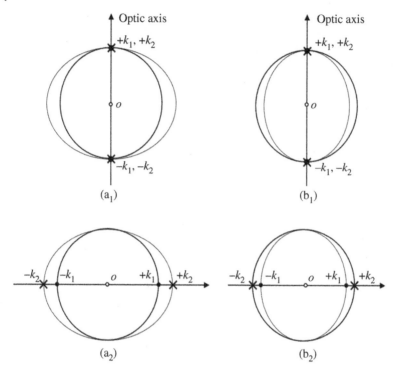

Figure B.5 Propagation in uniaxial media (a_1) along the optical axis, positive medium, (a_2) off the optical axis, positive medium, (b_1) along the optical axis, negative medium, (b_2) off the optical axis, negative medium.

Here, $x = D_x/\sqrt{2\varepsilon_0 W_e}$, and y and z are analogous expressions containing D_y and D_z. We see that Equation B.23 is the equation for an ellipsoid. Every point on the surface of the ellipsoid represents the effective index of the wave in that direction. For simplicity, we will henceforth assume a uniaxial medium, simplifying Equation B.23 to

$$\frac{x^2 + y^2}{n_o^2} + \frac{z^2}{n_e^2} = 1.$$

From this we can see that the first term represents a circle, which means that the entire xy-plane has the same index of refraction which we will call n_o, the ordinary index. The z-axis has a different index of refraction n_e, the extraordinary index. If $n_e > n_o$, the medium is referred to as positive, and if $n_e < n_o$ the medium is called negative. The $\vec{k}$ surfaces for each case may be seen in Figure B.5. Since x, y, and z are all defined in terms of components of $\vec{D}$, this ellipse shows us that any vector representing a polarization of $\vec{D}$ will point to a specific point on the surface of the

Figure B.6 Index ellipsoid of a positive
uniaxial medium.

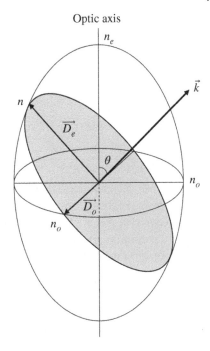

ellipsoid. This point represents the index of refraction for that polarization. Further, the index ellipsoid allows us to systematically determine which polarizations are associated with a given propagation direction (Figure B.6).

1. Draw the ray representing the $\vec{k}$ vector with its tail at the origin.
2. Draw a plane that includes the origin and is normal to the $\vec{k}$ vector.
3. This plane will intersect with the ellipsoid, creating an ellipse.
4. The semi-major and semi-minor axes of this ellipse represent the two eigenpolarizations for that direction of propagation.

Consider a wave traveling at some angle θ with respect to the optical axis, with $\theta = 0$ representing propagation along the z-axis and $\theta = \frac{\pi}{2}$ representing propagation in the xy-plane. The semi-major axis of the ellipse will always be at some angle $\theta - \frac{\pi}{2}$ with respect to the z-axis. The semi-minor axis remains fixed in the xy-plane, and is not dependent on θ. This polarization that remains fixed always travels with index of refraction n_o, and is called the ordinary ray. The other polarization, which is called the extraordinary ray, will experience a different index of refraction depending on the propagation direction of the wave. It is possible to find the index of refraction for the ordinary ray as a function of θ. The point n representing this index of refraction is at an angle of $\theta - \frac{\pi}{2}$ with respect to the z-axis.

The projection of the semi-major axis into the xy plane is given by

$$x^2 + y^2 = \left(n \sin\left(\theta - \frac{\pi}{2}\right)\right)^2 = (n \cos\theta)^2.$$

Similarly, the projection of the semi-major axis onto the z-axis is

$$z = n \cos\left(\theta - \frac{\pi}{2}\right) = n \sin\theta.$$

Substituting these into the ellipsoid expression, we get

$$\frac{(n \cos\theta)^2}{n_0^2} + \frac{(n \sin\theta)^2}{n_e^2} = 1 \Rightarrow \frac{\cos^2\theta}{n_0^2} + \frac{\sin^2\theta}{n_e^2} = \frac{1}{n^2}.$$

In this form, it is easy to see that the value of the index of refraction will depend on the angle of incidence. For waves traveling in the xy-plane, $\theta = \frac{\pi}{2}$ and the expression reduces to $n = n_e$. Thus the ordinary ray travels with index of refraction n_0 and the extraordinary ray travels with index n_e.

Waves traveling along the z-axis are at $\theta = 0$, so the expression reduces to $n = n_0$. Thus, the ordinary and extraordinary wave both travel with the index n_0. In other words, the two polarizations do not have different indices of refraction, and birefringence does not occur in this direction. This is the same conclusion reached in Section B.6.3, where we showed that for waves traveling along the optical axis, the wavenumbers/phase velocities of the two eigenpolarizations were equal. Further, we see that the xy-plane is the only planar section of the index ellipsoid that produces a circle rather than an ellipse. Thus, not only do the polarizations along the semi-major and semi-minor axes have index n_0, but *every* polarization in the xy-plane does as well. Put simply, this means that, for waves traveling along the optical axis, the medium supports every possible polarization, not just the two eigenpolarizations. This corresponds to what happens in isotropic media.

Waves incident at any angle $0 < \theta < \frac{\pi}{2}$ will experience an index of refraction with an intermediate value n between n_0 and n_e known as the effective index. The ordinary ray, as always, travels with index n_0, while the extraordinary ray will travel with the effective index n.

B.6.5 Phase and Group Velocity in Anisotropic Media

Since $\bar{\bar{\epsilon}}$ is no longer represented by a scalar quantity in anisotropic media, $\vec{E}$ and $\vec{D}$ are, in general, not parallel. We know that

$$\vec{S} = \vec{E} \times \vec{H},$$

which implies that $\vec{S}$ is perpendicular to $\vec{E}$, but also that, in source-free media, Maxwell's equations give us

$$\nabla \cdot \vec{D} = \vec{k} \cdot \vec{D} = 0,$$

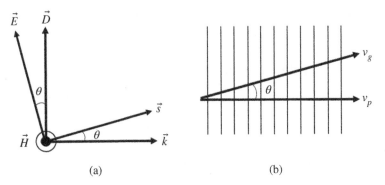

Figure B.7 Spatial walk-off as represented by non-parallel nature of (a) field vectors and (b) phase and group velocities.

which implies that $\vec{D}$ is perpendicular to $\vec{k}$. But the fact that $\vec{E}$ and $\vec{D}$ are not necessarily parallel tells us that $\vec{S}$ and $\vec{k}$ are also not necessarily parallel. This affects the phase and group velocities, since v_p is associated with the propagation direction of the wavefronts, $\vec{k}$, but the group velocity (and thus transmission of information) is determined by the direction of power flow, $\vec{S}$. This phenomenon is referred to as *spatial walk-off*, and is depicted in Figure B.7 where $\vec{E} \cdot \vec{D} = |\vec{E}||\vec{D}| \cos \theta$. Note that when $\vec{D}$ and $\vec{E}$ are parallel, $\theta = 0$, so v_p and v_g travel in the same direction. This only occurs if $\vec{E}$ is an eigenvector of $\overline{\overline{\epsilon}}$, i.e.

$$\overline{\overline{\epsilon}}\vec{E}_i = \lambda_i \vec{E}_i. \tag{B.24}$$

We saw earlier that the eigenvalues λ_i are the principal permittivities and that their corresponding eigenvectors lie along the principal axes. This tells us that as long as a wave propagates along one of the principal axes of a medium, spatial walk-off will not occur, so the phase and group velocities will be parallel to each other.

Figure B.7 represents the most general case, and must be used when considering biaxial media. However, uniaxial and isotropic media may be seen as special cases in which the constraints on the system are relaxed. If we consider a uniaxial medium, we see that ϵ_x is a repeated eigenvalue with multiplicity two. Thus, the corresponding eigenbasis has dimension two, representing an eigenplane – an entire plane which is invariant under the linear transformation represented by $\overline{\overline{\epsilon}}$. In other words, any vector in this plane (the *xy*-plane by convention) is an eigenvector of $\overline{\overline{\epsilon}}$, so any wave traveling in this plane will not exhibit spatial walk-off. The *z*-axis typically corresponds to the optical axis, and is orthogonal to the *xy*-plane. Waves traveling along this axis are also eigenvectors, so no spatial walk-off occurs in this direction either. Any vector in neither of these categories, i.e. vectors at some oblique angle to the optical axis, are not eigenvectors, so they will exhibit spatial walk-off. Similar reasoning may be applied to an isotropic medium, in which

ε is an eigenvalue of multiplicity three. Thus every vector in $\mathbb{R}^3$ is an eigenvector of $\overline{\overline{\varepsilon}}$, so no wave will ever exhibit spatial walk-off in an isotropic medium.

B.6.6 Birefringence and Spatial Walk-off in $\vec{k}$ Surfaces

A useful way to visualize spatial walk-off is through the use of $\vec{k}$ surfaces. Imagine once again the projection of the three-dimensional surfaces into the xy-plane, given by Equation B.20. Recall that this is the dispersion relation for the anisotropic medium. It is possible to find the group velocity by taking the derivative of this ω with respect to k. In multiple dimensions, the group velocity is given by

$$\vec{v}_g = \nabla_{\vec{k}}\omega(\vec{k}),$$

where $\omega(\vec{k})$ can be found from the dispersion relation. The gradient of a surface defines a tangent plane which may be described by a vector normal to its surface. When this plane and vector are projected into the xy-plane, it results in a line tangent to the 2D curve and a vector normal to the curve.

It is immediately evident in Figure B.8 that spatial walk-off is absent on a circular (spherical) $\vec{k}$ surface and in general present on an elliptical (ellipsoidal) $\vec{k}$ surface. The only propagation directions for which the ellipse does not exhibit spatial walk-off are along the major and minor axes, where the tangent line is perpendicular to the $\vec{k}$ vector. In general the tangent line on the ellipse will not be perpendicular to the propagation vector (spatial walk-off), and if we overlay the circle and the ellipse, we see that any given propagation direction will intersect each curve at a different point (birefringence).

In biaxial materials there are two special directions where the circle and the ellipse overlap and birefringence does not occur. These are the two optical axes (or

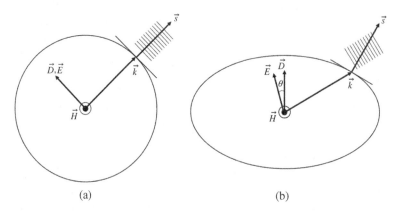

(a)   (b)

Figure B.8 (a) Circle, no walk-off. (b) Ellipse, walk-off present.

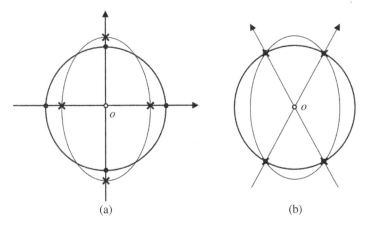

(a)   (b)

Figure B.9   (a) Biradials (principal axes): No spatial walk-off. (b) Binormals (optical axes): No birefringence.

binormals) of the medium. There are also three directions in which spatial walk-off does not occur. These directions are called the secondary optical axes (or biradials) and they lie along the principal axes of the medium. In a biaxial medium, the binormals and biradials never overlap, meaning there is no direction where birefringence and spatial walk-off may be simultaneously avoided. Figure B.9 demonstrates this in two dimensions by using the projection onto one of the principal planes, but since there is no symmetry about any axis of a biaxial medium, this representation is analogous for any of the three principal planes. The eccentricity of the ellipse will be different in each case, but the circle and ellipse will always intersect at four distinct points.

A uniaxial medium is a special case in which the two optical axes coincide not only with each other, but also with one of the principal axes. Since a uniaxial medium is symmetric about its optic axis, the xz- and yz-planes are identical. The optical axis exists in both of these planes, and it can be seen that the circle and the ellipse intersect at two points along the optical axis. Since the optic axis is also a principal axis, the line tangent to the ellipse at these points is perpendicular to the propagation vector. Therefore, waves propagating along the optical axis experience neither birefringence nor spatial walk-off.

The xy-plane is different from the other two principal planes. In this plane, both curves are circles of different radii. For every direction of propagation, the tangent line is perpendicular to the $\vec{k}$ vector, so spatial walk-off never occurs in this plane. However, there are also no points where the two curves intersect, so there are also no directions for which birefringence does not occur. These results may be observed for a positive uniaxial medium in Figure B.10. In a negative medium, the semi-axis corresponding to n_e would be shorter than the one corresponding to n_o.

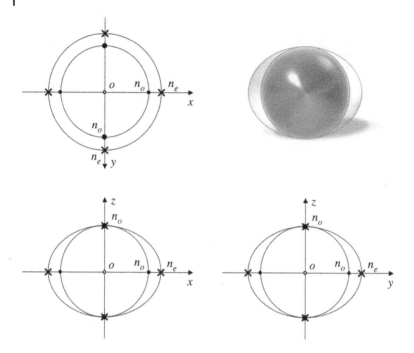

Figure B.10 Orthographic projection of $\vec{k}$ surface for a positive uniaxial medium.

This has no impact on the conditions we have just discussed, so the same results also apply to negative media.

An isotropic medium is so simple that it does not even warrant a diagram. Both of the 3D curves that comprise the $\vec{k}$ surface are identical spheres. Therefore they intersect at every point, and every tangent plane is orthogonal to its corresponding $\vec{k}$ vector. Thus, neither birefringence nor spatial walk-off occur in an isotropic medium.

B.7 Gyrotropic Media

B.7.1 Gyrotropic Susceptibility Tensor

Until now, we have assumed that the magnetic effects on a medium were negligible. However, in many media this assumption cannot be made. Ferromagnetic materials (permanent magnets) have an intrinsic magnetic dipole moment per unit volume. Diamagnetic and paramagnetic materials are not normally magnetized, but a magnetic dipole moment is induced in the presence of an external magnetic field. In the ionosphere, the presence of Earth's magnetic field will also induce such effects. A medium which is normally isotropic, but obtains anisotropy due to the presence of a magnetic field is called a (non-reciprocal) *gyrotropic medium*.

Magnetization changes the form of the susceptibility tensor, giving rise to several unique effects. One such effect is Faraday rotation, which is a significant effect at GNSS frequencies in the ionosphere. If we return to our derivation of susceptibility from Section B.2.2, this time assuming the presence of a static magnetic field, we now have

$$m_e \left(\frac{d^2 \vec{r}}{dt^2} + \Gamma \frac{d\vec{r}}{dt} + \omega_0^2 \vec{r} \right) = -e \left(\vec{E} + \frac{d\vec{r}}{dt} \times \vec{B} \right). \tag{B.25}$$

Since the time-varying magnetic field produced by the electromagnetic wave itself is typically very small compared with the external static magnetic field (and particularly so in the ionosphere), we will assume that only the static magnetic field contributes to the force on the electron. The axis defined by the direction of the magnetic field is denoted as the *axis of gyration* of the medium. Let's assume that the axis of gyration is in the z-direction. The magnetic field is thus given by

$$\vec{B} = \hat{z} B_0.$$

Consider an electromagnetic wave propagating in the $+z$-direction. The electric field of this wave is given by

$$\vec{E} = \begin{bmatrix} \underline{E}_x \\ \underline{E}_y \\ \underline{E}_z \end{bmatrix} e^{j(kz - \omega t)}.$$

The displacement vector will be of the same form

$$\vec{r} = \begin{bmatrix} \underline{x}_0 \\ \underline{y}_0 \\ \underline{z}_0 \end{bmatrix} e^{j(kz - \omega t)}.$$

Substituting all of these equations back into our differential equation, we get:

$$m_e \left(-\omega^2 \vec{r} - j\omega \Gamma \vec{r} + \omega_0^2 \vec{r} \right) = -e \left(\vec{E} - j\omega \vec{r} \times \hat{z} B_0 \right). \tag{B.26}$$

Considering the x-component first and re-arranging a bit, we obtain:

$$\left[m_e \left(\omega_0^2 - \omega^2 - j\omega \Gamma \right) \right] \underline{x}_0 - j\omega e B_0 \underline{y}_0 = -e \underline{E}_x.$$

Similarly, the y and z-components give:

$$\left[m_e \left(\omega_0^2 - \omega^2 - j\omega \Gamma \right) \right] \underline{y}_0 + j\omega e B_0 \underline{x}_0 = -e \underline{E}_y,$$

$$\left[m_e \left(\omega_0^2 - \omega^2 - j\omega \Gamma \right) \right] \underline{z}_0 = -e \underline{E}_z.$$

If we divide through by m_e and define the *cyclotron* frequency as

$$\omega_c = -\frac{eB_0}{m_e},$$

we get the following system of equations in matrix form:

$$
\begin{bmatrix}
(\omega_0^2 - \omega^2 - j\omega\Gamma) & j\omega\omega_c & 0 \\
-j\omega\omega_c & (\omega_0^2 - \omega^2 - j\omega\Gamma) & 0 \\
0 & 0 & (\omega_0^2 - \omega^2 - j\omega\Gamma)
\end{bmatrix}
\begin{bmatrix} x_0 \\ y_0 \\ z_0 \end{bmatrix}
= -\frac{e}{m_e}
\begin{bmatrix} E_x \\ E_y \\ E_z \end{bmatrix}.
$$

We wish to solve for $\vec{r}$, so we must find the inverse of the matrix on the left hand side, which we will denote $\mathbf{A}$. The inverse of $\mathbf{A}$ is given by

$$\frac{1}{\det(\mathbf{A})}\mathbf{C}^{\mathrm{T}},$$

where $\mathbf{C}$ is the cofactor matrix, given by

$$
\mathbf{C} =
\begin{bmatrix}
+\begin{vmatrix} a_{22} & a_{23} \\ a_{32} & a_{33} \end{vmatrix} & -\begin{vmatrix} a_{21} & a_{23} \\ a_{31} & a_{33} \end{vmatrix} & +\begin{vmatrix} a_{21} & a_{22} \\ a_{31} & a_{32} \end{vmatrix} \\[2mm]
-\begin{vmatrix} a_{12} & a_{13} \\ a_{32} & a_{33} \end{vmatrix} & +\begin{vmatrix} a_{11} & a_{13} \\ a_{31} & a_{33} \end{vmatrix} & -\begin{vmatrix} a_{11} & a_{12} \\ a_{31} & a_{32} \end{vmatrix} \\[2mm]
+\begin{vmatrix} a_{12} & a_{13} \\ a_{22} & a_{23} \end{vmatrix} & -\begin{vmatrix} a_{11} & a_{13} \\ a_{21} & a_{23} \end{vmatrix} & +\begin{vmatrix} a_{11} & a_{12} \\ a_{21} & a_{22} \end{vmatrix}
\end{bmatrix},
$$

where a_{ij} is the ij^{th} element of $\mathbf{A}$. The cofactor matrix for our system of equations earlier is

$$
\mathbf{C} =
\begin{bmatrix}
\left[\omega_0^2 - \omega^2 - j\omega\Gamma\right]^2 & -j\omega\omega_c\left[\omega_0^2 - \omega^2 - j\omega\Gamma\right] & 0 \\
j\omega\omega_c\left[\omega_0^2 - \omega^2 - j\omega\Gamma\right] & \left[\omega_0^2 - \omega^2 - j\omega\Gamma\right]^2 & 0 \\
0 & 0 & \left[\omega_0^2 - \omega^2 - j\omega\Gamma\right]^2 - \omega^2\omega_c^2
\end{bmatrix}.
$$

Transposing this matrix merely swaps the (1,2) and the (2,1) element. They only differ by a negative sign.

$$
\mathbf{C}^{\mathrm{T}} =
\begin{bmatrix}
\left[\omega_0^2 - \omega^2 - j\omega\Gamma\right]^2 & j\omega\omega_c\left[\omega_0^2 - \omega^2 - j\omega\Gamma\right] & 0 \\
-j\omega\omega_c\left[\omega_0^2 - \omega^2 - j\omega\Gamma\right] & \left[\omega_0^2 - \omega^2 - j\omega\Gamma\right]^2 & 0 \\
0 & 0 & \left[\omega_0^2 - \omega^2 - j\omega\Gamma\right]^2 - \omega^2\omega_c^2
\end{bmatrix}.
$$

Since most of the entries of $\mathbf{A}$ are zero, the determinant is simple to compute:

$$\det(\mathbf{A}) = \left[\left(\omega_0^2 - \omega^2 - j\omega\Gamma\right)^2 - \omega^2\omega_c^2\right]\left(\omega_0^2 - \omega^2 - j\omega\Gamma\right).$$

Taking the inverse of this expression and multiplying it by $\mathbf{C}^{\mathrm{T}}$ results in the inverse of our original matrix:

$$
\mathbf{A}^{-1} =
\begin{bmatrix}
\dfrac{\omega_0^2 - \omega^2 - j\omega\Gamma}{\left(\omega_0^2 - \omega^2 - j\omega\Gamma\right)^2 - \omega^2\omega_c^2} & \dfrac{j\omega\omega_c}{\left(\omega_0^2 - \omega^2 - j\omega\Gamma\right)^2 - \omega^2\omega_c^2} & 0 \\[4mm]
\dfrac{-j\omega\omega_c}{\left(\omega_0^2 - \omega^2 - j\omega\Gamma\right)^2 - \omega^2\omega_c^2} & \dfrac{\omega_0^2 - \omega^2 - j\omega\Gamma}{\left(\omega_0^2 - \omega^2 - j\omega\Gamma\right)^2 - \omega^2\omega_c^2} & 0 \\[4mm]
0 & 0 & \dfrac{1}{\omega_0^2 - \omega^2 - j\omega\Gamma}
\end{bmatrix}.
$$

Therefore, we can write the system of equations earlier as

$$
\vec{r} = -\frac{e}{m_e}\mathbf{A}^{-1}\vec{E}.
$$

Knowing that $\vec{P} = -Ne\vec{r}$ and $\vec{P} = \varepsilon_0 \overline{\overline{\chi}}\vec{E}$, we obtain

$$
\frac{Ne^2}{\varepsilon_0 m_e}\mathbf{A}^{-1}\vec{E} = \overline{\overline{\chi}}\vec{E}.
$$

This gives us an expression for the susceptibility tensor $\overline{\overline{\chi}}$ in gyrotropic media.

$$
\overline{\overline{\chi}} =
\begin{bmatrix}
\chi_{11} & j\chi_{12} & 0 \\
-j\chi_{12} & \chi_{11} & 0 \\
0 & 0 & \chi_{33}
\end{bmatrix}
$$

with

$$
\chi_{11} = \frac{Ne^2}{\varepsilon_0 m_e}\frac{\omega_0^2 - \omega^2 - j\omega\Gamma}{\left(\omega_0^2 - \omega^2 - j\omega\Gamma\right)^2 - \omega^2\omega_c^2},
$$

$$
\chi_{12} = \frac{Ne^2}{\varepsilon_0 m_e}\frac{\omega\omega_c}{\left(\omega_0^2 - \omega^2 - j\omega\Gamma\right)^2 - \omega^2\omega_c^2},
$$

$$
\chi_{33} = \frac{Ne^2}{\varepsilon_0 m_e}\frac{1}{\omega_0^2 - \omega^2 - j\omega\Gamma}.
$$

We see that this matrix is no longer symmetric nor Hermitian, which means that medium is lossy and non-reciprocal, as discussed before. Non-reciprocal media can be exploited to create devices such as isolators and circulators.

B.7.2 Propagation in Gyrotropic Media

The dielectric tensor of a gyrotropic medium with a z-directed axis of gyration is

$$
\overline{\overline{\varepsilon}} = \varepsilon_0\left(\overline{\overline{I}} + \overline{\overline{\chi}}\right) = \varepsilon_0
\begin{bmatrix}
1 + \chi_{11} & j\chi_{12} & 0 \\
-j\chi_{12} & 1 + \chi_{11} & 0 \\
0 & 0 & \chi_{33}
\end{bmatrix}
=
\begin{bmatrix}
\varepsilon_{11} & j\varepsilon_{12} & 0 \\
-j\varepsilon_{12} & \varepsilon_{11} & 0 \\
0 & 0 & \varepsilon_{33}
\end{bmatrix}.
$$

For a plane-wave in a source-free medium, we have seen that

$$\nabla \times \left(\nabla \times \vec{E} \right) = -\mu \bar{\bar{\varepsilon}} \frac{\partial^2 \vec{E}}{\partial t^2} \quad \Rightarrow \quad \vec{k} \times \left(\vec{k} \times \vec{E} \right) + \omega^2 \mu \bar{\bar{\varepsilon}} \vec{E} = 0.$$

Assuming a wave traveling in the $+z$-direction (i.e. let $k_x = k_y = 0$ and $k_z = k$), we get the following system of equations

$$\begin{bmatrix} \omega^2 \mu \varepsilon_{11} - k^2 & j\omega^2 \mu \varepsilon_{12} & 0 \\ -j\omega^2 \mu \varepsilon_{12} & \omega^2 \mu \varepsilon_{11} - k^2 & 0 \\ 0 & 0 & \omega^2 \mu \varepsilon_{33} \end{bmatrix} \begin{bmatrix} \underline{E}_x \\ \underline{E}_y \\ \underline{E}_z \end{bmatrix} = \begin{bmatrix} 0 \\ 0 \\ 0 \end{bmatrix}. \tag{B.27}$$

Nontrivial solutions occur when the determinant of the square matrix equals zero.

$$\left[\left(\omega^2 \mu \varepsilon_{11} - k^2 \right)^2 - \left(\omega^2 \mu \varepsilon_{12} \right)^2 \right] \omega^2 \mu \varepsilon_{33} = 0. \tag{B.28}$$

Since $\omega \mu \varepsilon_{33}$ is a nonzero quantity, the only solution to this equation is

$$\left(\omega^2 \mu \varepsilon_{11} - k^2 \right)^2 - \left(\omega^2 \mu \varepsilon_{12} \right)^2 = 0 \Rightarrow \omega^2 \mu \varepsilon_{11} - k^2 = \pm \omega^2 \mu \varepsilon_{12}.$$

Solving for k, we get

$$k = \omega \sqrt{\mu \left(\varepsilon_{11} \pm \varepsilon_{12} \right)}.$$

This tells us that there are two different values of k and therefore two phase velocities and two indices of refraction. This implies birefringence. If we substitute this value of k back into the Equation B.27, we see that

$$\underline{E}_x = \pm j \underline{E}_y.$$

Therefore the polarizations corresponding to the two different phase velocities are right-hand and left-hand circular. The indices of refraction are respectively

$$n_R = \sqrt{\varepsilon_{11} + \varepsilon_{12}}, \tag{B.29}$$

$$n_L = \sqrt{\varepsilon_{11} - \varepsilon_{12}}. \tag{B.30}$$

This is called *circular birefringence*.

We know that any linearly polarized wave can be decomposed into one left-hand and one right-hand circularly polarized wave. Using the result we just proved, we see that these two constituent circular polarizations will travel at different phase velocities through a gyrotropic medium such as the ionosphere. This can be interpreted as a rotation of the linear polarization's angle of polarization. To show this (derivation from [266, pp. 186, 187]), let's consider a wave linearly polarized in the x-direction. This polarization can be split into two circular polarizations. In Jones vector notation, this is:

$$\begin{bmatrix} 1 \\ 0 \end{bmatrix} = \frac{1}{2} \begin{bmatrix} 1 \\ -j \end{bmatrix} + \frac{1}{2} \begin{bmatrix} 1 \\ j \end{bmatrix}.$$

Expressing each polarization as a wave traveling in the +z-direction gives

$$\vec{E} = \frac{1}{2} \begin{bmatrix} 1 \\ -j \end{bmatrix} e^{j(k_R z - \omega t)} + \frac{1}{2} \begin{bmatrix} 1 \\ j \end{bmatrix} e^{j(k_L z - \omega t)}.$$

We are only concerned with the amplitude, so we may ignore the time dependence of this wave. Consider its amplitude after it has traveled a distance d in the z-direction:

$$\vec{E}(x, y, z) = \frac{1}{2} \begin{bmatrix} 1 \\ -j \end{bmatrix} e^{jk_R d} + \frac{1}{2} \begin{bmatrix} 1 \\ j \end{bmatrix} e^{jk_L d}$$

$$= e^{j(k_R + k_L)d/2} \left(\frac{1}{2} \begin{bmatrix} 1 \\ -j \end{bmatrix} e^{j(k_R - k_L)d/2} + \frac{1}{2} \begin{bmatrix} 1 \\ j \end{bmatrix} e^{-j(k_R - k_L)d/2} \right).$$

Let's define the quantities:

$$\psi \equiv \frac{1}{2} \left(k_R + k_L \right) d,$$

$$\theta \equiv \frac{1}{2} \left(k_R - k_L \right) d. \tag{B.31}$$

Therefore:

$$\vec{E} = e^{j\psi} \left(\frac{1}{2} \begin{bmatrix} 1 \\ -j \end{bmatrix} e^{j\theta} + \frac{1}{2} \begin{bmatrix} 1 \\ j \end{bmatrix} e^{-j\theta} \right)$$

$$= e^{j\psi} \begin{bmatrix} \frac{1}{2} \left(e^{j\theta} + e^{-j\theta} \right) \\ \frac{1}{2j} \left(e^{j\theta} - e^{-j\theta} \right) \end{bmatrix}$$

$$= e^{j\psi} \begin{bmatrix} \cos \theta \\ \sin \theta \end{bmatrix}.$$

This tells us that a linearly polarized wave is rotated by an angle θ after traveling a distance d in a gyrotropic medium. This is precisely the phenomenon of Faraday rotation. Using Equation B.31, we can express this angle in terms of the two circular waves' indices of refraction:

$$\theta = \frac{1}{2} \left(k_R - k_L \right) d = \frac{1}{2} \left(\frac{2\pi}{\lambda_0} n_R - \frac{2\pi}{\lambda_0} n_L \right) d = \left(n_R - n_L \right) \frac{\pi d}{\lambda_0}.$$

Here, λ_0 is the vacuum wavelength. If we remove the dependence on the distance traveled, we get a quantity called the specific rotary power:

$$\delta = \left(n_R - n_L \right) \frac{\pi}{\lambda_0}.$$

Since n_R and n_L will in general be functions of frequency, the amount of rotation that occurs is also dependent on frequency. This effect is called *rotary dispersion*, and accounts for the dependence of Faraday rotation on frequency.

Faraday rotation effects are strong at GNSS frequencies. Because of the spatiotemporal variability of the ionosphere, it is impossible to predict with adequate precision the amount of Faraday rotation for a linear polarized wave traversing the ionosphere. As a result, GNSS links employ circular (rather than linear) polarized waves.

Bibliography

1 D. J. Griffiths. *Introduction to Electrodynamics*. Cambridge University Press, Cambridge, United Kingdom; New York, NY, 4 edition, July 2017. ISBN 978-1-108-42041-9.

2 GPS: The Global Positioning System. URL http://www.gps.gov/.

3 C. DeMets, R. G. Gordon, and D. F. Argus. Geologically Current Plate Motions. *Geophysical Journal International*, 181(1):1–80, April 2010. ISSN 0956540X, 1365246X. doi: 10.1111/j.1365-246X.2009.04491.x.

4 E. D. Kaplan and C. J. Hegarty. *Understanding GPS: Principles and Applications*. Artech House, Inc., Boston, MA, 2nd edition, 2006. ISBN 1-58053-894-0.

5 ITU-R Recommendation V.431-8. Nomenclature of the Frequency and Wavelength Bands Used in Telecommunications, August 2015.

6 IEEE Standard Letter Designations for Radar-Frequency Bands. *IEEE Std 521-2002 (Revision of IEEE Std 521-1984)*, pages 1–10, January 2003. doi: 10.1109/IEEESTD.2003.94224.

7 J. Sanz Subirana, J. M. J. Zornoza, and M. Hernández-Pajares. *GNSS Data Processing*, volume I: Fundamentals and Algorithms. ESA Communications, 2013.

8 R. Gold. Optimal Binary Sequences for Spread Spectrum Multiplexing (Corresp.). *IEEE Transactions on Information Theory*, 13(4):619–621, October 1967. ISSN 0018-9448. doi: 10.1109/TIT.1967.1054048.

9 J. A. A. Rodriguez. *On Generalized Signal Waveforms for Satellite Navigation*. Doctoral Thesis, University of Munich, June 2008.

10 A. B. Carlson and P. B. Crilly. *Communication Systems: An Introduction to Signals and Noise in Electrical Communication*. McGraw-Hill Education, 5th edition, February 2009.

11 Günter Seeber. *Satellite Geodesy*. Walter de Gruyter, Berlin; New York, 2nd edition, 2003. ISBN 978-3-11-017549-3.

Tropospheric and Ionospheric Effects on Global Navigation Satellite Systems, First Edition.
Timothy H. Kindervatter and Fernando L. Teixeira.
© 2022 The Institute of Electrical and Electronics Engineers, Inc. Published 2022 by John Wiley & Sons, Inc.

12 United States Coast Guard. GPS Frequently Asked Questions (FAQ). URL https://www.navcen.uscg.gov/?pageName=gpsFaq.

13 International GNSS Service. IGS Products. URL http://www.igs.org/products/data.

14 US Coast Guard Home. URL https://www.uscg.mil/.

15 L. Wang. Introduction to GPS: Theory and Applications, CE5441 Lecture Notes, The Ohio State University. March 2017.

16 Global Positioning Systems Directorate. Navstar GPS Space Segment/ Navigation User Interface Specification. IS-GPS-705D, September 2013.

17 Global Positioning Systems Directorate. Navstar GPS Space Segment/User Segment L5C Interface Specification. IS-GPS-200H, September 2013.

18 Global Positioning Systems Directorate. Navstar GPS Space Segment/User Segment L1C Interface Specification. IS-GPS-800D, September 2013.

19 B. Hofmann-Wellenhof, H. Lichtenegger, and E. Wasle. *GNSS–Global Navigation Satellite Systems: GPS, GLONASS, Galileo, and More.* Springer, Wien; New York, 2008. ISBN 978-3-211-73012-6.

20 G. Xu. *GPS: Theory, Algorithms, and Applications.* Springer-Verlag, Berlin; New York, 2nd edition, 2007. ISBN 978-3-540-72714-9.

21 M. Petovello. GNSS Solutions: Calculating Time Offsets. *Inside GNSS*, pages 32–37, May/June 2013.

22 O. Montenbruck, P. Steigenberger, and A. Hauschild. Broadcast versus Precise Ephemerides: A Multi-GNSS Perspective. *GPS Solutions*, 19(2):321–333, April 2015. ISSN 1080-5370, 1521-1886. doi: 10.1007/s10291-014-0390-8.

23 R. Shankar. PHYS 200 relativity Notes, Yale University, 2006. URL https://oyc .yale.edu/sites/default/files/relativity_notes_2006_7.pdf.

24 Z. Hećimović. Relativistic Effects on Satellite Navigation. *Tehnički vjesnik Technical Gazette*, 20(1):195–203, 2013. ISSN 1848-6339.

25 S. Y. Zhu and E. Groten. Relativistic Effects in GPS. In *GPS-Techniques Applied to Geodesy and Surveying* (ed. E. Groten and R. Strauss), Lecture Notes in Earth Sciences, pages 41–46. Springer-Verlag, Berlin, Heidelberg, 1988. ISBN 978-3-540-50267-8.

26 R. Nelson and T. Allan Ely. Relativistic Transformations for Time Synchronization and Dissemination in the Solar System. In *Proceedings of the 38th Annual Precise Time and Time Interval Systems and Applications Meeting*, pages 305–317, Reston, Virginia, December 2006.

27 B. W. Parkinson and J. J. Spilker, editors. *Global Positioning System: Theory & Applications.* American Institute of Aeronautics and Astronautics, Washington, DC, 1st edition, January 1996. ISBN 978-1-56347-106-3.

28 N. Ashby. Relativity in the Global Positioning System. *Living Reviews in Relativity*, 6(1):1, January 2003. ISSN 1433-8351. doi: 10.12942/lrr-2003-1.

29 L. L. Wells. Field Test Results on the Use of Translated GPS for TRIDENT I. *Navigation*, 34(2):115–123, June 1987. ISSN 2161-4296. doi: 10.1002/j.2161-4296.1987.tb01494.x.

30 R. D. Van Nee. Multipath Effects on GPS Code Phase Measurements. *Navigation*, 39(2):177–190, June 1992. ISSN 2161-4296. doi: 10.1002/j .2161-4296.1992.tb01873.x.

31 T. S. Rappaport. *Wireless Communications: Principles and Practice.* Prentice Hall, Upper Saddle River, NJ, 2nd edition, January 2002. ISBN 978-0-13-042232-3.

32 F. K. Brunner and M. Gu. An Improved Model for the Dual Frequency Ionospheric Correction of GPS Observations. *Manuscripta Geodaetica*, 16(3):205–214, January 1991.

33 R. Dietrich, R. Dach, G. Engelhardt, J. Ihde, W. Korth, H.-J. Kutterer, K. Lindner, M. Mayer, F. Menge, H. Miller, C. Müller, W. Niemeier, J. Perlt, M. Pohl, H. Salbach, H.-W. Schenke, T. Schöne, G. Seeber, A. Veit, and C. Völksen. ITRF Coordinates and Plate Velocities from Repeated GPS Campaigns in Antarctica – an Analysis Based on Different Individual Solutions. *Journal of Geodesy*, 74(11–12):756–766, March 2001. ISSN 0949-7714, 1432-1394. doi: 10.1007/s001900000147.

34 J. Van Sickle. GEOG 862 - GPS and GNSS for Geospatial Professionals. *Pennsylvania State University Department of Geography*, August 2017. URL https://www.e-education.psu.edu/geog862/.

35 D. Haw. SP3 FORMAT. *National Geodetic Survey*, November 2017. URL https://www.ngs.noaa.gov/orbits/SP3_format.html.

36 S. Bancroft. An Algebraic Solution of the GPS Equations. *IEEE Transactions on Aerospace and Electronic Systems*, 21(1):56–59, 1985.

37 M. Mironovova and H. Havlis. Calculation of GDOP Coefficient. *2011 Konference STC, Czech Technical University in Prague: Faculty of Electrical Engineering*, 2011.

38 R. B. Langley. Dilution of Precision. *GPS World*, pages 52–59, May 1999.

39 H. S. Hopfield. Two-Quartic Tropospheric Refractivity Profile for Correcting Satellite Data. *Journal of Geophysical Research*, 74(18):4487–4499, August 1969. ISSN 2156-2202. doi: 10.1029/JC074i018p04487. URL http://agupubs .onlinelibrary.wiley.com/doi/10.1029/JC074i018p04487.

40 H. S. Hopfield. Tropospheric Effect on Electromagnetically Measured Range: Prediction from Surface Weather Data. *Radio Science*, 6(3):357–367, 1971. ISSN 1944-799X. doi: 10.1029/RS006i003p00357. URL http://agupubs .onlinelibrary.wiley.com/doi/abs/10.1029/RS006i003p00357.

41 H. Barrell and J. E. Sears Jr. The Refraction and Dispersion of Air for the Visible Spectrum. *Philosophical Transactions of the Royal Society of London, Series A: Mathematical and Physical Sciences*, 238(786):1–64, February 1939.

doi: 10.1098/rsta.1939.0004. URL https://royalsocietypublishing.org/doi/abs/10.1098/rsta.1939.0004.

42 D. E. Kerr, S. A. Goudsmit, and L. B. Linford. *Propagation of Short Radio Waves*. IET, 1987. ISBN 978-0-86341-099-4.

43 E. K. Smith and S. Weintraub. The Constants in the Equation for Atmospheric Refractive Index at Radio Frequencies. *Proceedings of the IRE*, 41(8):1035–1037, 1953.

44 J. E. Allnutt. *Satellite-to-Ground Radiowave Propagation*. The Institution of Engineering and Technology, London, 2nd edition, May 2011. ISBN 978-1-84919-150-0.

45 C. A. Levis, J. T. Johnson, and F. L. Teixeira. *Radiowave Propagation: Physics and Applications*. Wiley, 2010. ISBN 978-0-470-54295-8.

46 A. E. Niell. Global Mapping Functions for the Atmosphere Delay at Radio Wavelengths. *Journal of Geophysical Research: Solid Earth*, 101(B2):3227–3246, February 1996. ISSN 2156-2202. doi: 10.1029/95JB03048. URL http://agupubs.onlinelibrary.wiley.com/doi/abs/10.1029/95JB03048.

47 M. P. M. Hall and C. M. Comer. Statistics of Tropospheric Radio-Refractive-Index Soundings Taken Over a 3-Year Period in the United Kingdom. *Proceedings of the Institution of Electrical Engineers*, 116(5):685–690, May 1969. ISSN 2053-7891. doi: 10.1049/piee.1969.0138. URL https://digital-library.theiet.org/content/journals/10.1049/piee.1969.0138.

48 A. von Engeln, G. Nedoluha, and J. Teixeira. An Analysis of the Frequency and Distribution of Ducting Events in Simulated Radio Occultation Measurements Based on ECMWF Fields. *Journal of Geophysical Research: Atmospheres*, 108(D21), November 2003. ISSN 01480227. doi: 10.1029/2002JD003170. URL http://doi.wiley.com/10.1029/2002JD003170.

49 ITU-R Recommendation P.834. Effects of Tropospheric Refraction on Radiowave Propagation, 2017.

50 C. F. Bohren and D. R. Huffman. *Absorption and Scattering of Light by Small Particles*. John Wiley & Sons, 2008.

51 S. Prahl and S. Jacques. Classroom Resources, 2017. URL https://omlc.org/classroom/index.html.

52 M. Kerker. *The Scattering of Light and Other Electromagnetic Radiation*. Elsevier, 1969. ISBN 978-0-12-404550-7. URL https://linkinghub.elsevier.com/retrieve/pii/C20130061956.

53 H. C. van de Hulst, *Light Scattering by Small Particles*. Dover Publications, 1981. ISBN 978-0-486-64228-4.

54 B. J. Finlayson-Pitts and J. N. Pitts. Chapter 11 - Analytical Methods and Typical Atmospheric Concentrations for Gases and Particles. In B. J. Finlayson-Pitts and J. N. Pitts, editors, *Chemistry of the Upper and Lower*

Atmosphere, pages 547–656. Academic Press, San Diego, CA, January 2000. ISBN 978-0-12-257060-5. URL http://www.sciencedirect.com/science/article/pii/B9780122570605500137.

55 M. Jonasz and G. R. Fournier. Chapter 5: The Particle Size Distribution. In M. Jonasz and G. R. Fournier, editors, *Light Scattering by Particles in Water*, pages 267–445. Academic Press, Amsterdam, January 2007. ISBN 978-0-12-388751-1. URL http://www.sciencedirect.com/science/article/pii/B9780123887511500053.

56 M. Willis. Gaseous Attenuation Propagation Tutorial. URL http://www.mike-willis.com/Tutorial/gases.htm.

57 J. H. van Vleck. The Absorption of Microwaves by Oxygen. *Physical Review*, 71(7):413–424, April 1947. doi: 10.1103/PhysRev.71.413. URL https://link.aps.org/doi/10.1103/PhysRev.71.413.

58 J. H. van Vleck. The Absorption of Microwaves by Uncondensed Water Vapor. *Physical Review*, 71(7):425–433, April 1947. doi: 10.1103/PhysRev.71.425. URL https://link.aps.org/doi/10.1103/PhysRev.71.425.

59 S. B. Bayram and M. V. Freamat. Vibrational Spectra of N2: An Advanced Undergraduate Laboratory in Atomic and Molecular Spectroscopy. *American Journal of Physics*, 80(8):664–669, July 2012. ISSN 0002-9505. doi: 10.1119/1.4722793. URL https://aapt.scitation.org/doi/10.1119/1.4722793.

60 ITU-R Recommendation P.676-11. Attenuation by Atmospheric Gases, September 2016.

61 ITU-R Recommendation P.840-7. Attenuation Due to Clouds and Fog, December 2017.

62 P. L. Rice and J. W. Herbstreit. Tropospheric Propagation. In L. Marton, editor, *Advances in Electronics and Electron Physics*, volume 20, pages 199–259. Academic Press, January 1965. URL http://www.sciencedirect.com/science/article/pii/S0065253908604977.

63 International Radio Consultative Committee. *Recommendations and Reports of the CCIR, 1986*. International Telecommunication Union, 1986.

64 T. Moulsley and E. Vilar. Experimental and Theoretical Statistics of Microwave Amplitude Scintillations on Satellite Down-Links. *IEEE Transactions on Antennas and Propagation*, 30(6):1099–1106, November 1982. ISSN 0018-926X. doi: 10.1109/TAP.1982.1142964.

65 J. Saastamoinen. Contributions to the Theory of Atmospheric Refraction. *Bulletin Géodésique (1946–1975)*, 105(1):279–298, September 1972. ISSN 0007-4632. doi: 10.1007/BF02521844. URL https://doi.org/10.1007/BF02521844.

66 J. Saastamoinen. Atmospheric Correction for the Troposphere and Stratosphere in Radio Ranging Satellites. In *The Use of Artificial Satellites for Geodesy*, pages 247–251. Geophysical Monograph Series, Volume 15,

American Geophysical Union, March 2013. URL http://agupubs.onlinelibrary
.wiley.com/doi/10.1029/GM015p0247.

67 P. J. Mohr, B. N. Taylor, and D. B. Newell. CODATA Recommended Values of
the Fundamental Physical Constants: 2010. *Journal of Physical and Chemical
Reference Data*, 41(4):043109, December 2012. ISSN 0047-2689, 1529-7845.
doi: 10.1063/1.4724320. URL http://aip.scitation.org/doi/10.1063/1.4724320.

68 NOAA US Department of Commerce. NWS JetStream - Layers of the Atmosphere. URL https://www.weather.gov/jetstream/layers.

69 United States National Aeronautics and Space Administration. *U.S. Standard
Atmosphere, 1976*. National Oceanic and Atmospheric Administration, 1976.

70 Public Domain Aeronautical Software. The U.S. Standard Atmosphere 1976.
URL http://www.pdas.com/atmos.html.

71 J. K. Hargreaves. *The Solar-Terrestrial Environment: An Introduction to
Geospace - the Science of the Terrestrial Upper Atmosphere, Ionosphere, and
Magnetosphere*. Cambridge University Press, Cambridge England; New York,
NY, revised ed. edition, May 1995. ISBN 978-0-521-42737-1.

72 B. Zolesi and L. R. Cander. The General Structure of the Ionosphere.
In *Ionospheric Prediction and Forecasting*, (ed. B. Zolesi and L. R. Cander)
Springer Geophysics, pages 11–48. Springer-Verlag, Berlin, Heidelberg, 2014.
ISBN 978-3-642-38429-5.

73 Sun Fact Sheet, NASA Space Science Data Coordinated Archive. URL https://
nssdc.gsfc.nasa.gov/planetary/factsheet/sunfact.html.

74 A. B. Balantekin and N. Takigawa. Quantum Tunneling in Nuclear Fusion.
Reviews of Modern Physics, 70(1):77–100, January 1998. ISSN 0034-6861,
1539-0756. doi: 10.1103/RevModPhys.70.77.

75 R. Garner. Understanding the Magnetic Sun. *NASA*, January 2016. URL
http://www.nasa.gov/feature/goddard/2016/understanding-the-magnetic-sun.

76 H. W. Babcock. The Topology of the Sun's Magnetic Field and the 22-YEAR
Cycle. *The Astrophysical Journal*, 133:572, March 1961. ISSN 0004-637X. doi:
10.1086/147060. URL https://ui.adsabs.harvard.edu/abs/1961ApJ...133..572B.

77 S. Tsuneta. Structure and Dynamics of Magnetic Reconnection in a Solar
Flare. *The Astrophysical Journal*, 456:840, January 1996. ISSN 0004-637X. doi:
10.1086/176701.

78 T. Yokoyama, K. Akita, T. Morimoto, K. Inoue, and J. Newmark. Clear
Evidence of Reconnection Inflow of a Solar Flare. *The Astrophysical Journal
Letters*, 546(1):L69, 2001. ISSN 1538-4357. doi: 10.1086/318053.

79 J. Qiu, H. Wang, C. Z. Cheng, and D. E. Gary. Magnetic Reconnection and
Mass Acceleration in Flare-Coronal Mass Ejection Events. *The Astrophysical
Journal*, 604(2):900, 2004. ISSN 0004-637X. doi: 10.1086/382122.

80 J. T. Gosling, J. Birn, and M. Hesse. Three-Dimensional Magnetic Reconnection and the Magnetic Topology of Coronal Mass Ejection Events.

Geophysical Research Letters, 22(8):869–872, April 1995. ISSN 1944-8007. doi: 10.1029/95GL00270.

81 K. Davies. *Ionospheric Radio*. The Institution of Engineering and Technology, London, UK, December 1990. ISBN 978-0-86341-186-1.

82 A. Aitta. Iron Melting Curve with a Tricritical Point. *Journal of Statistical Mechanics: Theory and Experiment*, 2006(12):P12015, 2006. ISSN 1742-5468. doi: 10.1088/1742-5468/2006/12/P12015.

83 J. H. Nguyen and N. C. Holmes. Melting of Iron at the Physical Conditions of the Earth's Core. *Nature, London*, 427(6972):339–342, January 2004. ISSN 00280836.

84 G. A. de Wijs, G. Kresse, L. Vočadlo, D. Dobson, D. Alfè, M. J. Gillan, and G. D. Price. The Viscosity of Liquid Iron at the Physical Conditions of the Earth's Core. *Nature*, 392(6678):805–807, April 1998. ISSN 0028-0836. doi: 10.1038/33905.

85 B. A. Buffett. Earth's Core and the Geodynamo. *Science (New York, N.Y.)*, 288(5473):2007–2012, June 2000. ISSN 1095-9203.

86 NASA Content Administrator. 2012: Magnetic Pole Reversal Happens All The (Geologic) Time. *NASA*, April 2015. URL http://www.nasa.gov/topics/earth/features/2012-poleReversal.html.

87 IGRF Geomagnetic Model Map Spreadsheet Tool, National Oceanic and Atmospheric Administration. URL https://www.ngdc.noaa.gov/geomag/magfield-wist/.

88 E. Thébault et al. International Geomagnetic Reference Field: The 12th Generation. *Earth, Planets and Space*, 67:79, May 2015. ISSN 1880-5981. doi: 10.1186/s40623-015-0228-9.

89 M. B. McElroy. Ionosphere and Magnetosphere - Magnetosphere — Atmospheric Science. *Encyclopedia Britannica*, August 2012. URL https://www.britannica.com/science/ionosphere-and-magnetosphere/Magnetosphere.

90 L. R. Lyons and D. J. Williams. Wave-Particle Interactions. In *Quantitative Aspects of Magnetospheric Physics* (ed. L. R. Lyons and D. J. Williams), *Geophysics and Astrophysics Monographs*, pages 133–228. Springer, Dordrecht, 1984. ISBN 978-90-481-8391-3 978-94-017-2819-5.

91 D. N. Baker, S. G. Kanekal, V. C. Hoxie, M. G. Henderson, X. Li, H. E. Spence, S. R. Elkington, R. H. W. Friedel, J. Goldstein, M. K. Hudson, G. D. Reeves, R. M. Thorne, C. A. Kletzing, and S. G. Claudepierre. A Long-Lived Relativistic Electron Storage Ring Embedded in Earth's Outer Van Allen Belt. *Science*, 340(6129):186–190, April 2013. ISSN 0036-8075, 1095-9203. doi: 10.1126/science.1233518.

92 H. Rishbeth. Basic Physics of the Ionosphere. In *Propagation of Radiowaves* (ed. L. Barclay). The Institution of Electrical Engineers, 2nd edition, 2003.

93 T. Obayashi and A. Nishida. Large-Scale Electric Field in the Magnetosphere. *Space Science Reviews*, 8(1):3–31, March 1968. ISSN 0038-6308, 1572-9672. doi: 10.1007/BF00362569.

94 S. S. Fernández de Córdoba. 100km Altitude Boundary for Astronautics, FAI Astronautic Records Commission. URL http://www.fai.org/page/icare-boundary.

95 D. J. Griffiths. *Introduction to Quantum Mechanics*. Pearson Prentice Hall, Upper Saddle River, NJ, 2nd edition, April 2004. ISBN 978-0-13-111892-8.

96 H. Lohninger. Ionization Energies of Diatomic Molecule, general chemistry, May 2011. URL http://www.vias.org/genchem/chembonding_14777_04.html.

97 Introduction to the Atmosphere, National Weather Service Online Weather School (JetStream). URL https://www.weather.gov/jetstream/atmos_intro.

98 H. Rishbeth and O. Garriott. *Introduction to Ionospheric Physics*. Academic Press, 1969.

99 S. Matsushita. *Physics of Geomagnetic Phenomena*. Elsevier, October 2013. ISBN 978-0-323-14121-5.

100 V. C. A. Ferraro. Some Remarks on Ambipolar Diffusion in the Ionosphere. *Journal of Atmospheric and Terrestrial Physics*, 26:913–917, 1964.

101 J. Jackson, E. Schmerling, and J. Whitteker. Mini-Review on Topside Sounding. *IEEE Transactions on Antennas and Propagation*, 28(2):284–288, March 1980. ISSN 0018-926X. doi: 10.1109/TAP.1980.1142318.

102 D. Bilitza, R. Benson, B. Reinisch, and X. Huang. Alouette and ISIS Topside Sounder Measurements -A Data Source for Improvements of the IRI Model in the Topside Ionosphere. In *38th COSPAR Scientific Assembly*, volume 38, page 5, 2010. URL http://adsabs.harvard.edu/abs/2010cosp...38.1500B.

103 N. Wakai and N. Matuura. Operation and experimental results of the Ionosphere Sounding Satellite-b. *Acta Astronautica*, 7(8):999–1020, August 1980. ISSN 0094-5765. doi: 10.1016/0094-5765(80)90097-1. URL http://www.sciencedirect.com/science/article/pii/0094576580900971.

104 MIT Haystack Observatory: Incoherent Scatter Radar. URL https://www.haystack.mit.edu/atm/mho/instruments/isr/index.html.

105 Compilation of basic technical specifications of the EISCAT facilities. URL https://eiscat.se/scientist/document/technical-specifications/.

106 C.-F. Enell. EISCAT Metadata and Data Formats, EISCAT Scientific Association. 2017.

107 R. Schunk and A. Nagy. *Ionospheres: Physics, Plasma Physics, and Chemistry*. Cambridge University Press, Cambridge, UK; New York, 2nd edition, September 2009. ISBN 978-0-521-87706-0.

108 J. V. Evans. Theory and Practice of Ionosphere Study by Thomson Scatter Radar. *Proceedings of the IEEE*, 57(4):496–530, April 1969. ISSN 0018-9219. doi: 10.1109/PROC.1969.7005.

109 H. Akbari, A. Bhatt, C. La Hoz, and J. L. Semeter. Incoherent Scatter Plasma Lines: Observations and Applications. *Space Science Reviews*, 212(1–2):249–294, October 2017. ISSN 0038-6308, 1572-9672. doi: 10.1007/s11214-017-0355-7.

110 R. S. V. Raman, J. P. St-Maurice, and R. S. B. Ong. Incoherent Scattering of Radar Waves in the Auroral Ionosphere. *Journal of Geophysical Research: Space Physics*, 86(A6):4751–4762, September 2012. ISSN 0148-0227. doi: 10.1029/JA086iA06p04751.

111 M. P. Sulzer. RADAR | Incoherent Scatter Radar. In *Encyclopedia of Atmospheric Sciences* (ed. G. R. North, J. A. Pyle, and F. Zhang), pages 422–428. Elsevier, 2015. ISBN 978-0-12-382225-3.

112 B. E. Cherrington. The Use of Electrostatic Probes for Plasma Diagnostics—A Review. *Plasma Chemistry and Plasma Processing*, 2(2):113–140, June 1982. ISSN 0272-4324, 1572-8986. doi: 10.1007/BF00633129.

113 R. L. Merlino. Understanding Langmuir Probe Current-Voltage Characteristics. *American Journal of Physics*, 75(12):1078–1085, December 2007. ISSN 0002-9505, 1943-2909. doi: 10.1119/1.2772282.

114 Yu I. Stozhkov, N. S. Svirzhevsky, and V. S. Makhmutov. Cosmic Ray Measurements in the Atmosphere. 2001. URL https://cds.cern.ch/record/557167/files/p41.pdf.

115 H. Friedman. Ultraviolet and X Rays from the Sun. *Annual Review of Astronomy and Astrophysics*, 1:59–96, 1963.

116 A. K. Saha and S. Ray. Some Features of the E2-Layer Observed at the Ionosphere Field Station, Haringhata, Calcutta. *Journal of Atmospheric and Terrestrial Physics*, 7:107–IN8, July 1955. ISSN 0021-9169. doi: 10.1016/0021-9169(55)90114-8.

117 J. M. Young, C. Y. Johnson, and J. C. Holmes. Positive Ion Composition of a Temperate-Latitude Sporadic E Layer as Observed during a Rocket Flight. *Journal of Geophysical Research*, 72(5):1473–1479, March 1967. ISSN 2156-2202. doi: 10.1029/JZ072i005p01473.

118 H. E. Hinteregger. EUV Fluxes in the Solar Spectrum below 2000 Å. *Journal of Atmospheric and Terrestrial Physics*, 38(8):791–806, August 1976. ISSN 0021-9169. doi: 10.1016/0021-9169(76)90020-9.

119 G. H. Munro. Diurnal Variations in the Ionosphere Deduced from Satellite Radio Signals. *Journal of Geophysical Research*, 67(1):147–156, January 1962. ISSN 2156-2202. doi: 10.1029/JZ067i001p00147.

120 H. Rishbeth and C. S. G. K. Setty. The F-Layer at Sunrise. *Journal of Atmospheric and Terrestrial Physics*, 20(4):263–276, April 1961. ISSN 0021-9169. doi: 10.1016/0021-9169(61)90205-7.

121 M. R. Torr and D. G Torr. The Seasonal Behaviour of the F2-Layer of the Ionosphere. *Journal of Atmospheric and Terrestrial Physics*, 35(12):2237–2251, December 1973. ISSN 0021-9169. doi: 10.1016/0021-9169(73)90140-2.

122 W. K. Lee, H. Kil, Y.-S. Kwak, Q. Wu, S. Cho, and J. U. Park. The Winter Anomaly in the Middle-Latitude F Region during the Solar Minimum Period Observed by the Constellation Observing System for Meteorology, Ionosphere, and Climate. *Journal of Geophysical Research: Space Physics*, 116(A2):A02302, February 2011. ISSN 2156-2202. doi: 10.1029/2010JA015815.

123 N. Jakowski and M. Förster. About the Nature of the Night-Time Winter Anomaly Effect (NWA) in the F-Region of the Ionosphere. *Planetary and Space Science*, 43(5):603–612, May 1995. ISSN 0032-0633. doi: 10.1016/0032-0633(94)00115-8.

124 H. Rishbeth and I. C. F. Müller-Wodarg. Why Is There More Ionosphere in January than in July? The Annual Asymmetry in the F2-Layer. *Annales Geophysicae*, 24(12):3293–3311, December 2006.

125 G. H. Millward, H. Rishbeth, T. J. Fuller-Rowell, A. D. Aylward, S. Quegan, and R. J. Moffett. Ionospheric F2 Layer Seasonal and Semiannual Variations. *Journal of Geophysical Research: Space Physics*, 101(A3):5149–5156, March 1996. ISSN 2156-2202. doi: 10.1029/95JA03343.

126 T. J. Fuller-Rowell. The "Thermospheric Spoon": A Mechanism for the Semi-annual Density Variation. *Journal of Geophysical Research: Space Physics*, 103(A3):3951–3956, March 1998. ISSN 2156-2202. doi: 10.1029/97JA03335.

127 R. G. Ezquer, J. L. López, L. A. Scidá, M. A. Cabrera, B. Zolesi, C. Bianchi, M. Pezzopane, E. Zuccheretti, and M. Mosert. Behaviour of Ionospheric Magnitudes of F2 Region over Tucumán during a Deep Solar Minimum and Comparison with the IRI 2012 Model Predictions. *Journal of Atmospheric and Solar-Terrestrial Physics*, 107(Supplement C):89–98, January 2014. ISSN 1364-6826. doi: 10.1016/j.jastp.2013.11.010.

128 A. G. Kanatas and A. D. Panagopoulos. *Radio Wave Propagation and Channel Modeling for Earth–Space Systems*. CRC Press, June 2016. ISBN 978-1-4822-4971-2.

129 D. N. Anderson. A Theoretical Study of the Ionospheric F Region Equatorial Anomaly—I. Theory. *Planetary and Space Science*, 21(3):409–419, March 1973. ISSN 0032-0633. doi: 10.1016/0032-0633(73)90040-8.

130 H. Rishbeth. Dynamics of the Equatorial F-Region. *Journal of Atmospheric and Terrestrial Physics*, 39(9):1159–1168, September 1977. ISSN 0021-9169. doi: 10.1016/0021-9169(77)90024-1.

131 N. R. Thomson, C. J. Rodger, and R. L. Dowden. Ionosphere Gives Size of Greatest Solar Flare. *Geophysical Research Letters*, 31(6):L06803, March 2004. ISSN 1944-8007. doi: 10.1029/2003GL019345.

132 W. D. Reeve. *Sudden Frequency Deviations Caused by Solar Flares*. Reeve Observatory, Anchorage, Alaska, 2015. URL https://reeve.com/RadioScience/ Radio%20Astronomy%20Publications/Articles_Papers.htm.

133 H. Volland. On the Theory of the Sudden Enhancement of Atmospherics (SEA). *Journal of Atmospheric and Terrestrial Physics*, 28(4):409–423, April 1966. ISSN 0021-9169. doi: 10.1016/0021-9169(66)90096-1.

134 National Research Council. *Committee on the Societal and Economic Impacts of Severe Space Weather Events*. National Academies Press, Washington, DC, 2008.

135 S. Odenwald. The Day the Sun Brought Darkness, May 2017. URL http:// www.nasa.gov/topics/earth/features/sun_darkness.html.

136 T. Phillips. Solar Shield: Protecting the North American Power Grid, NASA Science Mission Directorate, 2010. URL https://science.nasa.gov/science-news/science-at-nasa/2010/26oct_solarshield.

137 A. T. Price. Daily Variations of the Geomagnetic Field. *Space Science Reviews*, 9:151–197, March 1969. ISSN 0038-6308. doi: 10.1007/BF00215632.

138 R. S. Lindzen and S. Chapman. Atmospheric Tides. *Space Science Reviews*, 10(1):3–188, 1969.

139 T. Nagata and S. Kokubun. A Particular Geomagnetic Daily Variation (Sqp) in the Polar Regions on Geomagnetically Quiet Days. *Nature*, 195(4841):555–557, 1962.

140 A. Nishida. The Origin of Fluctuation in the Equatorial Electrojet; a New Type of Geomagnetic Variations. *Annales Geophysicae*, 22:478–484, 1966.

141 C. R. Clauer and Y. Kamide. DP 1 and DP 2 Current Systems for the March 22, 1979 Substorms. *Journal of Geophysical Research: Space Physics*, 90(A2):1343–1354, February 1985. ISSN 2156-2202. doi: 10.1029/JA090iA02p01343.

142 A. Nishida. Geomagnetic Dp 2 Fluctuations and Associated Magnetospheric Phenomena. *Journal of Geophysical Research*, 73(5):1795–1803, March 1968. ISSN 2156-2202. doi: 10.1029/JA073i005p01795.

143 K. S. Brathwaite and G. Rostoker. DP2 Current System in the Ionosphere and Magnetosphere. *Planetary and Space Science*, 29(5):485–494, May 1981. ISSN 0032-0633. doi: 10.1016/0032-0633(81)90063-5.

144 A. Nishida. Coherence of Geomagnetic DP 2 Fluctuations with Interplanetary Magnetic Variations. *Journal of Geophysical Research*, 73(17):5549–5559, September 1968. ISSN 2156-2202. doi: 10.1029/JA073i017p05549.

145 J. Bartels, N. H. Heck, and H. F. Johnston. The Three-Hour-Range Index Measuring Geomagnetic Activity. *Terrestrial Magnetism and Atmospheric Electricity*, 44(4):411–454, 1939. ISSN 0096-8013. doi: 10.1029/ TE044i004p00411. URL http://onlinelibrary.wiley.com/doi/abs/10.1029/ TE044i004p00411.

146 National Geophysical Data Center. Solar and Terrestrial Physics. URL https://www.ngdc.noaa.gov/stp/GEOMAG/kp_ap.html.

147 J. Matzka. *Indices of Global Geomagnetic Activity.* German Research Centre for Geosciences. URL http://www.gfz-potsdam.de/kp-index.

148 T. Iyemori. Geomagnetic Data Service, University of Kyoto. URL http://wdc.kugi.kyoto-u.ac.jp/.

149 J. MacDougall, M. A. Abdu, I. Batista, P. R. Fagundes, Y. Sahai, and P. T. Jayachandran. On the Production of Traveling Ionospheric Disturbances by Atmospheric Gravity Waves. *Journal of Atmospheric and Solar-Terrestrial Physics*, 71(17):2013–2016, 2009.

150 G. Crowley, I. Azeem, A. Reynolds, T. M. Duly, P. McBride, C. Winkler, and D. Hunton. Analysis of Traveling Ionospheric Disturbances (TIDs) in GPS TEC Launched by the 2011 Tohoku Earthquake. *Radio Science*, 51(5):2015RS005907, May 2016. ISSN 1944-799X. doi: 10.1002/2015RS005907.

151 R. D. Hunsucker. Atmospheric Gravity Waves Generated in the High-Latitude Ionosphere: A Review. *Reviews of Geophysics*, 20(2):293–315, 1982.

152 Z. T. Katamzi and L. A. McKinnell. Observations of Traveling Ionospheric Disturbances Associated with Geomagnetic Storms. In *2011 XXXth URSI General Assembly and Scientific Symposium*, pages 1, August 2011. doi: 10.1109/URSIGASS.2011.6050928.

153 T. Tsugawa, A. Saito, and Y. Otsuka. A Statistical Study of Large-Scale Traveling Ionospheric Disturbances Using the GPS Network in Japan. *Journal of Geophysical Research: Space Physics*, 109(A6):A06302, June 2004. ISSN 2156-2202. doi: 10.1029/2003JA010302.

154 T. Tsugawa, Y. Otsuka, A. J. Coster, and A. Saito. Medium-Scale Traveling Ionospheric Disturbances Detected with Dense and Wide TEC Maps over North America. *Geophysical Research Letters*, 34(22):L22101, November 2007. ISSN 1944-8007. doi: 10.1029/2007GL031663.

155 C. Ferencz, G. Lizunov, F. Crespon, I. Price, L. Bankov, D. Przepiórka, K. Brieß, D. Dudkin, A. Girenko, V. Korepanov, A. Kuzmych, T. Skorokhod, P. Marinov, O. Piankova, H. Rothkaehl, T. Shtus, P. Steinbach, J. Lichtenberger, A. Sterenharz, and A. Vassileva. Ionosphere Waves Service (IWS) – a Problem-Oriented Tool in Ionosphere and Space Weather Research Produced by POPDAT Project. *Journal of Space Weather and Space Climate*, 4:A17, 2014. ISSN 2115-7251. doi: 10.1051/swsc/2014013.

156 S. T. Loi, I. H. Cairns, T. Murphy, P. J. Erickson, M. E. Bell, A. Rowlinson, B. S. Arora, J. Morgan, R. D. Ekers, N. Hurley-Walker, and D. L. Kaplan. Density Duct Formation in the Wake of a Travelling Ionospheric Disturbance: Murchison Widefield Array Observations. *Journal of Geophysical Research: Space Physics*, 121(2):1569–1586, February 2016. ISSN 21699380. doi: 10.1002/2015JA022052.

157 W. Calvert and J. M. Warnock. Ionospheric Irregularities Observed by Topside Sounders. *Proceedings of the IEEE*, 57(6):1019–1025, June 1969. ISSN 0018-9219. doi: 10.1109/PROC.1969.7146.

158 W. Calvert and C. W. Schmid. Spread-F Observations by the Alouette Topside Sounder Satellite. *Journal of Geophysical Research*, 69(9):1839–1852, May 1964. ISSN 2156-2202. doi: 10.1029/JZ069i009p01839.

159 G. G. Bowman. A Review of Some Recent Work on Mid-Latitude Spread-*F* Occurrence as Detected by Ionosondes. *Journal of Geomagnetism and Geoelectricity*, 42(2):109–138, 1990. doi: 10.5636/jgg.42.109.

160 B. H. Briggs. Observations of Radio Star Scintillations and Spread-F Echoes over a Solar Cycle. *Journal of Atmospheric and Terrestrial Physics*, 26(1):1–23, January 1964. ISSN 0021-9169. doi: 10.1016/0021-9169(64)90104-7.

161 X. He, S. Chen, and R. Zhang. A Lattice Boltzmann Scheme for Incompressible Multiphase Flow and Its Application in Simulation of Rayleigh–Taylor Instability. *Journal of Computational Physics*, 152(2):642–663, July 1999. ISSN 0021-9991. doi: 10.1006/jcph.1999.6257.

162 D. H. Sharp. An Overview of Rayleigh-Taylor Instability. *Physica D: Nonlinear Phenomena*, 12(1):3–18, July 1984. ISSN 0167-2789. doi: 10.1016/0167-2789(84)90510-4.

163 H. J. Kull. Theory of the Rayleigh-Taylor Instability. *Physics Reports*, 206(5):197–325, August 1991. ISSN 0370-1573. doi: 10.1016/0370-1573(91)90153-D.

164 A. J. Scannapieco and S. L. Ossakow. Nonlinear Equatorial Spread F. *Geophysical Research Letters*, 3(8):451–454, August 1976. ISSN 1944-8007. doi: 10.1029/GL003i008p00451.

165 E. Ott. Theory of Rayleigh-Taylor Bubbles in the Equatorial Ionosphere. *Journal of Geophysical Research: Space Physics*, 83(A5):2066–2070, May 1978. ISSN 2156-2202. doi: 10.1029/JA083iA05p02066.

166 C.-S. Huang, M. C. Kelley, and D. L. Hysell. Nonlinear Rayleigh-Taylor Instabilities, Atmospheric Gravity Waves and Equatorial Spread F. *Journal of Geophysical Research: Space Physics*, 98(A9):15631–15642, September 1993. ISSN 2156-2202. doi: 10.1029/93JA00762.

167 P. J. Sultan. Linear Theory and Modeling of the Rayleigh-Taylor Instability Leading to the Occurrence of Equatorial Spread F. *Journal of Geophysical Research: Space Physics*, 101(A12):26875–26891, December 1996. ISSN 2156-2202. doi: 10.1029/96JA00682.

168 M. A. Abdu, R. T. de Medeiros, and J. H. A. Sobral. Equatorial Spread F Instability Conditions as Determined from Ionograms. *Geophysical Research Letters*, 9(6):692–695, June 1982. ISSN 1944-8007. doi: 10.1029/GL009i006p00692.

169 M. A. Abdu. Outstanding Problems in the Equatorial Ionosphere–Thermosphere Electrodynamics Relevant to Spread F. *Journal of Atmospheric and Solar-Terrestrial Physics*, 63(9):869–884, January 2001. ISSN 1364-6826. doi: 10.1016/S1364-6826(00)00201-7.

170 M. Nicolet. The Collision Frequency of Electrons in the Ionosphere. *Journal of Atmospheric and Terrestrial Physics*, 3(4):200–211, May 1953. ISSN 0021-9169. doi: 10.1016/0021-9169(53)90110-X.

171 K. C. Yeh, H. Y. Chao, and K. H. Lin. Polarization Transformation of a Wave Field Propagating in an Anisotropic Medium. *IEEE Antennas and Propagation Magazine*, 41(5):19–33, October 1999. ISSN 1045-9243. doi: 10.1109/74.801511.

172 ITU-R Recommendation P.618-12. Propagation Data and Prediction Methods Required for the Design of Earth-Space Telecommunication Systems, July 2015.

173 ITU-R Recommendation P.531-13. Ionospheric Propagation Data and Prediction Methods Required for the Design of Satellite Services and Systems, September 2016.

174 J. A. Klobuchar. Ionospheric Effects on GPS. In *Global Positioning System: Theory and Applications*, Volume I (ed. J. J. Spilker Jr., P. Axelrad, B. W. Parkinson and P. Enge), pages 485–515. American Institute of Aeronautics and Astronautics, 1996. ISBN 978-1-56347-106-3.

175 C. Jiang and B. Wang. Atmospheric Refraction Corrections of Radiowave Propagation for Airborne and Satellite-Borne Radars. *Science in China Series E: Technological Sciences*, 44(3):280–290, June 2001. ISSN 1006-9321, 1862-281X. doi: 10.1007/BF02916705.

176 S. Bassiri and G. A. Hajj. Higher-Order Ionospheric Effects on the GPS Observables and Means of Modeling Them. *NASA STI/Recon Technical Report A*, 95:1071–1086, January 1993. ISSN 0065-3438. URL https://ui.adsabs.harvard.edu/abs/1993STIA...9581411B.

177 R. C. Moore and Y. T. Morton. Magneto-Ionic Polarization and GPS Signal Propagation through the Ionosphere. *Radio Science*, 46(1):RS1008, February 2011. ISSN 1944-799X. doi: 10.1029/2010RS004380.

178 M. M. Hoque and N. Jakowski. Higher Order Ionospheric Effects in Precise GNSS Positioning. *Journal of Geodesy*, 81(4):259–268, 2007.

179 M. Hernández-Pajares, J. M. Juan, J. Sanz, and R. Orús. Second-Order Ionospheric Term in GPS: Implementation and Impact on Geodetic Estimates. *Journal of Geophysical Research: Solid Earth*, 112(B8):B08417, August 2007. ISSN 2156-2202. doi: 10.1029/2006JB004707.

180 H. Li, Z. Wang, and J. An. GPS Differential Code Biases Considering the Second-Order Ionospheric Term. *GPS Solutions*, 21(4):1669–1677, October 2017. ISSN 1080-5370, 1521-1886. doi: 10.1007/s10291-017-0643-4.

181 K. Miya. *Satellite Communications Technology*. KDD Engineering, Tokyo, Japan, 1985.

182 J. Aarons. Global Morphology of Ionospheric Scintillations. *Proceedings of the IEEE*, 70(4):360–378, April 1982. ISSN 0018-9219. doi: 10.1109/PROC.1982.12314.

183 D. H. Zhang, Z. Xiao, M. Feng, Y. Q. Hao, L. Q. Shi, G. L. Yang, and Y. C. Suo. Temporal Dependence of GPS Cycle Slip Related to Ionospheric Irregularities over China Low-Latitude Region. *Space Weather*, 8(4):S04D08, April 2010. ISSN 1542-7390. doi: 10.1029/2008SW000438.

184 J. A. Klobuchar. Ionospheric Time-Delay Algorithm for Single-Frequency GPS Users. *IEEE Transactions on Aerospace and Electronic Systems*, AES-23(3):325–331, May 1987. ISSN 0018-9251. doi: 10.1109/TAES.1987.310829.

185 S. K. Llewellyn. *Documentation and Description of the Bent Ionospheric Model*. U.S. Department of Commerce, National Technical Information Service, 1973.

186 C.-M. Lee and K.-D. Park. Generation of Klobuchar Ionospheric Error Model Coefficients Using Fourier Series and Accuracy Analysis. *Journal of Astronomy and Space Sciences*, 28(1):71–77, 2011.

187 T. Bi, J. An, J. Yang, and S. Liu. A Modified Klobuchar Model for Single-Frequency GNSS Users over the Polar Region. *Advances in Space Research*, 59(3):833–842, February 2017. ISSN 0273-1177. doi: 10.1016/j.asr.2016.10.029.

188 G. Di Giovanni and S. M. Radicella. An Analytical Model of the Electron Density Profile in the Ionosphere. *Advances in Space Research*, 10(11):27–30, January 1990. ISSN 0273-1177. doi: 10.1016/0273-1177(90)90301-F.

189 J. R. Dudeney and R. I. Kressman. Empirical Models of the Electron Concentration of the Ionosphere and Their Value for Radio Communications Purposes. *Radio Science*, 21(3):319–330, May 1986. ISSN 1944-799X. doi: 10.1029/RS021i003p00319.

190 K. Rawer. Replacement of the Present Sub-Peak Plasma Density Profile by a Unique Expression. *Advances in Space Research*, 2(10):183–190, January 1982. ISSN 0273-1177. doi: 10.1016/0273-1177(82)90387-8.

191 S. M. Radicella and M. L. Zhang. The Improved DGR Analytical Model of Electron Density Height Profile and Total Electron Content in the Ionosphere. *Anali di Geofisica*, 38:35–41, March 1995.

192 G. Hochegger, B. Nava, S. Radicella, and R. Leitinger. A Family of Ionospheric Models for Different Uses. *Physics and Chemistry of the Earth, Part C: Solar, Terrestrial & Planetary Science*, 25(4):307–310, January 2000. ISSN 1464-1917. doi: 10.1016/S1464-1917(00)00022-2.

193 S. M. Radicella and R. Leitinger. The Evolution of the DGR Approach to Model Electron Density Profiles. *Advances in Space Research*, 27(1):35–40, January 2001. ISSN 0273-1177. doi: 10.1016/S0273-1177(00)00138-1.

194 R. Leitinger, M.-L. Zhang, and S. M. Radicella. An Improved Bottomside for the Ionospheric Electron Density Model NeQuick. *Annals of Geophysics*, 48(3):525–534, 2005.

195 P. Coïsson, S. M. Radicella, R. Leitinger, and B. Nava. Topside Electron Density in IRI and NeQuick: Features and Limitations. *Advances in Space Research*, 37(5):937–942, January 2006. ISSN 0273-1177. doi: 10.1016/j.asr .2005.09.015.

196 B. Nava, P. Coïsson, and S. M. Radicella. A New Version of the NeQuick Ionosphere Electron Density Model. *Journal of Atmospheric and Solar-Terrestrial Physics*, 70(15):1856–1862, December 2008. ISSN 1364-6826. doi: 10.1016/j.jastp.2008.01.015.

197 C. Liu, M.-L. Zhang, W. Wan, L. Liu, and B. Ning. Modeling M(3000)F2 Based on Empirical Orthogonal Function Analysis Method. *Radio Science*, 43(1):RS1003, February 2008. ISSN 1944-799X. doi: 10.1029/2007RS003694.

198 R. Leitinger, J. E. Titheridge, G. Kirchengast, and W. Rothleitner. Ein "einfaches" globales empirisches modell für die f-schicht der ionosphäre (engl. transl. a "Simple" *Global Empirical Model for the F Layer of the Ionosphere). Kleinheubacher Ber.*, 39:697–704, 1996.

199 W. B. Jones and R. M. Gallet. The Representation of Diurnal and Geographic Variations of Ionospheric Data by Numerical Methods. *Telecomm. J,* 29(5):129–147, 1962.

200 N. Wang, Y. Yuan, Z. Li, Y. Li, X. Huo, and M. Li. An Examination of the Galileo NeQuick Model: Comparison with GPS and JASON TEC. *GPS Solutions*, 21(2):605–615, April 2017. ISSN 1080-5370, 1521-1886. doi: 10.1007/s10291-016-0553-x.

201 J. R. Dudeney. *A Simple Empirical Method for Estimating the Height and Semi-Thickness of the F2-Layer at the Argentine Islands, Graham Land*, volume 88. British Antarctic Survey, Cambridge, 1974. ISBN 978-0-85665-027-7.

202 P. A Bradley and J. R Dudeney. A Simple Model of the Vertical Distribution of Electron Concentration in the Ionosphere. *Journal of Atmospheric and Terrestrial Physics*, 35(12):2131–2146, December 1973. ISSN 0021-9169. doi: 10.1016/0021-9169(73)90132-3.

203 J. R Dudeney. An Improved Model of the Variation of Electron Concentration with Height in the Ionosphere. *Journal of Atmospheric and Terrestrial Physics*, 40(2):195–203, February 1978. ISSN 0021-9169. doi: 10.1016/0021-9169(78)90024-7.

204 J. R. Dudeney. The Accuracy of Simple Methods for Determining the Height of the Maximum Electron Concentration of the F2-Layer from Scaled Ionospheric Characteristics. *Journal of Atmospheric and Terrestrial*

Physics, 45(8):629–640, August 1983. ISSN 0021-9169. doi: 10.1016/S0021-9169(83)80080-4.

205 M. Mosert de González and S. M. Radicella. On a Characteristic Point at the Base of the F2 Layer. *Advances in Space Research*, 10(11):17–25, January 1990. ISSN 0273-1177. doi: 10.1016/0273-1177(90)90300-O.

206 B. Arbesser-Rastburg. The GALILEO Single Frequency Ionospheric Correction Algorithm. *Third European Space Weather Week*, 13:17, 2006.

207 Y. Béniguel. Global Ionospheric Propagation Model (GIM): A Propagation Model for Scintillations of Transmitted Signals. *Radio Science*, 37(3):4-1–4-13, 2002. ISSN 1944-799X. doi: 10.1029/2000RS002393. URL https://agupubs .onlinelibrary.wiley.com/doi/abs/10.1029/2000RS002393.

208 ITU-R Recommendation P.2097-0. Transionospheric Radio Propagation - The Global Ionospheric Scintillation Model (GISM). URL https://www.itu.int:443/ en/publications/ITU-R/Pages/publications.aspx.

209 D. E. Winch, D. J. Ivers, J. P. R. Turner, and R. J. Stening. Geomagnetism and Schmidt quasi-normalization. *Geophysical Journal International*, 160(2):487–504, February 2005. ISSN 0956-540X. doi: 10.1111/j.1365-246X .2004.02472.x. URL https://academic.oup.com/gji/article/160/2/487/659348.

210 C. B. Haselgrove and J. Haselgrove. Twisted Ray Paths in the Ionosphere. *Proceedings of the Physical Society*, 75(3):357–363, March 1960. ISSN 0370-1328. doi: 10.1088/0370-1328/75/3/304. URL https://doi.org/10.1088 %2F0370-1328%2F75%2F3%2F304.

211 K. G. Budden. *The Propagation of Radio Waves: The Theory of Radio Waves of Low Power in the Ionosphere and Magnetosphere*. Cambridge University Press, Cambridge, August 1988. ISBN 978-0-521-36952-7.

212 C. J. Coleman. Ionospheric Ray-Tracing Equations and their Solution. *URSI Radio Science Bulletin*, 2008(325):17–23, 2008.

213 C. J. Coleman. Point-to-Point Ionospheric Ray Tracing by a Direct Variational Method. *Radio Science*, 46(5), October 2011. ISSN 1944-799X. doi: 10.1029/2011RS004748. URL http://agupubs.onlinelibrary.wiley.com/doi/10 .1029/2011RS004748.

214 S. Priyadarshi. A Review of Ionospheric Scintillation Models. *Surveys in Geophysics*, 36(2):295–324, March 2015. ISSN 1573-0956. doi: 10.1007/s10712-015-9319-1. URL https://doi.org/10.1007/s10712-015-9319-1.

215 H. G. Booker, J. A. Ratcliffe, D. H. Shinn, and W. L. Bragg. Diffraction from an Irregular Screen with Applications to Ionospheric Problems. *Philosophical Transactions of the Royal Society of London, Series A: Mathematical and Physical Sciences*, 242(856):579–607, September 1950. doi: 10.1098/rsta.1950.0011. URL http://royalsocietypublishing.org/doi/abs/10.1098/rsta.1950.0011.

216 J. A. Ratcliffe. Some Aspects of Diffraction Theory and their Application to the Ionosphere. *Reports on Progress in Physics*, 19(1):188–267, January 1956.

ISSN 0034-4885. doi: 10.1088/0034-4885/19/1/306. URL https://doi.org/10
.1088%2F0034-4885%2F19%2F1%2F306.

217 K. C. Yeh and C.-H. Liu. Radio wave scintillations in the ionosphere. *Proceedings of the IEEE*, 70(4):324–360, 1982.

218 D. Bilitza. International Reference Ionosphere 2000: Examples of Improvements and New Features. *Advances in Space Research*, 31(3):757–767, January 2003. ISSN 02731177. doi: 10.1016/S0273-1177(03)00020-6.

219 D. Bilitza, K. Rawer, L. Bossy, I. Kutiev, K.-I. Oyama, R. Leitinger, and E. Kazimirovsky. International Reference Ionosphere 1990, Technical Report NSSDC/WDC-A-R&S 90-22, National Space Science Data Center/World Data Center A for Rockets and Satellites, 1990.

220 D. Bilitza, D. Altadill, Y. Zhang, C. Mertens, V. Truhlik, P. Richards, L.-A. McKinnell, and B. Reinisch. The International Reference Ionosphere 2012–a Model of International Collaboration. *Journal of Space Weather and Space Climate*, 4:A07, 2014.

221 A. D. Danilov and N. V. Smirnova. Improving the 75 to 300 Km Ion Composition Model of the IRI. *Advances in Space Research*, 15(2):171–177, February 1995. ISSN 0273-1177. doi: 10.1016/S0273-1177(99)80044-1.

222 A. D. Danilov and A. P. Yaichnikov. A New Model of the Ion Composition at 75 to 1000 Km for IRI. *Advances in Space Research*, 5(7):75–79, January 1985. ISSN 0273-1177. doi: 10.1016/0273-1177(85)90360-6.

223 P. G. Richards, D. Bilitza, and D. Voglozin. Ion Density Calculator (IDC): A New Efficient Model of Ionospheric Ion Densities. *Radio Science*, 45(05):1–11, October 2010. ISSN 1944-799X. doi: 10.1029/2009RS004332.

224 V. Truhlik, D. Bilitza, and L. Triskova. Towards Better Description of Solar Activity Variation in the International Reference Ionosphere Topside Ion Composition Model. *Advances in Space Research*, 55(8):2099–2105, April 2015. ISSN 0273-1177. doi: 10.1016/j.asr.2014.07.033.

225 D. Bilitza, L. H. Brace, and R. F. Theis. Modelling of Ionospheric Temperature Profiles. *Advances in Space Research*, 5(7):53–58, January 1985. ISSN 0273-1177. doi: 10.1016/0273-1177(85)90356-4.

226 D. Bilitza. Implementation of the New Electron Temperature Model in IRI. *Advances in Space Research*, 5(10):117–121, January 1985. ISSN 0273-1177. doi: 10.1016/0273-1177(85)90193-0.

227 V. Truhlik, D. Bilitza, and L. Triskova. A New Global Empirical Model of the Electron Temperature with the Inclusion of the Solar Activity Variations for IRI. *Earth, Planets and Space*, 64(6):531–543, June 2012. ISSN 1880-5981. doi: 10.5047/eps.2011.10.016.

228 M. Friedrich, R. Pilgram, and K. M. Torkar. A Novel Concept for Empirical D-Region Modelling. *Advances in Space Research*, 27(1):5–12, January 2001. ISSN 0273-1177. doi: 10.1016/S0273-1177(00)00133-2.

229 A. D. Danilov, A. Yu. Rodevich, and N. V. Smirnova. Problems with Incorporating a New D-Region Model into the IRI. *Advances in Space Research*, 15(2):165–168, February 1995. ISSN 0273-1177. doi: 10.1016/S0273-1177(99)80042-8.

230 K. Rawer, D. Bilitza, S. Ramakrishnan, and N. Sheikh. Intentions and Build-up of the International Reference Ionosphere. In *Operational Modelling of the Aerospace Propagation Environment, AGARD Conf. Proc. 238*, volume 1, November 1978.

231 D. Bilitza. A Correction for the IRI Topside Electron Density Model Based on Alouette/ISIS Topside Sounder Data. *Advances in Space Research*, 33(6):838–843, January 2004. ISSN 0273-1177. doi: 10.1016/j.asr.2003.07.009.

232 D. Altadill, J. M. Torta, and E. Blanch. Proposal of New Models of the Bottom-Side B0 and B1 Parameters for IRI. *Advances in Space Research*, 43(11):1825–1834, June 2009. ISSN 0273-1177. doi: 10.1016/j.asr.2008.08.014.

233 T. L. Gulyaeva. Progress in Ionospheric Informatics Based on Electron-Density Profile Analysis of Ionograms. *Advances in Space Research*, 7(6):39–48, January 1987. ISSN 0273-1177. doi: 10.1016/0273-1177(87)90269-9.

234 D. Altadill, S. Magdaleno, J. M. Torta, and E. Blanch. Global Empirical Models of the Density Peak Height and of the Equivalent Scale Height for Quiet Conditions. *Advances in Space Research*, 52:1756–1769, November 2013. ISSN 0273-1177. doi: 10.1016/j.asr.2012.11.018.

235 V. N. Shubin, A. T. Karpachev, and K. G. Tsybulya. Global Model of the F2 Layer Peak Height for Low Solar Activity Based on GPS Radio-Occultation Data. *Journal of Atmospheric and Solar-Terrestrial Physics*, 104:106–115, November 2013. ISSN 1364-6826. doi: 10.1016/j.jastp.2013.08.024.

236 V. N. Shubin. Global Median Model of the F2-Layer Peak Height Based on Ionospheric Radio-Occultation and Ground-Based Digisonde Observations. *Advances in Space Research*, 56(5):916–928, September 2015. ISSN 0273-1177. doi: 10.1016/j.asr.2015.05.029.

237 H. G. Booker. Fitting of Multi-Region Ionospheric Profiles of Electron Density by a Single Analytic Function of Height. *Journal of Atmospheric and Terrestrial Physics*, 39(5):619–623, May 1977. ISSN 0021-9169. doi: 10.1016/0021-9169(77)90072-1.

238 E. Mechtley and D. Bilitza. Models of D-Region Electron Concentration, Rep. *IPW-WB1*, Inst. fiir phys. Weltraumforsch., Freiburg, Germany, 1974.

239 S. S. Kouris and L. M. Muggleton. Diurnal Variation in the E-Layer Ionization. *Journal of Atmospheric and Terrestrial Physics*, 35(1):133–139, January 1973. ISSN 0021-9169. doi: 10.1016/0021-9169(73)90221-3.

240 S. S. Kouris and L. M. Muggleton. Analytical Expression for Prediction of F0E for Solar Zenith Angles 0–78 degrees. *Proceedings of the Institution* *of Electrical Engineers*, 121(4):264–268, April 1974. ISSN 2053-7891. doi: 10.1049/piee.1974.0051.

241 C. M. Rush, M. PoKempner, D. N. Anderson, F. G. Stewart, and J. Perry. Improving Ionospheric Maps Using Theorctically Derived Values of f0F2. *Radio Science*, 18(1):95–107, January 1983. ISSN 1944-799X. doi: 10.1029/RS018i001p00095.

242 C. M. Rush, M. PoKempner, D. N. Anderson, J. Perry, F. G. Stewart, and R. Reasoner. Maps of f0F2 Derived from Observations and Theoretical Data. *Radio Science*, 19(4):1083–1097, July 1984. ISSN 1944-799X. doi: 10.1029/RS019i004p01083.

243 G. C. Reid. Production and Loss of Electrons in the Quiet Daytime D Region of the Ionosphere. *Journal of Geophysical Research*, 75(13):2551–2562, 1970.

244 V. N. Shubin, A. T. Karpachev, V. A. Telegin, and K. G. Tsybulya. Global Model SMF2 of the F2-Layer Maximum Height. *Geomagnetism and Aeronomy*, 55(5):609–622, September 2015. ISSN 0016-7932, 1555-645X. doi: 10.1134/S001679321505014X.

245 D. Bilitza, D. Altadill, V. Truhlik, V. Shubin, I. Galkin, B. Reinisch, and X. Huang. International Reference Ionosphere 2016: From Ionospheric Climate to Real-Time Weather Predictions. *Space Weather*, 15(2):418–429, 2017.

246 D. Bilitza, R. Eyfrig, and N. M. Sheikh. A Global Model for the Height of the F2-Peak Using M3000 Values from the CCIR Numerical Map. *ITU Telecommunication Journal*, 46:549–553, September 1979.

247 D. Bilitza and B. W. Reinisch. International Reference Ionosphere 2007: Improvements and New Parameters. *Advances in Space Research*, 42(4):599–609, 2008.

248 S. K. Llewellyn and R. B. Bent. Documentation and Description of the Bent Ionospheric Model. Technical report, Atlantic Science Corp., Indian Harbour Beach, FL, 1973.

249 B. W. Reinisch and X. Huang. Redefining the IRI F1 Layer Profile. *Advances in Space Research*, 25(1):81–88, January 2000. ISSN 0273-1177. doi: 10.1016/S0273-1177(99)00901-1.

250 C. Scotto, M. Mosert de Gonzalez, S. M. Radicella, and B. Zolesi. On the Prediction of F1 Ledge Occurrence and Critical Frequency. *Advances in Space Research*, 20(9):1773–1775, January 1997. ISSN 0273-1177. doi: 10.1016/S0273-1177(97)00589-9.

251 S. Ramakrishnan and K. Rawer. Empirical Electron Density Profiles from Incoherent Scatter Data. *Annales de Geophysique*, 30:347–350, May 1974. ISSN 0003-4029.

252 D. Bilitza. Electron Density in the D-Region as given by the IRI, in *International Reference Ionosphere–IRI 79,* World Data Center A for Solar-Terrestrial Physics, Report UAG-82, Boulder, Colorado, 1981.

253 D. Bilitza, V. Truhlik, P. Richards, T. Abe, and L. Triskova. Solar Cycle Variations of Mid-Latitude Electron Density and Temperature: Satellite Measurements and Model Calculations. *Advances in Space Research,* 39(5):779–789, January 2007. ISSN 0273-1177. doi: 10.1016/j.asr.2006.11.022.

254 K. Rawer, J. V. Lincoln, and R. O. Conkright, International Reference Ionosphere-IRI 79, World Data Center A for Solar-Terrestrial Physics, Report UAG-82, Boulder, Colorado, 1981.

255 L. Scherliess and B. G. Fejer. Radar and Satellite Global Equatorial F Region Vertical Drift Model. *Journal of Geophysical Research: Space Physics,* 104(A4):6829–6842, 1999.

256 M. A. Abdu, J. R. Souza, I. S. Batista, and J. H. A. Sobral. Equatorial Spread F Statistics and Empirical Representation for IRI: A Regional Model for the Brazilian Longitude Sector. *Advances in Space Research,* 31(3):703, 2003. doi: 10.1016/S0273-1177(03)00031-0.

257 Y. Zhang, L. J. Paxton, D. Bilitza, and R. Doe. Near Real-Time Assimilation in IRI of Auroral Peak E-Region Density and Equatorward Boundary. *Advances in Space Research,* 46(8):1055–1063, October 2010. ISSN 0273-1177. doi: 10.1016/j.asr.2010.06.029.

258 E. A. Araujo-Pradere, T. J. Fuller-Rowell, and M. V. Codrescu. STORM: An Empirical Storm-Time Ionospheric Correction Model 1. Model Description. *Radio Science,* 37(5):1–12, 2002.

259 C. J. Mertens, X. Xu, D. Bilitza, M. G. Mlynczak, and J. M. Russell. Empirical STORM-E Model: I. Theoretical and Observational Basis. *Advances in Space Research,* 51(4):554–574, February 2013. ISSN 0273-1177. doi: 10.1016/j.asr.2012.09.009.

260 C. J. Mertens, X. Xu, D. Bilitza, M. G. Mlynczak, and J. M. Russell. Empirical STORM-E Model: II. Geomagnetic Corrections to Nighttime Ionospheric E-Region Electron Densities. *Advances in Space Research,* 51(4):575–598, February 2013. ISSN 0273-1177. doi: 10.1016/j.asr.2012.09.014.

261 J. A. Kong. *Theory of Electromagnetic Waves.* Wiley, 1975. ISBN 978-0-471-50190-9.

262 W. C. Chew. *Waves and Fields in Inhomogenous Media.* Wiley-IEEE Press, Piscataway, NY, February 1999. ISBN 978-0-7803-4749-6.

263 E. Hecht. *Optics.* Addison-Wesley, Reading, MA, 4th edition, August 2001. ISBN 978-0-8053-8566-3.

264 E. M. Purcell and D. J. Morin. *Electricity and Magnetism.* Cambridge University Press, Cambridge, 3rd edition, January 2013. ISBN 978-1-107-01402-2.

265 R. P. Feynman. *The Feynman Lectures on Physics*, volume 1. Basic Books, New York, The New Millennium Edition, January 2011. ISBN 978-0-465-02382-0.

266 G. R. Fowles. *Introduction to Modern Optics*. Dover Publications, New York, 2nd edition, June 1989. ISBN 978-0-486-65957-2.

267 F. L. Teixeira. On Aspects of the Physical Realizability of Perfectly Matched Absorbers for Electromagnetic Waves. *Radio Science*, 38(2), 2003.

268 M. Schönleber, D. Klotz, and E. Ivers-Tiffée. A Method for Improving the Robustness of Linear Kramers-Kronig Validity Tests. *Electrochimica Acta*, 131:20–27, 2014.

269 K. Sainath, F. L. Teixeira, and Donderici. Robust Computation of Dipole Electromagnetic Fields in Arbitrarily Anisotropic, Planar-Stratified Environments. *Physical Review E*, 89(1):013312, 2014.

270 K. Sainath and F. L. Teixeira. Tensor Green's Function Evaluation in Arbitrarily Anisotropic, Layered Media Using Complex-Plane Gauss-Laguerre Quadrature. *Physical Review E*, 89(5):053303, 2014.

271 K.-Y. Jung, B. Donderici, and F. L. Teixeira. Transient Analysis of Spectrally Asymmetric Magnetic Photonic Crystals with Ferromagnetic Losses. *Physical Review B*, 74(16):165207, 2006.

272 M. Born, E. Wolf, A. B. Bhatia, P. C. Clemmow, D. Gabor, A. R. Stokes, A. M. Taylor, P. A. Wayman, and W. L. Wilcock. *Principles of Optics: Electromagnetic Theory of Propagation, Interference and Diffraction of Light*. Cambridge University Press, Cambridge; New York, 7th edition, October 1999. ISBN 978-0-521-64222-4.

273 K. Sainath and F. L. Teixeira. Spectral-Domain Computation of Fields Radiated by Sources in Non-Birefringent Anisotropic Media. *IEEE Antennas and Wireless Propagation Letters*, 15:340–343, 2016.

Index

Tropospheric and Ionospheric Effects on Global Navigation Satellite Systems, First Edition.
Timothy H. Kindervatter and Fernando L. Teixeira.
© 2022 The Institute of Electrical and Electronics Engineers, Inc. Published 2022 by John Wiley & Sons, Inc.